THE
GREEN
INDUSTRIAL
REVOLUTION

THE GREEN INDUSTRIAL REVOLUTION

Energy, Engineering and Economics

BY

WOODROW W. CLARK II

GRANT COOKE

AMSTERDAM • BOSTON • HEIDELBERG • LONDON
NEW YORK • OXFORD • PARIS • SAN DIEGO
SAN FRANCISCO • SINGAPORE • SYDNEY • TOKYO

Butterworth-Heinemann is an imprint of Elsevier

Acquiring Editor: Kenneth P. McCombs
Editorial Project Manager: Chelsea Johnston
Project Manager: Preethy Simon
Designer: Maria Inês Cruz

Butterworth Heinemann is an imprint of Elsevier
The Boulevard, Langford Lane, Kidlington, Oxford OX5 1GB, UK
225 Wyman Street, Waltham, MA 02451, USA

Library of Congress Cataloging-in-Publication Data
Application submitted

British Library Cataloguing-in-Publication Data
A catalogue record for this book is available from the British Library

ISBN: 978-0-12-802314-3

For information on all Butterworth–Heinemann publications
visit our website at http://store.elsevier.com

Working together
to grow libraries in
developing countries

www.elsevier.com • www.bookaid.org

DEDICATION

This book is dedicated to Woody's wife, Andrea Kune-Clark and Grant's children and grandchildren. The Green Industrial Revolution will have a far greater, and beneficial, impact on their lives than anything we can imagine.

ACKNOWLEDGMENTS

In 2013, the world pumped 36 billion tons of carbon dioxide into the air, adding to atmospheric and environmental degradation that has continued since the First Industrial Revolution. We are reaching a tipping point with our fragile planet, and how the world responds, or does not respond, to climate change will have an unprecedented impact on the course of human history.

The changes that will be made to mitigate global warming and equally disastrous environmental changes will create unimaginable social, economic, and political change. With almost 7 billion people now on this fragile planet, the sands of time are marking our existence if we continue to use the environment and atmosphere as garbage cans.

Fortunately, there is a growing trend toward a more sustainable and environmentally sensitive way of life. This trend is being accelerated by new wondrous technologies that make renewable, carbonless energy a viable replacement for the fossil fuel energy generation of the twentieth century.

In examining and explaining these developments, which offer hope to our damaged planet, Clark and Cooke called upon international scholars and experts. Our goal was to provide the reader with a perspective that was global for we quickly realized that the United States was far behind the rest of the world in entering what we call the Green Industrial Revolution.

There are many to be thanked for helping the authors in their labors. Ken McCombs, the senior editor at Elsevier's Science and Technology Books committed to the project early, pushed the authors and others into a comprehensive examination of the topic. We are very grateful for Ken's hard work and efforts on our behalf and future generations. We all have families that will be impacted by the Green Industrial Revolution. And we all want that sooner than later.

We are also grateful to the many scholars, practitioners, business, government, and experts that contributed the case studies that are found in the Appendix. While we knew that others beside ourselves saw the emergence of the Green Industrial Revolution, we were amazed at the interest shown by those wanting to contribute to our project.

There are many other scholars, academics, and scientists whose work influenced our writings. We have done our best to note their opinions and give credit where credit is due. We want to particularly acknowledge for his insights as to how history's primary industrial phases evolved, Al Gore and his team, along with the hundreds of scientists who have contributed to the United Nations committees and teams, and the thousands of others whose work on climate change has begun to make a difference by creating the Green Industrial Revolution. Without their credible work, climate change would not be such a significant concern.

While intelligence, experience, and insight can help in the prescient recognition of a megatrend like the Green Industrial Revolution, hard work is what captures the transformative ideas and hones them into a creative and multi-disciplinary coherent viable plans (technologies, public policies and economics) for a carbon free future around the world.

This hard work is impossible without the support of family, and the authors would like to thank their respective wives, Andrea Kune-Clark and Susan Cooke, for their forbearance and patience as time and the research went on as the manuscript came together.

CONTENTS

Introduction   *xv*

1. **The Endgame**   **1**

   From San Diego on...   1
   The GIR's Key Components   6
   Government Support and Financing Is Required   8
   The GIR Cannot Come Soon Enough   9
   References   10

2. **Industrial Development**   **13**

   The First Industrial Revolution   13
   The Second Industrial Revolution   16
   Not Enough Supply, too Much Demand   19
   The Green Industrial Revolution   21
   Sustainability Is the Key   25
   Renewable Energy and Distributed on-Site Power   25
   Government Support Is Crucial   29
   The Next Economics: Social Capitalism   30
   References   34

3. **Big Oil's Impact**   **37**

   OIL: An Ancient Product   38
   The Oil Industry   41
   Developing an Oil Industry   44
   The First Oil Well Was Drilled in China in 347   46
   OPEC and the Oil Embargo   51
   OPEC Pricing and the Politics of the Persian Gulf   55
   Subsidies and Politics   59
   Massive Oil Spills   59
   Big Oil's Influence in the United States   62
   Oil's Reserves and Its Future   66
   References   69

4. Coal, Natural Gas, and Nuclear Power   **71**

Coal   72
Natural Gas   76
Hydraulic Fracturing   78
Nuclear Power   85
The Most Polluted Place on Earth: Russia's Mayak Nuclear Reactor   91
References   92

**5. Climate Change, Science and Technology, and Economics
Are the Forces Behind the GIR**   **95**

What Is Climate Change?   101
Top 10 Warmest Years (1880-2013)   102
Pioneering Climate Change Research   106
United Nations: The Epic Step Forward   108
AR5: Key Findings   109
There Are So Many More Human Beings—And More Coming   110
Science and Technology   111
Additive Manufacturing   113
LED Lighting   114
Hydrogen Power and Fuel Cell Storage   115
Economics   116
Marketing Green   117
Green Jobs   119
References   119

6. Renewable Technologies   **123**

What is Electricity?   123
Renewable Energy   125
Wind Power Generation   126
Development of China's Wind Industry   130
China's Wind Power   131
Solar: Energy from the Sun   134
China's Solar Valley City   136
US Solar   137
Water as Energy: From Hydroelectric to Ocean Waves and the Run of Rivers   139
Geothermal   141
Biomass, Recycled and Reusable Generation   143
For the Planet's Benefit and Human Health: Renewables Have to Come First   144
Renewable Energy Systems Are Protecting the Environment and Changing
Local Communities   145
References   147

7. Storage Technologies — **149**

V2G Power Storage in Electric Cars — 154
Hydrogen: A Breakthrough Technology — 156
References — 162

8. Smart, Green Grids — **163**

Europe's Parallel Lines — 166
China's Leading Smart Grid — 167
References — 171

9. Emerging Green Industrial Revolution Technologies — **173**

A Revolution in Lighting Technologies and Peak Demand
Response — 174
Cool Roofs Will Offset Carbon — 177
Nanotechnology: "Really" Small Things — 178
Regeneration Braking: From Trains to Cars to Trains and Back Again — 179
Combined Heat and Power — 181
Heat Pumps and Seawater Heat Pumps — 182
High-Speed Rail and Maglev Trains Have Become Realities — 183
Biofuel: A Transitional Energy Power — 184
Algae as a Biofuel Source — 185
GIR Fuel from Plants — 186
Waste to Energy — 187
Commercializing Emerging Technologies — 188
References — 189

10. China: The Twenty-First-Century Green Powerhouse — **191**

Emerging World Leader in Environmental Sustainability — 194
China's Energy Needs are Massive — 196
China's Solar Valley City — 197
Wind Power — 198
The Green Technology Culture — 200
References — 203

**11. The Green Industrial Revolution Is Spreading Around
the World** — **205**

Asia Leading — 205
South Korea — 206
Europe — 208
America — 211
Canada — 215

Central and South America   216
India   218
Southeast Asia   218
Australia and New Zealand   219
Middle East and North Africa (MENA)   220
References   223

12. Economics of the Green Industrial Revolution   225
True Cost of Oil   229
Free-Market Economics Has Failed   232
China's Central Planning Model   235
The Feed-in-Tariff (FiT) Model   237
FiT: California Style   239
Paying to Mitigate Climate Change   240
Green Jobs: The GIR's Results   243
Private Investment Is Needed   246
Google Invests Over $1 Billion in Green Tech   247
References   250

**13. Smart Sustainable Communities: The Way We Will Live
in the Future   253**
Los Angeles: The Car-Centric 2IR City   254
Sustainable Communities Are the Answer   255
Denmark   257
Agile Systems   259
Sustainability Starts at Home   261
Creating Sustainable Communities   265
Three Smart Green Sustainable Communities   266
References   273

14. A Smart Green Future   275
To the End of the Oil Age   277
Grid Parity   279
Energy Deflation   280
Zero Marginal Cost   282
References   285

Appendices 1, 3, 5, 9 and 16 will appear online and not in the print-book

Appendix 1 Macro Economic Monetary Reform: Greening Capital
and Finance   e1
Appendix 2 The Economic Implications of Green Industrial
Revolution (GIR) in Central and Eastern Europe: The Case of Poland   287
Appendix 3 Implementing the Green Industrial Revolution   e23
Appendix 4 The Green Revolution Applied in Everyday Life   313
Appendix 5 The Green Industrial Movement (Revolution)
in South Asia   e33
Appendix 6 Reforming the Energy Vision in New York State:
Clean Coalition Comments on Matter 14-00581 / Case 14-M-0101
Before the New York State Public Service Commission
(July 18, 2014)   333
Appendix 7 The World Is Round and Green: It Needs
Strong Medicine   347
Appendix 8 Creating a Cradle to Cradle World: Executive Summary
Cradle to Cradle Products Innovation Institute   357
Appendix 9 Pilot Study: Certified Products Program: Impacts of the
Cradle to Cradle (C2C) Certified Products Program: Shaw
Product Analysis   e57
Appendix 10 The Formidable Fight for Fuel   369
Appendix 11 Smart Green Energy Communities: A Definition
and an Analysis of Their Economic Sustainability in Europe   377
Appendix 12 Potential of Offshore Wind in the
Republic of Mauritius   393
Appendix 13 A Case of Community Involvement
in Wind Turbine Planning   419
Appendix 14 eHealth for Sustainable Health Care in Serbia   491
Appendix 15 UC Davis West Village Energy Initiative
Annual Report 2012-2013   515
Appendix 16 Qualitative Economics: Micro-Economics
Interactionism   e87
Appendix 17 Achieving Fossil-Free Homes through Residential
PACE Financing   535
Appendix 18 Smart Explorer's Wheel: Accelerating Innovation
Integration in the Green Industrial Revolution   551
Index   559

INTRODUCTION

Climate change is real, and we are all impacted by it. Greenhouse gases from carbon emissions cover our planet, forcing drastic changes in weather patterns that create destructive super storms never before witnessed or experienced. Decades of failing to curb the world's dependency on fossil fuels have made the planet hotter and more polluted. It has killed people and stolen their livelihoods. The world's poorest nations are the most vulnerable as they face increased risk of drought, crop failure, poverty, and disease.

Coastal cities and island nations are at the highest risk and the planet's food production faces increasing complications (see Case #12 in Appendix about Mauritius Island). Deserts and dry regions are becoming dryer, and wet regions wetter. Many parts of the planet, like the tropics, are experiencing heat waves and tropical super cyclones and the irreversible loss of biodiversity. Coral reefs that surround and protect vast miles of islands and coastal beaches are dying. A hotter planet threatens to roll back decades of sustainable growth, and the science is clear.

A major US problem is real estate development since buildings generate large amounts of greenhouse gases. Case #17 in the Appendix provides an overview of this issue from an individual's perspective.

Scientific reports and national and international security and military resources all conclude that humans are the cause of climate change. Oceans are warming, sea levels are rising, and the global mean temperature is increasing. Recent predictions are that global mean temperature will rise more than 2 degrees Fahrenheit in less than three decades. Melting Arctic and Antarctic ice caps will cause rising ocean tides and swells resulting in more global damage and loss of lives.

This grim reality of a planet in environmental crisis may be the legacy of the world's most prosperous and careless era (see Case #10 in Appendix about the human costs). The extraordinary wealth of the last century has been supercharged by fossil fuels that were, and still are, indiscriminately burned for cheap energy, while their emissions create a suffocating blanket for our atmosphere. The tragedy is that the fossil fuel industry was supported with land grants and massive rail systems to transport their coal, oil, nuclear waste, and natural gas. A century later, this industry, which is history's wealthiest, still receives massive US tax subsidies.

IT STARTED WITH INDUSTRY

The First Industrial Revolution was a turning point in human history. Great Britain led that eighteenth century revolution in machine-based manufacturing. The Age of Enlightenment offered powerful ideas and pushed art, literature, science, and democracy into new frontiers. Powered by coal, Watt's steam engine drove textile and manufacturing industries that were the envy of the world.

The Second Industrial Revolution saw the tremendous power and flexibility of fossil fueled internal combustion engines. The internal combustion engine, along with the commercialization of oil, created a previously unimaginable world of machines and personal transportation. Thomas Edison with his electricity and then Alexander Graham Bell with the telephone revolutionized the daily lives of ordinary people and led to telecommunication centers, huge data server farms, and complex electrical networks, all of which required vast amounts of energy.

The Green Industrial Revolution started to emerge at the end of the twentieth century. Initially proclaimed as occurring in northern Europe, it actually had started and took hold in Asia. Japan led the way in the 1970s with its TOTO Corporation, which created technologies and systems to control and save water. Then at the beginning of the twenty-first century, China leapfrogged into this new Green Industrial Revolution driven by unprecedented economic growth, urbanization, and infrastructure needs.

The Green Industrial Revolution is a far more significant and life changing than either of the previous industrial periods, for there is so much more at stake. While the planet has been getting hotter and smokier, the world has been getting more crowded. Each day, precious resources get scarcer. Today there are 7 billion people living on the planet, and by 2053, the UN predicts that there will be 10 billion people (see Case #6 in Appendix on these and other changes).

Compounding the problem is the rise of a middle class in developing nations. People in emerging nations want to get out of poverty. They want the things that developed nations have—nice clothes, nutritious food (including animal protein for their children), and large, air-conditioned, electrified homes as well as education and a future for themselves and their children. They also want the things that most citizens of developed nations take for granted: washing machines, cell phones, refrigerators, televisions, and cars (see Case #5 in Appendix about micro-cities in India).

Add it up, and the world will soon be resource-constricted, particularly since the planet is running out of fossil fuels. This alone threatens to shake the very foundation of our existence and future for our children and grand children. But the environmental degradation and the collapse of various parts of our planet's ecosystem are adding a heightened urgency. The planet cannot delay halting this degradation and exploitation.

THE SOLUTION FROM COMMERCIAL TECHNOLOGIES

It is a dangerous time—a point at which global warming and environmental degradation may become irreversible. Critical decisions must be made on a global level for the good of the planet. It is also an age of opportunity, and the Green Industrial Revolution with its renewable energy and storage system technologies can provide those opportunities and solutions.

A new era of sustainability and carbonless energy generation is at our doorstep. A push for renewable energy and a carbonless lifestyle will become history's largest social and economic megatrend, with the potential of extraordinary benefits in the form of economic revival, innovation, emerging technologies, and significant job growth for those nations capable of fast entry.

Countries around the world are starting to realize the need to replace fossil fuels with renewable energy generation. Denmark has been one of the leaders (see Case #13 in Appendix). Most of the European Union nations have environmental and renewable energy policies that move them aggressively into the GIR. See Case #11 in the Appendix, which highlights Italy and other EU nations. Poland (Case #2 in Appendix) and Serbia (Case #14 in Appendix) are doing just that along with other central and east EU nations. Developing nations like Brazil, Chile, and Thailand are adding renewable energy wisely. Island countries like England, Singapore, and Mauritius (Case #12 in Appendix) are adding more carbonless energy generation to their economies. More and more nations are slowly realizing that their futures are not rooted in carbon-based fuels and they are finding better solutions.

These solutions are taking the form of remarkably innovative technologies that are exponentially more stunning, numerous, and revolutionary. Toyota led the way with its Prius hybrid car that used regenerative brakes to create and then store energy. Now Toyota is the number one car company in the world. The same pattern is seen with inventions that emerged from previous industrial periods. Amazing technologies—from tiny nano-crystals to 200 mph trains propelled by magnetic force—are being designed by scientists and engineers who are changing human history. We look at emerging green technologies that will provide a sense of what can be done in the face of truly monumental societal, financial, and policy challenges.

The genius of mechanical engineers and inventors is showcased in tech-nologies like Scotland's Pelamis, an offshore wave energy converter that uses the energy of ocean waves to generate electricity. Equally innovative is a Davis, California company's Fast Ox, a refrigerator-size blast furnace that can create energy out of organic waste or junk like old IPods.

To store renewable energy, innovative ideas like the Cash-Back Cars, which uses plugged-in cars for storage, are on the horizon. New adaptations and uses for old technologies like flywheels and technologically advanced batteries like the one Tesla Motors developed are part of the storage solu-tions. Hydrogen fuel cells for storage and power is another major break-through and will quickly become a reality when California launches its hydrogen highway in 2016.

Technologies that share power like smart grid systems are moving toward the mainstream. China has deployed over 250 million smart meters and created a new grid city to showcase its science.

At the foundation of the Green Industrial Revolution are sustainable and green communities. Singapore and China have launched the Tianjin Eco-City as a model for a sustainable eco-city. West Village in northern Califor-nia is the largest Net Zero Energy community in the United States and a model for future residential construction (see the Report as Case #15 in Appendix). Aalborg, Denmark, was probably the first modern community to focus on sustainable growth, and it continues as a model for northern European cities.

Many of these new technologies need to move more quickly from the laboratory into the mainstream. When we talk about huge infrastructure changes like replacing coal with renewable generation, there are tremendous costs. It is clear to us that the 2008 world financial meltdown signaled an end to the free-market, modern economic theory that served the world during the twentieth century. This supply-side, deregulated, free-market econom-ics worked well in the last part of the twentieth century, but this form of neo-classical economics will not be able to address the problems of a twenty-first century world that is threatened by irreversible environmental degradation. Case #1 and then Case #16 in the Appendix demonstrate how economics has changed around the world to a new economic paradigm.

Except for the United States, developed nations around the world have started to grapple with this new reality, and we explore various economic and financial alternatives and methods that are being implemented as a new economic model emerges. At the core of this new economic paradigm is how to pay to mitigate climate change, and it recognizes that for too long business greed and carbon-based corporations have been using the atmosphere as a garbage can. That cost is due.

Nations like China and Denmark, and organizations like the UN, DARA, and the World Bank are developing strategies that can move the world into the Green Industrial Revolution. Equally as important, the developed world needs to reduce its energy dependency on the Middle East, a geopolitical region in turmoil and the channel for the most massive wealth transfer of all time—from the world's oil consumers to the oil suppliers. This transfer of wealth is creating political instability that threatens world security, drains precious financial resources, and keeps nations from focusing on crucial domestic issues.

For example, as this book goes to press in fall 2014, Russia is using the natural gas dependency of Europe and Eastern Europe to leverage its incursion into the Crimea and the Ukraine. There has been a series of not-so subtle threats to decrease or withhold natural gas to European nations if they retaliate against Russia for its violation of Ukrainian sovereignty. It is a massive display of extortion and power, and provokes European comparisons of Stalin's post-World War II land grab.

While at press time we do not know how it will be resolved, and the European Union is acknowledging that Russian natural gas comes at huge political and economic price. European nations are calling for the development of renewable, non-carbon energy generation. In the end, it may be that Russian aggression has pushed Europe and the developed world deeper and sooner into the Green Industrial Revolution than the threat of climate change and environmental degradation.

CHAPTER 1

The Endgame

The endgame for the carbon-intensive, utility-controlled centralized power generation era has started. Powered by the oil-fueled internal combustion engine, this era is slowly giving way to a revolutionary new industrial and economic model powered by renewable green energy (Clark and Bradshaw, 2004). Instead of being generated in monolithic plants with huge fossil-fueled turbines and passed along rigid one-way power lines, this energy will come from many small-scale renewable processes. Electricity will flow through a smart and flexible grid and be controlled via the Internet. Like spilled water settling on a tabletop, power will be able to scale laterally.

This new process is called "distributed," or on-site, energy generation (Andersen and Lund, 2007: 289). Unlike carbon sources such as coal, oil, gas, and tar sand, which are elite and come from special finite sources, distributed renewable energy comes from common sources that cover every inch of the planet. Solar and wind energy, geothermal heat under the ground, biomass from garbage, small hydro, and ocean tides and waves are all easily harnessed sources to generate electricity (Lund and Clark, 2002: 473). These never-ending sources are at the core of an economic and industrial revolution that will transform the way we live. It is called the green industrial revolution, GIR for short, and it will emerge as the largest megatrend in history (Clark and Cooke, 2011).

FROM SAN DIEGO ON...

In San Diego, California, the sun is powering over 5000 buildings. The sunny Southern California coastal region has nearly 40 megawatts of rooftop solar power installed on homes and commercial and government buildings. With one megawatt able to power 750-1000 homes, San Diego is powering the equivalent of 8000 homes, making this city of three million America's ground zero when it comes to per capital use of solar energy (Environmental California Study, 2012).

Children in Ghana's mud huts study their school lessons after dark by the light of a shiny LED light, powered by a solar collector about twice the size of pack of cards. The inexpensive reading light, which was invented by a

California university professor, comes from a group called Unite to Light that is bringing these small solar power lights to rural Africa. The group is intent on furthering children's education and health by replacing the noxious kerosene lamps used by Africa's rural poor. Besides Africa, Unite to Light is supplying solar-based lighting to 33 developing nations on 4 continents (Unite-to-Light, 2013).

In Norway, a hydrogen highway stretches 375 miles or 600 kilometers from Oslo to Stavanger. It opened in 2009 with the first hydrogen station in Stavanger followed by one in Porsgrunn. A modified Toyota Prius can travel 200 km using just 2 kilograms of hydrogen. StatoilHydro retails hydrogen for $6.28 per kilo for a fuel cost of around 7.4 cents a km (11.8 cents a mile). In Norway, regular gasoline costs around $1.40 a liter and a normal Prius gets around 22 km to the liter (6.4 cents a km, 10.3 cents a mile). While slightly more expensive than regular gas, the vehicles are odorless, noiseless, and free of CO_2 emissions. Japan, Sweden, Denmark, Germany, and California in the United States have built hydrogen-refueling stations in major cities with plans for more. To provide vehicles for the hydrogen highways, car companies are quietly producing hydrogen fuel cell cars that electrolyze renewable energy sources into hydrogen for refueling at home, office, or public buildings (Norway Hydrogen Highway, 2009).

The Chinese have taken cutting-edge German technology with some originally created in and then licensed from the United States and implemented the first commercial high-speed magnetic levitation train. Called maglev, this carbonless technology uses controlled tension from magnets to form an electromagnetic suspension system for lift and propulsion. The train connects downtown Shanghai and Pudong International Airport, making the 30 km trip in just 7 minutes. Other trains that can travel up to 1000 km, or about 620 miles per hour, are in development. These maglev trains will travel in underground airless tubes.

A Canadian cement company is running their fleet of trucks on goopy oil made from CO_2-consuming algae. Ontario's St. Marys Cement is using algae's photosynthesis to convert the CO_2 produced by the cement manufacturing process into a biofuel. The CO_2-consuming algae are being harvested and dried using waste heat from the plant. Then, the algae are burned as fuel inside the plant's cement kilns as well as being used as fuel to power the cement trucks.

As the green industrial revolution takes hold, it will result in a complete restructuring of the way energy is generated, supplied, and used. It will be the largest industrial and economic megatrend of the postmodern age and create an era of huge potential and opportunity. Remarkable innovations in science and energy will lead to sustainable and carbonless economies powered by advanced technologies like hydrogen fuel cells and nonpolluting technologies like wind and solar. Small community-based and on-site renewable energy generation will replace central plant utilities, and smart green grids will deliver energy effortlessly and efficiently to offices, homes, plug-in autos, and even intelligent appliances.

This new era encompasses changes in technology, economics, manufacturing, and consumer lifestyles. The transition will be more comprehensive than the steam-driven first industrial revolution of the 18th and 19th centuries. It will have more impact than the oil-powered industrial era of the 20th century. This monumental shift is already under way in parts of the world, where renewable energy, sustainable communities, smart grids, and environmentally sound technologies are being implemented (Clark and Cooke, 2011).

Nations are starting to make historically significant decisions in light of this new era. For example, Scotland, which has chaffed under English rule for centuries, has decided to push for complete autonomy. The Scots will go to the polls in fall 2014 to decide on secession from England. Some of this political drive for independence is based on the belief that they can decouple from England's finance services-based economy and move to their own knowledge-based economy supported by green energy generation. The Scots point out that they are nearly energy-independent with a mix of North Sea oil and natural gas and a growing renewable energy industry with unlimited potential.

Scotland already generates 40 percent of their electricity via renewable energy and will be able to offset a decline in North Sea oil with more renewable energy. The Scots expect to exploit onshore and offshore wind resources near Glasgow and Edinburgh as well as the emerging wave and tidal power potential of the Highlands and Hebrides areas. Scotland is uniquely situated for tidal and wave power generation. Government, business, and science are supporting this push. They believe that a large continuous supply of renewable energy that can satisfy domestic use as well as be exported will allow them to develop a robust economy.

Germany has been making major policy changes based on solar energy. The Nordic countries of northern Europe—Denmark, Finland, Iceland, Norway, and Sweden—are committed to developing renewable energy. Government policies have been adopted, and financing is in place to nurture the green industry, both of which are creating new businesses and related jobs (Lund and Østergaard, 2010).

Asian nations are rapidly turning to renewable energy generation and related technologies. Japan is developing small hydrogen cells that can provide power for individual buildings. It has led the way in developing electric and hybrid autos, purchasing the patents for the regenerative braking system from the United States in the 1980s. After Toyota developed the extraordinarily successful Prius, and hybrid technology became commercially viable, they licensed the technology back to United States and other automakers (Adams and Funaki, 2009).

Although China is the world's largest carbon emitter, it is rapidly developing renewable energy sources along with extraordinary technologies like high-speed trains that are propelled up to 200 mph by magnetic force (China Coal, 2012).

China is also considering new policies that will address the horrendous smog and miserable air quality of major cities like Beijing. Speculation is spreading that some sort of restriction on carbon-fueled autos will be developed, forcing new car owners to electric or possibly hydrogen-propelled vehicles (Clark and Isherwood, 2010).

In recognition that a significant new trend toward more efficient vehicles is overwhelming them, US automakers are revolutionizing their cars. Ford, for example, has introduced a major change in its iconic F-150 pickup series. The new 2014 full-size trucks will replace portions of heavy steel with aluminum, shaving off 500 pounds of mass. This could yield as much as five mpgs increase in mileage, and considering that the F-150 is the most popular vehicle sold in the United States, this mileage increase represents huge gasoline savings. Other automakers have now developed drivetrain technologies like direct-injection engines, advanced transmissions and drive gears, and more efficient hybrids and plug-ins.

One of the leaders of this new green auto movement is Tesla Motors, a San Jose, California, company. Founder Elon Musk, a South African born Canadian-American engineer and businessman who struck it rich with PayPal, the Internet finance system, created Tesla to prove that an electric car could compete with the world's luxury gas-powered sedans. The result was the Model S, a plug-in, high-performance sedan that by 2014 has sold over 30,000 cars globally. This extraordinary car can equal the best of the gas-powered sedans and was Norway's top seller for 2 months in 2013. It has rapidly become the "go-to" vehicle of Silicon Valley's hip venture capitalist and entrepreneurs (Tesla GigaFactory, 2014).

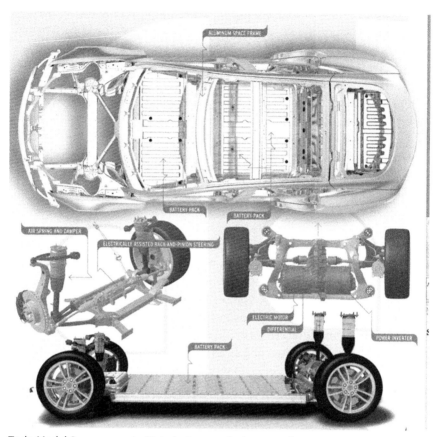

Tesla Model 3 components. Note battery packs between frames. *Source: CarandDriver. com December 2012. http://www.caranddriver.com/tesla/model-s.*

Though the United States has many breakthrough technologies, it is lagging the rest of the world in moving into the green industrial revolution. Tremendous economic and political forces connected to the established oil, utility, and carbon-based industries are at play in America to restrict the adoption of renewable energy generation and green and clean technologies. The oil industry and billionaire oil businessmen like the Koch brothers have spent millions of dollars attacking climate change science, environmental groups, and United Nation's research and conclusions.

Despite pressure from oil and carbon-based industries and the lack of federal support, the green industrial revolution is budding in the United States. Aware that other countries like Germany have created new green jobs, US states, cities, and local communities have started to act on their own to nurture the GIR benefits (Gipe, 2010–2014).

California, where innovation and environmental concerns are historically prominent political themes, is leading the United States. The state has a rich tradition of innovative research and venture capital-driven economy and just cannot seem to resist developing cutting-edge technologies. Tesla Motors is a prime example. Innovative new discoveries in technologies and products are emerging from the state's growing green industry.

These technologies range from biofuels to electricity-assisted bicycles and include a broad spectrum of new building materials like porous concrete that lets rainwater drip back into the water table. The University of California Davis campus includes a center devoted to revolutionizing facility lighting, and the Lawrence Berkeley National Laboratory's Molecular Foundry is working on windows that take advantage of electrochromic nanotechnology to adjust lighting levels according to ambient conditions via sensors and the Internet. The Lawrence Livermore National Laboratory, also in California, has had a long history of working with energy and environmentally sound technologies (Clark et al., 2003; Cooper and Clark, 1996; Yano, 1997).

THE GIR'S KEY COMPONENTS

Just as coal and steam engines were central to the development of the first industrial revolution, today's green industrial revolution relies on core energy components. Briefly, those components are the following:

- *Renewable energy.* Clean, never-ending energy is available throughout the planet; existing technologies just need to be applied in a systemic way. Humans have been using water and wind power for thousands of years, and these sources are abundant and the energy is easily gathered. The sun is the most energy-intensive object in our galaxy and its offers far more energy than humans will ever need. Hydrogen is another abundant energy source, as is tidal action. Turning waste—organic and inorganic—into energy isn't a complicated process and can be adapted to almost every community. The more common renewable sources and their technologies are wind, solar, geothermal, biomass, and ocean waves. Not so common renewable sources include hydrogen cells, magnetic levitation, algae, and bacterial or microbial.

- *Energy storage.* Most renewable energy sources, particularly wind and solar, are called "intermittent" because the sun is not always shining and the wind does not always blow. To make renewable energy sources work smoothly, devices that can store the energy and release it when needed are critical. These storage devices can take the natural form of large salt formations, or artificial ones like batteries, fuel cells, or flywheels. Innovation is exploding in energy storage, including the use of hybrid and plug-in autos. Cost-effective energy storage is the holy grail of the renewable energy dynamic.
- *Flexible energy sharing.* The old centralized, one-way power lines of the traditional utility system must change. The old grids are inefficient and dysfunctional. Smart grids are necessary to maximize distributed energy from many small sources. These Internet-like grids need to be scalable and distribute electricity flexibly, moving energy multiple ways among users. The idea is that even though you are not at home, your solar panels are generating power. For example, while you are gone, your neighbor is editing video in his home office while washing his clothes and brewing coffee, plus recharging his plug-in auto, all of which is using more electrons than his solar panels can produce. The smart grid can seamlessly divert your excess power to him, track it, and bill it.
- *Integrated transportation.* Mobility and transportation are critical functions that cannot be done without in our modern world. Yet, transportation contributes massive amounts of greenhouse gases that need to be eliminated. Autos, buses, and other forms of vehicular transportation have to switch from fossil fuels to green and environmentally sensitive energy. The international transportation industry has started this transition, and it will only accelerate, especially if China moves forward with its nonfossil fuel auto regulations. Automakers are under pressure to increase gas mileage in their vehicles, and the results are remarkably innovative concepts and technologies, including hybrids, electric cars, and hydrogen-powered buses and vehicles. A new program called the cash-back car is being tested in the United States and has great promise. On a small scale, this new program connects hybrids and plug-in autos to a local grid. The cars are charged when needed, and if not, the cars can push electricity back into the grid. The system is easily regulated by computer, which can keep tabs on the kWhs coming and going. The system can calculate how much electricity each car uses and puts back, making them ideal

storage devices. The concept is simple and efficient and has enormous potential. It can be scaled to be a major part of the energy storage dynamic. Plus it adds one more incentive to owning an electric car.

These components are technologically accessible and individually useful. However, the real value is in connecting them together in a seamless and environmentally clean infrastructure. Connected, they allow us to conceive of a new energy and economic model that is independent of fossil fuels (Lund and Clark, 2008: 60).

GOVERNMENT SUPPORT AND FINANCING IS REQUIRED

If clean renewable energy with a limitless supply is available, how come the world is still addicted to fossil fuels with their destructive emissions and dwindling supplies? The answer is fairly simple—power and money. The fossil fuel interests, including huge international oil companies, monopolistic utilities, and sovereign oil-rich nations, have rallied enough political support to delay and minimize the emergence of an international carbonless economy. Even today with its staggering budget deficits, the United States still provides special tax subsidies to the oil industry—the richest corporations in world history. An estimated $4 billion a year in tax breaks is extended to companies like Chevron and Exxon at the expense of developing alternative energy generation.

Our connection to oil and carbon goes all the way back to the oil seeps of Pleistocene era, or about 12,000 years ago. Ancient Mesopotamia prospered by trading bitumen (a sludge-like mixture of hydrocarbons) to the Egyptians. Bitumen was used as mortar for the extraordinary Tower of Babel.

The modern oil industry has a tradition in the United States that goes back to the mid-1800s. John D. Rockefeller launched Standard Oil in 1870, and the industry was destined to be the most powerful and wealthiest in world history. As a nation, the United States owes much of its extraordinary wealth to the oil industry and the subsequent internal combustion engine that propelled it to 20th-century superpower status.

This power and wealth continues, and America still ignores the calls by the national community to curb carbon emissions and to provide leadership in the battle against climate change.

Changing something as enormous and moribund as a fossil fuel infrastructure to a carbonless one will take government leadership, regulations, and sophisticated financial tools. The science and the technology are there; what is lacking, particularly in the United States, is the will and commitment.

THE GIR CANNOT COME SOON ENOUGH

The planet is becoming hotter, smokier, and more crowded. It groans as the 7th billion person is born, and the UN is predicting 10 billion by midcentury. There are too many of us, and all natural resources are finite and limited as the United Nations Intergovernmental Panel on Climate Change pointed out (UNIPCC, 2014).

Climate change is real. Today, the planet is being threatened by melting ice caps and ocean acidification, and we are losing critical environments. Yet with climate change impacting daily lives, the world's leaders are unable to stop the damage. In 2009, 192 countries came together in Copenhagen to address global warming. Even with our scientists saying that the planet's climate has begun a significant shift because of the emissions of industrial produced carbon dioxide, methane, and nitrous oxide, world leaders could not come to grips with the reality.

The world's rising temperature is causing freakish storms and creating havoc with the Earth's water cycles and hydrology. Ecosystems are imperiled and mass extinctions of plant and animal life are being threatened. From Uganda's mountain gorillas to California's native trout, the world's animal species are endangered by the detritus of cheap carbon energy and unsustainable lifestyles.

Humans must stop using the environment as a massive garbage can. Not stopping is expensive and it is killing people. The humanitarian organization DARA estimates that 5 million deaths occur each year from air pollution, hunger, and disease as a result of climate change caused by carbon-intensive economies. Over 100 million deaths will occur in the next decade. That toll will likely rise if current patterns of fossil fuel use continue (DARA, 2012).

According to DARA's 2012 report, developing countries will bear the brunt of it, since the world's poorest nations are the most vulnerable to increased risk of drought, water shortages, crop failure, poverty, and disease. Economically, the world's poorest nations could see an average 11 percent loss in GDP by 2030 due to climate change. Agriculture and fisheries, two areas that most poor nations are dependent on, face losses of more than $500 billion per year by 2030. The effects of climate change are already costing the global economy about $1.2 trillion a year, and this could double by 2030 if global temperatures are allowed to rise.

The green industrial revolution is about climate change mitigation, renewable energy, smart grids, health, education, and environmental sensitivity. A clean, healthy planet is within the reach of science and emerging technologies. A new wave of business enterprises and green jobs are available. Profound and innovative new green discoveries are now routine. Worldwide, communities are talking about sustainability, but the effort requires governmental support and leadership.

Despite the lack of collective action by world leaders to address global warming and put a stop to the growth of carbon-based energy consumption, there are signs of an alternative. While frustratingly slow for those concerned about an environmental catastrophe, a new green economy based on distributed energy from renewable sources is emerging. As this green industrial revolution takes hold, it is destined to be the most significant social and economic transition in world history.

REFERENCES

Adams, L., Funaki, K., 2009. Japanese experience with efforts at the community level towards a sustainable economy. Chapter #15. In: Clark II, W.W. (Ed.), Sustainable Communities, Springer Press, New York.

Andersen, A.N., Lund, H., 2007. New CHP (combined heat and power) partnerships offering balancing of fluctuating renewable electricity productions. J. Cleaner Prod. 15, 288–293, Elsevier Press.

Environment California Study, January 24, 2012.

China Coal, 2012. http://thinkprogress.org/climate/2012/02/22/430441/coal-consumption-in-china/.

Clark II, W.W., Cooke, G., 2011. Global Energy Innovation: Why America Must Lead. Preager Press, Westport.

Clark II, W.W., Isherwood, W., Winter 2010. Utilities Policy Journal. Special Issue on China: environmental and energy sustainable development (from Asian Development Bank Report in 2007).

Clark, W.W., 2003. with William Isherwood, J. Ray Smith, Salvador Aceves, and Gene Berry (Lawrence Livermore National Laboratory) and Ronald Johnson, Deben Das, Douglas Goering, and Richard Seifert (University of Alaska Fairbanks) "Remote Power Systems with Advanced Storage Technologies for Alaskan Villages", Energy Policy, Elsevier.

Cooper, J.F., Clark II, W.W., 1996. Zinc/air fuel cell: an alternative to clean fuels in fleet electric vehicle applications. Int. J. Environ. Conscious Des. Manufact. 5 (3–4), 49–54.

DARA, 2012. http://daraint.org/climate-vulnerability-monitor/climate-vulnerability-monitor-2012/.

Gipe, P. Feed-in-tariff monthly reports, 2010–2014. http://www.wind-works.org/FeedLaws/RenewableTariffs.qpw.

Lund, H., Clark II, W.W., 2002. Management of fluctuations in wind power and CHP: comparing two possible danish strategies. Energy Policy 27 (5), 471–483, Elsevier Press.

Lund, H., Clark II, W.W., 2008. Sustainable energy and transportation systems introduction and overview. Utilities Policy 16, 59–62, Elsevier Press.

Lund, H., Østergaard, P.A., 2010. Climate change mitigation from a bottom up community approach: a case in Denmark. Chapter 14, In: Clark II, W.W. (Ed.), Sustainable Communities Design Handbook, Elsevier Press, Amsterdam.

Norway Hydrogen Highway, 2009. http://www.greenmuze.com/climate/cars/1149-norways-hydrogen-highway.html.

Telsa GigaFactory, 2014. http://blogs.marketwatch.com/energy-ticker/2014/02/26/teslas-gigafactory-what-elon-musk-didnt-say/.

UNIPCC, March 31, 2014. www.unipcc.org/WGIIAR5 Phase I Report Launch.

Unite-to-Lite, 2013. www.unite-to-light.org.

Yano, G., 1997. Press Release. Laboratory moves toward $100 million agreement to develop zinc-air fuel cell technology. Lawrence Livermore National Laboratory. NR-97-09-02.

CHAPTER 2

Industrial Development

Human history is marked by extraordinary leaps in technology. When a technological window opens, it triggers remarkable social, political, and scientific advances. Archeologists point to the ancients rise as they discovered fire and entered the Stone Age. Suddenly, stones provided the basis for new technology applications that increased survival—prey was easier to kill and the odds were better against predators.

So it went through history. Technology responded to human needs, and the wheel was discovered, making transportation and the distribution of goods easier. Oxen were taught to pull a plow; horses tamed to be ridden. Stunning advances in technology, science, and engineering allowed the great civilizations of the Egyptians, Mayans, and Chinese to flourish. Superior battle technology aided Hannibal as he conquered Italy; later, it made the Venetians the rulers of the Mediterranean Sea.

Breakthroughs in technology opened the door to the glories of the European Renaissance, an era when art, science, and politics raced ahead after the despair of the Dark Ages.

THE FIRST INDUSTRIAL REVOLUTION

In the middle of the fifteenth century, a German named Johannes Gutenberg, and two friends Andreas Dritzehn and Andreas Heilmann, became interested

Historic image of Gutenberg's original press.

in a crude invention that pressed images on paper from a flat woodcut. Gutenberg began to work with a new form of type and an oil-based ink that was durable and lasting, and he used paper as well as vellum, a high-quality parchment.

When he had mastered the process, Gutenberg printed a beautifully executed bible in 1455. Copies sold for 30 florins, or about 3 years worth of wages. Through various iterations of his bible, Gutenberg refined his technique for mechanical movable-type printing. His inventions were astounding, revolutionary in their impact on society.

From Gutenberg's small shop in Mainz, Germany, printing spread to hundreds of cities throughout Europe, helping to stimulate the Renaissance and later Europe's scientific revolution of the sixteenth and seventeenth centuries. Historians estimate that by the end of the sixteenth century, European printing presses had produced 150 million books (Febvre and Martin, 1976).

These early printers unlocked the human mind on an unprecedented scale. New ideas emerged beyond the cold, rock walls of monasteries and the monks who laboriously hand copied scripts. Major Greek and Latin texts that had survived the Inquisition and the book burnings of the Middle Ages were printed and poured over. Literacy was spreading as economies grew and readership leaped beyond clergy and royal clerks.

Ideas crossed borders, and literacy broke the monopoly of the elite and bolstered a middle class. Mass communication had arrived, and ideas were emerging and suddenly accessible. Another advance in technology had pushed civilization forward.

Since the printing press was a European phenomenon, the rest of the world fell behind in science and social development. The rapid exchange of new ideas led to the Age of Enlightenment, the extraordinary era of cultural, social, and scientific advancement led by Isaac Newton and Charles Darwin.

In the early 1700s, as the Age of Enlightenment emerged in Great Britain, a young Scottish inventor and mechanical engineer named James Watt started to tinker with the design of a crude steam engine that was used for pumping water from mines. In a flash of insight, he realized that the engine wasted most of its steam. As a remedy, he changed the steam so that it condensed in a chamber separate from the piston. Watt's change radically improved the power, efficiency, and cost-effectiveness of steam engines. Then, he added rotary shafts and gears and a new world rose.

In 1775, Watt entered a partnership with Matthew Boulton, a businessman who quickly recognized the commercial potential of this new energy-generating source. Boulton and Watt were enormously successful with their steam engine, providing the power for a change in the way people lived, worked, and played.

By the end of the Napoleonic Wars in 1815, wood was scarce in Britain. Most of the nation's magnificent oak trees had been cut into planks and masts for the naval warships that held France's Napoleon at bay until his eventual defeat at Waterloo. However, Britain had an abundance of coal, which contained twice as much energy as wood. Coal quickly displaced wood as the fuel for the steam engine. Soon, it was used to produce heat for industrial processes, to drive engines, and to create propulsion, as well as to warm buildings.

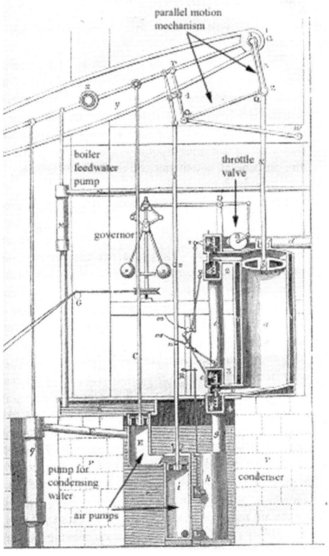

Illustration of the Boulton-Watt double-acting engine. *Adapted from an engraving.*

The combination of Gutenberg's printing press, which made the quick dissemination of innovation and ideas possible, and Watt's coal-driven steam engine, which created a whole new energy source, triggered an explosion in industrial activity. Chemical energy from coal was transferred to thermal energy and then to mechanical energy. The steam engine powered industrial machinery and steam locomotives. Suddenly, the world needed people who understood combustion engines. New education, business, and employment sectors were created to support these new technologies.

The first industrial revolution (1IR) surged. It was a turning point in human history, and soon, all of Western civilization was entrenched in this new industrial age, driven by steam, and surrounded by innovative ideas, which were distributed with the help of the printing press.

The first industrial revolution hastened the drive toward large-scale manufacturing. At first, machines started to replace manual labor, horsepower, and wind and waterpower. This transition eventually reached North America, hastened by the westward expansion and the 1849 discovery of gold in California. American communities, which were based on trade and agriculture and dependent on tools and animals, began to rely more and more on machines and engines.

THE SECOND INDUSTRIAL REVOLUTION

America had its own version of the Enlightenment driven by the brilliant Ben Franklin. A Founding Father of the United States, Franklin was a prolific scientist and inventor. In 1752, this pudgy, pear-shaped physicist ventured out to a muddy field near Philadelphia. As rain clouds gathered and lightning sparked the dark sky, Franklin calmly fitted his kite with a key and then sailed the kite into the lightning. Luckily, he was spared electrocution as the lightning traveled down the kite lines to his Leyden jar. He had captured electricity in a jar, and his famous kite experiment ushered in the era of electricity.

Franklin's experiment led to a series of ideas, inventions, and breakthroughs. People began to understand, harness, and commercialize electricity, which had been studied since 1600. Along with the discovery and nascent efforts to harness electricity, energy generation shifted to oil and the internal combustion engine.

In 1851, Samuel Kiers, an American from Pennsylvania began selling kerosene to local coal miners, calling it carbon oil. Petroleum was seeping into some salt mines that Kier owned, fouling the salt, and he wanted to find a use for the oil. He invented a process to distil the crude and a lamp that burned the oil efficiently. Kier founded America's first oil refinery in

Pittsburgh in 1853. Taking the crude oil from his salt mines, he used a still to produce the first commercial illuminating oil from petroleum.

Kiers' efforts to rid his salt mines of petroleum led to commercial oil drilling and production. With oil, the internal combustion engine became the force for modern industrialization.

In its simplest form, an internal combustion engine harnesses a small intense explosion to drive a shaft or gears, like wheels on a car or a turbine or propeller. A small amount of high-energy fuel like gasoline is placed in a small, enclosed space. Then, air is mixed in and the mixture is ignited releasing energy in the form of expanding gas. This miniexplosion applies force to a part of the engine, usually to pistons, turbine blades, or a nozzle. This force pushes the engine part, converting chemical energy into useful mechanical energy.

A Belgian engineer, Jean J. Lenoir, developed the first commercialized internal combustion engine. He developed a single-cylinder, two-stroke engine in 1860. His engine burned a mix of coal gas and air, which was ignited by a "jumping spark" ignition system. In 1863, the hippomobile with a hydrogen gas-fueled one-cylinder internal combustion engine made a test drive in France from Paris to Joinville-le-Pont in about three hours. The auto had a top speed of six miles per hour. Leaping to the conclusion that oil and gas would soon be the new energy source, the Parisian news-paper *Cosmos* pronounced the steam age over (Georgano, 1990).

By the end of the nineteenth century, the first of the world's major auto manufacturer was founded in Germany. Wilhelm Maybach designed an engine built at Daimler Motoren Gesellschaft, following the specifications of Emil Jellinek. As part of the deal, Jellinek required that the engine to be named *Daimler-Mercedes* after his daughter. In 1902, automobiles with that engine were put into production by DMG (Georgano, 1990).

As the 2IR took hold, the internal combustion engine, fueled by various types of carbon and oil-based gases, became the driving force behind this new age of extraordinary machines—cars, trains, boats, space ships, etc.—that soon defined modern life.

Manufacturing and jobs grew as the assembly line made it possible to mass-produce goods and products. Except for some brutal wars in Europe, and the relentless genocide of the world's aboriginal natives, the planet's overall population expanded. With this expansion came ever more products and increased dependence on fossil fuels.

Along with machines, electricity, and transportation came the telephone—a technology that revolutionized the daily lives of ordinary people. There is some controversy surrounding its invention, including a claim that an Italian, Antonio Meucci, was the actual inventor, rather than

Alexander Graham Bell. However, it was Bell who spoke the first complete sentence from one telephone to another. This first of eventually several trillion calls was transmitted on 10 March 1876. To his assistant Thomas Watson, Bell said across the line, "Watson, come here; I want you." The commercialization of the telephone was an iconic example of American entrepreneurship.

This revolutionary technology provided the same explosion in communications as had Gutenberg's printing press. It was a new world of analog communications. Ideas, concepts, and images now driven by electricity could travel faster than ever before. Once more, science and knowledge exploded exponentially.

The second industrial revolution's (2IR's) commercialization of fossil fuels opened the world to the wonders of the personal transportation vehicle. At first, fossil fuels allowed for the transition from an agrarian society to an urban one and provided a way to make electricity. But when used to power a car, fossil fuels allowed urban workers to leave their city apartments and settle in the suburbs. This led to the need to construct highways and build housing developments, and America's car culture took root. Not only did the car become the means to get to work and how we measured success, but also, ironically, it became a symbol of rebellious freedom, like the spontaneous romantic exhilaration of Jack Kerouac's *On the Road*.

The US automobile culture with powerful internal combustion engines pushed fossil fuel as the main source of energy for the 2IR. *Source: Popular Hotroddding. com. http://www.popularhotrodding.com/features/1301phr_top_41_hottest_muscle_cars_in_ your_garages/photo_24.html.*

Unfortunately, the Western world's improving lifestyle—and the human passion for autos—has been dependent on fossil fuels. At one time, the coal, oil, and natural gas that powered this new economic model and the prosperity it entailed seemed relatively cheap, inexhaustible, and presumably harmless. More and bigger homes and buildings were built, more concrete poured, more fossil fuels extracted and burned. Frankly, there was not much to stop this social and economic juggernaut.

As industrialization led to urbanization, and urbanization to suburbanization, America built highways and thousands of miles of freeways that circled and interlaced our cities. The more concrete that was poured, the faster the suburbs grew and the American lifestyle was forever changed and thoroughly dependent on fossil fuels.

By the middle of the 1990s, the world's undeveloped countries followed and soon were building highways and suburbs that sprawled along concrete ribbons, creating congestion, generating pollutants, and producing an atmospheric overhang of smog. Societies around the world evolved to require greater and greater amounts of energy for light, heat, locomotion, mechanical work, and communications and then for smartphones, computers, televisions, microwaves, washing machines, coffee makers, and all the other technology and gadgets of modern living.

In fact, world energy consumption is predicted to grow by 53% from 2008 to 2035, according to the US Energy Information Administration. Much of the growth in energy consumption comes from developing nations like Brazil, Russia, India, and China—known as the BRIC nations. Energy use has exploded in these nations and predicted to increase 85% by 2030.

Power demand in the United States is going up 15-20% in the next decade because of the continuing electrification of our society. In the past, and probably in the near future, these energy demands will primarily come from fossil fuel—coal, natural gas, and oil.

To keep its energy-intensive lifestyle going, the United States uses about 19 million barrels of oil a day, or about 7 billion barrels of oil a year. In 2011, this worked out to about 22% of the world's total consumption of oil, which was about 32 billion barrels in 2011, according to the US Energy Information Administration (USEIA, 2011). This high level of consumption is not sustainable.

NOT ENOUGH SUPPLY, TOO MUCH DEMAND

For over 100 years, oil and gas discoveries have made fortunes for those optimistic and smart enough to understand the Earth's geology and exploit it or, in the case of the Middle East, lucky enough to live on top of enormous hydrocarbon deposits.

Now, scientists believe that the world's oil and natural gas supplies have peaked and are rapidly declining. As Hubbert (1956), the Shell Oil geophysicist, observed in his startling prediction first made in 1949, the fossil fuel era will be of very short duration. In 1956, he predicted that US oil production would peak about 1970 and then decline. At the time, he was scoffed at; now, Hubbert looks extraordinarily prescient.

While oil supply decreases, demand increases. In early 2011, China released customs data that showed that oil imports rose 18% in 2010. Platts, an oil industry research company, reported in October 2012 that China's oil consumption averaged 9.8 million barrels per day. Platts calculated that China's apparent oil demand is up 9.1 times year to year (China Oil Demand, 2013).

Driving this consumption is China's adoption of the automobile by its emerging middle class. Once a nation where everyone commuted by bicycle, China now has 60 million cars on the road, with 12-18 million more new cars predicted annually. To solve its demand problem, China state-controlled oil and gas companies have been buying massive amounts of oil and gas from around the world (BP, 2011).

However, as China prepares for its 13th Five-Year Plan (from 2016 to 2020), there is speculation that the nation will allow only electric and hydrogen fuel vehicles to be sold. That would be monumental for China's own auto manufacturing, and it would force other nations to convert rapidly to non-fossil fuel transportation (Hong, 2013).

India is not far behind China in oil consumption. India consumes nearly 3 million barrels per day as car sales have jumped (Asian Age, 2010). India is expected to be the fourth largest car market in the next 3 years (Jafry and Silvers, 2014).

The rest of the world's undeveloped societies are modernizing and using more oil for cars, trucks, airplanes, and boats. For example, Saudi Arabia is the world's largest oil producer. However, it is also the sixth largest oil consumer and internal consumption is growing rapidly. Increasing population and fuel subsidies are pushing internal demand growth to 7-9% per year (Luft and Korin, 2012: 33).

Meanwhile, as more and more oil and other fossil fuels are burned, pollution increases and emissions—called greenhouse gases (GHGs)—grow, spreading around the world and causing climate change. The buildup of GHGs has marched in lock step with the expansion of fossil fuel use since the 1700s. While primarily composed of CO_2, GHGs also include methane

(CH_4) and nitrous oxide (N_2O). The gases float upward into the atmosphere and wrap themselves like a blanket around the Earth. As more and more are added, the blanket gets thicker and warmer.

Unlike empty beer cans, plastic bags, or the other garbage that pile up along our roads and rivers, GHGs pile up out of sight, in the Earth's atmosphere above our heads. Visualize all the CO_2 that is released from cars, from coal and gas-burning power generation, and from the burning and clearing of forests and the deforestation of regions like Brazil or Indonesia.

As a consequence, the planet is getting hotter. Each day, the Earth is gaining huge amounts of extra energy from GHG emissions. NASA climate scientist James Hansen said the current increase in global warming is equivalent to exploding 400,000 Hiroshima atomic bombs per day 365 days per year (Economist, 2012: 3). That is 278 atomic bombs worth of energy every minute or more than four explosions per second. To be clear, this is the extra energy being gained each day. The results are devastating to the land areas and the atmosphere that is breathed daily around the world. Further, severe changes to weather patterns are created by the differences in temperatures and the impacts from evaporation as hot air streams hit cold water.

THE GREEN INDUSTRIAL REVOLUTION

In the midst of this fossil-fueled, internal combustion, GHG suffocating age, the green industrial revolution, or GIR, is starting to emerge.

Social scientists argue that industrial revolutions are triggered by the confluences of a new energy generating technology and a new form of communication technology that provides rapid dissemination of new ideas to accelerate the adoption of inventions. In the First Industrial Revolution, it was the steam engine and the printing press, and in the second, it was the internal combustion engine and analog communications. For the Green Industrial Revolution, renewable energy has combined with digital communications.

While Jeremy Rifkin (Rifkin, 2004) coined the term, the third industrial revolution, in his work, *The European Dream*, the larger paradigm of the GIR is more descriptive as of what is has already emerged in the EU and Asia. While Europe was and is environmentally conscious and aware of being "green," the GIR actually began in Japan and South Korea many years before it emerged in Europe.

As a small and densely populated island nation of 130 million people, Japan has a tradition of "no waste" that dates back to the Middle Ages. For centuries, Japan relied on its own natural resources for energy and development. Natural resources were exploited, but because 70-80% of Japan is mountainous or forested, development of land for commercial, farm, and residential use was limited. Even today, human waste is recycled for fertilizer. It is no wonder that for three decades, Japan leads in the creation of photovoltaic and other renewable energy systems. Its concerns for water conservation led to the success of TOTO, one of the greatest and most efficient water use companies in the world.

By the 1980s, Japan and South Korea were concerned with the need to become energy-secure, and, as a result, they developed national policies and programs to reduce their growing dependency on foreign fuels. These countries realized after World War II and the Cold War that their futures were not rooted in the same carbon-intensive economies that had built the United States and western Europe.

Decades later, Japan is once again struggling with an energy crisis, created by a devastating earthquake and tsunami in the northeast coastal region that destroyed one of the key Fukushima nuclear power plants. From this tragedy, Japan may leap even further ahead in developing a carbonless economy as it expands renewable energy generation to compensate for the loss of nuclear power. Other Asian nations are rapidly developing large-scale renewable energy generation as well.

Despite all the activity in Europe and Asia, few Americans (outside of a small circle of scholars and a handful of prescient venture capitalists and investment bankers) saw this new global megatrend looming. Even many people within the green industry have remained oblivious. For the most part, America's dependence on fossil fuels has clouded its ability to see that the carbon-based 2IR is ending. Today, the corporations and people vested in fossil fuels and related products from the 2IR are holding America back, preventing it from competing and advancing into the new green future.

The GIR, with its extraordinary new technologies and the promise of thousands of new green jobs, is trying to come to America. It is hampered by the lack of a national energy policy and a political process that is beholden to the fossil fuel industry. Big oil has been America's "elephant in the room" for over 100 years, exploiting the nation's resources, pushing the country into a dependence on foreign oil producers who are politically destabilizing, and not aligned with our national interests.

The United States remains in the 2IR, when in January 2010, the US Supreme Court institutionalized the problem. In *Citizens United v. Federal Election Commission*, the high court ruled that large corporations, with unlimited financial resources, are to be considered as "individuals." This means that they have no limits of freedom of speech in terms of financial and political influences. The powerful interests that buttressed America's lavish carbon-intensive lifestyle are using their enormous resources to influence public opinion and politics, trying to keep America desperately clinging to an era that the rest of the world is leaving.

The planet is threatened by an environmental and climate catastrophe of unimaginable proportions. Population is the ticking time bomb. The United Nations predicts that we will increase from today's 7 billion people to 10 billion by 2050. In other words, we will add 3 billion people in less than 40 years. China will add 320 million people for a total of 1.4 billion, India will add 600 million for a total of about 1.5 billion, while the United States will add 120 million for a total of about 400 million residents by midcentury (United Nations, 2010).

All natural resources, particularly fossil fuels, are finite. Experts are warning that there are not enough resources and that we are inviting environmental collapse. In the 2014 Quarterly Spring Report, the US Department of Defense said,

> The impacts of climate change may increase the frequency, scale, and complexity of future missions, including defense support to civil authorities, while at the same time undermining the capacity of our domestic installations to support training activities. Our actions to increase energy and water security, including investments in energy efficiency, new technologies, and renewable energy sources, will increase the resiliency of our installations and help mitigate these effects.
>
> **US DOD (2014): VI**

This new industrial revolution features fast-as-light communication of the digital age with its Internet access to almost all scientific knowledge and the Facebook- and Twitter-led social networking that has truly created Marshall McLuhan's "global village." This digital age will intersect with renewable and sustainable sources for power. Smart grids, intelligent machines, and additive manufacturing will augment it. This emerging worldwide GIR is being led by the Asian nations, particularly China. The United States is lagging far behind.

In major historical irony, the communications tools of this new GIR helped overthrow the notorious despots who ruled the countries that controlled the world's oil supply. The Arab Spring, which has changed the political reality of the Middle East, was made possible by the instant communications of the social networks and Facebook, in particular.

The GIR has the potential to be more significant and life changing than either the 1IR or the 2IR. It may also turn out to be the planet's only real chance for survival. With an estimated 10 billion inhabitants by midcentury, there is so much more at stake.

Despite the wild claims by the oil and natural gas industries that there is an abundant supply, the reality is that the world is running out of fossil fuel, particularly oil. This alone threatens to shake the very foundation of human existence. Adding a heightened sense of urgency is the environmental degradation and the collapse of various parts of our planet's ecosystem, like the Brazilian watershed and the Arctic.

Fortunately, in some parts of the world, the GIR has begun. Parts of Asia and Europe have been moving into it for over three decades, developing sustainable, energy-independent communities. South Korea has urban regions that are already energy-independent and carbon-neutral (Clark, 2000).

Japan was heading in this direction as well but got redirected toward nuclear power stations and plants in the 1970s. However, after the March 2011 nuclear disaster at Fukushima, the Japanese government is replacing nuclear power with renewable energy systems for building complexes and individual homes (Adams and Funaki, 2009).

Meanwhile, a large-scale effort is under way in China where the nation has overtaken other countries in the new GIR. In 2008, the Climate Group, an international think tank, reported China's rapid gains in the race to become the leader in developing renewable energy technologies via its 12th Five-Year Plan. This plan that started in March 2011 committed the nation to spending the equivalent of over three trillion dollars in funding for renewable energy (Climate Group, 2008; Clark and Isherwood, 2010; Wen, 2012; Sun et al., 2013).

Germany through its feed-in tariff (FiT) program was the number one producer and installer of solar panels for homes, offices, and large open areas from 2006 to 2009. In 2010, Italy then copied the FiT and held that distinction of world leader in solar panel installation. China took the lead in 2011 and continues as the number one solar panel and photovoltaic manufacturer and installer. Japan is now leading in auto manufacturing, jumping ahead of the competition with its hybrids (Gipe, 2014).

Other European nations like Spain and the Nordic countries are pursuing policies to achieve energy-independent through renewable energy. They are succeeding. Denmark has made extraordinary advances already. The Danes have a program that includes local plans and financing to develop on-site energy-renewable power systems. By 2015, several Danish cities will be energy-independent with renewable energy power and smart green grids with the whole nation 100% using renewable energy by 2025 (Lund and Østergaard, 2010; Lund and Hvelplund, 2012).

SUSTAINABILITY IS THE KEY

The decline in natural resources and fossil fuels and increasing climate change plus an accelerating population are pushing us closer to environmental catastrophe. If global energy policies do not change, political and social tensions will mount over the supplies and locations of fossil fuels as they become scarcer and more expensive. The decline in fossil fuel and rise in climate change will exacerbate the difficulties in feeding the world's expanding population.

The way out is by embracing the GIR and its promise of sustainable communities, renewable and distributed energy, and smart grids. Asian and European nations have set the pace for sustainable and secure communities with their own renewable energy sources, storage devices, and emerging technologies.

Sustainable and green communities represent an improved new design for how we can live, particularly in urban areas. They can integrate renewable energy generation and storage technologies with non-fossil fuel transportation. They can focus on environmentally sensitive business development, green job creation, and healthy social activities. Scientists describe this as sustainable development or the integration of a community's energy and infrastructure requirements, economic needs, and social activities for the protection and preservation of the environment. Business and new commerce are stimulated by this interaction, which in turn provides economic reasons for pursuing and creating sustainable communities.

Most modern cities have the potential to implement some, if not all, sustainable activities. With a little guidance, most communities can have locally distributed renewable energy, clean water, recycled garbage and waste, and efficient community transportation systems that run on renewable energy sources for power. This generation must create a sustainable lifestyle that is free from the carbon-intensive, fossil fuel-based, inefficient centralized energy generation of the past (Most Polluted Cities, 2013).

RENEWABLE ENERGY AND DISTRIBUTED ON-SITE POWER

Renewable energy generation is the foundation for a sustainable community and the heart of the GIR. Basically, renewable energy is a source of energy that is not carbon-based and will not run out. The most common are systems that create power from wind, sun, or water. Yet these renewable power system are intermittent and they also need storage systems. The integration of these systems is a key component to sustainable communities (Clark and Eisenberg, 2008).

Other renewable systems use digestive processes that can convert waste into biomass and other systems that can recycle waste for fuel generation. Still other renewable sources include geothermal, run-of-the-river systems, which include small hydroelectric turbines to create electricity and do not involve creating dams. Other systems can convert bacteria and algae into energy.

Wind generation is fairly straightforward. Wind has been used as a power source for thousands of years. Today, a large propeller is placed in the path of the wind. The force of the wind turns the propeller and a gear coupling interacts with a turbine, which generates electricity. The concept of wind generation may be ancient, but technological advances have transformed it. New generation wind turbines are extraordinarily sophisticated machines that use new materials like carbon fiber. They are strong, quiet, and cost-efficient.

Wind farms harness the energy of dozens, even hundreds, of wind turbines. Turbines can be installed on land or offshore. Small ones can be placed on rooftops or in highway medians to capture airflow.

Solar generation systems capture sunlight, including ultraviolet radiation, via solar cells made from silicon. Passing sunlight through silicon creates a chemical reaction that generates a small amount of electricity. A photovoltaic, or PV, reaction is at the core of solar panel systems. A second process uses sunlight to heat liquid, which is then converted to electricity. A number of communities are now installing solar "concentrated" systems in which the sun is captured in heat tubes and used for heating and cooling of homes, buildings, and central power plants. Solar, like wind, can be installed on homes and building complexes closer to the end user, instead of on distant farms. That is why, on-site renewable power generation is the key to becoming energy-independent with solar and wind power as well as other renewable energy sources.

Biomass is biological material from living or recently living organisms such as plants that can be used to generate energy. This remarkable chemical process converts plant sugars like corn into gases like ethanol or methane. The gases are burned or used to generate electricity. The process is referred to as "digestive" and it is not unlike an animal's digestive system. Abundant and seemingly unusable plant debris—rye grass, wood chips, weeds, grape sludge, almond hulls, and the like—can then be used to generate energy. Algae can be grown in ponds and harvested and used for biomass fuel.

Geothermal power is created from heat stored in the Earth. This heat originates from the formation of the planet, from radioactive decay of minerals, and from solar energy absorbed at the Earth's surface. It has been used for space heating and bathing since ancient Roman times but is now better known for generating electricity. In 2007, geothermal plants worldwide had the capacity to generate about 10 GW, or 10 billion watts of power, and in practice generated enough power to meet 0.3% of the global electricity demand (GEA, 2014).

In the last few years, engineers have developed remarkable devices such as geothermal heat pumps, ground source heat pumps, and geoexchangers that gather ground heat to provide heating for buildings in cold climates. Through a similar process, they can use ground sources for cooling buildings in hot climates. More and more communities with concentrations of buildings, like colleges, government centers, and shopping malls, are turning to geothermal systems.

In 2013, New Zealand completed the world's largest binary geothermal power plant! Designed by Mighty River Power, the 100-MW Ngatamariki geothermal power station is the country's third geothermal project. The plant is located near Taupo on the country's North Island. *Source: http://inhabitat.com/the-worlds-largest-binary-geothermal-power-plant-opens-in-new-zealand/.*

Ocean and tidal waves have power that can be harnessed to create usable energy. That is the concept behind the revolutionary SeaGen tidal power system, the world's first large-scale commercial tidal stream generator, installed in Strangford Narrows in Northern Ireland. (OPT, 2014).

America, particularly the Pacific coastline, is equally suitable for producing massive amounts of energy with the right technology. Ocean power technologies vary, but the primary types are *wave power* conversion devices, which bob up and down with passing swells; *tidal power* devices, which use strong tidal variations to produce power; *ocean current* devices, which look like wind turbines and are placed below the water surface to take advantage of the power of ocean currents; and *ocean thermal energy conversion devices*, which extract energy from the differences in temperature between the ocean's shallow and deep waters.

Run-of-the-river systems generate electricity without the large water storage required of traditional hydroelectric dams. Run-of-the-river systems are ideal for streams or rivers with consistent water levels or minimum loss of water flow during the dry season. In most cases, power turbines are mounted along the river, and as the water flows, the turbines generate electricity. This is being done in Europe and Asia, where the flowing river water generates considerable amounts of energy without harming the surrounding land or changing the natural elements in the water. These systems do minimal destruction to pristine environments and could easily be adapted to large inland rivers (Run of River, 2014).

Fuel cells are electrochemical cells that change a source fuel into an electrical current. They generate electricity through reactions between a fuel and an oxidant, triggered in the presence of an electrolyte. The reactants flow into the cell, and the reaction products flow out of it, while the electrolyte remains within it. Fuel cells are energy storage devices that can operate continuously as long as the necessary reactant and oxidant flows are maintained (Clark et al., 2002; Cooper and Clark, 1996).

Fuel cells are different from conventional electrochemical cell batteries in that they consume reactant from an external source, which must be replaced (Clark, 2007; Clark, 2008; Clark et al., 2006; Clark et al., 2005). Many combinations of fuels and oxidants are possible. A hydrogen fuel cell uses hydrogen as its fuel and oxygen as its oxidant. Other fuels include hydrocarbons and alcohols and chlorine and chlorine dioxide.

Bacterial, or microbial, fuel cells use living, nonhazardous microbial bacteria to generate electricity. BP has made a $500 million investment in this futuristic process, which is now being developed by researchers at the University of California, Berkeley, and the University of Illinois, Urbana.

Researchers envision small household power generators that look like aquariums but are filled with water and microscopic bacteria instead of fish. When the bacteria inside are fed, the power generator—referred to as a "biogenerator"—would produce electricity. Ironically, the funding for this technology comes from the same company that caused the April 2010 oil spill in the Gulf of Mexico that killed 11 people, damaged the Gulf waters, and polluted the coastland while destroying fishing and tourist businesses.

GOVERNMENT SUPPORT IS CRUCIAL

The huge carbon energy industries were launched with government support. Even today, these megacorporations receive vast sums in subsidies from various governments. Similar government support needs to go into the sustainable technologies of the GIR. The EU has consistently taken this position, including its efforts to create a hydrogen highway throughout the EU (Clark, 2004a; Clark, 2004b). Incentives that are now going to the carbon-generating industries—$4 billion annually in the United States to the oil companies—must be reduced and applied to renewable energy generation. This tax shift has been very successful in other industries and can be designed so there is little or no additional tax burden on consumers (Clark and Demirag, 2005; Fortune, April 2012).

In the 1990s, energy deregulation took hold in America but was called "privatization" or "liberalization" in Europe. It failed because the policies turned into market manipulation, fraud, and loss of service (Clark and Bradshaw, 2004).

To move the GIR forward will require modification to these deregulation rules and new policies. Central power plants are still the norm and regulations are needed to oversee supplies, costs, and delivery of energy. German and Danish central power plants, for example, have significant government involvement through partial ownership or by appointed directors. The concept has begun in the United States and it is critical in China where large renewable power systems are partly state-owned (Borden and Stonington, 2014).

Local energy generation is critical. While some fossil fuels, like coal, oil, and gas, are still cheap and used in central power plants, they are the major global atmospheric polluters. If a carbon tax or some other method that calculated the damage to the environment and public health were added, the costs would be higher than renewable power systems. That public policy and economic strategy is beginning in the Europe and China today.

China and the United States rank at the top of the emission polluters, and in both cases, coal is the major problem. If the human and environmental impacts of coal were calculated into its true costs, then the real cost of coal energy generation for power would soar.

The 2IR pushed the operation of large fossil-based power plants in the early part of the twentieth century and then nuclear power plants in the century's last half. The plants had to be powerful to withstand the degradations over the vast distribution of a central-powered grid system. At each conversion from alternating current (AC) to direct current (DC), electricity loses some power, but for the fossil or nuclear fueled plants, there is so much power coming out at the beginning that it did not matter at the end. This system resulted in the loss of efficiency in transmission over power lines as well as the constant need for repairs and upgrades.

Not so in the case of the environmentally friendly renewable systems. For best results, community energy systems need nearby renewable power generation connected to "smart" distribution systems, so electricity does not have to travel far and lose efficiencies. An alternative is to hook into a transmission line. This way the local grid is added to the existing distribution system and the transmission line can act as a battery for the renewable energy that needs storage. A system like this is similar to the Internet where there is no single area for control over data, or in this case power, rather it is spread out and localized.

The sooner the world evolves into the GIR, the sooner the planet will begin to heal. How quickly this happens will be heavily influenced by the United States, the world's leading democracy. The United States can no longer afford oil wars, considering their financial impact and their costs in human lives and injuries. Nor can the country afford another environmentally crippling deep ocean oil spill, such as the April 2010 BP spill in the Gulf of Mexico, nor too many devastating storms like Hurricane Sandy.

THE NEXT ECONOMICS: SOCIAL CAPITALISM

It is not all about money—or is it?

In light of the October 2008 world financial meltdown, which even in 2014 continues with the monetary crisis in Europe, it seems silly to think that the supply-side, deregulated, free-market economics so passionately espoused by President Ronald Reagan and Prime Minister Margaret Thatcher in the 1980s will work for a twenty-first-century world threatened by irreversible environmental damage.

The 2008 economic implosion from trillions of dollars in hedge funds, subprime mortgages, credit swaps, and related marginal derivatives nearly pushed the Western world's financial structure into the abyss. It underlined what happens when governments ignore their responsibility to govern.

Market economists and others had argued that there was no need for regulation. Government would act as "the invisible hand."

In the end, the worst financial disaster since the Great Depression was a testament to the venal side of free-market capitalism—greed, stupidity, carelessness, and total disregard for risk management. These are not behaviors that can be repeated if the planet is going to survive climate change and its impact on the Earth and its inhabitants.

The GIR must develop an economy that fits its social and political structures, similar to the 1IR and 2IR. The first one replaced an agrarian, draft animal-powered economy with one powered by steam engines and combustion machine-driven manufacturing, an evolution that was accelerated by colonial expansion. The second created a fossil fuel-powered economy that extracted natural resources in an unregulated, consumer-fed, free-market capitalist society.

As the GIR grows, the world will become much more interdependent. What happens in one part of the world, be it weather, pollution, politics, or economics, impacts other regions. For example, the dramatic change in the Egyptian government in early 2010 has affected the rest of the Middle East and will result in global changes of oil and gas supplies. The result might well be the forced end of the 2IR as continuous Middle East turmoil forces developed nations to push for energy independence with renewable energy sources.

There is historical precedence for this forced transition. The Arab oil embargo of the early 1970s pushed Europe and Asia toward social policies that eventually led to the beginnings of the GIR. Energy independence, climate change, and environmental protection became serious political issues. Both these regions have been developing economic forms of what has become known as social or collective capitalism, an economic view that includes sustainable growth, health and educational issues, environmental concerns, and climate change mitigation, along with interest in diverse populations, gender equality, and democratic processes. The essence of social capitalism is that there are some social and political problems so complex and overriding that free markets and deregulation cannot address them.

Social and environmental factors—sustainable communities, climate change mitigation, and environmental protection—are growing in importance and will soon demand far greater international cooperation and agreement. Rampant economic growth and individual accumulation of wealth are being replaced by social and environmental values that benefit the larger community. For example, the European Union is pushing for limits on the salaries of corporate executives.

Without a national policy and investment, countries cannot address their basic infrastructures. Without government consensus, there can be no action, no improvement, no resources, and certainly no response to environmental degradation. For example, the United States' inability to develop a national energy policy that addresses climate change is often cited as a monumental failure of its free-market and deregulated economic model. Energy and infrastructure, the argument goes, are extraordinarily important national issues, just as important as defense or entitlement programs. To address these basic systems for the greater good, a nation needs to have plans, which are outlined and offered by the central government.

The People's Republic of China, not the United States, is showing real global leadership in responding to climate change. More than anything, China demonstrates how important the role the government plays in overseeing, directing, and supporting the economics of technologies and creation of employment. China's economic system is the prototype of social capitalism. Since the 1949 revolution, the Chinese have moved away from communism toward economic development through a series of five-year plans, now being referred to as guidelines.

In the post-Mao era, China moved aggressively into a market capitalism system, but one where state institutions were owned in part by the Chinese government and shared in joint ventures with foreign companies. Companies wanting to do business in China had to keep their profits there for reinvestment and have at least 49% of the company owned by the Chinese government. Today's new Chinese government leadership will put more controls on industrial growth and more focus on mitigating climate change (Lo, 2011).

Europeans adjusted their economies to fit the requirements of the GIR early on. Both the Scandinavians and the Germans realized that the move away from fossil fuels to renewable energy distribution would require more than neoclassical free-market economics could deliver. While the Danes and the other Scandinavians shifted national energy resources toward renewable energy power by national consensus, the Germans developed the innovative feed-in tariff process.

Germany's FiT was part of their 2000 Energy Renewable Sources Act, formally called the Act of Granting Priority to Renewable Energy Sources. This remarkable policy was designed to encourage the adoption of renewable energy sources and to help accelerate the move toward grid parity, making renewable energy for the same price as existing power from the grid (Morris, 2014).

Creating an economy that can move the world into the GIR is an exceptionally complex process. Various governments and states are approaching the problem differently. The European FiT program and China's direct government subsidies have been the most successful. Some US states, such as California with its newly designed Renewable Auction Mechanism (RAM), have developed possible improvements over the European FiT. But the RAM is much more limited and available only in California.

So the GIR is not all about the money. It is about climate change mitigation, renewable energy, smart grids, and health, education, and environmental sensitivity. But achieving the benefits—a wave of new technologies, business enterprises, and green jobs—will require substantial public support.

Yet with global warming and climate change impacting daily lives, the United States is still mired in fossil fuel dependence without the political will to break loose and move forward.

As the world's leading democracy, the United States must also lead the GIR. The United States must establish a national energy policy with funding that makes sense. One that can move the entire country rapidly from the fossil fuels and wholesale resource extraction of that dominated the twentieth century to a renewable energy based system for the twenty-first century. The old, "dirty" economy was dependent on fossil fuels and internal combustion engines, as well as heavy manufacturing based on cheap oil and massive infrastructures to support energy and transportation. The GIR is about using renewable energy to power local communities where renewable power and smart grids can monitor power and increase efficiencies.

Humanity's lust for fossil fuels may be our ruin. In the 6 June 2012 issue of *Nature,* Anthony Barnosky of the University of California, Berkeley, along with 21 coauthors publicly worried about the same thing. The planet may be on the verge of a "tipping point" they cautioned, which is a point of environmental decline that cannot be reversed (Climate Change, 2013).

The authors called for accelerated cooperation to reduce population growth and per-capita resource use. Fossil fuels need to be replaced with sustainable sources. We need to develop more efficient food production and distribution systems, and more protection for land and ocean is needed. Barnosky wrote, "Humans may be forcing an irreversible, planetary-scale tipping point that could severely impact fisheries, agriculture, clean water and much of what Earth needs to sustain its inhabitants."

These scientists are among those urging the world to join the GIR before it is too late.

REFERENCES

Adams, L., Funaki, K., 2009. Japanese experience with efforts at the community level towards a sustainable economy. In: Clark II, W.W. (Ed.), Sustainable Communities. Springer Press, New York.

Asian Age, 2010. India Car Sales Jump 21 Percent. www.asianage.com/business/india-car-sales-jump-21-cent-504.

Borden, E., Stonington, J., 2014. Germany's Energiewende. In: Global Sustainable Communities Design Handbook: Green Design, Engineering, Health, Technologies, Education, Economics, Contracts, Policy, Law and Entrepreneurship. Elsevier Press, New York.

BP report, 2011. China the Fuel for Growth. Issue 1. www.bp.com/sectiongenericarticle.do?categoryId=9037009&contentId=7068199.

China Oil Demand, 2013. http://news.yahoo.com/platts-report-chinas-oil-demand-1500502.

Clark II, W.W., 2000. Developing and Diffusing Clean Technologies: Experience and Practical Issues. In: OECD Conference, Seoul, Korea, 2000.

Clark II, W.W., 2004a. Hydrogen: the pathway to energy independence. Utilities Policy. Elsevier.

Clark, W.W. II., 2004b. Innovation for a Sustainable Hydrogen Economy, Boosting Innovation from Research to Market. Brussels, Belgium: European Union. Spring 2004. pp. 65–67.

Clark II, W.W., 2007. The green hydrogen paradigm shift. Co-Generation and Distributed Generation Journal 22 (2), pp. 6–38.

Clark II, W.W., 2008. The green hydrogen paradigm shift: energy generation for stations to vehicles. Utilities Policy Journal. Elsevier Press.

Clark II, W.W., Bradshaw, T., 2004. Agile Energy Systems: Global Solutions to the California Energy Crisis. Elsevier Press, London, UK.

Clark II, W.W., Demirag, I., 2006. US financial regulatory change: the case of the California energy crisis. Special Issue, Journal of Banking Regulation 7 (1/2) pp. 75–93.

Clark II, W.W., Eisenberg, L., 2008. Agile sustainable communities: on-site renewable energy generation. Utilities Policy 16 (4), 262–274.

Clark II, W.W., Isherwood, W., 2010. Report on energy strategies for Inner Mongolia Autonomous Region. Utilities Policy Journal 18, pp. 3–10. http://dx.doi.org/10.1016/j.jup.2007.07.003.

Clark II, W.W., Paulocci, E., Cooper, J., 2002. Commercial Development of energy—environmentally sound technologies for the auto-industry: the case of fuel cells. Journal of Cleaner Production 11, pp. 427–437.

Clark II, W.W., Rifkin, J., et al., 2005. Hydrogen energy stations: along the roadside to a hydrogen economy. Utilities Policy 13, 41–50 Elsevier Press.

Clark II, W.W., Rifkin, J., et al., 2006. A green hydrogen economy. Special Issue on Hydrogen Energy Policy 34, 2630–2639 Elsevier, Fall 2006.

Climate Change, 2013. University of Oslo. http://davis.patch.com/articles/are-humans-bringing-earth-to-an-irreversible-tipping-point.

Climate Group, 2008. China's Clean Revolution. www.guardian.co.uk/environment/2008/aug/01/renewableenergy.climatechang.

Cooper, J.F., Clark II, W.W., 1996. Zinc/air fuel cell: an alternative to clean fuels in fleet electric vehicle applications. International Journal of Environmentally Conscious Design & Manufacturing 5 (3-4), 49–54.

Economist: Special Report—The Melting North. June 16, 2012. p. 3.

Febvre, L., Martin, H.-J., 1976. The Coming of the Book: The Impact of Printing 1450-1800. New Left Books, London.

Georgano, G.N., 1990. Cars: Early and Vintage, 1886-1930. Grange-Universal, London, p. 39.

Geothermal Energy Association (GEA), 2014. http://www.geo-energy.org/data/2014.

Gipe, P., 2014. Feed-in-Tariff Monthly Reports. http://www.wind-works.org/FeedLaws/RenewableTariffs.qpw.

Hong, L., 2013. Developing an analytical approach model for Offshore Wind in China, PhD Thesis.

Hubbert, M.K., 1956. Nuclear Energy and the Fossil Fuels 'Drilling and Production Practice'. (PDF). In: Spring meeting of the Southern District, Division of Production. American Petroleum Institute. San Antonio, Texas (Shell Development Company, June 1956), pp. 22–27.

Jafry, N., Silvers, G., 2014. Micro Cities: the case of India. In: Clark II, W.W. (Ed.), Global Sustainable Communities Design Handbook: Green Design, Engineering, Health, Technologies, Education, Economics, Contracts, Policy, Law and Entrepreneurship. Elsevier Press, New York, NY.

Wen, L., 2012. The Integration of Sustainable Transport in Future Renewable Energy Systems in China. PhD Thesis.

Lo, V., 2011. China's Role in Global Economic Development. Speech by Chairman of Shui On Land, given at Asian Society, Los Angeles, CA, April 25, 2011.

Luft, G., Korin, A., 2012. The American Interest, The Folly of Energy Independence. p. 33.

Lund, H., Østergaard, P.A., 2010. Climate Change mitigation from a bottom up community approach: a case in Denmark. In: Clark II, W.W. (Ed.), Sustainable Communities Design Handbook. Elsevier Press, New York.

Lund, H., Hvelplund, F., 2012. The economic crisis and sustainable development: the design of job creation strategies by use of concrete institutional economics. Energy 43, 192–200, Elsevier Press, New York, NY.

Morris, C., 2014. Energiewende—Germany's community-driven since the 1970s. In: Global Sustainable Communities Design Handbook: Green Design, Engineering, Health, Technologies, Education, Economics, Contracts, Policy, Law and Entrepreneurship. Elsevier Press, New York, NY.

Most Polluted Cities, 2013. http://www.stateoftheair.org/2013/city-rankings/most-polluted-cities.html.

Ocean Power Technologies (OPT), 2014. info@oceanpowertech.com.

Rifkin, J., 2004. European Dream. Penguin Patnam, New York.

Run of River, 2014. http://www.runofriverpower.com/.

Sun, X., Li, J., Wang, Y., Clark II, W.W., 2013. China's Sovereign Wealth Fund Investments in overseas energy: the energy security perspective. Energy Policy Journal 65, 654–661. http://dx.doi.org/10.1016/j.enpol.2013.09.056.

United Nations (UN), 2010. Revision of World Population Prospects. www.un.org/esa/population.html.

US Department of Defense (US DOD), 2014. Quadrennial Defense Review, "Executive Summary". Secretary of Defense, Washington, DC, p. VI.

USEIA International Energy Outlook, 2011. 20554.135.7/forecasts/ieo/.

CHAPTER 3

Big Oil's Impact

In one of history's great ironies, today's gigantic oil industry owes its existence to the slaughter of whales—one of the world's most spectacular animals.

Creatures that left dry land some 50 million years ago, whales were the unchallenged masters of the oceans until the late 1700s. At their peak, they numbered in the millions and inhabited every sea. These marvelous ocean-dwelling mammals ranged from the huge blue whale (the world's biggest animal) at around 30 meters or 100 feet and 180 tons to the pygmy sperm whale at 3 meters.

Though humans have hunted whales since ancient times, it was not until around 1700 that commercial whaling took hold. Originally a food source, whales became prized for their oil, which slowly replaced tallow candles as a light source. Gallons of whale oil were used as an illuminant in lamps and as candle wax. At the same time, the world's great cities added street lamps to facilitate the growing crowds of nighttime revelers. By the end of the 18th century, the world's whale population was overhunted and in serious decline. Whale oil became scarce and expensive, and the world looked for an alternative.

Kerosene, or paraffin as it is known in the United Kingdom, is a thin clear liquid made from hydrocarbons. It started as a substitute in lamps and then later in heaters to replace coal. Al-Razi, a 9th-century Persian physician and chemist referred to a simple lamp in his notes that used crude mineral oil. He called it "naffatah" and described a rough distilling process using crude oil and bitumen. Al-Razi used clay as an absorbent, repeating the distillation process until the liquid was clear and safe to light.

By the mid-1800s, with whale oil becoming scarce, a Canadian geologist, Abraham Gesner invented a coal distilling process that produced a clear fluid that burned well in lamps. He called it "kerosene," a contraction for "keroselaion" or wax oil.

Others followed Gesner and distilled kerosene from coal and carbon-based ore. Eventually, Samuel Kiers used the concepts to develop his carbon oil from the petroleum that seeped into his Pennsylvania salt mines. When Kiers added an efficient lamp for his distilled crude oil, and followed it up with the first US oil refinery, the world's multitrillion dollar oil industry was born.

OIL: AN ANCIENT PRODUCT

Oil is a product of ancient life millions of years ago. It was made deep in the earth, the crushed remains of rotting organic matter that clumped together in large low spots in rock. While a few dinosaurs contributed to these organic-rich stews, most of the matter came from plankton, tiny microscopic one-cell plants and animals that blanketed the prehistoric seas. Similarly, prehistoric algae were thrown into the mix from large freshwater lakes.

As these tiny creatures died, masses of them sank to the bottom of the seas and lakes. Over time, they were covered in sediment, encapsulated like a stew in huge Dutch ovens. As more and more sediment covered the deposits, pressure and heat built up. Being organic matter, these stews were a thick mix of hydrogen, carbon, and oxygen. With heat, these compounds broke down into hydrocarbons. It took millions of years with just the right temperature and pressure for these materials to cook into oil. Too much sediment on top of the remains would overcook the matter and turn oil into natural gas or unusable material. Too little sediment meant no oil at all (History of Oil, 2014).

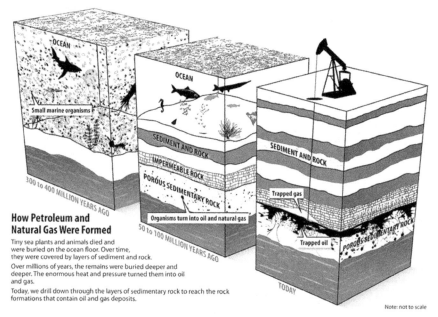

How Petroleum and Natural Gas Were Formed

Tiny sea plants and animals died and were buried on the ocean floor. Over time, they were covered by layers of sediment and rock.

Over millions of years, the remains were buried deeper and deeper. The enormous heat and pressure turned them into oil and gas.

Today, we drill down through the layers of sedimentary rock to reach the rock formations that contain oil and gas deposits.

Note: not to scale

Source: Oring, CNG Fuel Systems, LLC. http://www.oringcngfuelsystems.com/where-does-cng-come-from/.

Finding oil has always been a wild and exciting explorative process. While most of it is buried deep, some oil has always been right here, sitting on the Earth's surface.

Prehistoric humans first contact with oil came in the form of smallish sludge or tar lakes filled with a black, oily viscous material. This is bitumen, a naturally occurring by-product of decomposed organic material. Animals and early humans would blunder into these sticky, watery-looking lakes and not be able to get out. In the United States, California's La Brea Tar Pits are one place where a smelly oily bitumen lake can be seen and smelled. They are now a natural history museum that traces the pit's existence back to the end of Pleistocene era, or about 12,000 years ago.

These bitumen seeps were known to the earliest dwellers of the region, and in 1875, William Denton, a college professor, first discovered fossil remains. Over the years, paleontologists have painstakingly excavated, categorized, and exhibited over one million ice age fossils. The fossils represent 650 species, including mammoths, bears, sloths, smilodon, camels, and condors (the La Brea Tar Pits, Page Museum).

Seeps or pits of bitumen occurred in other parts of the prehistoric world. This sticky tar was used by the Neanderthals to fasten wooden hafts to sharp edge tools some 40,000 years ago. Other ancients used this asphalt or tar as a sealant or adhesive. It was used as mortar for buildings, as incense, and as decorative applications on pots or human skin. Bitumen was used to waterproof canoes and reed boats and in the mummification process toward the end of ancient Egypt's New Kingdom. In fact, the word from which mummy is derived "mūmiyyah" means bitumen in Arabic. In Mesoamerica, bitumen stains were found on human remains, perhaps as a ritual pigment or from the axes used to dismember the bodies.

Ancient Mesopotamia prospered by trading bitumen to the Egyptians and others in North Africa. This "land between the rivers"—between the Tigris and the Euphrates rivers—was dotted with all sorts of petroleum deposits, including bitumen seeps, crude oil springs, and even bituminous rock, which released crude oil when heated.

The great king of Babylon, Nebuchadnezzar II, used bitumen as mortar for the extraordinary Tower of Babel. A great builder, he fashioned a bridge over the Euphrates using piers of burnt bricks cemented and coated with bitumen. He also constructed large sewers lined with a mixture of bitumen, clay, and gravel. He laid down the first paved streets by setting stone slabs in bitumen mortar (Bilkadi, 1984).

The Chinese discovered bitumen and natural gas as a by-product of digging for salt. Early records show that the Chinese used bitumen and natural gas before the start of the Han dynasty around 400 B.C. Of all the historic uses of bitumen, or natural asphalt or tar, nothing was as important as the discovery that this substance would burn. By the time of the prophet Muhammad, around 600 A.D., crude oil was being used as fuel by the Iranians. There is evidence that around the same time, the Syrians invented an incendiary mix of petroleum, quick lime, and sulfur, which oddly they called "Greek fire." The ancient Chinese called the natural gas deposits fire wells (Bilkadi, 1984).

The Byzantines adapted Greek fire to open warfare around 675 A.D. They mounted pumps in the bows of their warships and spewed the flaming mixture onto their enemies' vessels. By 850, the Arabs were using it in the Indian Ocean to protect their ships against pirates, and their infantry had archers that shot grenades of the fire at the enemy. They used it in Asia, where the Chinese adopted it.

In the ninth century, Muslim forces were using large quantities of flammable petroleum products. This was the result of distillation, an amazing technological advance. Early distillation consisted of pouring boiling bitumen through a stretched animal hide over a kettle. As the hide was wringed, a condensed liquid would drip into the kettle. An Egyptian scientist al-Mas'udi discovered a crude process similar to today's "cracking" technique. His process used two superimposed jars separated by a sieve. The upper jar, filled with bitumen, was heated with a fire, and the oily distillate allowed to drip through the screen into the bottom jar buried in damp sand.

The Muslims also invented an advanced distillation technique, called *taqtir,* which used a long glass column capped with a water-cooled condenser. A devise known an "alembic" was originally used to distill olive oil. These Muslim scientists were exceptional in developing refining techniques. By the 12th century, kerosene or white naphtha could be purchased in the streets of Damascus. A distilling industry had developed in Damascus and Egypt.

But it was the medical applications, not the warfare technologies, that brought bitumen to western Europe. The Europeans learned of bitumen's healing process from Muslim physicians around the 12th century. Unfortunately, at about the same time, science went into a severe decline, in both Europe and the Middle East. During this decline, bitumen, kerosene, and crude oil were forgotten. It took until the middle of the 19th century when the first modern oil well was drilled that the vast potential uses for petroleum began to be recognized. Eventually, oil would become the very lifeblood of the second industrial revolution.

THE OIL INDUSTRY

According to *IBIS*, which publishes global business analysis, the international oil and gas exploration and production industry will generate about $5 trillion in revenue in 2014, compared to $3.03 trillion in 2007. From 2009 to 2014, the industry had an annualized growth of 11.9 percent. *IBIS* reports that the gain in revenue over the 5 years is from higher oil and gas prices well in excess of the general inflation rate. In 2012, the industry's estimated profit was about $3.37 trillion, or about and 5.1 percent of global GDP (IBIS, 2014).

Climate Progress' 2012 report estimated that the big five oil companies in the United States—BP, Chevron, ConocoPhillips, ExxonMobil, and Shell—combined to make more than $1 trillion from 2001 to 2011. They earned $375 million in profits per day in 2011, or about $261,000 per minute (ClimateProgress, 2012a,b).

How did today's international petroleum industry become a juggernaut—the largest, most profitable and politically powerful industry in human history? Wealthy and powerful enough so that the industry overlaps borders and its executives try to skirt national sovereignty.

The oil industry is unique in how commercial and government interests intersect and sometimes collide. The industry is roughly split between state-owned and publicly owned megacompanies. State-owned companies like Saudi Aramco and Russia's Rosneft dominate exploration and production. Saudi Aramco is the world's wealthiest and largest petroleum company. It controls an estimated 25 percent of the world's petroleum reserves, with annual revenues of over $2 trillion (Oil Industry Giants, 2009). PetroChina is the largest of China's three state-controlled oil giants. The company produces more oil than ExxonMobil and could someday vie with Gazprom as a regional gas power. This portion of the business—the suppliers—is referred to as "upstream."

In contrast, huge publicly owned international companies like Exxon and Royal Dutch dominate refining and marketing, which is called the "downstream" part of the business. Since its modern beginnings, the industry has shamelessly leveraged its wealth to influence governments and political leaders for tax advantages and subsidies—currently about $4 billion annually in the United States alone. It is an industry that has been unrelenting in its pressure on governments to allow expanded drilling in sensitive ecosystems and to relax environmental protections. It is an industry with a product so critical that nations go to war and countries explode into chaos to control it.

Today's petroleum industry extends to every ocean, landmass, and nook and granny in, or under, the modern world. The industry knows where every likely drop of oil is, no matter how difficult to recover. The petroleum industry includes exploration, extraction, refining, transporting, and, of course, marketing and sales. Fuel oil and gasoline are the prime products, though oil is the raw material for many chemical products—pharmaceuticals, solvents, fertilizers, pesticides, and plastics.

Oil in the form of gasoline, diesel, or jet fuel represents almost all of the energy consumed by the transportation industries and the major slice of the world's consumed energy. Coal, a fossil fuel cousin of oil, is the predominate fuel burned by the world's utilities to generate electricity. For example, North America uses oil for about 40 percent of its energy needs.

Overall, the world consumes about 90 million barrels of oil per day, and over 30 billion barrels per year, making it by the far the world's largest industry (Morrigan, 2010a,b). Much of the oil industry's power comes from the mutual dependency of the state-owned producers and the publicly owned megacompanies. While usually cooperative, this relationship is always tense.

In July 2012, this complex and sometimes conflicted government and oil company relationship was on display in a classic Russian power struggle. The vast wealth and political intrigue sometimes indulged in by the oil industry came to light when billionaire Russian oil tycoon Mikhail Fridman tried to defy Russian President Vladimir Putin (Levine, 2012). The US magazine *Forbes* listed Fridman as the 43rd richest man in the world and was one of the few remaining Russian oligarchs from the Boris Yeltsin era. He was the chief executive officer of TNK-BP, the Russian partner of BP. Fridman was intent on leveraging TNK-BP's position in the Russian oil patch and catapulting himself into being the largest single shareholder of the British company.

As the chart below indicates, the European Union is very dependent on oil and gas from Russia including through its shipping and pipelines.

According to *Foreign Policy*, what was at stake besides the numerous blocks of stock and billions of dollars were partnerships and rights to oil and gas drilling in the Arctic and other unidentified places around the world (Levine, 2012).

In this clash of oil oligarch and Russian president, the oligarch lost. According to *Foreign Policy*, he seemed to have misjudged Putin, who takes a proprietary interest in the Russia oil patch and has defeated past oligarchs who threatened to destabilize it or who have double-crossed him. In the end, Rosneft, the state-owned oil company bought the shares from BP, as well as swallowed up Fridman's shares of the parent company, TNK-BP. Unlike some of the other oligarchs Putin has dispatched, Fridman appears to have been stripped of his CEO's position but avoided jail and retained most of his wealth.

More significant than oil and government conflicts among the Russians, the petroleum industry has given the Middle East a position of global power that routinely disrupts geopolitical stability. Mostly ruled by centuries-old monarchs or despots, the Middle East holds a fraction of the world's population. However, since 1995, the United States and its allies have fought three major wars—Desert Storm, Iraq, and Afghanistan—and have had numerous smaller military operations in the region.

In all, the United States and its allies have spent well over a trillion dollars since Desert Storm trying to maintain Middle East political stability. In fact, by the fall of 2012, the National Priorities Project "Cost of War" calculator estimates the US expenditure at $1.5 trillion for just the Iraq and Afghanistan wars (Cost of National Security, 2014). Ensuring that the oil industry remained intact and commercially viable had as much to do with these wars as political ideology or religion.

More instability is ahead as the region's population matures. Middle East nations have a disproportionate number of young people who are now connected to social media. They are restless and pushing for modernity and greater participation in their countries' politics. The story is the same whether it is Egypt, Libya, Syria, or the other Middle East nations. Most importantly, work and the necessary jobs are not there. Throughout region, private sector development is severely hampered. Without jobs, incomes are low, and frustrations high. Young men cannot marry without a job, increasing frustrations and challenging political stability. These were the dynamics of the 2011 Arab Spring and they are still very much in play.

Creating greater problems, water and food are in tight supply, and as the region's population grows, it will only get worse. Additionally, the maturing population wants more cars and gasoline at cheap prices. A restless and growing population adds to the problems of authoritative governments. The response has been to raise oil prices to the purchasing nations, bring in more revenue, and increase subsidies and cheap oil to the resident populations. The Middle East nations are now using more of their own oil—Saudi Arabia is the sixth largest oil user in the world—and rising prices on the oil their export.

DEVELOPING AN OIL INDUSTRY

Except for the early Chinese salt drilling activities, oil wells are a relatively modern technology. Marco Polo witnessed the mining of oil seeps in Persia in 1264, and by the 1500s, oil for street lamps was collected in Poland's Carpathian Mountains. There are reports that Persia had hand-dug wells of 35 m by 1594, and oil was being extracted from sand around 1735 in Alsace, France. In the early 1800s, oil was appearing as an undesirable by-product in the salt wells of Pennsylvania in the United States.

The Russian engineer F. N. Semyenov is said to have drilled the first modern oil well in 1848 on the Absheron peninsula northeast of Baku in 1848. A few years later, wells of 50 m were being drilled in Bóbrka, Poland. Colonel Edwin L. Drake is generally credited with the first US commercial oil well. His well went down about 22 m near Titusville in Pennsylvania. Records indicate it started producing oil on August 28, 1859 (Extreme Oil, 2004a,b).

Without autos, the market for "rock oil" as it was called was mostly for medicine, and at $40 a barrel, it was expensive. However, this sleepy specialty niche business soon changed. In the United States, George Bissell contracted for Drake's well and built the Pennsylvania Rock Oil Company of New York around it. This was a key moment for the commercial oil industry and John D. Rockefeller's Standard Oil of Ohio quickly followed Bissell into the oil patch.

In 1870, Rockefeller put down his $1 million to fund Standard Oil in 1870 and launched what became the most powerful and wealthiest industry in world history. He and Standard Oil were at the core of the second industrial revolution's dependency on oil—the lifeblood of the modern era and basis for the growth of the United States and other Western nations. Not only did he change how the world uses energy, but also he changed the politics behind the fortunes (Chernow, 1998).

A shrewd and ruthless businessman, Rockefeller bought up competing refineries, schemed with the railroads, pressured politicians, and drove competitors out of business. Gradually by underselling, differential pricing, and secret transportation rebates among other tricks, Standard Oil gained almost complete control of the oil and kerosene business.

By the 1890s, Standard Oil was responsible for 90 percent of the US refining capacity. It had gained an aura of invincibility, always prevailing against competitors, critics, and political enemies. It was the richest, biggest, most feared company in the world, consistently racking up profits year after year. Its vast American empire included 20,000 domestic wells, 4000 miles of pipeline, 5000 tank cars, and over 100,000 employees (Extreme Oil, 2004a,b).

To further control pricing, Rockefeller created one of his most important innovations. In the early 1880s, he issued certificates against oil stored in Standard Oil's pipelines. Speculators soon traded the certificates, creating the first oil futures market, which effectively set spot market prices from then on. The National Petroleum Exchange opened in Manhattan in late 1882 to facilitate the oil futures trading (Extreme Oil, 2004a,b).

Rockefeller revolutionized the petroleum industry, and as kerosene and gasoline grew in importance, his wealth soared. He became an industrial oligarch, one of the most powerful men in the United States. He was described as the world's richest man and the first American billionaire. As Standard Oil's rapacious ways continued, the company came under the attack by journalists and politicians. In 1880, the *New York World* branded Standard Oil as "the most cruel, impudent, pitiless, and grasping monopoly that ever fastened upon a country" (Segall, 2001).

By 1909, the United States was using the Sherman Antitrust Act to dismantle the huge industrial monopolies that had emerged in the last part of the 19th century and were dominating the economy and politics. In 1911, the Supreme Court of the United States found Standard Oil Company of New Jersey in violation of the Sherman Antitrust Act. The court ruled that the trust engaged in illegal monopoly practices and ordered it to be broken up into 34 new companies.

Originally, kerosene was the most important product of Standard Oil, but it was being replaced by gasoline. In 1903, a young Michigan engineer named Henry Ford started Ford Motor Company, and the great modern fascination with the auto began. Though others had built automobiles, Ford introduced mass production with the assembly line. His introduction of the Model T automobile revolutionized transportation.

With the Model T, Ford brought the automobile to the American heart-land. Suddenly, gasoline became important and the petroleum industry gathered momentum. It was the auto, and more precisely the Model T, that turned oil into a gigantic international industry.

As the trenches of World War I scarred Europe, oil became a critical military asset. As the battles raged, military transportation shifted to oil-powered naval ships, new horseless army vehicles such as trucks and tanks, and even military airplanes. During the war, the use of oil increased so rapidly that a severe shortage developed in 1917-1918.

By the middle of the 20th century, the oil industry started to achieve real growth in scale and economic importance. The first hint came in 1938, when Standard Oil ventured into Saudi Arabia. With the Saudi's permission, Standard Oil drilled near the country's eastern province, striking a gusher near Dhahran that yielded commercial quantities of crude oil.

As World War II erupted, the internationalization of oil played a vital role, providing the allies with a much-needed and appreciated supply advantage. The Allied forces' access to oil was considered a crucial factor in their victory over the Axis powers in World War II. At the same time, scientific discoveries and inventions fueled a growing market for petroleum products in plastics, synthetics, and other industries.

THE FIRST OIL WELL WAS DRILLED IN CHINA IN 347

While the ancients in Mesopotamia were salvaging bitumen and tar from the seeps in the desert, the Chinese were developing a salt industry, which started along the coast about 5000 years ago. Not only was salt a vital food supplement, but also the early Chinese were using it as a preservative. While the people along the coast harvested salt from seawater, those in the Sichuan region discovered salt formations along the Yangtze River. Sichuan salt farmers began digging wells around 2500 years ago to tap into this huge brine aquifer (Kuhn, 2008).

The drilling technique included a bamboo rig, drill bits of iron, and a bamboo pipe. The pipe was lifted by hand or a wheel and then dropped into the hole crushing rock and gravel. This pulverizing process continued until the loose dirt was removed by a length of hollow bamboo fitted with a leather foot valve. The Chinese kept digging deeper wells, and by the 3rd century AD, salt wells were being drilled to 140 m deep. While it may sound crude, the process is similar to a modern oil rig without the benefit of modern machining methods.

The Chinese then used bamboo pipelines to connect gas wells with the salt springs. In both China and Japan, there were early allusions to the use of natural gas for lighting and heating. In the 7th-century Japanese records, the reference to "burning water" is thought to be a reference to oil or natural gas. In 1088, the Chinese scientist and statesman Shen Kuo of the Song dynasty used the term "Shiyou" or rock oil for petroleum, which remains the current term. Remarkably prescient, Shen Kua wrote *Dream Pool Essays* in which he predicted that fossil fuels such as oil and natural gas would become widely used for energy. At the time, the Chinese were using oil for weaponry, medicine, lubricants, ink, and lighting.

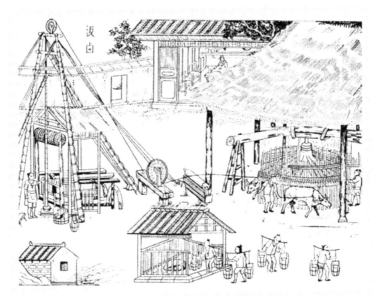

Ancient sketch from the "The Annals of Salt Law of Sichuan Province" shows the structure of the ancient derrick. Oil and natural gas came from salt formations.

About the same time, the Chinese drillers replaced their solid bamboo pipes with a thinner and more flexible cable. This breakthrough allowed them to dig even deeper, drilling down to 400 m. By 1835, a Shanghai well exceeded 1000 m, or about twice the depth of wells being drilled in the United States and Russia.

Originally, sun and air were used to evaporate the water from the brine, but later, wood was burned to accelerate the drying process. In the 16th century, as the salt production grew in volume and importance, the Chinese developed extensive networks of bamboo pipelines to transport the brine and natural gas. With wood becoming scarce and natural gas a residue of the salt wells, the Chinese switched to burning natural gas. The gas heated large salt pans, and the adaption of natural gas pushed Zigong's salt production to industrial scale. As the wells were drilled down deeper, they started tapping into the brine and gas from the Jialingjiang group Triassic formations (Kuhn, 2008).

By the mid-1800s, annual salt production was estimated at 150,000 tons, and the lucrative salt industry was a major economic driver. Toward the end of the 1800s, the Chinese invented the "Kang Pen" drum, a major technological advancement. The drum sat on top of a wellhead. Pressure in the drum allowed both brine and gas to be produced together and then separated. One bamboo pipeline would take away the brine and another the gas. It is estimated that over 2000 years, the Sichuan salt industry has drilled 130,000 brine and gas wells. The area continues to be a major salt producer, and many of the historical wells are still in production.

In the late 1950s, vast oil reserves were discovered in the Songhua Jiang-Liao basin in northeast China. A few years later, the Daqing oil field in Heilongjiang Province was discovered. These two discoveries pushed the nation into the oil industry. The first drilling machines came from Russia, but China quickly started providing its own drilling equipment.

Refining and plant equipment came from Europe and Japan as China grew its petroleum and petrochemical industries. The Arab oil embargo hit Japan particularly hard, and China started to export oil to Japan as its oil industry ramped up production. However, the industry was slow to evolve as various economic, political, and nation-building issues took precedence.

Deng Xiaoping's "open-door" policy at the end of the 1970s opened up the Chinese economy. The open-door policy loosened the political and economic constraints holding back modernization of the oil industry. Foreign participation was legitimized leading to quick development of financial, technical, and economic ties with overseas suppliers. The international markets were tapped to provide the capital to help the petroleum industry develop.

The 1990s internationalization of the oil industry was driven by the growing adoption by China of foreign business practices and standards. Since China was a net exporter of oil, the requirements of finance, technology, and trade with the rest of the world forced reform on the industry. The acquisition of foreign technology led to foreign partnerships. Gradually, the state-owned firms that controlled the petrochemical industry opened up to foreign influence and investment.

From the 1950s onward, the firms in China's petrochemical industry were state-owned. The State Council headed by the premier decides on overall industrial strategies, major investments, and import quotas and approves large projects. The State Planning Commission is responsible for the National Economic Plan and sets production targets and collects taxes.

In the mid-1980s, the State Council wanted to reduce competition for key resources and work more directly with the petrochemical industry. The outcome was Sinopec, or China Petrochemical Corporation, which is by far the dominant petrochemical firm. Sinopec controls the production and refining of all petroleum products. A giant organization, Sinopec controls more than 70 subsidiaries and processes about 90% of China's crude oil, gas, and petrochemical products. This includes 38 refineries, 21 basic organic chemical facilities, 15 synthetic fiber facilities, 5 synthetic rubber plants, 3 synthetic resin plants, 13 chemical fertilizer facilities, 5 construction companies, 7 research and design institutes, and 6 trade and sales companies. Sinopec is directly accountable to the State Council and the State Planning Commission. There are two other major state-owned corporations, China National Offshore Oil Corporation, which regulates and controls offshore drilling, and China National Petroleum Corporation, which produces about 97 percent of China's domestic crude oil (China's Demand for Oil, 2011).

By the early 1990s, China exported hundreds of different petrochemical products, including crude oil that had risen as an export from about 7 million tons in 1973 to 20 million by the mid-1980s. Most of this oil was shipped to Japan. However, by the mid-1990s, the export/import equation had reversed with China's growing industrialization and development. As internal needs increase and domestic supply declines, China has had to import a much larger amount of crude.

Originally importing oil from its Southeast Asian neighbors like Malaysia and Indonesia, China's growing demand has forced it to pull heavily from the international market. By 2012, China is now the world's second largest oil importer, importing over 200 million tons of oil in 2010 (China's Demand for Oil, 2011).

China's oil needs now overwhelm its domestic capabilities. Dependence on foreign oil has had a radical impact on its exploration and acquisition polices and left it vulnerable to market fluctuations and more susceptible to international oil conflicts. In response, China has been investing in foreign lands that are rich in oil and creating an internal oil reserve for emergencies. Domestic oil production supplies only two-thirds of the nation's oil needs, and it is estimated that China will require 600 million tons of crude oil by 2020 (China's Integration, 2011).

Since 2004, China has been stockpiling oil in a massive storage program that includes sites at three different providences. The stockpiling plan calls for 90 days worth of oil. Along with the domestic reserves, there is a push for more oil on all levels. Offshore drilling rigs in the South China Sea are increasing, using FPSO, a floating production, storage, and off-loading system.

The big three state-owned companies in China's oil industry—China National Offshore Oil Corp, China National Petroleum Corp, and Sinopec—have all invested in exploration and development in countries that has oil fields but lack capital. They have invested billions of dollars in African countries like Angola, Nigeria, and the Sudan and signed agreements to extract oil from the Melut Basin and Darfur.

In the Middle East, CNPC invests heavily in Iran and Iraq, signing an agreement with National Iranian Oil Company to develop the North Azadegan oil field. The company also has large investments in projects in Kazakhstan and Turkmenistan and is looking to invest more in Russia. In South America, Chinese companies are large investors in oil-rich Venezuela.

China has also been pushing to be a major player in the development of Canadian and American oil shale projects. Investment consultant Dealogic estimates that Chinese oil companies have invested nearly $40 billion since 2008 in North American oil projects, including proposed takeovers of Calgary-based Nexen and projects with Texas-based Chesapeake Energy Corporation (Schneyer, 2012).

China's energy needs are immense. The nation is vast, and large numbers of people have moved to urban areas and are needing various infrastructure elements from water and energy to transportation and waste. The chart below shows the demand for energy, mostly from fossil fuels, has risen dramatically.

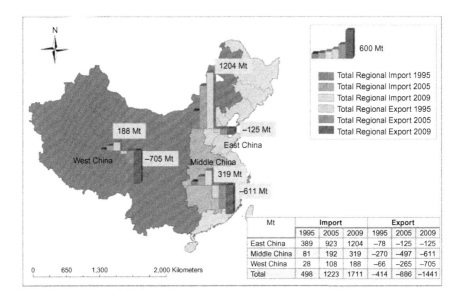

Mt	Import			Export		
	1995	2005	2009	1995	2005	2009
East China	389	923	1204	−78	−125	−125
Middle China	81	192	319	−270	−497	−611
West China	28	108	188	−66	−265	−705
Total	498	1223	1711	−414	−886	−1441

Transportation has been the most profound as well as environmentally challenging in its greenhouse gases and carbon emissions. The Chinese middle class has grown and demands for services, products, and governmental support have risen at double digit rates. The entire population is vast with huge numbers of people seeking to improve their lifestyles and create a modern, diverse, and technology-laden economy. Renewable energy is a continuing and increasing area for China's five-year plans as well as a government focus for investment and finance.

But with strategic plans and large sums of government investment and support, China's environmental and climate difficulties are being controlled and mitigated. Vast regions of China are pursuing and installing renewable energy systems.

OPEC AND THE OIL EMBARGO

As the major American and European oil companies prospered during the post-World War II period, they retained 65 percent or more profit from the revenue that was produced on someone else's property. In reaction, the

The provinces with top ten renewable energy power capacity(2012)

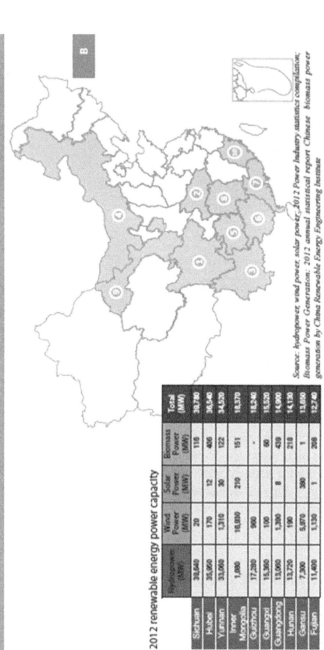

2012 renewable energy power capacity

	Hydropower (MW)	Wind Power (MW)	Solar Power (MW)	Biomass Power (MW)	Total (MW)
Sichuan	39,640	20		116	39,780
Hubei	35,950	170	12	406	36,540
Yunnan	33,060	1,310	30	122	34,520
Inner Mongolia	1,000	16,000	210	151	18,370
Guizhou	17,280	900		-	18,240
Guangxi	15,360	100		60	15,520
Guangdong	13,060	1,300	8	439	14,900
Hunan	13,720	190		218	14,130
Gansu	7,300	5,670	380	1	13,050
Fujian	11,400	1,130	1	208	12,740

Source: hydropower, wind power, solar power, 2012 Power Industry statistics compilation; Biomass Power Generation: 2012 annual statistical report Chinese biomass power generation by China Renewable Energy Engineering Institute suppliers, mostly undeveloped Arab nations, pressed to protect their interests and increase their leverage on production, prices, and concession rights. In 1960, the oil-exporting nations formed a cartel called the Organization of Petroleum Exporting Countries. OPEC's founding nations were Iran, Iraq, Kuwait, Saudi Arabia, and Venezuela. Other nations joined, including Qatar, Indonesia, Libya, United Arab Emirates, Algeria, Nigeria, Ecuador, Angola, and Gabon. OPEC's goal was to show a united front in price nego- tiations with the giant oil companies, which were colliding among themselves.

The industry expanded as oil became the lifeblood of the industrialized world. Cheap oil was flowing into the United States and Europe from the Middle East. Each day, oil became more and more of a critical commodity for the developed nations that relied on it for vital parts of their economies, particularly for the distribution of goods—ships, trucks, boats, and planes— and in vital areas like fertilizers for farming, drugs for medicines, and syn- thetics like plastic and nylon.

The cooperation between the Western oil companies and OPEC sup- pliers evaporated in a flash in October 1973. OPEC declared an oil embargo against the nations that had supported Israel during the Yom Kippur War with Egypt, Syria, and Jordan. It was one of the most significant geopolitical events of the second half of the 20th century, hitting Europe and the United States particularly hard. After 20 years of growth and prosperity, the Western nations were in an inflationary spiral struggling with the heavy demand for raw materials. Consumer prices were rising rapidly and inflation was increasing even more. Oil demand had been increasing throughout the industrial world and it was outstripping supply. Oil industry profits were surging and OPEC was getting stronger and determined to increase its profit share.

The Yom Kippur War was so named because Egyptian and Syrian forces attacked Israel on October 6, 1973 as Israel celebrated the Jewish national holy day. Egyptian forces swarmed into Israel from across the Suez Canal, while at the same time, Syrian troops flooded the Golan Heights in a surprise offensive. Israeli recovered from the initial attacks and countered with US help, reversing the Arab gains. Before the November cease fire, OPEC struck against the West by imposing an oil embargo on the United States while increasing prices by 70 percent to western Europe nations. Overnight, the price of a barrel of oil rose from $3 to $5.11, and in January 1974, OPEC raised it further to $11.65 (Extreme Oil, 2004a,b).

The shock to the Western nations was immediate. Having gotten used to a steady flow of cheap Middle Eastern oil, they were suddenly at the mercy of Arab nationalism. OPEC was standing up to the American oil companies that had once held their countries in a vice grip.

As the energy crisis hit, it pushed the United States and much of the Western world into a severe recession. Gasoline lines went around city blocks in the United States and gas prices quadruple from about 30 cents to $1.20 a gallon. Before the embargo was lifted in March 1974, the United States and the West had suffered a severe recession, and the New York Stock Exchange lost $97 billion in share value.

A second oil crisis hit at the end of the decade. When the Iranian Revolution swept across Iran in 1979, it replaced the Shah of Iran's autocracy with a new fundamentalists theocracy led by the Ayatollah Khomeini. Massive protests and disruptions crippled the Iranian oil sector, effectively removing Iranian oil from the market. This triggered a spike in prices and widespread panic buying of oil and gasoline until Saudi Arabia was able to bridge the production decline. In the 1980s, the Iran-Iraq War again curtailed Iran as well as Iraq's oil production. At the height of this second crisis, oil reached a high of $15.85 per barrel.

After this second crisis, a 20-year price decline started as exploration techniques and recovery technology vastly improved. Exporters like Mexico, Nigeria, and Venezuela raised production. The vast supplies of Russia were uncovered, pushing Russia into the world's number one producer. North Sea and Alaskan oil entered the market. The price of oil continued to drop despite the 1990s Iraqi invasion of Kuwait and the West's retaliation in Operation Desert Storm. By the end of the 1998, oil had dropped to less than $10 a barrel and the industrialized nations were riding a wave of cheap energy and prosperity.

From 1998's low of $10 per barrel, the world price of oil has steadily risen. Demand has increased as the emerging nations, particularly the giants like Brazil, China, and India, have modernized and gained economic strength. Globalization, the most important economic force of the end of the 20th century, has gained momentum in the 21st century. The result is an ever-interdependent global economy with energy as the main driver.

Now as the world's economy struggles to recover from the deepest recession since the 1930s, oil hovers around $105 a barrel. High prices have propelled oil and the oil industry to the forefront of economics and politics, and the geopolitical implications are critical. From Asia, to the

Middle East, to the United States, big oil has become intertwined in power struggles, international conflicts, and presidential campaigns. The clash between Russian President Vladimir Putin and oil oligarch Mikhail Fridman is a high-profile example of the power and the politics behind the oil industry.

OPEC PRICING AND THE POLITICS OF THE PERSIAN GULF

The multitrillion dollar petroleum industry, which includes oil and natural gas, is the largest and wealthiest in the world. Likewise, so are the major companies that pump the crude oil and refine it. The data for the chart below are from the Forbes magazine 2014 list of the world's largest oil companies. "Big oil" has gone from the Standard Oil monopoly in America at the beginning of the 20th century to an international juggernaut. It's impossible to underestimate the impact that this industry has on geopolitics. Note that 6 of the top 10 companies are state-owned or state-controlled; these companies often are tools of their nations' foreign policy (Forbes, 2014).

1. Saudi Aramco (12.5 million barrels per day)	Saudi Aramco is by far the biggest energy company in the world, generating more than $1 billion a day in revenues
2. Gazprom (9.7 million barrels per day)	Russia's Gazprom is the world's largest producer of natural gas. Controlled by the Kremlin, Gazprom provides Russia with a prime lever for projecting power in the region. Gazprom's profits are more than $40 billion a year
3. National Iranian Oil Co. (6.4 million barrels per day)	Iran has been forced to curtail oil production due to international sanctions but remains a huge oil and gas producer. To skirt sanctions, Turkey and India have reportedly been paying for Iranian oil with gold. The Strait of Hormuz remains the world's most significant choke point for oil. Iran has threatened to close the Strait if attacked
4. ExxonMobil (5.3 million barrels per day)	ExxonMobil has $40 billion in annual profits on $400 billion in sales. Exxon needs to make giant projects like the one with Russia's state-controlled oil giant Rosneft to grow
5. PetroChina (4.4 million barrels per day)	PetroChina is the largest of China's three state-controlled oil giants. The company produces more oil than ExxonMobil and could someday vie with Gazprom as a regional gas power

Continued

6. BP (4.1 million barrels per day)	Formerly known as British Petroleum, BP is in a turnaround, selling assets, settling lawsuits, and promising improvements. BP may not maintain its 4.1 million barrels per day for long; it is selling its 50% stake in Russian venture TNK-BP, which provides a quarter of production
7. Royal Dutch Shell (3.9 million barrels per day)	Shell is hoping to start drilling for oil in Alaska's Chukchi Sea. For years since leasing offshore blocks from the federal government, Shell has been perfecting its drilling plan and preparing the *Kulluk* floating drilling rig
8. Pemex (3.6 million barrels per day)	Production from Mexico's biggest field, Cantarell, has plunged from 2 million bbl per day to roughly 600,000. State-owned Pemex is working to replace that shortfall with other fields. Reforming Pemex to allow foreign investment is a significant issue
9. Chevron (3.5 million barrels per day)	Chevron bought Atlas Petroleum in 2010 for $4.3 billion to gain acreage in the Marcellus and Utica shale formations. Some expect a bigger deal to come
10. Kuwait Petroleum Corp. (3.2 million barrels per day)	Kuwait's oil company was formed in 1934 by Chevron and BP and nationalized in 1975. Kuwait's fields suffered greatly by fires set by Iraq's forces in 1990. Kuwait's biggest field, Burgan, continues to be operated by Chevron

Economies have historically evolved through a process that starts with the extraction of natural resources for initial wealth, then to a period of product manufacturing and distribution that builds a middle class, and on to a diversified, service-based economy that is reliant on processes like finance, innovation, and management to drive a robust mature economy. The evolutions of the United States and now China are examples.

The United States is rapidly moving away from the wealth of resource extraction, referred to as the "dirty" economy to a "clean" economy based on knowledge and service companies. After a long reign at the top of the US economic ladder, the "dirty" rich are being displaced. David Callahan, author of *Fortunes of Change: The Rise of the Liberal Rich and the Remaking of America*, points out that of the top US 20 companies on the Fortune 500 list in 1960, 16 were engaged in heavy industry (companies like US Steel and DuPont) or resource extraction (companies like Texaco and Mobil). By 2009, only six such companies were in the top 20 (Callahan, 2010).

Even in Texas, the US historical center where fossil fuel fortunes were made, things are changing. Fewer than half the of the state's billionaires on the Forbes 400 list of the wealthiest Texans made their money in oil or energy, a major departure from the past. Two of the wealthiest Texans are not oil magnates: one is Alice Walton, heir to the Wal-Mart fortune, and the other is Michael Dell, founder of Dell computers.

Energy costs are important at every phase of the economic evolution, but they are critical to the highly sophisticated and technology-laden mature economies and economies like China that are rapidly growing. Sadly, this critical component is little understood and greatly misused in politics, particularly in the US election campaigns where candidates continually call for oil independence.

For example, in the 2012 US presidential campaign, Mitt Romney, the Republican candidate, argued strongly and emotionally that the United States could achieve energy independence by drilling for oil on American soil. While clearly a campaign "sound bite," this claim echoed a long-held belief by most Americans and one continually pushed by the oil industry. However, this assertion is based on the misconception that the exporters can choose the destination of their oil exports for commercial or political purposes. The oil business simply does not work that way.

Basically, the oil industry does not know or care where a barrel of oil originates. Oil is a "fungible" commodity, like the currency markets. This means that a barrel—like a dollar, yen, euro, or pound—is exchangeable, or replaceable, in whole or in part, by another barrel (or dollar, yen, euro, or pound). Imagine a huge swimming pool. Producers pour oil in; consumers take oil out. Except in the rare cases where an individual producing country has a direct contract with a consuming nation for a tanker or two, most oil is bought on the world's spot market. The whole process is enabled by the large international oil companies that determine what happens to the oil when it enters the global market (American Interests, 2012).

Supply is regulated by the oil producers and is withheld or increased based on keeping the "fair price" intact. For the most part, OPEC sets the fair price of oil based on what is needed to satisfy OPEC nations' social and political needs. With the Arab Spring, this fair price rose based on the threat of revolution, not economics, as the gulf regimes blanketed their citizens with gifts and gasoline subsidies.

So when the price of oil spikes, it spikes for everyone. For example, the 2011 Libyan upheavals caused major supply problems. While America imports hardly any oil from Libya, when the Libyan oil disruptions took place, America was just as affected by the resulting $25 per barrel price increase as China was as one of Libya's major oil purchasers.

OPEC controls price by keeping the level of production low, which is not surprising since it is a cart and cartels aim to maximize profit by holding back supply. OPEC controls almost 80 percent of the world's conventional oil reserves, yet its members produce only 36 percent of global supply, which is why in the past three decades, global GDP went up six times, while OPEC's crude production increased only by 15 percent. OPEC supports its fair price by keeping production below what its reserves allow (American Interests, 2012).

In summary, OPEC decides on a price based on the needs of the cartel's members and then adjusts supply to maintain that price. This price is constantly rising, jumping from $25 per barrel in 2004, to $90 in 2010, to $100 in 2011. Another contribution to OPEC's rising oil price is the increasing use of oil by the cartel members. Saudi Arabia, for example, is now the world's sixth largest oil consumer. The cartel members increasing domestic consumption and dwindling supply is why oil analyst Charles Maxwell said in the February 2011 issue of Barron's magazine that based on simple supply and demand, the price of oil will climb to $300 a barrel by 2020 (Strauss, 2011).

Why are the world's economies so vulnerable to high oil prices? Unfortunately, it has to do with the unbridled expansion of petroleum-based transportation. Virtually all cars, trucks, railroads, airplanes, and shipping run on petroleum products—mostly gas or diesel. There are no other energy commodities to compete with oil's dominance in these critical sectors; this in turn allows OPEC and the oil industry to maintain ever-higher prices.

The desire for personnel transportation, or for autos, drives the petroleum industry. As developing countries like Brazil, China, and India become more prosperous, their citizens want motorcycles, scooters, cars, and other forms of modern transportation. Luxury cars are enormous status symbols, and just as many rich Brazilians, Indians, and Chinese want the new Audi A8 or the Mercedes S-Class as do rich Americans or Germans. Even in cities like London, Paris, or New York, some of the most traffic-impacted cities in the worlds, expensive cars are part of the social fabric.

Unfortunately, given geopolitics, it will stay that way until the oil is gone, or an alternative energy source, such as hydrogen, comes into the market place in sufficient quantity to offer an alternative fuel. While science is on the verge of developing alternative energy commodities like biofuels and hydrogen, their adoption by the market place will require major political and commercial transformations.

SUBSIDIES AND POLITICS

Jesse "Big Daddy" Unruh, a 1960 era, bigger-than-life California politician and the state's 54th Assembly speaker, remarked, "Money is the mother's milk of politics" (Cannon, 1969). While true at all levels of politics throughout the ages, it has probably never been truer than in today's relationship between the petroleum industry and geopolitics.

Big oil impacts almost all global politics, and its influence, power, and the corresponding river of cash that it creates are barely hidden or challenged. At the top of the list is Saudi Aramco with its $1 billion a day in revenues. Not only is the vast wealth of this state-owned company the driving force for the well-being of the Saudis, but also it is the critical component in maintaining the House of Saud, the kingdom's monarchy.

The same is true for the Iranian theocrats, the government in Venezuela, and Putin's Russia and in almost all nations with large state-owned oil companies. Governments of all kinds, either freely elected or the playgrounds of tyrants, crave and require the wealth that petrodollars bring. This is as true for the Persian Gulf nations trying to dampen revolutionary zeal with gifts of food and cheap gasoline, as it is for London, the financial capital of the world, that takes in every shade of petrodollars by the boatload, and turns (some say launders) them into pale conservative investments for a cut of the action.

MASSIVE OIL SPILLS

More than any other industry, the oil industry has had an extraordinarily ruinous impact on the environment. Besides releasing over 39 billion tons of CO_2 per year into the atmosphere, the industry has done huge damage through a series of oil spills. The worst spill was in 1991 during the Arabian Gulf War, when retreating Iraqi soldiers set Kuwait's oil fields on fire. Here are the 13 worst oil spills in history, according to Mother Nature Network (Moss, 2010):

1. Arabian Gulf/Kuwait When: January 19, 1991 Where: Persian Gulf, Kuwait Amount spilled: 380-520 million gallons	The worst oil spill in history was deliberate. During the Gulf War, Iraqi forces opened valves at an offshore oil terminal and dumped oil from tankers. A 4 inch thick oil slick spread across 4000 square miles in the Persian Gulf

Continued

2. Gulf of Mexico oil spill
When: April 22, 2010
Where: Gulf of Mexico
Amount spilled: An estimated 206 million gallons

The Gulf oil spill is officially the largest accidental spill. It began when an oil well a mile below the surface of the Gulf blew out, causing an explosion on BP's Deepwater Horizon rig that killed 11 people. BP made several unsuccessful attempts to plug the well, but oil flowed—possibly at a rate as high as 2.5 million gallons a day—until the well was capped on July 15, 2010. Oil gushed from the broken well for more than 85 days, oiled 572 miles of the Gulf shoreline, and killed hundreds of birds and marine life. The long-term effects of the oil and the 1.82 million gallons of dispersant used on this fragile ecosystem remain unknown, but experts say they could devastate the Gulf coast for years to come

3. Ixtoc 1 oil spill
When: June 3, 1979
Where: Bay of Campeche off Ciudad del Carmen, Mexico
Amount spilled: 140 million gallons

Like the Gulf oil spill, this spill didn't involve a tanker, but rather an offshore oil well. Pemex, a state-owned Mexican petroleum company, was drilling an oil well when a blowout occurred, the oil ignited, and the drilling rig collapsed. Oil began gushing into the Gulf of Mexico at a rate of 10,000-30,000 barrels a day for almost an entire year before workers were finally able to cap the well

4. Atlantic Empress oil spill
When: July 19, 1979
Where: Off the coast of Trinidad and Tobago
Amount spilled: 90 million gallons

This Greek oil tanker was caught in a tropical storm off the coast of Trinidad and Tobago when it collided with the Aegean Captain. The damaged ship started losing oil and continued to leak it into the ocean while it was towed. The oil tank finally sunk into deep water on August 3, 1979, where the remaining cargo solidified

5. Kolva River oil spill
When: August 6, 1983
Where: Kolva River, Russia
Amount spilled: 84 million gallons

A poorly maintained pipeline caused this massive oil spill. The pipeline had been leaking for eight months, but a dike contained the oil until sudden cold weather caused the dike to collapse. Millions of gallons of accumulated oil were released that spread across 170 acres of streams, fragile bogs, and marshland

6. Nowruz oil field spill
   When: February 10, 1983
   Where: Persian Gulf, Iran
   Amount spilled: 80 million gallons

The oil spill was the result of a tanker collision with an oil platform. The weakened platform was closed, and it collapsed upon impact, spewing oil into the Persian Gulf. The ongoing war between Iran and Iraq prevented the leak from being capped quickly

7. Castillo de Bellver oil spill
   When: August 6, 1983
   Where: Saldanha Bay, South Africa
   Amount spilled: 79 million gallons

The Castillo de Bellver caught fire about 70 miles northwest of Cape Town and drifted in the open sea until it broke in two 25 miles off the coast. The ship's stern sank along with the 31 million gallons of oil. The bow section was towed and deliberately sunk later

8. Amoco Cadiz oil spill
   When: March 16, 1978
   Where: Portsall, France
   Amount spilled: 69 million gallons

The massive Amoco Cadiz was caught in a winter storm that damaged the ship's rudder. The ship put out a distress call, but while several ships responded, none were able to prevent the ship from running aground. On March 17, the gigantic supertanker broke in half, sending its 69 million gallons of oil into the English Channel. The French later sunk the ship

9. ABT Summer oil spill
   When: May 28, 1991
   Where: About 700 nautical miles off the coast of Angola
   Amount spilled: 51-81 million gallons

This ship exploded off the coast of Angola, discharging massive amounts of oil into the ocean. Five of the 32 crew members on board died as a result of the incident. A large slick covering an area of 80 square miles spread around the tanker and burned for three days before the ship sank on June 1, 1991. Subsequent efforts to locate the wreckage were unsuccessful

10. M/T Haven tanker oil spill
    When: April 11, 1991
    Where: Genoa, Italy
    Amount spilled: 45 million gallons

This oil tanker exploded and sank off the coast of Italy, killing six people and leaking its remaining oil into the Mediterranean for 12 years. The source of the explosion was thought to be the ship's poor state of repair—supposedly the Haven was scrapped after being hit by a missile during the Iran-Iraq War but was put back into operation

11. Odyssey oil spill
    When: November 10, 1988
    Where: Off the coast of Nova Scotia, Canada
    Amount spilled: 40.7 million gallons

This large oil spill occurred about 700 nautical miles off the coast of Newfoundland and spilled more than 40 million gallons of oil into the ocean

Continued

12. The Sea Star oil spill When: December 19, 1972 Where: Gulf of Oman Amount spilled: 35.3 million gallons	The South Korean supertanker, Sea Star, collided with a Brazilian tanker, the Horta Barbosa, off the coast of Oman on the morning of December 19, 1972. The vessels caught fire after the collision and the crew abandoned ship. Although the Horta Barbosa was extinguished in a day, the Sea Star sank into the Gulf on December 24 following several explosions
13. The Torrey Canyon oil spill When: March 18, 1967 Where: Isles of Scilly, the United Kingdom Amount spilled: 25-36 million gallons	The Torrey Canyon was one of the first big supertankers, and it was also the source of one of the first major oil spills. Although the ship was originally built to carry 60,000 tons, it was enlarged to a 120,000 ton capacity, and that's the amount the ship was carrying when it hit a reef off the coast of Cornwall The spill created an oil slick measuring 270 square miles, contaminating 180 miles of coastland. More than 15,000 sea birds and enormous numbers of aquatic animals were killed before the spill was finally contained Toxic solvent-based cleaning agents were used by Royal Navy vessels to try to disperse the oil, but that didn't work very well and instead caused a great deal of environmental damage. It was then decided to set fire to the ocean and burn away the oil by dropping bombs

BIG OIL'S INFLUENCE IN THE UNITED STATES

In the United States, the connection between big oil and government is just as rampant as other nations, but the process is different. In the United States, big oil wages a war against any form of government regulations or policies that restricts drilling or access to oil deposits. While at the same time challenging, anything that questions the superiority of fossil fuel is the key energy source for the nation.

Sometimes, the oil industry's influence takes the form of sheer political "quid per quo" in the form of campaign contributions. For example, Mitt Romney, the 2012 Republican presidential candidate, held a fundraising event in Texas in August 2012. Two days after the oil industry gave him $7 million for his campaign, he announced the most pro-oil, antienvironment energy policy statement in the US history.

While Romney's behavior was extraordinarily overt, even for a pro-oil Republican politician, most of big oil's influence and political pressure is more circumspect. Probably the most pernicious form of favoritism to the oil companies is the ongoing tax subsidies. According to the Organization for Economic Cooperation and Development's 2009 *Inventory of Estimated Budgetary Support and Tax Expenditures for Fossil Fuels*, the US taxpayer subsidizes America's oil companies for about $4-5 billion a year (OECD, 2009).

The US taxpayer subsidies to the oil industry go back over a hundred years. They were originally intended to encourage the exploration and the development of rudimentary technology. Now, the subsidies continue with reduced tax rates on oil field leases and equipment and capital investments. Senator Robert Menendez, Democrat of New Jersey said, "The flow of revenues to oil companies is like the Deepwater Horizon gusher at the bottom of the Gulf of Mexico: heavy and constant." He added, "There is no reason for these corporations to shortchange the American taxpayer" (NYTimes, 2010).

The Deepwater Horizon oil rig was an ultradeep water rig positioned off the US waters in the Gulf of Mexico. Owned by Transocean and leased to BP, the rig drilled the deepest oil well in history at a vertical depth of 35,050 feet or 10,685 meters in 2009. Six months later in spring of 2010, while drilling at the Macondo Prospect, a blowout exploded on the rig sending a fireball up that was visible from 35 miles away. Eleven men were killed, and the fire could not be stopped. The rig toppled into the sea, leaving the well gushing at the seabed. It was the largest offshore oil spill in the US history. According to the US Geological Survey, scientists concluded that between 504,000 and 798,000 gallons of oil a day had been billowing out of the mud a mile beneath the sea.

At the time of this unprecedented natural disaster, the rig was registered in the Marshall Islands. Though it was drilling in the US waters for the US oil deposits, registering the rig in the Marshall Islands allowed Transocean to significantly reduce its American taxes. In another tax-avoidance maneuver, Transocean, which was originally a Houston, Texas, company, moved its corporate headquarters to the Cayman Islands in 1999 and to Switzerland in 2008 (Kocieniewski, 2010).

The Times reported that while Transocean owned the rig, BP was leasing it, using a major US tax break to the oil industry to write off 70 percent of the Deepwater Horizon rent. Reporting on the aftermath of the oil spill, the New York Times cited a letter that BP had sent to the US Senate Finance Committee documenting that BP received a deduction of more than $225,000 a day under the lease.

While the pro-oil groups like the American Petroleum Institute claim that any reduction in tax subsidies will cost jobs, others involved say that it's only the industry's political power that preserves the tax breaks.

"We're giving tax breaks to highly profitable companies to do what they would be doing anyway," Sima Gandhi told the New York Times in 2010. Gandhi, a policy analyst at the Center for American Progress, continued, "That's not an incentive; that's a giveaway" (Kocieniewski, 2010).

Perhaps the clearest, most reasoned analysis of the US oil and gas industry tax subsidies and the potential benefit of removing them came from Alan Krueger, the chief economist and assistant secretary for economic policy at the US Department of Treasury.

In October 2009, Krueger gave a report on the US Treasury's oil and gas tax subsidies to the American Tax Policy Institute. Krueger was reporting on the potential elimination of the subsidies as the Obama administration's put together the fiscal year 2010 budget. Because of congressional recalcitrance, none of the subsidies were eliminated.

Krueger explained that the current US tax law provided several credits and deductions specific to the oil and gas industry. The administration proposed to repeal three deductions that were available only to nonintegrated oil and gas firms. (Nonintegrated companies receive all revenue from production at the wellhead and not from refining.) Eliminating these tax preferences would raise revenues about $10.3 billion from 2010 to 2019. The administration also proposed to repeal three tax preferences available for integrated and nonintegrated firms. Elimination of these would result in about $20.3 billion from 2010 to 2019 (Krueger, 2009).

The elimination of tax preferences for the oil and gas industry, Krueger argued, would be consistent with the principle that tax policy should be neutral across industries. He maintained that these preferences resulted in inefficiency and reduced economic growth and pointed out that maintaining neutrality in economic policy was a principle that went back to George Washington, the first US president, who said in his farewell address, "even our Commercial policy should hold an equal and impartial hand: neither seeking nor granting exclusive favours or preferences..." (Krueger, 2009).

Krueger pointed out that oil and natural gas prices do not reflect the environmental harm caused by the release of greenhouse gases in the atmosphere associated with oil and gas production and consumption. In addition, the price of oil does not reflect the risks associated with the US oil dependency or the costs of traffic congestion. He also noted that these tax subsidies lead to inefficiency by encouraging an over investment of domestic resources in this industry. Removing them, he said, would improve the industry's overall economic efficiency.

With about $40 billion at stake, Krueger probably anticipated that his remarks would be met with criticism and ridicule from the oil and gas industry. Probably as a preempted move, Krueger included in his report an analysis of the economic impact on prices, oil and gas production, and employment if the government went ahead with removing the favored tax treatment. He and his staff found these impacts relatively nominal, for example, as follows:

- *Impact on prices:* Removing the subsidies would have no significant effect on world oil prices and without a change in world oil prices; the US consumer would experience no impact at the pump.
- *Impact on domestic output:* Since the price of oil will not change, consumers would not change their demand for petroleum products. Krueger estimated that because of supply elasticity, the decrease in domestic production would be less the one-half of 1 percent (0.05 percent).
- *Impact of employment:* With no change in domestic oil output, there would be no appreciable change in employment.

In conclusion, Krueger noted that "current tax subsidies for the oil and gas industry encourage the overproduction of oil and natural gas, they divert resources from other, potentially more efficient investments. . .Removing these subsidies is also consistent with the recent G-20 agreement to phase out fossil fuel subsidies. Furthermore, removing these subsidies will have a very small effect on the price of oil and gas, the production of oil and gas, and domestic jobs. In fact, removing these subsidies would actually make the U.S. economy more efficient by reducing distortions in the tax code. The possibility to promote our broader energy goals at no long-run cost and at a very low short-run cost—in terms of prices, productivity, and jobs—makes removing these subsidies sound economic and public policy" (Krueger, 2009).

Krueger's report and those of other economists and objective tax analysts failed to sway a Republican-controlled US Congress. Though Obama's administration continued to try to eliminate the subsidies, Congress steadfastly rejected the cuts.

To protect subsidies and guard against the incursion of clean and alternative energy sources, the oil industry' spends hundreds of millions of dollars in the United States on political campaigns, lobbying, and misinformation campaigns. According to a 2012 report by ThinkProgress, the oil industry spent $105 million in lobbying Congress in 2011, based on disclosures from the lobby industry. The oil and gas industry spent 90 percent of its $50 million in political contributions on Republicans (ClimateProgress, 2012a,b).

In addition, over $150 million was spent on 2012 election ads, including attack ads by the Americans for Prosperity, a group funded by Charles and David Koch, two conservative, pro-oil billionaires, that launched a bogus ad campaign claiming that clean energy stimulus dollars went overseas. Other ads by American Petroleum Institute chief Jack Gerard claimed that oil production on federal land was down. On the contrary, oil production on federal lands was increased by 240 million barrels by the Obama administration compared to the Bush administration (ClimateProgress, 2012a,b).

The industry also actively lobbies to defeat clean energy, currently trying to eliminate the modest tax incentives for wind energy. This effort seems to be led by two Koch-related groups, the American Energy Alliance and the Americans for Prosperity.

These massive amounts of lobbying money seem to be working for the pro-fossil fuel industries. According to ThinkProgress, the current House of Representatives is the most antienvironment in congressional history, averaging at least one antienvironment vote per day to eliminate or undermine pollution protections. To further the cause, the Paul Ryan Republican budget plan proposed an additional $2.3 billion in annual tax cuts to oil companies (ClimateProgress, 2012a,b).

OIL'S RESERVES AND ITS FUTURE

Unlike the sun, oil is not a renewing fount of energy. Like copper, zinc, or gold, there is a finite supply locked under the earth. Unlike these hard minerals that have internationally agreed upon measuring standards, the oil industry's supply numbers are unreliable at best and potentially false or fraudulent at worst. Like everything in the oil industry, supply numbers—and its future—are intertwined with big money and even bigger politics. No less an expert than Sadad al-Husseini, a former executive of Saudi Aramco, described reserve reporting data in 2007, as "Reserves are confused and in fact inflated. Many of the so-called reserves are in fact resources. They are not delineated, they are not accessible, and they are not available for production" (Strahan, 2007).

Oil data in the public domain are greatly inflated, falsified, and misreported. Al-Husseini's comments fairly describe the reporting problem. Much of the supply is not free-flowing black liquid that is easily discovered, then cheaply drilled into, and then piped up like sucking on a straw in a milk shake. The days of easy oil are long gone. Today's oil reserves are mere gaps or shadows on a computer screen miles beneath the earth or thousands of feet below the surface of a wild and unforgiving sea. Scientists use best-guessed algorithm to estimate location and volume, which makes all reported oil reserves extremely difficult to measure and very suspect.

In March 2010, the Smith School of Enterprise and the Environment at England's Oxford University put out a report saying that the "capacity to meet projected future oil demand is at tipping point." The Oxford report, published in the journal Energy Policy, noted that the demand has started to outstrip supply and that the age of cheap oil is over. The report highlighted that in the past, political and financial objectives have led to the exaggerated reporting of oil reserve estimates. Further, the report "suggests that the current oil reserve estimates should be downgraded from between 1150-1350 billion barrels to between 850-900 billion barrels" (University of Oxford, 2010).

Despite the difficulties inherent in measuring just how much oil in available for practical production, scientists are beginning to conclude that "peak oil" either has happened or is happening now. Peak oil means the maximum rate of oil production, after which the rate of production declines. While there may still be oil remaining in the ground, it will become increasingly difficult and more costly to produce.

Richard Sears, a geophysicist and former vice president for exploration and deepwater technical evaluation at Shell, appeared at the 2010 Technology, Entertainment and Design (TED) Conference in Long Beach, California. In his talk on the future of energy, Sears said that there is only 30-50 years left before a broad gap opens between worldwide oil supply and demand. Sears then held up a pincushion of the globe with red thumbtacks. "This is it," he said. "This is all the oil in world. Geologists have a pretty good idea of where it is" (Sears, 2010).

Sears made the point that throughout the world, oil supplies are in decline, for example, as follows:
- Kuwait's oil supplies have been in decline since 1970.
- The US oil supplies have been in decline since 1971, hence the need to protect those resources around the world.
- Iran's oil production has been in decline since 2008, while demand is soaring.
- Indonesia's oil supplies have been in a free fall since 1991, and this former OPEC exporter is now an importer.

- The European North Sea oil reserves have been declining since 1999, and the decline is accelerating. The United Kingdom is no longer an oil exporter.
- Norway's oil production has been in decline since 2001, and for environmental reasons, it has limited and cutback on offshore drilling.
- Mexico's oil production has been dropping since 2005.

Complicating oil's declining supply issues is the ever-increasing demand. Under current use, the demand for oil is projected to increase about 1 percent a year until 2030, or from today's 90 million barrels per day to about 105 million barrels per day. This means the world has to develop about 64 mbpd of gross capacity, or about six times the production of Saudi Arabia by 2030 to supply the projected business-as-usual demand. Unfortunately, the industry is estimating that 20 percent of this production will be based on undiscovered oil fields. The reality is that neither scientists nor industry experts know if there will be enough oil to supply this 20 percent projected crude production.

The era of cheap and abundant oil is over. As global production peaks, oil is harder to access and the quality is lower raising refinery costs. Huge investments will be needed to develop these obscure reserves, and more complex drilling and recovery methods will be required to maintain global production capacity. While the energy and economic investment required in producing the remaining oil increases, the energy yield from the reserves is decreasing. In financial terms, the EROI, or energy return on investment, is decreasing. The current EROI on oil is significantly lower than the past EROI for oil, and the future EROI will be even lower (Morrigan, 2010a,b).

Eventually, the financial and energy cost of producing oil will exceed the marginal profit and energy gained. This is very similar to the paradox of creating ethanol from corn. It takes as much or more energy to manufacture ethanol from corn than it produces. Instead of corn, ethanol should come from higher starch plants like switchgrass or sugarcane.

The same is true for "unconventional" oil sources. Unconventional oil includes extraheavy oil, oil sands, oil shale, coal to liquids (CTL), and gas to liquids (GTL). Despite the many claims of the oil industry that these unconventional sources are extensive, it is unlikely that unconventional oil resources will be able to replace conventional oil supplies in the future. According to the International Energy Agency (the IEA) estimates, the global supply of unconventional oil will increase to 8.8 mbpd in 2030, with the biggest increase coming from Canadian oil sands. These projected unconventional oil supplies are less than 7 percent of 2030s projected global demand.

Unconventional oil resources cost at least 2-3 times more to produce than conventional oil, making the EROI of these unconventional oil resources much lower than that of conventional oil. Unconventional oil resources also have much greater environmental impacts associated with them, including higher CO_2 emissions (Morrigan, 2010a,b).

Global conventional oil production is peaking or has already peaked. Subsequently, global conventional oil production will likely decline, perhaps severely. Since oil is used in the production and transportation of fossil fuels, it has a major impact on the second industrial revolution energy generation systems. A decline in oil production would make it harder and more expensive to use coal, gas, and unconventional oil and nuclear energy resources. Global peak oil may mean global peak fossil fuel energy generation. A decline in oil production would also create havoc for the world's transportation industry, which has not yet embraced alternative energy sources.

Throughout the second industrial revolution, humans used fossil fuels to dominant the global ecological, environmental, and climate systems while pushing the limits of the planet's carrying capacity. Cheap and abundant oil fueling the power of the internal combustion engine helped create the most extraordinary era in human history. Society, culture, and technology have all advanced to unprecedented levels. Yet, it created ruthless and pitiless environmental destruction that may have forced the planet to a critical tipping point. Human dependency on cheap oil and other fossil fuels may be the worst thing that ever happened to humanity and the planet Earth. Truly, it may have been the proverbial "Deal with the Devil."

REFERENCES

Bilkadi, Z., November 1984, Bitumen—a history, Saudi Aramco World. http://www.saudiaramcoworld.com/issue/198406/bitumen.-.a.history.htm.

Callahan, D., 2010. Fortunes of Change: the Rise of the Liberal Rich and the Remaking of America. Wiley, Hoboken, NJ.

Cannon, L., 1969. Ronnie and Jesse: A Political Odyssey. Doubleday, New York.

Chernow, R., 1998. Titan: the Life of John D. Rockefeller, Sr. Random House, New York, N.Y. p. 258.

China's Demand for Oil to Grow 6.2% in 2011: PetroChina The China Perspective, January 24, 2011. http://thechinaperspective.com/articles/china039sdemandforoilt.

China's Integration into the World, 2011. Economy. Whalley, J. World Scientific Publishing Co. Pte, Ltd., Singapore

Cost of War Calculator, 2014. https://www.nationalpriorities.org/cost-of/.

Extreme Oil, 2004a. WNET, New York. http://www.pbs.org/wnet/extremeoil/history/1850.html.

Extreme Oil, 2004b. WNET, New York. www.pbs.org/wnet/extremeoil/history/index.html.

Hirst, K., Bitumen, A Smelly but Useful Material of Interest. http://archaeology.about.com/od/bcthroughbl/qt/bitumen.htm.

The History of the Oil Industry, 2014. Oil Through the Ages. http://www.sjvgeology.org.

IBIS, June 2014. Report Snapshot. http://www.ibisworld.com/industry/global/global-oil-gas-exploration-production.html.

Kocieniewski, D., July 3, 2010. As Oil Industry Fights a Tax, It Reaps Subsidies. http://www.nytimes.com/2010/07/04/business/04bptax.html.

Remarks of Alan B. Krueger Chief Economist and Assistant Secretary for Economic Policy United States Department of the Treasury to the American Tax Policy Institute Conference Washington, DC. October 15, 2009. http://tcgasmap.org/media/Krueger%20Address%20to%20American%20Tax%20Policy%20Institute.pdf.

Kuhn, O., July 2, 2008. Ancient Chinese Drilling. E&P Magazine. http://www.epmag.com/Production-Drilling/Ancient-Chinese-drilling_4266.

Leber, R., July 2012a. What Five Oil Companies Did With Their $375 Million in Daily Profits. ClimateProgress. http://thinkprogress.org/climate/2012/07/24/574161/what-five-oil-companies-did-with-profits/.

Leber, R., October 24, 2012b. Three Ways Big Oil Spends Its Profits to Defend Oil Subsidies and Defeat Clean Energy. ClimateProgress. http://thinkprogress.org/climate/2012/10/24/1064231.

Levine, S., July 25, 2012. The Last Free Oligarch. Foreign Policy. http://www.foreignpolicy.com/articles/2012/07/25/the_last_free_oligarch.CC4.pdf.

Luft, G., Korin, A., 2012. The folly of energy independence. The American Interest (July/August), 33–40.

Morrigan, T., 2010a. Peak Energy, Climate Change, and the Collapse of Global Civilization, Global Climate Change: The Peak Oil Crisis. http://www.global.ucsb.edu/climateproject/papers/pdf/Morrigan_2010_Energy.

Morrigan, T., 2010b. Peak Energy, Climate Change, and the Collapse of Global Civilization, Global Climate Change: The Peak Oil Crisis. http://www.global.ucsb.edu/climateproject/papers/pdf/Morrigan_2010_Energy_CC4.pdf.

Moss, L., July 10, 2010. The 13 Largest Oil Spills in History. Mother Nature Network. http://www.mnn.com/earth-matters/wilderness-resources/stories/the-13-largest-oil-spills-in-history.

Oil Industry Giants, 2009. Saudi Aramco. http://www.arabianoilandgas.com/article-6051-oil_industry_giants_saudi_aramco/#aram.

Organization for Economic Cooperation and Development's 2009 Inventory of Estimated Budgetary Support and Tax Expenditures for Fossil Fuel, 2009. http://www.oecd.org/tad/environmentandtrade/inventoryofestimatedbudgetarysupportandtaxexpendituresforfossilfuels.htm.

Page Museum. La Brea Tar Pits. http://www.tarpits.org.

Schneyer, J., Sheppard, D., October 24, 2012. Analysis: Canada's Tough Stance on Foreign Buyers Won't Cut Oil Growth. Reuters. http://www.reuters.com/article/2012/10/24/us-energy-petronas-idUSBRE89N06320121024.

Sears, R., 2010. Planning for the end of oil. In: Presentation to the 2010 Technology Entertainment and Design Conference, February.

Segall, G., 2001. John D. Rockefeller: Anointed with Oil. Oxford University Press, London, UK, p. 14, ISBN.

Strahan, D., 2007. Oil reserves over-inflated by 300bn barrels—al-Husseini. The Last Oil Shock. http://www.davidstrahan.com/blog/?p=68.

Strauss, L., February 12, 2011. Whatever Happens in Egypt, Oil Will Hit $300 by 2020. http://online.barrons.com/article/SB50001424052970204098404576130370708044708.html#articleTabs_article%3D1.

World Oil Reserves at "Tipping Point", March 24, 2010. University of Oxford. http://www.ox.ac.uk/media/news_stories/2010/100324.html.

World's 25 Biggest Oil Companies, 2014. http://www.forbes.com/pictures/mef45glfe/1-saudi-aramco-12-5-million-barrels-per-day-3/.

CHAPTER 4

Coal, Natural Gas, and Nuclear Power

Most of the world's energy comes from oil, gas, and coal sources that generate carbon and emit greenhouse gases, adding to the planet's environmental degradation. In fact, the Environmental Defense Fund estimates that 65% of global warming comes from energy generation (www.EDF.org), which is not surprising since about 85% of the world's energy is generated by carbon-emitting fossil fuels.

Liquid crude makes up about 35% of the total generated energy, coal is 28%, natural gas is 23%, and nuclear power is 5%. (See table.)

Data shown below are the latest available as of December 2012(World Energy Statistics (The US Energy Information Agency)):

Since these data were released, China became the world's number one consumer of energy as well as the largest emitter of carbon and greenhouse

World energy consumption by fuel, 2008	505 quadrillion Btu
• Liquid oil	• 35%
• Coal	• 28%
• Natural gas	• 23%
• Nuclear	• 5%
• Others	• 10%

1. Liquids from biomass, crude oil, coal, and natural gas
2. Biomass, geothermal, hydropower, wind, solar, and other miscellaneous fuel

World energy consumption, 2009	483 quadrillion Btu
By top five countries	
• The United States	• 20%
• China	• 19%
• Russia	• 6%
• India	• 4%
• Japan	• 4%

gases. However, on a per capita basis, the United States still emits the most carbon and greenhouse gases, and China, unlike the United States, has plans for renewable energy power systems as well as financing them as the chart below indicates:

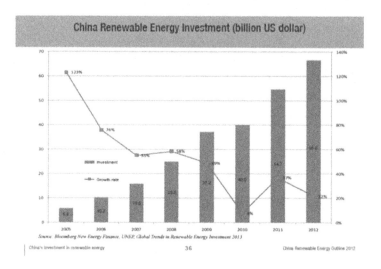

COAL

Like oil, coal is an ancient resource. Where oil is organic matter decomposed and then trapped in pools or pockets, coal is organic matter that clumped together and then was pressed by the weight of sand and dirt into carbon. Most of the Earth's coal was formed about 300 million years ago. At the time, enormous ferns and other prehistoric plants covered much of the swamp-like planet. As the plants died, they were covered with water and slowly decomposed. Without oxygen, this decomposition eroded away the hydrogen, leaving a material rich in carbon. Sediment was compressed on top of this carbon-heavy matter creating coal.

For centuries, the world has been discovering and exploiting coal, burning it for heat and energy. Despite heavy use, there are still significant coal deposits left in China, the United States, Canada, and Australia, with smaller deposits in Europe. China possesses the world's most extensive coal supplies. Because there may be a hundred years or more worth of supply, bituminous coal is a valuable commodity. Unfortunately, burning it for energy is a major contributor of carbon emissions and a big reason that the Earth is getting hotter.

Consider China's fossil fuel and renewable energy sources in 2012.

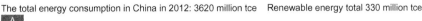

2012 China's total primary energy consumption

The total energy consumption in China in 2012: 3620 million tce   Renewable energy total 330 million tce

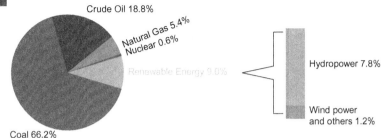

China's energy and power generation mtx   China Renewable Energy Outline 2012

For years, scientists have recorded the harsh effects—carbon emissions, acid rain, and ocean acidification—of burning coal. In response, the coal industry has been trying to develop technology that will mitigate or reduce the amount of carbon dioxide and other GHGs that are released from coal power plants. The result is described as "clean" coal. The clean coal process includes technologies, called CCS, that capture and store the carbon. Technologies like precapture, oxyfuel combustion, and postcapture have been tried.

Germany's Vattenfall's Schwarze Pumpe plant is an example of the oxyfuel process, probably the most popular example of carbon-capture technology. Another technology under development is the integrated gasification combined cycle or IGCC. This process is burden with the problem that for every ton of coal burned, almost 3 tons of carbon dioxide is created, which means that for every train bringing coal to a CCS coal plant, three trains would be needed to remove the CO_2.

Babcock and ThermoEnergy are American companies that are developing a system called the Zero Emission Boiler System (ZEBS). This system claims near 100% carbon capture and according to company information virtually no air emissions. The process uses pressurized oxycombustion. This high-pressure process strives to capture carbon dioxide in a clean, pressurized form that can be isolated or reused while simultaneously capturing the carbon dioxide and other harmful emissions (Thermoenergy, 2014).

Another process called underground coal gasification (UCG) converts coal into a gas within the mining process. Oxidants are injected into underground coal seams, and then, the gas is brought to the surface by wells. A proposed project in the United Kingdom would use this process to create pure streams of hydrogen and carbon dioxide. The hydrogen is then used as an emissions-free fuel to run an alkaline fuel cell while the carbon dioxide is captured (Thermoenergy, 2014).

Most scientists doubt that coal for energy generation can ever be "clean." As with any fossil fuel-burning process, emissions will be released no matter how capable the technology or how precise the plant is operated. In addition, developing a utility-scale coal mine creates enormous environmental damage. Huge quantities of rock and dirt must be removed to expose the underlying coal seams, causing damage to ecosystems and the environment. Removing the dirt and rock also creates a liquid waste called slurry that is deposited in open lagoons. When it floods, this waste (which contains heavy metals and carcinogenic compounds) flows into groundwater, affecting water for drinking and irrigation (Care2, 2010).

A 2014 study by the Society for Freshwater (Hector et al., 2014) reveals the environmental and personal health dangers from removing mountain tops for coal in the US state of West Virginia's Appalachian Mountain region.

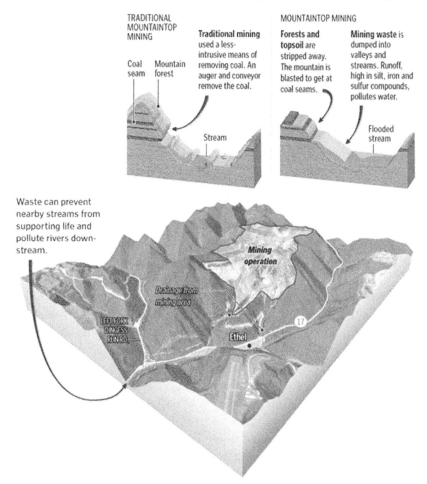

Patterson Clark, Gene Thorp, April Umminger/The Washington Post

Not only is coal mining environmentally destructive, but also actually creating clean coal plants may not be feasible. While there are promising CCS technologies that do have the potential to reduce and even mitigate carbon dioxide and other GHG emissions released in the burning process, this technology is very expensive and probably cannot be delivered in time to avoid dangerous climate change. According to industry sources, adding CCS technology would double the cost of plants, and the earliest chances for large-scale implementation are not expected before 2030. To avoid the worst impacts of climate change, global greenhouse gas emissions have to start falling after 2015 (Clark et al., 2014; Care2, 2010).

CCS technology also takes large amounts of energy to operate, as much as 10-40% of the energy produced by a power station. Wide-scale adoption of CCS would erase the efficiency gains of the last 50 years and increase resource consumption by one-third. The CCS process calls for storing CO_2 underground, which is risky. Even low leakage rates could undermine any climate mitigation efforts (Clark et al., 2014; Care2, 2010).

The United States faces extreme problems today as 82% of its energy is from fossil fuels, as shown below (Unger, 2013).

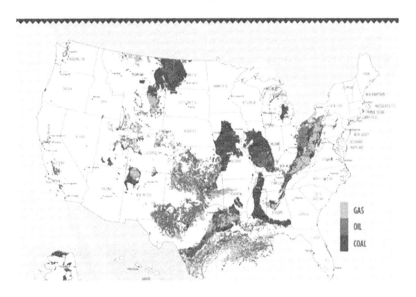

Despite significant environmental damage and an increased threat to climate, coal exploitation and energy generation are increasing. Most countries that burn coal do not consider the overall costs of the process. Like with most fossil fuels, these costs are vastly understated compared to a comprehensive analysis that includes health, environment, and cleanup. Instead of spending enormous amounts of money on trying to "clean" coal, it would be far more efficient and environmentally sound to develop renewable energy sources and efficient green grids.

NATURAL GAS

Like oil and coal, natural gas is a product of the Paleozoic era, within the late Carboniferous period. Massive amounts of plant and other organic life died and decomposed forming large deposits of carbon.

Under the right conditions, the heat of the planet's core turned the carbon solids into liquid. Being a liquid, oil sought the deep cavities between the layers of sediment to puddle up into reservoirs. Natural gas was created as the oil in crevices was heated and then compressed by the weight from massive geologic shifts as mountains formed, or by earthquakes deep in the core, or by the shearing and twisting of tectonic plates.

Historically, few wells are dug specifically for natural gas; most of it is extracted as a by-product of petroleum drilling. Once the natural gas was recovered, some fuels that have primarily automotive purposes were extracted by condensation or absorption. The remaining gas was captured and then piped directly for commercial and residential applications. The western hemisphere, Europe, and parts of Africa contain the largest natural gas deposits. The gas is moved by pipelines or converted into a liquid called LNG and shipped or stored in tanks.

The use of natural gas has dramatically increased in the last decade because of changes in exploration and new drilling techniques, especially hydraulic fracturing, or commonly known as "fracking." Fracking is a controversial drilling technique that uses various toxic chemicals to blast through underground levels of shale or rock. Once below the shale line, the well can be turned in various directions—this process is sometimes called lateral drilling. (See appendices for studies and report on fracking dangers to the environment.)

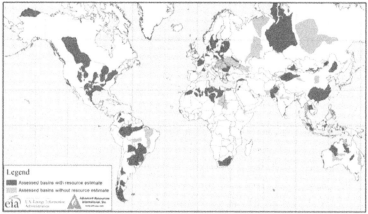

Source: United States basins from U.S. Energy Information Administration and United States Geological Survey; other basins from ARI based on data from various published studies

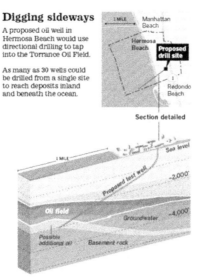

Digging sideways

A proposed oil well in Hermosa Beach would use directional drilling to tap into the Torrance Oil Field.

As many as 30 wells could be drilled from a single site to reach deposits inland and beneath the ocean.

Source: Proposed Settlement: City of Hermosa and Gas Drilling, EB Energy Company, 2013

Compared to coal or oil, using natural gas for energy produces fewer emissions. However, natural gas, like the other fossil fuels, is not a clean or green fuel, since it emits particulates and causes carbon pollution and greenhouse gases. Nonetheless, increasing numbers of proponents of natural gas point to its cheaper costs and less pollution than oil and coal. However, natural gas industry supporters do not include the external costs related to the deterioration of the environment and subsequent health costs. In addition, fracking for natural case is causing increased concern that precious water tables and underground water supplies are being impacted by the toxic chemicals used by the drilling companies.

The supporters of natural gas push it as a clean fuel for vehicles. However, natural gas, like coal, and other fossil fuels create pollutants and GHGs from their emissions in both extraction and usage.

Despite the environmental destruction, the United States is undergoing a boom in natural gas. In fact, the International Energy Association (IEA) claims that America will be the biggest oil and gas producer in the world for the next few decades (CNNMoney, 2012).

According to IEA World Energy Outlook, the US resurgence in oil and gas production coupled with making the transport sector more efficient are reshaping the nation's energy market. In fact, the United States may overtake Saudi Arabia to become the world's biggest oil producer before 2020. While the United States currently imports 20% of its total energy needs, it could become a net exporter of oil around 2030.

Over three years, from 2008 to 2011, the US crude oil production jumped 14%, according to the US Energy Information Administration. Natural gas production is up by about 10% over the same period. Unfortunately, much of the increase in production comes from fracking and the injection of toxic chemicals to batter through shale levels.

The IEA projects that the US natural gas prices will rise to $5.5 per million British thermal units (MBtu) in 2020, from around $4.40 per MBtu in 2014, driven by rising domestic demand rather than a forecast increase in exports to Asia and other markets (CNNMoney, 2012).

North America's new role in the world energy markets will accelerate a change in the direction of international oil trade toward Asia and underscore the importance of securing supply routes from the Middle East to China and India. The IEA said it expects global energy demand to increase by more than a third by 2035, with China, India, and the Middle East accounting for 60% of the growth, and more than outweighing reduced demand in developed economies.

If this happens, the IEA estimates that oil prices would go up to $125 per barrel by 2035, from around $100 per barrel at present, but they could be much higher if Iraq fails to deliver on its production potential. Politics and civil war are critical when discussing Iraq; however, a stable Iraq is set to become the second largest oil exporter by the 2030s, as it expands output to take advantage of demand from fast-growing Asian economies.

The IEA points out that new fuel economy standards in the United States and efforts by China, Japan, and the European Union to reduce demand could have a significant impact on oil and natural gas production. According to the IEA, there is a large potential to improve energy efficiency, and about four-fifth of the potential in the building sector and more than half in industry remain untapped. Growth in demand over the years to 2035 would be halved and oil demand would peak just before 2020, if governments took action to remove barriers preventing the implementation of energy efficiency measures that are already economically viable, the IEA noted.

HYDRAULIC FRACTURING

No form of energy production is without risk. Conventional oil and gas wells leak. Nuclear power stations melt down. Coal-fired power stations emit lots of mercury and greenhouse gases. And wind farms kill migrating birds and disfigure the landscape. Like any other technology, drilling for natural gas imposes its own costs and benefits.

While there are conventional methods of extracting natural gas as a by-product of drilling for oil, hydraulic fracturing is becoming the most common method of tapping huge natural gas deposits miles below the Earth's surface. At the same time, it is becoming controversial with communities in New York, California, and even Texas banning the practice. The process dates back to 1947, and up to 95% of all oil and gas wells drilled today are hydraulically fractured, accounting for 43% of the total US oil production and 67% of the natural gas production, according to a report prepared by the National Petroleum Council for the US Department of Energy.

Fracking as it is commonly called is heralded as a way to tap oil and gas reserves that were once considered too costly to extract. The process involves drilling a pipe horizontally into an underground oil- or natural gas-bearing formation and pumping slurry into the formation at high pressure to release the trapped hydrocarbons. The concerns focus directly on the slurry injected into the well, which is a mix of toxic chemicals. The exact mix is bitterly contested, with oil companies claiming that the formulas for the mix are trade secrets, and environmentalists convinced that the chemicals are toxic and carcinogenic.

Halliburton Company in the late 1940s, a major oil exploration and drilling company that remains a huge player in the field, pioneered the fracking process in the United States. The company's former CEO, Richard Cheney, became George W. Bush's vice president. During George W. Bush's presidential administration, eager oil and gas companies leaped into the extraction of the natural gas accompanied by political support, particularly a highly controversial and soundly criticized 2004 report from the Environmental Protection Agency. The EPA report declared that hydraulic fracturing posed no threat to drinking water. This led to hydraulic fracturing being excluded from the Safe Drinking Water Act. The SDWA exempted underground injections of hydraulic fracturing fluids from EPA's original authority under SDWA, which was needed to protect drinking water sources. The industry is constantly trying to convince the public that its new technologies are safe. They are not as can be seen in the issues from the chart below:

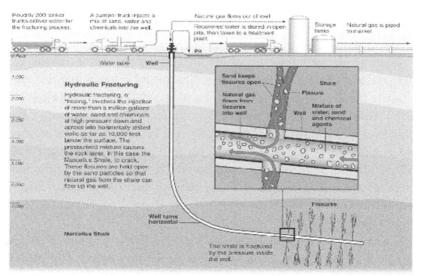

Graphic by Al Granberg

There are simply too many factors in fracking, from the actual drilling to its processing, storage, piping, and transportation (LNG, train or trucks) to see this sort of energy as either cost-effective or good for the environment. In fact, now, there is evidence of fracking destroying water in the ground and polluting other connecting water sources. Data in 2014 are now looking at the impact on land itself from fracking since there are statistically more small earthquakes in areas and communities that rarely had them before the oil and gas wells were dug.

Aside from evidence that fracking causes environmental and health damage that could be irreversible, the original statements and data on the magnitude of oil and gas, for example, in California, have been proved wrong. See appendices for data and evidence.

Controversy surrounds the original and highly politicized EPA decision, and the push to repel it continues in the US Congress. In 2009, the Fracturing Responsibility and Awareness of Chemicals Act, called the FRAC Act, was introduced. The bill would have repealed the exemption for hydraulic fracturing in the SDWA and regulated the oil and natural gas recovery process under the Underground Injection Control Program. Under the FRAC Act, the energy industry would be required to disclose the chemicals mixed with the water and sand that are pumped underground in the fracking process. Again, the oil and gas interests opposed the bill, claiming that the chemicals used were protected as proprietary or trade secrets. Without knowing the identity of the proprietary components, regulators cannot test for their presence. This prevents government regulators from establishing baseline levels of the substances prior to hydraulic fracturing and documenting changes in these levels, thereby making it impossible to determine whether hydraulic fracturing is contaminating the environment with these substances. The natural gas industry opposed the legislation. As of 2014, Congress had not yet passed a bill that requires disclosure of the fracking chemicals or established a process to regulate the activity.

Fracking is being used extensively in the gas-producing Northeast. The Marcellus Shale natural gas field lies under the Appalachian Mountains of New York, Pennsylvania, Ohio, Maryland, and West Virginia. It is described as a supergiant reserve containing a rumored 500 trillion cubic feet of natural gas and has attracted major extraction companies looking for huge profits and anxious politicians looking for a quick fix for the energy needs of New York City and other east coast population centers.

With the natural gas extraction, many economically depressed rural communities are finding new business as the gas companies roll in. Jobs are created, supplies are purchased, restaurants are active, and it looks like prosperity is on the rebound.

However, the problem with fracking is simple. Shale will clog up a drill and bit system, so hydraulic fracturing is used to blast through the shale with enormous amounts of sand and fluids under extreme pressure. This blasting takes place thousands of feet underground. While shale is nearly impossible to drill through, it will shatter. So, the process is much like breaking a glass window with a high-pressure fire hose. It can take up to 350,000 gallons of water slurried with sand and "trade secret" chemicals to frack a gas well (Walsh, 2011).

This "cocktail" of toxic chemicals is the greatest concern. While the drilling companies refuse to disclose the chemicals, other scientists claim it may include, among other things, barium, strontium, benzene, glycol ethers, toluene, 2-(2-methoxyethoxy) ethanol, and nonylphenols. A well can be fracked multiple times, using up to 4 million gallons of water. Estimates are that gas companies are on track to drill 32,000 new gas wells each year, contaminating 100 billion gallons of water a year. None of the states involved have the capacity to filter this amount of contaminated water, nor are they sure of what toxic fluids are used in the process. Even local medical personnel do not know how to treat gas drillers exposed to the fracking fluids because they are unsure of the chemicals (Walsh, 2011).

Hydraulic fracturing may have devastating environmental consequences. Not only does the huge amount of contaminated water overwhelm filtering systems, but also the fluids are usually stored in reservoirs dug near the well. The fluids seep into the ground and eventually find the tiny fissures and cracks in the shale caused by the pressurized hydraulic fracturing process. Eventually, these toxins could make their way into the local water supply systems.

No one mentioned these consequences of gas exploration in a shale formation; and no one worried about them until cattle started being quarantine, and water faucets and riverbeds caught on fire. Then, criticisms of the SDWA exemption began, and it was discovered that the EPA had done no scientific studies before making the 2005 SDWA exemption. In April 2010, Pennsylvania banned Cabot Oil from further drilling until they repaired the contaminated wells of 14 homes in Dimock Township. EOG Resources and CC Forbes have been banned from drilling pending investigations into a well blowout that spewed 35,000 gallons of fracking fluids onto campers and residents.

The 28 cattle that were quarantined were located in north-central Pennsylvania on a ranch that is right on top of the Marcellus Shale Formation. The state Department of Agriculture said the action was its first livestock quarantine related to pollution from natural gas drilling.

Carol Johnson, who along with her husband owns the farm, said she noticed in early May 2010 that fluids pooling in her pasture had killed the grass. She immediately notified the well owner, East Resources Inc. "You could smell it. The grass was dying," she said. "Something was leaking besides ground water." The well has since been shut down (Walsh, 2011).

Unfortunately, hydraulic fracturing is not limited to the Northeast. It is also used to break through the shale layer underneath California's south San Joaquin Valley. This area has oil reserves that have been drilled since the early 1900s. The oil is referred to as "heavy," which means it takes a significant amount of steam and fluids to extract. The fracking was started several years ago, and there are numerous sludge and slurry reservoirs dotting the area, which is some of the world's most fertile farmland. Farmers and nut growers are starting to get nervous about fracking creating fissures in the shale and bedrock and worried that the toxins from the reservoirs might leak into the water table.

Fracking is also finally drawing critics in the Rocky Mountain states where recent tests by the EPA in Wyoming and other areas have led to concern about contamination of well water used for cattle grazing. A 2010 study published in the International Journal of Human and Ecological Risk Assessment was coauthored by Dr. Theo Colburn, head of the Endocrine Disruption Exchange (TDEX) in Colorado.

Titled "Natural Gas Operations from a Public Health Perspective," the studies' researchers demonstrated that toxic chemicals are used during both the fracturing and the drilling phases of gas operations. They found that fracking not only pollutes the groundwater system wherever the process is undertaken but also pollutes the air. The chemicals used in fracking range from methanol to tetramethylammonium chloride, with over a thousand other chemicals used in natural gas operations across the United States.

The report noted that natural gas drilling operations also uses chemicals. ". . . From the first day the drill bit is inserted into the ground until the well is completed, toxic materials are introduced into the borehole and returned to the surface along with produced water and other extraction liquids. In the western U.S. it has been common practice to hold these liquids in open evaporation pits until the wells are shut down, which could be up to 25 years. These pits have rarely been examined to ascertain their chemical contents outside of some limited parameters (primarily metals, chlorides, and radioactive materials). Our data reveal that extremely toxic chemicals are found in evaporation pits and indeed, these and other similar sites may need to be designated for Superfund cleanup. In the eastern U.S., and increasingly in the west, these chemicals are being re-injected underground, creating yet another potential source of extremely toxic chemical contamination."

In other words, what ends up in evaporation pits in the West will in other parts of the country be injected underground. The study also showed that there may be long-term health effects that are not immediately recognized and that waste evaporation pits may contain numerous chemicals on the US Environmental Protection Agency (EPA) Superfund list.

The TEDX also issued a statement at the time of the studies' release that warned that human development can be affected in the womb by low levels of toxins and warns that the danger is even greater because ". . . government standards for chemical safety are deeply flawed" (Dearing, 2010).

In the United States, public pressure is mounting so that the EPA, under the Obama administration, is conducting a new study and efforts are being made to put hydraulic fracturing back under the Safe Water Drinking Act regulations and EPA control.

All sides have an urgent interest in better regulation. While fracking was mostly about drilling for gas, the US Congress and local lawmakers have treated it as primarily an environmental issue. But as it moves into the oil patch, fracking is set to become an issue of national energy security, and the environmental aspects will be much lower down in Congress' list of concerns. Now is the time for EPA and environmental groups to compromise for a stronger regulatory framework.

In the meantime, hydraulic fracturing in the Allegheny Mountain region and elsewhere may cause irreparable environmental damage. The fluids used are clearly toxic, and since they are unknown, local water treatment plants cannot test for their presence. Given the close proximity of the hydraulic fracturing processes to major eastern population centers, the danger is mounting and threatens the water supply of New York City as well as Philadelphia and New Jersey.

The Scientific American journal reported after the annual 2012 American Association for the Advancement of Science (AAAS) meeting in Vancouver, Canada, that resistance to hydraulic fracturing in the United States has risen (Fischetti, 2012). Citizens and politicians are worried that fracking deep shales to extract natural gas can contaminate groundwater, trigger earthquakes, and release methane, another potent greenhouse gas, into the atmosphere.

In a particularly ironic case, the worry has extended to even the natural gas fields in Texas. Denton, Texas, is an upscale community north of Dallas, with new parks, a golf course, and miles of grassy soccer fields. The community's prosperity comes from the huge gas reserves underneath its streets. The gas fields have produced billion dollars in mineral wealth and pumped more than $30 million into city bank accounts (Schmall, 2014).

Once a farm center, Denton is now at the center of the fracking controversy in gas-friendly Texas. The leaders of Denton, where many wealthy oil executives live, have temporarily halted all fracking, while they consider an ordinance to permanently ban the practice. Unlike other communities that have embraced the lucrative drilling boom made possible by hydraulic fracking, Denton would be the first city in the state to outlaw the practice. If the city council rejects the ban, it will go to voters in November 2014.

Although in the heat of oil and gas country, Denton has tried to preserve much of its agricultural past. Historic downtown streets lined with nineteenth-century buildings open up to expansive fields with greenhouses and grazing cattle. However, drilling is never far away, with some 275 active gas wells piercing the earth.

Fracking began in Denton around 2000, and about the same time, the population started to increase. More graduates from the University of North Texas and from Texas Woman's University chose to stay in town and set up small businesses. Then, emerging concerns over the proximity of new wells to residential areas came to a head in 2009, with an organized protest against five wells planned in a meadow across from a city park.

A community group proposed tighter fracking rules and even won a series of temporary bans on new drilling permits. At the same time, drillers defied city rules that required them to line wastewater pits and prohibited them from burning off, or "flaring," waste gas in residential areas. Residents were so angry that they called for an outright ban.

While the worries and concerns about fracking continue to mount in the United States, the practice is fiercely opposed in Europe. In Europe, environmental groups have more political power than in the United States and higher population densities magnify the possible damaging effects of the drilling practice. Some countries have banned fracking outright; others have imposed onerous regulations that effectively make the practice illegal, though they are reconsidering fracking in light of the standoff with Russia over Ukraine (Johnson, 2014).

European nations that have banned or imposed a moratorium on fracking include France, Bulgaria, Germany, the Czech Republic, Spain, Switzerland, Austria, Italy, Northern Ireland, Ireland, Romania, and the Netherlands. The United Kingdom has rejected shale gas technology, but has not passed an outright ban of fracking.

Outside of Europe, nations like Argentina, Australia, New Zealand, South Africa, and Canada have imposed at least partial bans. See the appendix for a complete list of nations that have bans or moratoriums (www.keeptapwatersafe.org).

NUCLEAR POWER

In 1954, the Obninsk Nuclear Power Station reactor was turned on and connected to Russia's grid, launching worldwide hopes for an environmentally sound, inexpensive, and endless supply of cheap energy. Located about 110 km southwest of Moscow, Obninsk had a single reactor nicknamed "peaceful atom" and generated 6MW of electrical power with a thermal output of 30 MW. Obninsk was a prototype design and lasted until 2002 when it was decommissioned.

Obninsk was followed by Calder Hall in England in 1956, and a half-century later, the International Atomic Energy Agency reports that there are 433 nuclear power plants operating in 30 different countries. The United States has the most with 104 power plants, followed by France with 58, Japan with 55 (all but two are shut following the Fukushima disaster), and Russia with 33. China has 15 with 25 under construction and 51 more in the planning stages (IAEA).

Most nuclear power plants are memorable for their large cylindrical cooling tower or towers that jut into the air emitting a plume of white vapor. The cooling tower vents water vapor from the nonradioactive side of the plant and is usually separated from the nuclear reactor.

Large cylindrical cooling tower, or towers, of a nuclear reactor. The cooling tower vents water vapor from the nonradioactive side of the plant. *(Source: http://www.freeenterprise. com/energy-environment/construction-first-new-nuclear-reactor-more-30-years-approved. Freeenterprise. "Construction of First New Nuclear Reactor in More Than 30 Years Approved." February 9, 2012.)*

At the heart of the power plant is the nuclear reactor; in its center is the core where the nuclear reactions, or fissures, take place. Inside the core are the fuel rods and assemblies, the control rods, the moderator, and the coolant. Outside the core are the turbines, the heat exchanger, and usually the cooling tower system.

The fuel rods are about 3.5 m long and about a centimeter in diameter. Inside each fuel rod are hundreds of pellets of uranium fuel stacked end to end. The rods are grouped into large bundles called fuel assemblies, which are then placed in the reactor core (World Nuclear Association, 2013).

Also in the core are control rods. These rods have pellets inside that are made of very efficient neutron capturers, usually in the form of cadmium. These control rods are connected to machines that can raise or lower them in the core. When they are fully lowered into the core, fission cannot take place. However, when they are pulled out of the reactor, fission can start again anytime a stray neutron strikes a ^{235}U atom, thus releasing more neutrons and starting a chain reaction.

Another component of the reactor is the moderator, which serves to slow down the high-speed neutrons "flying" around the reactor core. If a neutron is moving too fast, and thus is at a high-energy state, it passes right through the ^{235}U nucleus. It must be slowed down to be captured by the nucleus and to induce fission. Water is the most common moderator.

The job of the coolant is to absorb the heat from the reaction. The most common coolant used in nuclear power plants today is water. In actuality, in many reactor designs, the coolant and the moderator are one and the same. The coolant water is heated by the nuclear reactions going on inside the core. However, this heated water does not boil because it is kept under intense pressure, thus raising its boiling point above the normal 100 °C.

The conversion to electrical energy takes place indirectly, as in conventional thermal power plants. The heated water, or pressurized steam, is then usually fed to a steam turbine. After the steam turbine has expanded and partially condensed the steam, the remaining vapor is condensed in a condenser. The condenser is a heat exchanger, which is connected to the cooling tower. The water is then pumped back into the nuclear reactor and the cycle begins again.

Since nuclear fission creates radioactivity, a protective shield surrounds the reactor core. This containment absorbs radiation and prevents radioactive material from being released into the environment. Additionally, most reactors are covered with a dome of concrete to protect against external impacts (World Nuclear Association, 2013).

The following figure shows that

The Diablo Canyon, California, nuclear power plant. The two light-colored dome structures are the reactors. The front building with the vertical strips contains heat turbine engines for electricity generation and transmission of the power to the grid. Each reactor has pressure vessels for the nuclear fuel, control rods, moderator, coolant, and containment.

The plants nuclear fission reactors produce heat through a moderated nuclear fission of fissile materials deployed at the critical mass level.

This section of California is vulnerable to earthquakes. A key public concern is whether the plant is sufficiently earthquake-proof. Originally built in 1968, it was retrofitted in 1990s to withstand a 7.5 magnitude of earthquake. Its design includes safety features and intentional redundant seismic monitors to sense earth movements and to shut it down. Its design can divide the fissile material's mass level to be well below a critical mass during any significant ground motion. The Diablo Canyon plant's design is considered one of the most reliable in use today, and it serves as the standard for modern reactor designs.

It is worth noting that nuclear power plants are not the only abundant energy. The United States has ventured into a variety of better energy resources. The Integral Fast Reactor was built, tested, and evaluated during the 1980s and then retired under the Clinton administration in the 1990s due to nuclear nonproliferation policies of the administration. Recycling spent fuel is the core of its design and it therefore produces only a fraction of the waste of current reactors.

One of the key challenges to nuclear power plants includes the high start-up cost, which is typically higher with a longer investment return period than other comparative power generators.

Another issue includes the disposal of the nuclear waste. Recycling both used fuel and wastewater is imperative to reduce the high-level hazard and the volume of waste. To address this challenge, the reprocessing of uranium and plutonium is undertaken for reuse in a nuclear power plant. Furthermore, when old reactors close down, the safe and secure storage for numerous fuel rods has attracted intense international interest. Several countries including France, Russia, and China have dedicated efforts to recycle used fuel.

While nuclear power originally was thought to be limitless and perfectly safe, a steady series of disasters has forced nations to reject or rethink nuclear power as a viable source of energy.

Table below lists the 10 worst nuclear disasters as complied by Time Magazine.

Fukushima Daiichi, 11 March 2011	An 8.9 magnitude earthquake and subsequent tsunami overwhelmed the cooling systems of an aging reactor along Japan's northeast coastline. The accident triggered explosions at several reactors in the complex, forcing a widespread evacuation in the area around the plant
Tokaimura nuclear accident, 30 September 1999	At the time, it was the worst nuclear accident in Japan's history, occurring in a uranium reprocessing facility in Tokaimura, northeast of Tokyo. The incident took place while workers were mixing liquid uranium
Tomsk-7 explosion, 6 April 1993	The accident in the Siberian city of Tomsk took place after a tank exploded while being cleaned with nitric acid. The explosion released a cloud of radioactive gas drifting from the Tomsk-7 Reprocessing Complex
Goiânia accident, 13 September 1987	More than 240 people were exposed to radiation when a junkyard dealer in Goiânia, Brazil, broke open an abandoned radiation therapy machine and removed a small highly radioactive cake of cesium chloride. Children attracted to the bright blue of the radioactive material touched it and rubbed it on their skin, resulting in the contamination of several city blocks, which had to be demolished

Continued

Chernobyl, 26 April 1986	The Chernobyl disaster is considered to be the worst nuclear power plant disaster in history. On the morning of 26 April 1986, reactor number four at the Chernobyl plant exploded. More explosions ensued, and the fires that resulted sent radioactive fallout into the atmosphere. Four hundred times more fallout was released than had been by the atomic bomb of Hiroshima
K-431 Chazhma Bay, 10 August 1985	During refueling in Vladivostok, Russia, the Echo II class submarine suffered an explosion, sending a radioactive cloud of gas into the air. Ten sailors were killed in the incident and 49 people were observed to have radiation injuries
Three Mile Island, 28 March 1979	The partial meltdown of the Three Mile Island Unit 2 nuclear power plant was the serious accident in the history of the US nuclear power plant operations, although it led to no deaths or injuries
Yucca Flat, 18 December 1970	After the Baneberry test, involving the detonation of a 10 kiloton nuclear device underneath Yucca Flat in Nevada, the plug sealing the shaft from the surface failed and radioactive debris vented into the atmosphere. Eighty-six workers at the site were exposed to radiation
Thule accident, 21 January 1968	A cabin fire aboard a B-56 forced the crew of the American bomber to abandon the craft before they could carry out an emergency landing. The bomber then crashed onto sea ice near the Thule Air Base in Greenland, causing the nuclear payload to rupture, which resulted in widespread radioactive contamination
Palomares incident, 17 January 1966	A US B52 bomber collided with KC-135 tanker during midair flight refueling over the coast of Spain. The tanker was completely destroyed in the incident, while the B52 broke apart, spilling four hydrogen bombs from its broken fuselage. The nonnuclear weapons in two of the bombs detonated on impact with the ground, contaminating of a 490 acre area with radioactive plutonium. One of the devices was recovered from the Mediterranean Sea

Fukushima Daiichi, Japan. March 11, 2011. *(Source: Tech & innovation Daily. Com. November, 2013. http://www.techandinnovationdaily.com/2013/11/19/two-innovations-to-prevent-another-fukushima-nuclear-disaster/)*

THE MOST POLLUTED PLACE ON EARTH: RUSSIA'S MAYAK NUCLEAR REACTOR

Russia's Chelyabinsk province in the Ural Mountains, north of Siberia, is generally acknowledged by the scientific community as the most polluted spot of Earth. This region with over 1.3 million people was the home of "Mayak," one of the former Soviet Union's main military production centers, which included nuclear weapon manufacturing. Accidents, nuclear waste disposal, and day-to-day operation of the Mayak reactor and radio-chemical plant contaminated a vast area of the province.

In the early 1950s, there were so many occurrences of death and disease from the nuclear waste dumping in the Techa River that 22 villages along the riverbanks in a 50 kilometer zone downstream from Mayak were evacuated. In 1957, a nuclear waste storage tank accident released radiation double the amount released by the Chernobyl accident. This accident was kept secret and 10,700 people were evacuated. The severe environmental contamination of this region led to dramatic increases in cancer rates, birth defects, and sterility. Over the decades, there have been a 21% increase in the incidences of cancer and 25% increase in birth defects, and 50% of the population of childbearing age is sterile (Russia: Living and Dying in the Shadow of Mayak).

The Earth is a remarkably fragile planet, and a quick look around provides a vision of human carelessness, resource-extraction greed, and disregard for vulnerable and fragile biosystems. Besides environmental concerns, there are miserable examples of societies that have placed too much economic and political dependence on fossil fuel extraction and wealth that they have failed to develop.

Social scientists describe this as the "resource curse." This is the paradox that countries rich in minerals or petroleum tend to grow more slowly and have lower living standards than other nations. Typical examples are the Niger Delta, the Orinoco Belt in Venezuela, and the Iraqi Marshes. However, the United States has its own example of the resource curse in Louisiana where oil and gas riches stunted the state's development, leaving it far behind other states with fewer natural resources.

Despite great fossil fuel resources, Louisiana has created no Silicon Valley or North Carolina Triangle Research Park.

When Louisiana realized it needed new sources of tax revenue to make up for declining oil receipts, its best idea was to expand offshore drilling. Natural gas is being pushed by corporate interests for the same reason. As Thomas Friedman, the economist with the New York Times, noted in November 2012 when he quoted Fatih Birol, the chief economist for the International Energy Agency, "a golden age for gas is not necessarily a golden age for the climate" if the result is that renewables are either ignored or unfunded. Then, Friedman noted that Maria van der Hoeven, executive director for the IEA, had urged governments to keep in place subsidies and regulations that encourage investments in wind, solar, and other renewables "for years to come" so that they remain competitive.

REFERENCES

Environmental Defense Fund, 2014. Energy, The Challenge in Numbers. http://www.edf.org/energy/energy-challenge-numbers.
U.S. Energy Information Agency, http://www.eia.gov/totalenergy/.
Thermoenergy, 2014. http://www.thermoenergy.com/clean-combustion/zero-emissions-power-production.
Care2, 2010. The Many Problems with Clean Coal. http://www.care2.com/causes/the-many-problems-with-clean-coal.html.
Unger, D., 2013. The Big Picture, Christian Science Monitor.
Hector, B., Schenk, J., Clark II., W.W., Saavedra, A., 2014. The Universal Ecolabel. Int. J. Appl. Sci. Tech. 4.
Hermosa Beach City Council Report. Confidentiality Settlement between City of Hermosa Beach and E&B Natural Resources Management based on Contract with Macpherson Oil Company and Windward Associates 2012. Hermosa Beach, California.
EB Energy Company, 2013. Proposed Settlement: City of Hermosa and Gas Drilling.
Bloomberg, 2014. New Energy Finance. Business Council of Sustainable Energy.

Thompson, M., 2012. U.S. to Become Biggest Oil Producer—IEA. CNNMoney. Mark Thompson. http://money.cnn.com/2012/11/12/news/economy/us-oil-production-energy/.

Walsh, B., 2011. Could Shale Gas Power the World? Time. http://www.time.com/time/magazine/article/0,9171,2062456,00.html.

Dearing, S., 2010. Oil and Natural Gas Fracking has a Price Tag for Human Health. Digital Journal (http://digitaljournal.com/article/297727).

Fischetti, M., 2012. The Scientific American Journal.

Schmall, E., 2014. Texas Town in Revolt: City that has long drawn lifeblood from drilling might ban fracking. U.S. News.

Johnson, K., 2014. Russia's Quiet War Against European Fracking. The Week.

Keep Tap Water Safe, 2014. http://keeptapwatersafe.org/global-bans-on-fracking/.

Nuclear Power Plant Information, 2010. International Atomic Energy Agency, http://www.iaea.org/pris/.

World Nuclear Association, 2013. Nuclear Power Reactors. http://www.world-nuclear.org/info/inf32.html.

Time, 2013. 12 Worst Nuclear Disasters. http://content.time.com/time/photogallery/0,29307,1887705_2255451,00.html/.

Radio Free Europe, 2014. Russia: Living and Dying in the Shadow of Mayak. http://www.rferl.org/content/article/1063825.html.

CHAPTER 5

Climate Change, Science and Technology, and Economics Are the Forces Behind the GIR

Human-induced climate change since the 1960s has increased the frequency and intensity of heat waves and thus also likely exacerbated their societal impacts. In some climatic regions, extreme precipitation and drought have increased in intensity and/or frequency with a likely human influence.

The World Bank (2012).

Butterflies in California's vast Central Valley are emerging earlier in the spring. Magellanic penguin chicks along the Atlantic coast of Argentina are starving because rainstorms in Antarctica are becoming more common. Bangalore, India, once had 400 lakes; now, the New Indian Express newspaper reports that only 40 are left, and all of them polluted. Algae are dying or leaving the planet's coral reefs, turning these fertile rain forests of the sea into bleached limestone.

For years, scientists have warned us that human or anthropogenic activity has been changing the climate. Now, the evidence is becoming overwhelming (UN IPCC, 2014). The oceans are becoming more acidic, land masses are getting warmer, drinkable water is getting scarcer, and crops are failing. Since the planet is round, not flat, all of us suffer the consequences. Because of the jet stream, global wind patterns, the Earth's surface, and a complex variety of atmospheric factors, weather in one part of the world impacts the rest. When different weather patterns occur as they are doing more frequently because of the changes in climate, the results are fiercer tornadoes and hurricanes.

Hurricane Sandy is a case in point. Sandy, the largest Atlantic hurricane on record, slammed into the northeastern section of the United States on 2012 October 29. Propelled by twisting cyclonic winds and torrential rain, Hurricane Sandy crushed coastal New Jersey and New York, killing 253 people, destroying homes and businesses, and wrecking havoc and destruction for over 72 hours. Airports were abandoned, and millions of people were threatened as New York City's subway tunnels sparked, then shorted, and became eerily quiet as they were filled with water. Throughout the city, water burbled up through street covers, and Battery Park and Lower Manhattan were flooded. The huge brightly colored scrolling banners that ring Times Square went black. No honking taxis prowled the streets, and the homeless scattered as the rain fell in sheets. New York, the city that never sleeps, went dark and quiet for days.

Heavy winds, rains, and storm surges ravaged southern New Jersey. The shoreline that attracted millions each summer was a twisted mess of board-walk timbers and fallen trees. Eighty percent of Atlantic City, New Jersey's gambling and casino mecca, was underwater. Small towns along the shore were ruined as water surged and large boats from exclusive marinas were carried out to sea, thrown up on land, or left perched on buildings when the water withdrew. The few people who did not evacuate ahead of the storm tried to escape in anything that would float.

Sandy was the worst hurricane in US recorded history. Its wind diameter was estimated at 1100 miles, or 1800 km, stretching across the eastern portion of the United States and far into Canada. Twenty-four states were impacted including the entire eastern seaboard from Florida to Maine. The storm stretched west across the Appalachian Mountains to Michigan and Wisconsin. The damage in the United States alone was estimated at over $100 billion (Sandy Hurricane, October, 2012).

Massive though Hurricane Sandy was, it paled in comparison with Supertyphoon Haiyan that ripped through the Philippines in November 2013. The sheer magnitude of the typhoon was unprecedented as the archipelago was shattered with 250 miles per-hour sustained winds, and water surged over 16 foot barriers. The scale of the destruction and damage was shocking. President Benigno Aquino III declared the devastation a national calamity (Economist, November 16, 2014).

Some parts hit by Typhoon Haiyan were remote; however, the government said that more than 2300 people were killed and 11 million were affected. Roads and villages were destroyed, trees felled, crops flattened, power lines and houses blown away, and about 600,000 were made homeless. Cost estimates were well over $15 billion.

Typhoon Haiyan may be the strongest storm in recorded history, and scientist and politicians are blaming climate change. Naderev Saño, the Philippines representative at the 2013 Warsaw Climate Change Conference, was convinced that the severity of the storm was the result of climate change. "The trend we now see is that more destructive storms will be the new norm," he told reporters (Economist, November 16, 2014).

The Earth is getting warmer, at a much more rapid rate than scientists had predicted just a few years ago. Climate change is real, it is here now, and it is having a serious effect on the planet's weather. For example, Climate Central reported that 2012 was the third straight hurricane season with 19 named storms on the East Coast of the United States. Hurricane records go back to 1851, and 2010, 2011, and 2012 were the busiest on record except for 2005 and 1933. Scientists think one reason for this increase in storm activity comes from the warming of the Atlantic sea surface temperatures (Economist, November 16, 2014).

Hurricanes are exceptional because of their size, but they are not the only results of climate change. In February 2011, Pope Benedict XVI opened his bedroom windows to take a peek at Rome blanketed in snow. The freak snowstorm forced the closure of schools and the Coliseum. It covered palm trees, Baroque churches, and ancient Roman ruins across the city, which usually has temperate climates. Italian newspapers reported snowdrifts of 5-6 feet outside the city. Without snow plows, the Italian National Civil Protection Agency handed out 4000 shovels so residents could help with the cleanup.

From 2010 to 2012, Pakistan struggled with unprecedented flooding, and Mongolia and Texas in the United States, suffered from torturous droughts. Texas cattle ranchers were rushing steers to slaughter before they died from dehydration. Also in 2010, Western Russia had its hottest summer on record, with 500 wildfires around Moscow. In 2011, East Africa was ravaged by famine, the result of a drought that was linked to the Indian Ocean warming. The Western Amazon region had its worst drought, with record low water levels in Rio Negro. The spring of 2011 was the hottest and driest ever in France. A year later, the summer of 2012 was the wettest April-to-June period ever recorded in the United Kingdom. The rain blighted much of the Queen's Diamond Jubilee river pageant on the Thames (Sandy Remembered, November 2012).

Roads melted in Oodnadatta, an outback town in South Australia in summer 2013, and fires swept across the Tasman Peninsula. Australia, one of the countries most vulnerable to global warming, is getting ever hotter. The 2013 heat wave set new records (40.3 °C) for the highest national temperature ever recorded. To measure these new recordings, Australia's Bureau of Meteorology added new colors purple and pink to its weather map. The new colors denote temperatures once considered off the scale (World Bank, Climate Change, 2014).

Dramatically unusual weather is becoming the new "normal" as global warming changes our climate. Throughout the world, climate scientists are becoming concerned about the impact that excessive amounts of greenhouse gas and subsequent climate change are having on the world's weather. According to the 2011 Arctic Report Card, the Arctic Circle, one of the world's most sensitive environmental areas, is undergoing profound changes from global warming.

Produced by a team of 121 international scientists under the auspices of the National Oceanic and Atmospheric Administration, the report says that the rapid Arctic climate change may already be influencing weather and climate patterns in the Northern Hemisphere—including drafts in California.

These climate scientists say that the Earth's northern polar region is entering a new warmer era and the ice cap is melting at an unprecedented rate. They warn that the new era has warmer air and water temperatures, less summer sea ice and snow cover, and changed ocean chemistry. In 2011, the average annual air temperatures over the Arctic Ocean were 2.5 degrees greater than the 1981-2010 baseline and that ocean acidification from the increased absorption of carbon dioxide was rising. One scientist described the Arctic in summer now "as a giant slushie."

As Arctic sea ice declines in thickness, it alters the flow of heat between the Arctic Ocean and the air. As the sea ice declines, the corresponding dark ocean surface absorbs more incoming solar radiation. This has a dramatic influence on Arctic air temperatures and can alter atmospheric circulation. Warmer air weakens the high-altitude winds that circle the North Pole, causing a decline in the "polar vortex." This provides more chances for Arctic air to flow south into the United States and Europe, causing major changes in weather (Economist, January 12, 2013).

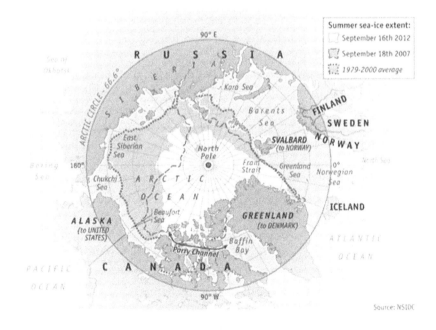

Source: NSIDC

As the Arctic region warms twice as fast as the rest of the planet, one irony stands out. While the ice melts, the world's major oil companies refuse to admit the role their products and activities play in this environmental death dance. They are doing their best not to be seen as profiting from the environmental destruction contributed to by their activities. The companies have poured millions of dollars into disinformation, politics, and public relations campaigns arguing against climate change. At the same time, they are pushing forward with major explorations to harvest the vast natural resources of the Arctic region, even as the ice continues to recede at a faster rate.

Carbon emissions have a dramatic impact of the world's great oceans. In a troubling discovery in April 2014, scientists studying the ocean waters off California, Oregon, and Washington found evidence that the ocean's increasing acidity threatens the base of the food chain (Pacific Ocean Acidity, 2014). Scientists with the National Oceanic and Atmospheric Administration in Seattle led the research. They found that the growing acidity in the ocean is dissolving the shells of a key species of a tiny sea snail, known as pteropods.

The scientists found that in the waters off the West Coast of the United States, 53 percent of the tiny floating snails had shells that were severely dissolving, twice the estimate from 200 years ago (Pacific Ocean Acidity, 2014). As the base food in the food chain, pteropods are a prime food source for salmon, herring, mackerel, and related fish. In the fertile Pacific shoreline, a wide variety of other sea creatures such as whales, dolphins, and sea lions feed on the fish. In addition, the fish are eaten by millions of people each year.

According to scientists, the concentrations of carbon dioxide in the Earth's atmosphere have increased 25 percent since 1960 and are now at the highest levels in at least 800,000 years. Nearly a third of the carbon dioxide emitted by humans is dissolved in the ocean, which produces carbonic acid and makes the ocean more corrosive. Over the past 200 years, the ocean's acidity has risen almost 30 percent and is on track to rise by 70 percent by 2050. Carbonic acid is harmful to most species that have calcium shells, such as clams, oysters, and corals (Pacific Ocean Acidity, 2014).

The transition to the Green Industrial Revolution would not be so critical if it was simply a matter of delaying when the last drop of the world's oil will fall from the spigot. It is not just a desire for cheap fuel as much as it is a concern for what the use of fossil and carbon-based fuel is doing to the planet. As scientists have known for more than a decade now, different areas of the planet are getting hotter or colder each year where the climate had not changed in thousands of years.

Fly into Los Angeles, Mexico City, Beijing, or nearly any of the world's major cities and see the noxious and toxic layers of smog that surround these communities and cause health problems for their residents. The smog layer has been developing for decades, since fossil and carbon-based fuels became the major source of cheap energy for the development that has swept the globe. However, as the film *An Inconvenient Truth* documented, this process has accelerated extremely since the turn of the 21st century into what Al Gore refers to in the film as a "hockey stick." The world is round, so what happens to the atmosphere and ocean in one area impacts other regions.

That the residents of these smog-covered communities tolerate this oppressive environmental degradation is a tribute to human adaptability and denial, mixed with a heavy dose of political self-interest. Their adaptability is most likely coupled with the extraordinary desire for cheap personal transportation. It is also a tribute to the monumental skill of the advertising industry, which has convinced the world that a personal vehicle with a 300 horsepower engine that can exceed 80 miles per hour is something to be desired, bought, and celebrated.

Given the overwhelming social and psychological needs of human beings to emulate each other, it was inevitable, as nations struggled to rise from poverty to a middle-class prosperity, that the first thing their citizens did after adding animal protein to their diet was to buy a car. Of course, with almost seven billion people owning one billion cars on an environmentally fragile planet, eventually a price had to be paid. Unfortunately, that price, which started out as the smog blanket common to most major population centers, has now become global warming with potentially disastrous consequences to the Earth's climate.

The recognition that increasing greenhouse gas emissions are causing global warming and climate change is not new. It was initially advanced by a few visionary scientists as early as 1980s and introduced to the world stage with the United Nations Intergovernmental Panel on Climate Change. Unfortunately, this realization was sidetracked during the United States' environmentally insensitive George W. Bush administration. America, under Bush, refused to sign the Kyoto Protocol and derailed international momentum for a cooperative approach to address the global threat.

WHAT IS CLIMATE CHANGE?

Typhoon Haiyan and Hurricane Sandy's vast devastation demonstrates how climate impacts the way humans live. Massive disruptions from the weather on daily life have both short- and long-term consequences. For example, the drought in Texas had a huge impact on cattle prices and the floods in Pakistan caused the loss of infrastructure in a volatile political region. Russia's heat wave threatened its extensive wheat crop, sending food prices higher in nations that import Russian wheat, like Egypt. But what exactly is climate change, and where does it come from?

The simplest answer is that climate change is the permanent change in the world's weather patterns over time. It may be a change in weather conditions, or in average weather conditions, like more or fewer extreme weather events. Several factors cause climate change. For example, oceanic circulation, variations in radiation from the sun, movement in the Earth's tectonic plates, and eruptions of volcanoes can all cause significant changes in regional and global weather patterns. However, the most crucial factor in climate change is the warming of the Earth's environment, which scientists are now convinced comes from human impacts.

Day by day, the planet's climate is changing, and we humans have caused it. Climate change "significantly" increases the odds of freaky weather like the brutal Texas drought, which is the findings of the American Meteorological Society 2012 State of the Climate report (AMAP, 2009). The 2013 report from the US National Climatic Data Center noted that 2013 was the fourth warmest year globally since records began in 1880. This marks the 37th consecutive year above average, and 9 of the 10 warmest years in the 134-year period have occurred in the 21st century.

TOP 10 WARMEST YEARS (1880-2013)

The following table lists the global combined land and ocean annually averaged temperature rank and anomaly for the 10 warmest years on record (State of the Climate, 2013).

Rank 1 = warmest period of record: 1880-2013	Year	Anomaly (°C)
1	2010	0.66
2	2005	0.65
3	1998	0.63
4 (tie)	2013	0.62
4 (tie)	2003	0.62
6	2002	0.61
7	2006	0.60
8 (tie)	2009	0.59
8 (tie)	2007	0.59
10 (tie)	2004	0.57
10 (tie)	2012	0.57

The planet's climate has been changing since the First Industrial Revolution in the late 1700s, when humans started to move toward large-scale manufacturing. At first, machines started to replace manual labor, horsepower, and wind- and waterpower. Starting in Britain, this transition spread through Europe and eventually reached North America. Societies based on trade and agriculture that were dependent on tools and animals began to rely more and more on machines and engines.

The First Industrial Revolution of the 18th century was propelled by James Watt's steam engine. Watt's engine converted the chemical energy in wood or coal to thermal energy and then to mechanical energy. Its main purpose was to power industrial machinery and steam locomotives.

The Second Industrial Revolution began in the 19th century, when power and energy shifted from steam and coal to oil and the internal combustion engine. Oil was used in the mid-1800s, several years before electricity. It was burned as kerosene in lamps and small stoves, replacing whale oil, as populations of these mammals declined.

Manufacturing grew and the assembly line made it possible to massproduce automobiles. Except for some brutal wars in Europe, Russia, and Asia and the relentless genocide of the world's aboriginal natives, the overall planet's population expanded. With this expansion came greater distribution of products and increased dependence on fossil fuels.

Society evolved to require greater and greater amounts of energy for light, heat, locomotion, mechanical work, and communications. Later, we added smartphones, computers, televisions, microwaves, washing machines, coffee makers, and all the other technologies and gadgets that make modern living modern. Since the 19th century, these energy requirements have primarily come from fossil fuels that emit carbon dioxide.

Fossil fuels opened the world to the wonders of the personal transportation device but without regard to the impact on the environment or the consumer. At first, fossil fuels allowed for the transition from an agrarian society to an urban one and provided a way to make electricity. But when used to fuel a car, fossil fuels allowed urban workers to leave their city apartments and settle in the suburbs. This led to the construction of highways and suburban housing developments and the creation of the car culture.

Today's new cars are a stunning testament to the creativity and brilliance of human design and engineering. Up close, a glistening, Italian-styled Ferrari or a low-slung dangerous-looking Porsche and an impregnable Land

Rover are machines that take one's breath away. They are awash in expensive and luxurious leather, encapsulated in a metal shell that is cushioned by multiple protective airbags. The audio systems are transcendent and the navigation, talk-through, and computer systems are the best the electrical engineering geniuses can design. All would be better in terms of the environment and the vehicles' costs if these extraordinary machines did not run on gasoline.

Unfortunately, the world's improving lifestyle and the human passion for autos have been dependent on fossil fuels. At one time, the fossil fuels powered this prosperity seemed relatively cheap, inexhaustible, and harmless. More and bigger homes and buildings were built, more concrete was poured, and more fossil fuels were extracted and burned. Frankly, there wasn't much to stop this social and economic juggernaut.

As industrialization led to urbanization, and urbanization to suburbanization, developed nations built highways and thousands of miles of freeways that circled and interlaced their cities. The more concrete that was poured, the faster the suburbs grew and the modern lifestyle was forever changed and thoroughly dependent on fossil fuels. The undeveloped nations followed the developed nations, and soon, India, China, and South America were building highways and suburbs that sprawled along concrete ribbons, creating congestion, generating pollutants, and producing an atmospheric overhang of smog.

The buildup of greenhouse gases has marched in lock step with the expansion of fossil fuel use (NCDC, 2013). While primarily composed of carbon dioxide (CO_2), greenhouse gases (GHGs) also include methane (CH_4) and nitrous oxide (N_2O). Greenhouse gases cannot be touched or smelled. Unlike empty beer cans, plastic bags, or the other garbages that pile up along our roads and rivers, GHGs pile up out of sight, in the Earth's atmosphere above our heads. Visualize all the CO_2 that is released from cars, from coal- and gas-burning power generation, and from the burning and clearing of forests and the deforestation of regions like Brazil or Indonesia. The gases float upward into the atmosphere and wrap themselves like a blanket around the Earth. As more and more are added, the blanket gets thicker and warmer.

The impact of greenhouse gases has increased and spread across the Pacific as this chart indicates. The atmospheric and weather changes that have resulted from tsunamis and earthquakes illustrate how environmental problems in one part of the world impact other regions.

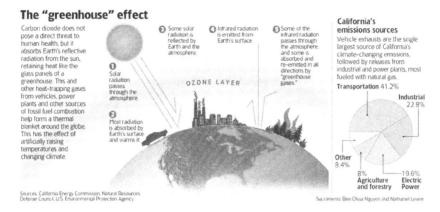

The "greenhouse" effect

Carbon dioxide does not pose a direct threat to human health, but it absorbs Earth's reflective radiation from the sun, retaining heat like the glass panels of a greenhouse. This and other heat-trapping gases from vehicles, power plants and other sources of fossil fuel combustion help form a thermal blanket around the globe. This has the effect of artificially raising temperatures and changing climate.

① Solar radiation passes through the atmosphere.

② Most radiation is absorbed by Earth's surface and warms it.

③ Some solar radiation is reflected by Earth and the atmosphere.

④ Infrared radiation is emitted from Earth's surface.

⑤ Some of the infrared radiation passes through the atmosphere, and some is absorbed and re-emitted in all directions by "greenhouse gases."

OZONE LAYER

California's emissions sources

Vehicle exhausts are the single largest source of California's climate-changing emissions, followed by releases from industrial and power plants, most fueled with natural gas.

Transportation 41.2%

Industrial 22.8%

Other 8.4%

8% Agriculture and forestry

19.6% Electric Power

Sources: California Energy Commission, Natural Resources Defense Council, U.S. Environmental Protection Agency

Sacramento Bee/Olivia Nguyen and Nathaniel Levine

As a greenhouse gas, methane is 23 times more damaging than carbon dioxide. Like CO_2, CH_4 is released through industrial processes and agriculture as well as through petroleum drilling, coal mining, and emissions from solid landfill sites. Perhaps the worst generator of methane is Bessie, the neighborhood milk cow. Livestock gas is high in CH_4 as well as N_2O. Once CH_4 is released into the atmosphere, it traps heat at a much greater rate than does CO_2. Livestock, and particularly cattle, release both CO_2 and CH_4 through belching as they chew their cud. Climate researchers estimate that the average cow releases about 600 liters of CH_4 per day (Food and Agriculture Organization of the United Nations, 2006). N_2O is released through livestock defecation.

According to the United Nations Food and Agriculture Organization's 2006 report, *Livestock's Long Shadow—Environmental Issues and Options,* the world's livestock sector generates 18 percent more greenhouse gas emissions (as measured in CO_2 equivalent) than does transportation. Livestock are also a major source of land and water degradation.

About 30 percent of the Earth's surface is now given to livestock production, including 33 percent of the global arable land used to produce feed for livestock, the FAO report notes. As forests are cleared to create new pastures, livestock production is a major driver of deforestation, especially in Latin America, where, for example, some 70 percent of former forests in the Amazon have been turned over to grazing.

From 1961 to 2010, worldwide emissions from livestock increased 51 percent with the majority of the increases coming from developing nations. Beef cattle produced more than half the emissions, followed by dairy cattle at 17 percent, sheep at 9 percent, pigs at 5 percent, and goats at 4 percent, according to a 2014 study released by the Proceedings of the National Academy of Sciences of the United States of America.

The largest increases came in Congo, the Central African Republic, and Oman (PNAS, 2014).

Unfortunately, the world's rising middle class wants animal protein—a "must have" perhaps only second to a car. Many people believe that a diet rich in animal protein is important for children's growth and mental development. With increased prosperity, people are consuming more meat and dairy products every year. Global meat production is projected to more than double, to 465 million tons in 2050, while milk output is set to climb to 1043 million tons in 2006. This rapid growth is extracting a huge environmental price. *Livestock's Long Shadow—Environmental Issues and Options* warns: "The environmental costs per unit of livestock production must be cut by one half, just to avoid the level of damage worsening beyond its present level" (Food and Agriculture Organization of the United Nations, 2006).

N_2O is also released as a by-product of warming temperatures. According to scientists studying the impact on global warming on the Arctic and the surrounding areas of permafrost, warming temperatures are causing N_2O to leak into the atmosphere. A 2009 study from the Arctic Monitoring and Assessment Program, a scientific body set up by the eight Arctic Rim countries, says that the Arctic is responsible for up to 9 percent of global N_2O emissions (AMAP, 2009).

Scientists estimate that there is 1.5 trillion tons of carbon locked inside icebound Earth, mostly in Alaskan and Russian permafrost areas. Since the age of mammoths—about 10,000 years ago—N_2O from the carbon has slowly seeped into the atmosphere via lakes and rivers. Over the last few decades, as the Earth has warmed, the icy ground has begun thawing more rapidly, accelerating the release of methane (NCDC, 2011).

PIONEERING CLIMATE CHANGE RESEARCH

No one has caused as violent an upheaval in the formerly staid science of climatology as James Hansen, director of the US National Aeronautics and Space Administration (NASA) Goddard Institute for Space Studies in New York City. Hansen joined NASA in 1967 and he gradually focused on planetary research that involved trying to understand anthropogenic (human made) impacts on the Earth's climate.

In 1988, the US Senate Committee on Energy and Natural Resource invited Hansen and other scientists to testify on climate and environmental concerns. They were not expecting the mild-mannered Midwesterner's blunt assessment (Shabecoff, 1988).

To the consternation of the US Senate Committee on Energy and Natural Resource and to the shock of an unaware national media, Hansen told the senators that, in the first 5 months of 1988, Earth had been warmer than any comparable period since measurements began 130 years ago. Hanson went further, making the claim that the higher temperatures were attributed to a long-expected global warming trend linked to pollution.

Until the hearing, scientists had been cautious about attributing rising global temperatures to the so-called greenhouse effect caused by pollutants in the atmosphere. Hansen was not shy and he told the senators that it was 99 percent certain that the warming trend was not a natural variation but was caused by a buildup of CO_2 and other gases in the atmosphere.

The New York Times captured Hansen's remarks. He said: "It is time to stop waffling so much and say that the evidence is pretty strong that the greenhouse effect is here." Hansen and the other scientists noted that humans, by burning fossil fuels and other activities, have altered the global climate in a manner that will affect life on Earth for centuries to come (Shabecoff, 1988).

Hansen's testimony, while not as politically charged as Galileo's defiance of Italian church doctrine in 1615, was a nevertheless stunning testimony from someone who worked for such a high-profile government agency as NASA. The testimony was not just controversial; it made him a lightning rod for skeptics, deniers, fossil fuel advocates, and protectors of 2IR technologies.

Despite this backlash, he managed to continue his work and further his advocacy to limit the production of greenhouse gases. On several occasions, Hansen criticized US public policy and has been particularly critical of the coal industry. In a 2007 testimony before the Iowa Utilities Board, he said coal contributes the largest percentage of CO_2 into the atmosphere and has called for phasing out coal power completely by the year 2030 (Iowa Coal, 2007).

While Hansen introduced climate change to an unaware America, other scientists across the world were equally concerned with the issue (Hansen, 2007). Reacting to international pressure that included the United States at the time, the United Nations established the Intergovernmental Panel on Climate Change (IPCC) in 1988. The IPCC's mandate was to review and assess the most recent scientific, technical, and socioeconomic information produced by scientists hoping to understand climate change (Hansen, 2007).

UNITED NATIONS: THE EPIC STEP FORWARD

The world's increased awakening to climate change—and the book you are reading—would not be possible without the United Nations Intergovernmental Panel on Climate Change.[1] Established in 1988, the UN IPCC was the stepchild of two UN organizations, the World Meteorological Organization (WMO) and the United Nations Environment Programme (UNEP). That the UN IPCC ended up sharing the 2007 Nobel Peace Prize with former US Vice President Al Gore is an extraordinary victory of science over politics and 2IR economic interest groups.

A main activity of the UN IPCC is publishing special reports on topics relevant to the implementation of the UN Framework Convention on Climate Change, an international treaty that acknowledges the possibility of harmful climate change.

The UN IPCC reports are based on science from worldwide sources and provide a clear view on the current state of climate change and its potential environmental and socioeconomic consequences. There have been five major UN IPCC reports published, with more scheduled. The reports, particularly the first two, were decried and criticized by a variety of business and political groups with interests rooted in the 2IR, or the "dirty" economy (UN IPCC Reports, 1990, 1995, 2001, 2007).

The UN IPCC First *Assessment Report* was published in 1990. Its executive summary brought a howl of criticism and calls of corrupt and biased science. The summary said that some scientists were certain that emissions resulting from human activities were substantially increasing the atmospheric concentrations of GHGs, resulting in additional warming of the Earth's surface. Further, the report argued that increased CO_2 was responsible for over half of the enhanced greenhouse effect. They predicted that global mean temperature would increase by about 0.3 °C per decade during the 21st century.

The report brought the potentially dramatic impacts of climate changes to world attention and kicked off the initial controversy. The 1995 *Second Assessment Report* (SAR) was equally insistent that GHGs and excessive CO_2 caused climate change. The UN IPCC *Third Assessment Report: Climate Change 2001* (TAR), published in 2001, continued to outline a world threatened by the reality of climate change. This report got close, but in

---

[1] Coauthor Woodrow Clark was an editor and author for the UN IPCC *Third Assessment Report* in 2001 and the author/editor for the 1999 UN FCCC's First Report on Climate Change, *Environmentally Sound Technology Transfer from Developed to Developing Nations*.

the end did not specifically blame human activity for climate change. That would come in the *Fourth Assessment Report* (AR4), issued in 2007 (UN IPCC Reports, 1990, 1995, 2001, 2007).

Lord Martin John Rees, the president of the Royal Society of London, summed up AR4 best when he said:

> This report makes it clear, more convincingly than ever before, that human actions are writ large on the changes we are seeing, and will see, to our climate. The UN IPCC strongly emphasizes that substantial climate change is inevitable, and we will have to adapt to this. This should compel all of us—world leaders, businesses and individuals—towards action rather than the paralysis of fear. We need both to reduce our emissions of greenhouse gases and to prepare for the impacts of climate change. Those who would claim otherwise can no longer use science as a basis for their argument (UK Scientists IPCC Reaction, 2007).

AR5: KEY FINDINGS

The fifth report, referred to as AR5, is being published in four parts from September 2013 to November 2014. It is the most comprehensive assessment of scientific knowledge on climate change since AR4 in 2007. AR5 was approved by nearly 200 nations at the end of weeklong meeting in Stockholm. The report said that they were more confident than ever that global warming was man-made and likely to get worse. It noted that sea levels will be much higher than previously thought, summers hotter, deluges much stronger, and drafts more severe.

The report's findings included the following:
- Sea levels may rise as much as 3 feet by century's end in the worst case.
- The earth will warm by at least 2 more degrees Fahrenheit (1.1 degrees Celsius) by midcentury.
- The Arctic will have summers that are ice-free by midcentury, and spring snow in North America will shrink by one quarter (UN IPCC, AR5, 2010a).

The second part of AR5, called *Climate Change 2014: Impacts, Adaptation, and Vulnerability*, was released in April 2014. "We live in an era of man-made climate change," said Vicente Barros, cochair of Working Group II. "In many cases, we are not prepared for the climate-related risks that we already face. Investments in better preparation can pay dividends both for the present and the future" (UN IPCC, AR5, April 2014).

The report sharply warned that climate change poses the greatest risk to most vulnerable populations. The gravest of those risks is to people living in low-lying coastal areas and on small islands because of storm surges, flooding, and rising sea levels. Poorer countries already struggling with food insecurity and civil conflict will have even harder times ahead. According to this report, climate change is likely to put added stress on natural and human systems with additional species loss. Food production will be impacted, the report said, from drought, flooding, and changing rainfall patterns. Crop yields are likely to decline, and food availability, severe price swings, and increasing civil unrest will hit countries already having difficulties meeting the basic needs of their citizens. Wheat and maize yields have already begun to be held back by climate change, and key crops like rice may be next.

The enormous and credible work done by the UN IPCC has opened the door for other scientists and scientific organizations. The research done by leading universities, the National Climate Data Center, and international meteorological organizations is mounting every day, tracing the environmental impacts of burning fossil fuels and a growing self-indulgent lifestyle patterned after the Western developed nations.

The World Bank 2012 report describes a planet warmed by 4 degrees Celsius by the end of the century. The scenarios are devastating, including

- the inundation of coastal cities;
- the increasing risks for food production potentially leading to higher malnutrition rates;
- many dry regions becoming dryer and wet regions wetter;
- unprecedented heat waves in many regions, especially in the tropics;
- substantially exacerbated water scarcity in many regions;
- increased frequency of high-intensity tropical cyclones;
- the irreversible loss of biodiversity, including coral reef systems.

In describing this ravaged world, the report makes the observation that a 4 degree hotter planet is "what scientists are nearly unanimously predicting by the end of the century, without serious policy changes" (WB, 2012).

THERE ARE SO MANY MORE HUMAN BEINGS—AND MORE COMING

Climate change and global warming would not have such an immediate and debilitating impact if fewer people inhabited the planet. Climate change is a matter of scale. A few tons of GHGs won't mean much to the ice cap. The primary concern comes from the accelerating growth in population and the exponentially increasing number of people entering the middle class.

There are many more human beings today, and people want more of everything. The growing middle classes of the developing world have their eyes squarely on America. They want animal protein and the American lifestyle, with all its extravagant, wasteful excesses and laborsaving devices. People moving up in class want gated communities and McMansions and fast cars to propel them from the suburbs to the inner city. Knowledge and the way out of poverty are a computer and an Internet connection away. Who can blame them?

There are 7 billion people on the planet now, and the UN is predicting that by 2053, there will be 10 billion (UN DESA, 2011). Since the Black Death ended in the 1400s, the world's population has experienced continuous growth. Most of the growth took place during the past two and half centuries. The world population reached 1 billion in 1804, 2 billion in 1927, 3 billion in 1960, 4 billion in 1974, 5 billion in 1987, and 6 billion in 1999. The highest rates of growth were seen after Word War II, in the decades of the 1950s-1970s.

The Earth's climate is changing dramatically each day, unlike anything seen in recorded time. Scientists have made it clear that human activity is the culprit. The planet's population is growing and the Earth's resources are being extracted faster and faster. The 2IR has locked us into a fossil-fueled and carbon-based economic and social structure, and not enough leaders are paying attention. Real progress is slow and nations like the United States continue to drag their heels. We live in an environment that is under assault, on a planet that has been thoughtlessly ravished by people far too careless with natural resources. The excesses of the carbon era have driven us to the brink and the clock is ticking.

The Green Industrial Revolution must come sooner around the world.

SCIENCE AND TECHNOLOGY

There may be no more compelling reason for the emergence of the GIR than the looming environmental devastation that could be caused by climate change and the Earth's warming. However, the consequences of climate change are not the only drivers of this megatrend. As pointed out in the last chapter, human history is marked by extraordinary leaps in science and technology. The late 20th century's revolution in computers and information technology is a prime example.

Science is never static; it is constantly moving forward, propelled by humanity's never-ending curiosity and lust for knowledge. These characteristics allow the species to adjust and survive and drive us further in the quest for discovery and understanding. As it evolves, the Green Industrial Revolution will become the next significant era in world history. As it takes hold globally, it will result in a complete restructuring of the way energy is generated, supplied, and used. It will be a revolutionary era of extraordinary potential and opportunity, with remarkable innovation in science and energy that will lead to sustainable and carbonless economies powered by advanced technologies like hydrogen fuel cells and nonpolluting technologies like wind and solar. Small community-based and on-site renewable energy generation will replace massive fossil fuel and nuclear-powered central plant utilities, and smart green grids will deliver energy effortlessly and efficiently to intelligent appliances. The world is on the cusp of this extraordinary and amazing era.

This green scientific and technological change is coming from all directions as more and more scientists turn their attention to the endless possibilities of this new age. For example, California's two iconic national laboratories, Lawrence Berkeley National Laboratory (LBNL) and Lawrence Livermore National Laboratory (LLNL) in Northern California, continue today to hotbeds for research and scientific breakthroughs in technology since World War II and now today. The addition of the National Renewable Energy Laboratory (NREL) in Golden, CO, has now places a focus on green technology. Much of this results from the labs shift away from nuclear and weapons research and development (LBNL, LLNL and NREL, 2014).

These breakthroughs are happening throughout the green industry, coming from chemistry, biology, and mechanical engineering. They range from more efficient methods for creating renewable energy—like the development of bigger and better carbon fiber blades for the next generation of wind turbines—to seawater source heat pumps that are being used in coastal cities like Dalian, China. They include the gasification of waste via small blast furnaces, the production of clean gasoline from straw, and ozone-injected water that cleans better than chemicals and needs fewer rinses.

Other discoveries are aimed at the consumer market. ENERGY STAR appliances are more efficient and use less electricity, and new smart meters that allow for a new agile grid system are being installed throughout the world. Italy is a leader in installing these new meters; they began the process as deterrent to electricity thefts. Eventually, the grid will be able to interact with smart appliances via these meters so that load can be balanced throughout the day.

Students at the University of Maryland developed an ultraefficient heat pump for clothes dryers. The pump is a major leap in efficient clothes drying technology and was the winning entry in a US Department of Energy contest. It combines compact heat exchangers, a brushless direct current (DC), and a vapor injection cycle to deliver energy savings of 59 percent compared with existing state-of-the-art US electric dryers. With an estimated 67 million US households that use electric clothes dryers, nationwide energy savings could amount to 21 gigawatt hours using this new two-stage pump (UMD, 2013).

These amazing new green technologies are reviewed in Chapter 7. However, before leaving this discussion, consider a brief look at three revolutionary technologies.

ADDITIVE MANUFACTURING

While there are millions of new "green" inventions emerging from the science and industrial laboratories, none may have the truly revolutionary impact of additive manufacturing. Also referred to as 3-D printing, this remarkable process uses a digital model to make three-dimensional solid objects. The process is similar to the way an inkjet printer works and uses an additive process to lay successive layers of material down in different shapes. Traditional manufacturing or machining techniques start with a large piece of material and use cutting or drilling to shape an object. The traditional method is referred to as subtractive processes.

A 3-D printer is a limited type of industrial robot that carries out an additive process under computer control. The technology has been around since the late 1980s, but didn't start to become viable until 2010. It has mushroomed into a multibillion dollar industry and is being used for creating prototypes and used in decentralized manufacturing. Applications for the 3-D process are rapidly expanding and include architecture, construction, industrial design, automotive, aerospace, military, engineering, civil engineering, dental and medical industries, biotech (human tissue replacement), fashion, footwear, jewelry, eyewear, education, geographic information systems, food, and many other fields.

This new technology will soon be a driving force in manufacturing and business development. It is far more flexible and economical than conventional manufacturing and requires fewer raw materials. Since the design is done in the software, each product can be different or customized, without retooling. 3-D printing has the potential to transform manufacturing because it lowers costs and reduces risks.

Not only will it make wondrous new products possible, but also it will increase energy efficiency, maximize natural resources, and lower product cost. It is certainly a critical component of the GIR. Yet reducing the use of energy through conservation and then providing new technologies that are efficient and "smart" are critical in reducing GHG, CO_2 emissions, and pollution from energy production and consumption for vehicles and buildings.

LED LIGHTING

In the 1960s, scientists developed a revolutionary, low-wattage light bulb that used light emitting diodes or LEDs for bulbs. This new process eliminated about 85 percent of the electricity used by conventional fluorescent lights. Early LEDs were used in electronic components that emitted low-intensity red light. Today's LEDs provide light across the visible, ultraviolet, and infrared wavelengths, with very high brightness. They are attractive and provide the same light as traditional bulbs.

LEDs are an extraordinary new generation of lighting. A 6 watt LED can provide the same amount of light as a standard 60 watt commercial overhead interior light. LEDs have a longer lifetime (measured in 8-10 years, instead of months), improved robustness, smaller size, faster switching, and greater durability and reliability. They are more expensive than traditional light bulbs, but the price is dropping rapidly as new manufacturers come to market and cost-effective ways are developed to finance lighting projects (Nularis, 2014).

Technological advances in lighting may seem mundane compared with some of the other emerging technologies, but lighting impacts every office, home, and room in the modern world. In the United States, facility lighting uses about 25 percent of the nation's generated electricity. LEDs can also be life-changing. In a mud-walled hut in a rural village in an undeveloped part of the world, lighting can offer a child a chance to study his or her lessons and slip the bonds of sustenance farming.

LED lighting has other remarkable advantages. It can be linked to sensors and controls for smart applications and, in essence, become a node on an intelligent network. These networks can turn off lights when people aren't around or dim them when its bright and natural light is plentiful. The sensors are also harvesting data about building temperature, occupancy, and the surroundings that can have all sorts of other applications (Nularis, 2014).

For example, a smart LED lighting retrofit on a city street has a number of benefits. The LED lights are saving money from electricity and maintenance, but the network can also be used to support revenue-generating services such as smart parking. If you add cameras, the system can issue parking tickets more efficiently. In a commercial application, these LED networks can help with security, let people into rooms, or manage conference room attendance or keep track of which department uses certain areas and resources (Nularis, 2014).

Navigant Research estimates that over the next 7 years, annual sales for occupancy sensors, photo sensors, and lighting networks gear for LED lighting applications will grow from $1.1 billion in 2013 to $2.7 billion by 2020 (Navigant, 2013).

Philips, the giant Netherlands-headquartered lighting company, estimates that by 2015, the worldwide LED lighting market will be between 75 and 80 billion euros—or roughly $100 billion—not including automotive light. They also predict that LED lamps for consumer application will fall to around $12.

HYDROGEN POWER AND FUEL CELL STORAGE

Hydrogen produces power without carbon emissions and may be the world's most significant new power source. In fact, there are those who think that the hydrogen fuel cell could be Watt's steam engine or the internal combustion engine of this era.

The fuel cell produces an electric current that can be directed to do work, such as powering an electric motor or illuminating a light bulb. Fuel cells can power an engine or an entire city. The real goal is to perfect a hydrogen-based fuel cell to take the place of the gasoline-based internal combustion engine. Hydrogen fuel cells use hydrogen as the fuel and oxygen as the oxidant.

The potential for a broad use of hydrogen, particularly in powering vehicles, is enormous.[2] Every major car company has developed a hydrogen fuel cell vehicle, and they are all planning to market them, starting in the Europe and Japan. In the United States, at least nine major car manufacturers

---

[2] See this chapter on storage devices for a detailed explanation of the workings and applications of a hydrogen fuel cell.

are introducing hydrogen fuel cell vehicles in 2016, under lease agreements rather than for sale. California is on the brink of successfully establishing a hydrogen highway that runs the entire length of the state.

Germany already has a hydrogen highway, so Daimler and other car makers already have hydrogen fuel cell cars on the road.

Hydrogen offers the promise of a zero-emission engine, where the only by-product created is a small amount of environmentally friendly water vapor. Hybrids and other green cars address these issues to a large extent, but hydrogen cars are the only ones that don't produce a single pollutant (Clark and Bradshaw, 2004).

Combined with a renewable energy source, fuel cell technologies, particularly hydrogen fuel cells, are a potentially revolutionary power source. They offer an exceptionally attractive alternative to oil dependency and fossil fuels (Clark et al., 2006). Scientists and manufacturers have a lot of work to do before fuel cells become a practical alternative to current energy production methods, but with worldwide support and cooperation, a viable hydrogen fuel cell-based energy system may be a reality starting with hydrogen fuel cell cars and their refueling stations (California ARB, 2014; HAA, 2014).

ECONOMICS

3-D printing, LED lighting, and hydrogen fuel cells have had, and will continue to have, huge commercial and economic viability. In fact, a study that included the essential GIR components—energy efficiency, clean energy, water recovery, waste-to-value, and environmental services—pegged the gross revenues in 2010 at $300 billion.

CleanTechnica, a renewable energy research company, projects that the global biofuel, wind, and solar markets will double in value from $248.7 billion in 2012 to $426.1 billion by 2022 (CleanTechnica, 2012).

With this much money being generated, it is not hard to understand why venture capital and risk money have invested heavily in these new industries and technologies. Investment money is very dynamic; it is constantly seeking a deal or an emerging technology that will be the GIR equivalent of Microsoft.

However, tracking venture capital investments in start-ups and young companies that are developing new technologies can be confusing because there is so much overlap. The venture capital industry refers to many of these technologies and companies as cleantech or greentech. However, the labels break down when they discuss investing in technologies like LEDs, sensors, and network gears, which would usually be described as IT infrastructure, though they are clearly GIR technologies.

Equally hard to label was the investment money that went into Duro-Last, the maker of a premium grade of reflective roofs. A Duro-Last single-ply roof membrane—called a "cool" roof—saves a huge amount of air conditioning in warm climates. As a major energy-efficient component that is revolutionizing the industry, it certainly deserves being described as a GIR technology. However, those who invested in the company probably considered the technology a superior roofing product.

Goldman Sachs, the world's foremost investment banker, describes this sector as the Clean Technology and Renewables Market and says via their website that "the firm has set a $40 billion target for financing and investing in clean technology companies" (Goldman Sachs, 2014).

The firm has investments in solar, wind, geothermal, energy-efficient, green transportation, and advanced biofuels. Interestingly, Goldman Sachs has published a very detailed report on the potential crisis in the world's fresh water supply. They note that over 60 percent of the Earth's freshwater supply is found in just 10 countries. "Severe water stress affects 3 billion people, two-thirds of whom reside in the BRICs (Brazil, Russia, India, China)" (Goldman Sachs, 2014). Water conservation and concerns are a growing part of what the GIR is addressing, particularly in the development of sustainable communities.

Goldman Sachs is one of many other investment banks putting money to work in the renewable field. One study by the GTM Research found that "global solar dominance is in sight as science trumps fossil fuels" (Evans, April 9, 2014).

MARKETING GREEN

While some companies and businesses have carelessly adopted tag lines claiming to be "sustainable" and "green" (which can be referred to as "green washing"), others are sincere in their efforts. For example, the world's luxury hotel industry has started to compete as green and sustainable hotels. Large luxury hotels are extraordinarily attractive environments, full of lavish furnishings, uniformed attendants, wonderful food, and dramatic interiors. In reality, they are minicommunities, providing the standard infrastructure elements of energy, transportation, water, waste handling, and telecommunications for a small, transient, and demanding client population. Hotels are energy hogs, constantly using energy and resources, even at night and during down times. Luxury hotels have amazingly complex inner workings that include kitchens, freight elevators, laundry rooms, parking structures, and enormous HVAC (heat, ventilation, and air conditioning) boilers, all of which wreak havoc on the environment. For example, the Hilton Hotel Complex in Agoura Hills, California, is now "zero emission" due to it conservation and efficiency programs (Hilton Hotel and Resort, 2014).

The worldwide economic downturn forced the hotel industry to rethink its operational costs and reevaluate its extravagant use of resources. Citigroup Inc. now states the "Age of Renewables has Begun" (Parkinson, April 1, 2014). Resources are expensive and the luxury hotel industry is turning to energy efficiency and green activities—like ozone laundry, retrocommissioning, and LED lighting—to hold costs down. A hotel aiming for sustainability uses resources wisely and conserves and preserves by saving water, reducing energy use, and cutting down on solid waste. Incorporating these waste reduction techniques into hotel operations is a win–win for the environment and the hotel. Not only will greener hotels save resources and reduce pollution, but also they will cut down on operating costs while increasing profit margins.

This extraordinary progress toward sustainability probably started with the eco-friendly hotels in countries with fragile environments, like Costa Rica and Australia. However, the big luxury hotels of Mumbai, Macau, and Dubai are taking sustainability seriously and are reaching out to architects and engineering firms for advice on LEED certification and renewable energy generation.

In the fiercely competitive luxury hotel market in San Francisco, sustainability has become a marketing tool. In this tourist-driven market, the luxury hotels are lining up to promote their greenness. Market studies are now convincing hotel owners that given the choice, high-end clients prefer a hotel that is environmentally responsible.

United Parcel Service, or UPS, the huge worldwide logistic company delivers about 16.3 million packages a day to over 220 countries. In 2012, it delivered a total of 4.1 billion packages. For decades, UPS has used diesel-powered large brown trucks. In fact, the company is often referred to as Big Brown because of the trucks and the distinctive rumble of their diesel engines.

Now, UPS is actively slashing its carbon emissions, first, by converting their fleet to alternative-fuel vehicles and, now, by using zero-emission, all-electric vehicles. The first 130 of the electric trucks went into service in California in 2013, and the company is marketing their green efforts, saying they are committed to electric vehicles as a way to clean the air and reduce carbon emissions. The trucks have a range of 75 miles and use a regenerative braking system like a Toyota Prius to augment the stored electricity in the battery (HEV, 2012).

IKEA, the huge Swedish furniture maker, also markets the company's environmental concerns. Many of IKEA's stores generate their own electricity with rooftop and parking lot solar installations. They have long offered LED lamps and energy-efficient appliances. Now, in a remarkable nod to how mainstream solar technology has become, the company is starting to sell solar panels directly to consumers. In fall 2013, IKEA rolled out a home residential solar package for UK customers. The cost is about $9,000 with installation arranged. Homeowners are expected to break even in 7 years.

GREEN JOBS

While the GIR is driving numerous economic benefits such as investment opportunities, significant new commercial markets, and lower energy costs, none may be as important as the creation of new jobs.

China, more so than any other nation has turned to the GIR for economic growth and job creation. In fact, China is now leading the world with economic and career innovations. Much of the nation's phenomenal economic growth in the last two decades has come from quickly grasping the potential for green industries.

The devastation of the planet's ecosystem may be the most important reason for the world to move quickly to embrace the Green Industrial Revolution. This new era, however, offers strong economic benefits and unlimited business potential.

REFERENCES

Arctic Monitoring and Assessment Program (AMAP), 2009. http://www.arctic.noaa.gov/reportcard/ArcticReportCard_full_report.pdf.
Arctic Map, NDSK, 2014. New society for the diffusion of knowledge. www.nsdk.org.uk.
World Bank, 2012. Turn down the heat. Why a 4 degree centigrade warmer world must be avoided. Report for the World Bank by the Potsdam Institute for Climate Impact Research and Climate Analytics. http://climatechange.worldbank.org/sites/default/files/Turn_Down_the_heat_Why_a_4_degree_centrigrade_warmer_world_must_be_avoided.pdf.
World Bank, Climate Change, 2014. http://climatechange.worldbank.org/sites/default/files/Turn_Down_the_heat_Why_a_4_degree_centrigrade_warmer_world_must_be_avoided.pdf.

California Hydrogen Refueling Program ARB, 2014. www.arb.ca.gov/hydrogen.

Clark II, W.W., Bradshaw, T., 2004. Agile Energy Systems: Global Solutions to the California Energy Crisis. Elsevier Press, London.

Clark II, W.W., Rifkin, J., et al., 2006. A Green Hydrogen Economy. Special Issue on Hydrogen, Energy Policy 34, 2630–2639, Elsevier.

CleanTechnica, 2012. http://cleantechnica.com/2013/03/12/biofuel-wind-and-solar-global-market-values-set-to-double-by-2012/.

Iowa Coal, 2007. http://www.columbia.edu/~jeh1/2007/IowaCoal_2071105.pdf.

Economist, November 16, 2014.

Economist, January 12, 2013.

Greenhouse Effect Map, 2012. California Energy Commission, Natural resources Defense Council, and US Environmental Protection Agency.

Evans, Ambrose, April 9, 2014. Global solar dominance in sight as science trumps fossil fuels, The Telegraph.

Food and Agriculture Organization of the United Nations, 2006a. http://www.fao.org/docrep/010/a0701e/a0701e00.htm.

Hansen, James, November 5, 2007. Direct testimony of James E. Hansen. State of Iowa, before the Iowa Utilities Board.

HEV, 2012. http://www.pressroom.ups.com/HEV/Related+Content/Documents/Fact+Sheets/UPS+Hybrid+Electric+Vehicle+Fleet.

Hilton Hotel and Resort, 2014. www.hilton.com/zero.emissions.

Hydrogen Association of America (HAA), 2014. http://www.hydrogencarsnow.com.

Lawrence Berkeley National Laboratory (LBNL), Lawrence Livermore National Laboratory (LLNL), and National Renewable Energy Laboratory (NREL), 2014. Websites for each Laboratory.

National Climate Data Center (NCDC), 2013. http://www.ncdc.noaa.gov/sotc/global/2013/13.

Navigant, 2013. http://www.navigantresearch.com.

NCDC, 2011. http://www1.ncdc.noaa.gov/pub/data/cmb/bams-sotc/2011/bams-sotc-2011-front-matter-and-abstract-lo-rez.pdf.

Nularis LED bulbs, 2014. http://www.nularis.com.

Parkinson, Giles, April 1, 2014. Citigroup Says the 'Age of Renewables' has begun, RenewEconomy.

Proceedings of the National Academy of Science, July 2014. http://www.smh.com.au/environment/climate-change/climate-scientists-have-a-real-beef-with-beef-20140723-zvvjd.html.

Rogers, Paul, April 30, 2014. Contra Costa Times. Climate change: pacific ocean acidity dissolving shells of key species. http://www.contracostatimes.com/news/ci_25664176/climate-change-pacific-ocean-acidity-dissolving-shells-key#.

Goldman Sachs, 2014. http://www.goldmansachs.com/our-thinking/focus-on/clean-technology-and-renewables/index.html.

Sandy Hurricane, October 2012. http://news.blogs.cnn.com/2012/10/29/hurricane-sandy-strengthens-to-85-mph/.

Sandy Remembered, November 2012. http://www.climatecentral.org/news/atlantic-hurricane-season-ends-sandy-will-be-long-remembered-15310.

Shabecoff, Philip, June 24, 1988. Global warming has begun, expert tells senate. New York Times. http://www.nytimes.com/1988/06/24/us/global-warming-has-begun-expert-tells-senate.html.

State of the Climate, 2013. http://journals.ametsoc.org/doi/abs/10.1175/2013BAMSStateoftheClimate.1.

UK Scientists' IPCC Reaction, February 2, 2007. BBC News. http://news.bbc.co.uk/2/hi/science/nature/6324093.stm.

UN IPCC, April 2014. Fourth report, WGII AR5, Phase 1 report launch.

UN Department of Economic and Social Affairs (UN DESA), 2011. World population prospects: the 2010 revision. http://esa.un.org/unpd/wpp/index.htm.

UN Intergovernmental Panel on Climate Change (UN IPCC), Reports, 1990, 1995, 2001, 2007 at www.ipcc.ch/reports.

UN IPCC, 2014. Fourth report, WGII AR5, Phase 1 report launch, April 2014.

UN IPCC AR5, 2010a. https://www.ipcc.ch/report/ar5/wg1/#.Uxyywlz20i0.

University of Maryland (UMD), 2013. http://www.umdrightnow.umd.edu/news/umd-team-wins-first-place-maxtech-competition.

World Bank (WB), 2012. http://climatechange.worldbank.org/sites/default/files/Turn_Down_the_heat_Why_a_4_degree_centrigrade_warmer_world_must_be_avoided.pdf.

CHAPTER 6

Renewable Technologies

The developed world uses a huge and disproportionate amount of energy, most of it being generated from fossil fuels. These nations grow their food with petrochemical fertilizers and carbon-based pesticides. Most construction materials are made from fossil fuels, as are the majority of our pharmaceutical products. Power, heat, light, and transportation are based on fossil fuels. All of which are creating critical levels of greenhouse gases, atmospheric changes, and environmental pollution.

The developed world wants a huge amount of electrons. The United States alone uses about 20 percent of the world's energy each year, consuming an average of about 14,000 kilowatt-hour (kWh) of electricity per person. As the global economy becomes more technical and knowledge-based, electricity consumption will grow significantly. A good example of the exponential electricity needs of a developing nation is Chile, whose coal, unfortunately, is in high demand around the world. This fast-growing South American country predicts that its electricity needs will double every decade until midcentury at least.

WHAT IS ELECTRICITY?

Electricity is the flow of electrons. Matter is made up of atoms, and a nucleus is the center of an atom. The nucleus contains positively charged particles called protons and uncharged particles called neutrons. Around the nucleus are negatively charged particles called electrons. An electron's negative charge is equal to the positive charge of a proton, and the number of electrons in an atom is usually equal to the number of protons. Often, protons and electrons are in balance, but when an outside force upsets the balance, an atom may gain or lose an electron. When electrons are "lost" from an atom, the movement of these electrons results in an electric current.

Thanks to Benjamin Franklin's flying kite experiment, science knows that electricity is a basic part of nature, mainly as part of cloud and storm formations. Under the right conditions, lightning occurs within these formations. Lightning is a massive electrostatic discharge between electrically charged areas in clouds or between a cloud and the Earth's surface. The flash, or strike, occurs when the charged areas temporarily equalize themselves. Lightning is always accompanied by thunder soon afterward. In fact, you can calculate the distance of the storm by the time between the lightning and thunder.

For whatever extraordinary reasons, electricity also occurs in fish. There are certain types of rays, eels, and catfish that have special organs that emit electrical discharges. The discharges are used to paralyze prey or for defense. South American electric eels (*Electrophorus electricus*) produce even electricity to power a dozen 40 watt light bulbs. These eels grow to 7 feet or more, and they are quite capable of killing a horse or man that stumbles into a swarm in shallow water.

In 1830, Michael Faraday, a British scientist figured out the basic method to create electricity. He discovered that electricity could be generated by moving a loop of wire, or copper disk, between the poles of a magnet. By the 1880s, electricity had emerged as a primary source of energy spurred on by new inventions. Mechanical transformers provided a way to generate electrical power from a centralized location. Alternating current power lines provided the method to transport electricity at very low costs across great distances by raising and lowering the voltage. These developments paved the way for the first power plants, which ran on coal or waterpower.

Today, electricity is generated by electromechanical generators, mostly driven by heat engines fueled by chemical combustion or nuclear fission, or by kinetic energy from sources such as flowing water. Coal is the world's main source of power for generating electricity, though other polluting sources of fossil fuels, like natural gas and oil shale, are increasing. According to the US Energy Information Administration's 2011 Annual Energy Outlook, about 85 percent of the US electricity is generated from fossil fuels, with about half of that coming from coal, due to government tax and finance incentives still in effect since the 1880s.

Over the last decade, natural gas generation has been increasing, and now, about 23 percent of the US electricity is from natural gas. That along with shale oil is promoted by the fossil fuel industry as the answer to America becoming energy-independent (Stansberry, 2012). Unfortunately, the permanent damage by the natural gas and shale oil industries to the environment and increased atmospheric emissions along with transportation, building, and security costs are never discussed or calculated. Few if any of the proponents of natural gas talk about its need to be shipped or piped from Canada to the US southern state processing plants and then transported, piped, or shipped to other regions for conversion into energy. These cost externalities also need to be included in the price of natural gas, which is supported by low government taxes, rebates, and other incentives (Clark, 2012).

Another 20 percent of the US electricity is generated by nuclear power and six percent by large-scale hydroelectric facilities. Wind and solar energy sources contribute only about three percent, nowhere near the potential amount that could be generated to meet American demand (US DoE, EIA, 2013).

Electricity is used primarily for lighting and for making buildings livable and comfortable. For example, heating and air conditioning use about 30 percent of the US's energy (US DoE, EIA, 2011). Data centers, the keystone to a knowledge-based economy, are another major and increasing user of energy. Data centers' energy use is growing quickly in China, India, and throughout the world.

Coal is the dominant energy generator in China, and its consumption is rapidly expanding with the dramatic increase in demand for cars, trucks, and buildings. According to energy consumption figures released by the Chinese government in 2012, coal consumption increased by 9.7 percent, the most year-over-year growth seen since 2005. Coal's role in generating electricity is due to its abundance and seemingly inexpensive extraction in China (Think Progress, 2012).

China's consumption of natural gas also saw a substantial increase of 12 percent in 2011 (Think Progress, 2012). The result is that China is now importing natural gas from other nations and creating pipelines from Russia. China has now become very dependent on foreign sources for its energy demands.

Economists do not include the cost of the environmental damage caused by digging and then using coal to generate its massive amounts of electricity. While coal-generated electricity may be cheap to the consumer, the damage that acid rain causes to the environment and human health is painfully expensive to remediate and never calculated. And nobody wants to spend money to offset the millions of tons of greenhouse gases (GHGs) that burning coal releases into the atmosphere. The cost of global warming is a hard concept to understand for people accustomed to cheap electricity.

RENEWABLE ENERGY

The core driver of the green industrial revolution is the move away from fossil fuel-generated electricity to renewable, environmentally friendly generation, transmission, and distribution. A degrading environment, global warming, severely damaging climate changes, dwindling fossil fuel supplies, and the need to end the continued dependence on the socially volatile and politically unstable Middle East, will force the adoption of alternative technologies and renewable energy sources. In some countries, like China and Denmark, this change is under way. Denmark has a goal of using 100 percent renewable energy by 2050. Today, Denmark is almost 50 percent there (Lund and Østergaard, 2010; Lund and Hvelplund, 2012).

Renewable energy is energy that is not carbon-based and will always be there. For example, the sun is shining during the day and the wind blows fairly constantly. Both can be used to generate energy but require some form of storage or feedback technology when the wind is not blowing and the sun is not shining. These renewable energy sources are called intermittent energy generation and need storage capabilities and integrated systems to provide baseload or round-the-clock power generation.

Renewable energy sources can be located in large areas often called "farms" or central plants where hundreds of solar panels or wind turbines are located and their energy must be transmitted hundreds of miles from a central plant to consumers. In the last decade, solar and wind systems have become distributed energies, because unlike the massive centralized fossil fuel power plants, they are spread out through many local sources. For example, the solar installations on numerous rooftops like in Germany or Arizona are on-site. These distributed energies are found in every inch of the world—the sun, the wind, wave and tidal action, the geothermal heat under the ground, and biomass like garbage and agricultural and forest waste. Other renewable sources include bacteria, algae, and hydrogen when it comes from renewable electrolyzed sources.

WIND POWER GENERATION

Wind has been used as a power source for tens of thousands of years. Ancient civilizations used wind power for sailboats, and this original technology had a major impact on the first windmills, which used sail-like panels to catch the wind.

The first documented use of windmills was in Persia around AD 500 to simplify tasks like grain grinding and water pumping. The designs used vertical sails made of reed bundles attached to a central vertical shaft by horizontal struts. The Chinese statesman Yehlu Chhu-Tshai constructed a windmill of similar design in China in 1219. One of the earliest and most scenic applications of wind power is the extensive use of water pumping machines on the island of Crete. Today, hundreds of these ancient sail-rotor windmills still pump water for crops and livestock.

The Europeans developed windmills, or "post-mills," around 1300. The earliest illustrations show a four-blade mill mounted on a central post. European mills used wooden cog-and-ring gears to translate the motion of the horizontal shaft to vertical movement to turn a grindstone. This gear was apparently adapted for use on post-mills from the horizontal-axis water wheel developed by Vitruvius, a Roman writer, architect, and engineer active in the first century BC.

The Dutch refined the design and created the tower mill around 1390. They affixed the standard post-mill to the top of a multistory tower, with separate floors devoted to grinding grain, removing chaff, and storing grain. On the bottom were the living quarters for the mill operator and his family. Both the post-mill and the later tower mill had to be oriented into the wind manually, by pushing a large lever at the back of the mill. The operator not only optimized windmill energy and power output but also protected the mill from damage by furling the rotor sails during storms.

It took 500 years and countless incremental improvements in efficiency to perfect the windmill sail. By the time the process was complete, windmill sails had all the major features recognized by designers as crucial to the performance of modern wind turbine blades. These mills were the power motors of preindustrial Europe.

The most important refinement of the fan-type windmill was the development of steel blades in 1870. Steel blades are lighter and could be crafted into more efficient shapes. They worked so well, in fact, that their high speed required a reduction (slowdown) gear to turn the standard reciprocal pumps at the required speed (Dodge, 2009).

Wind turbines translate the kinetic energy of moving air into mechanical energy and then transmit that energy into a generator that produces electrical power. Large-scale turbines have a horizontal-axis design. Two or three rotor blades are mounted atop a tower, similar to the blades of an airplane propeller. The movement of wind across the blades generates lift, spinning the shaft, which is connected to an electric generator. The output is a function of wind speed and the size of the turbines' rotors.

The first use of a large windmill to generate electricity was a system built in Cleveland, Ohio, in 1888 by Charles F. Brush. In 1891, Poul la Cour, a Danish scientist, developed the first electrical output wind machine to incorporate the aerodynamic design principles (low-solidity, four-bladed rotors incorporating primitive airfoil shapes) used in the best European tower mills.

By the close of World War I, the use of 25 kW electrical output machines had spread throughout Denmark, but cheaper and larger fossil fuel steam plants soon put the operators of these mills out of business.

A utility-scale wind energy conversion system was first attempted in 1931 in Russia. Experimental wind plants were also constructed in the United States, Denmark, France, Germany, and Great Britain during the period 1935-1970. These demonstration sites showed that large-scale wind turbines would work, but did not lead to a practical, large, electrical wind turbine. European developments continued after World War II, when temporary shortages of fossil fuels led to higher energy costs (O'Toole, 2010).

In the United States, the federal government's involvement in wind energy research and development began in earnest after the Arab oil embargo of 1973. Despite the speed with which it was initiated, political factors and the withdrawal of financial support halted its development before success could be achieved. However, by the mid-1980s, the commercial wind turbine market had evolved from the need for small (1 to 25 kW) machines for domestic and agricultural applications to intermediate-size (50 to 600 kW) machines for utility-interconnected wind farms (AWEA, 2011).

California installed the majority of wind turbines in the 1980s when an economic incentive program was offered by the state government. In California, more than 17,000 turbines, ranging in output from 20 to 350 kW, were installed as wind farms. At the height of development, these turbines had a collected rating of over 1,700 MW and produced more than 3 million MW-hours of electricity, enough to power a city of 300,000.

The U.S. wind farm market lagged and gradually declined after the 1980s. Fierce opposition to wind energy by the fossil fuel industry pressured Congress to drop the tax and financial incentives, triggering the decline (AWEA, 2011). While this was also the case in other parts of the world, national governments especially in Germany and the Nordic nations saw the need for renewable energy and resisted the fossil fuel interests (Lund and Clark, 2008). As a result of the Arab oil embargo, many of the nations of northern Europe and Asia developed policies to encourage wind energy. Wind turbine installations increased steadily through the 1980s and 1990s in these regions. The higher cost of electricity and excellent wind resources in northern Europe created a small, but stable, market for single, cooperative-owned wind turbines and small clusters of machines.

After 1990, most wind activity shifted to Europe and Asia. Driven by high utility power rates, cooperatives and private landowners in the

Netherlands, Denmark, and Germany installed first 50 kW; then 100 kW, 200 kW, and 500 kW; and finally 1.5 MW wind turbines. This impressive growth in installations now amounts to over 10,000 MW of European wind capacity and has helped support a thriving private wind turbine development and manufacturing industry (Gipe, 2014).

A key factor in Germany was the creation of their feed-in tariff (FiT) program that started in 1991 and tripled in size by the turn of the twenty-first century (Gipe, 2014). A FiT simply charges consumers more for power from which the additional funds can then be borrowed or loaned to build renewable energy systems (US DoE, IEA, 2010). For Germany, wind was a natural renewable resource.

In the United States, wind energy development resumed around the turn of century, buoyed by green power initiatives in Colorado and Texas. A variety of new wind projects were installed in Texas, northern Colorado, the upper Midwest, and California. The United States is rich in wind resources, and wind power generation has grown slowly, but steadily, in the twenty-first century (Clark and Sowell, 2002).

According to the American Wind Energy Association, today, wind power accounts for about 2.3 percent of the electricity generated in the United States, with an installed capacity of over 40,000 MW. Wind energy now produces enough electricity to power the equivalent of nearly nine million US homes. Use of wind, instead of fossil fuels, to generate power avoids 57 million tons of carbon emissions each year and reduces expected carbon emissions from the electricity sector by 2.5 percent. The Roscoe Wind Farm in Texas (with 780 MW of wind capacity) is the world's largest wind farm. With these new installations, the United States is on track to generate 20 percent of the nation's electricity from wind energy by 2030 (WWEA, 2011).

In the fall of 2010, England opened the world's largest offshore wind turbine farm. Thanet Wind Farm is located off England's southeast coast. It took more than two years to build and will generate 300 MW of electricity or about the amount required to power 200,000 homes. It has 100 turbines (each 377 feet tall) spread over 13.5 square miles. With this new capacity, Britain now generates approximately 5 GW of electricity from wind, about enough to power all the homes in Scotland. The British government has said it will support the renewable energy industry to achieve its goal of generating 15 percent of its energy from renewable sources by 2020. It gets about four percent of its electricity needs now from wind (REW, 2011).

Chile, along with other South American nations, has announced a major initiative for renewable energy, mostly from wind. Chile's economy is heavily dependent on the energy-intensive extraction of mineral resources but has no fossil fuels. It imports large quantities of natural gas from Bolivia and Argentina, two of its main South American rivals. Its electricity needs had been met through the use of hydroelectric projects on its rivers. However, there is a growing public opposition to any additional damming of Chile's pristine rivers, which are a main source of ecotourism and national pride.

In response, the Chilean government has announced a major effort to generate electricity with nonconventional renewable energy (NCRE). While lacking in fossil fuel resources, Chile is blessed with extraordinary wind resources, particularly in the northern desert area and along the Pacific coast. Chile has launched a major campaign to encourage the development of wind turbine farms, with the goal of obtaining 20 percent of the country's electricity from renewable sources by 2020 (PVResources, 2011).

The use of wind, one of the most ancient power sources, is growing rapidly as the world moves into the GIR. According to the World Wind Energy Association, in June 2010, 196 GW of electricity was being generated by wind power, which equated to about 2.5 percent of the world's energy production. Eighty countries are using wind power on a commercial basis. Several have achieved relatively high levels of wind power penetration by 2009: 20 percent of stationary electricity production in Denmark, 14 percent in Ireland and Portugal, 11 percent in Spain, and 8 percent in Germany (PVResources, 2011).

DEVELOPMENT OF CHINA'S WIND INDUSTRY

With the help of its central planning process and funding, China has leapfrogged other nations into the GIR. As part of China's social capitalism economic model, the Chinese leadership merged their interest in making money with their concern for protecting their society and the environment.

As the world's most populous nation, and one with a rapidly growing economy, China has huge energy needs that will continue to grow. Most of the nation's energy generation now comes from coal, which is causing extraordinary pollution problems. However, China is trying to address these problems with an ever-growing investment in renewable energy. In 2009, China led the United States and the other G-20 nations in annual clean energy investments and finance, according to a 2010 study by the Pew Charitable Trusts. In May 2010, it was reported that China used 34 percent of its stimulus funds ($586 billion) for clean technology energy generation. It will have the capacity for more than 100 GW of renewable energy installed and operating by 2020. Much of this will come from wind.

CHINA'S WIND POWER

The Danish wind turbine manufacturer Vestas saw early on that China and Asia were large emerging markets. In the early 1990s, Vestas agreed to China's business model and established a joint venture. The result was that the wind industry and supporting businesses grew rapidly. Jobs were created and people were hired to install, repair, and maintain the equipment. Today, China is a world leader in wind energy production, manufacturing, and installation.

Wind power is China's most economically competitive new energy source. The nation's wind industry emerged in 2005, after a decade of joint ventures and collaborations with northern European companies. Favorable government policies were key to doubling the country's wind power capacity each year (Clark and Isherwood, 2010). According to the Chinese Renewable Energy Industries Association, China has the world's largest installed wind turbine capacity. In 2012, China had a total wind power capacity of 75 GW, an increase of 21 percent, from a year earlier. China is predicted to reach 250 GW of installed wind capacity by 2020, almost 3.3 times more than current capacity and 42 times more than 2007 (Hong, 2013).

The wind industry is essential to achieving China's goals of secure and diversified energy production. The industry also contributes to economic growth and environmental and pollution control. China also plans to reduce emissions through the use of wind-generated power. If the Chinese wind power industry installs 250 GW by 2020, it will reduce GHG emissions by 500 million tons. China will also limit air pollution by reducing coal consumption, and CREIA predicts that at the same time, the country will generate more than 400 billion RMB in added value and create 500,000 jobs.

While engineering and new smart gird technologies are needed to integrate renewable energy into the China's electricity grid, the country is intent on a massive increase in wind generation. China is rich in wind energy resources, with a long coastline and large western open plains. Wind energy resources are particularly abundant in the southeast coastal regions and the islands off the coast. In the northern part of the country as well as western inland regions, mass wind energy systems have been built—now, the wind power must be transmitted to the coastal cities.

Offshore wind energy resources are plentiful, and in 2010, the first large offshore project was completed at Shanghai's Donghai Bridge. Thirty-four large 3 MW turbines, producing 100 MW, were installed. Analysts estimate that as much as 32,800 MW could be installed by 2020 (Hong, 2013).

Wind energy has enormous potential in China and could easily become a major part of the country's energy supply. Some scientists estimate that the total capacity for land-based and offshore wind energy could be as high as 2,500 GW.

China's wind turbine equipment manufacturing industry has developed rapidly by reaping the benefits of the green industrial revolution. Substantial new business and job growth have resulted with the development of new green technologies. Domestic wind turbine manufacturers now account for about 70 percent of China's supply market and are beginning to export their products. The largest manufacturers are Sinovel, Goldwind, and Dongfang Electric. China now leads the world and accounts for roughly a third of the global total, both in installed wind turbine capacity and in equipment manufacturing capability (Sun et al., 2013).

The state-owned power supply companies have developed the largest wind farms. These companies are given direction and funding to steadily increase their proportion of renewable energy. The CREIA reported that by the end of 2009, a total of 24 provinces and autonomous regions in China had their own wind farms and more than nine provinces had a cumulative installed capacity of more than 1,000 MW, including four provinces exceeding 2,000 MW. The Inner Mongolia Autonomous Region (IMAR) was the lead region, with newly installed capacity of 5,545 MW and a cumulative installed capacity of 9,196 MW (Clark and Isherwood, 2010).

At the UN's 2009 Copenhagen Conference on climate change, China committed that by 2020, it would meet 15 percent of the nation's energy demand with nonfossil fuels. Achieving this goal will require a huge increase in green energy development, including a much greater concentration on wind power.

Through its renewable energy land other policies, China has made a major commitment to wind energy. A major part of China's future efforts involves the creation of seven major scale wind power bases. Each wind base has potential for at least 10 GW of installed capacity.

The Chinese National Energy Bureau is developing these bases. They plan to create a total installed capacity of 138 GW by 2020 but only if the supporting grid network is established. A significant problem is that many of these bases are located in remote areas with a weak transmission grid and a long distance from China's main electricity load centers. There are also concerns about how to integrate large quantities of variable wind power into a grid built for coal-burning power stations.

Pricing is another important element. China's support mechanism for wind power has evolved from a price based on return on capital to a feed-in tariff, with variations based on differences in wind energy resources (Gipe, 2014).

The FiT system was introduced to China in 2009. The system divides the country into four categories of wind energy areas. This regional FiT policy seems to be a positive step in the development of wind power and is stimulating stronger economic growth, increasing manufacturing output, and adding jobs. Additionally, the Chinese see the need for trained workers for building, operating, and maintaining these new systems, so they have created engineering and science programs to train people to work in wind and other renewable technology industries (Gipe, 2014).

China faces several challenges when it comes to integrating large-scale wind-generated energy into its local and regional grid networks and infrastructures. Wind farms in China are located mainly in areas far from load centers and where the grid network has transmission issues and needs high maintenance (Hong, 2013). This causes a loss in efficiency, so the present design of the infrastructure grids places constraints on the development and use of wind power. This has become the biggest problem for the future development of wind power throughout the country. Nonetheless, the Global Wind Energy Council projects exceptional growth for China's wind power capacity. Wind power could account for 10 percent of total national electricity supply by 2020 and reach 16.7 percent in 2030. These figures do not take into account more local and regional wind farm systems as well as smaller systems that are integrated into buildings.

Wind energy, in contrast to fossil fuels, is plentiful, renewable, widely distributed and clean and produces no GHG emissions during operation. While there is some criticism of wind farms because of their visual impact, any effects on the environment are generally among the least significant of any power source.

Large-scale wind farms are not the only solution. Today's new technology allows for the installation of wind turbines in small communities. Even smaller systems can be placed on rooftops to capture the natural flow of air over buildings.

Advances in turbine construction have increased efficiency to the point that wind energy is quickly becoming the most cost-effective source of electrical power. In fact, one could make a strong case that it has already achieved this status. The actual life cycle cost of fossil fuels—which would include the external costs from mining, extraction, transportation, and use, as well as the costs from environmental and political impacts—is certainly far more than the current wholesale rates. The eventual depletion of fossil fuel energy sources will entail rapid escalations in price, which will require postponing actual costs that would be unacceptable by present standards. And this does not take into account fossil fuels' environmental and political costs, which are mounting every day.

The major technology developments that enabled wind power commercialization have already been made, but there will be more refinements and improvements. Based on the way other technologies have developed, the eventual push to full commercialization and deployment of wind power will happen in a manner that is unimaginable today (Hong, 2013).

SOLAR: ENERGY FROM THE SUN

The sun is the Earth's primary energy source. It radiates an enormous amount of power, about 170,000 terawatts (TW), and powers almost all natural processes that occur on the Earth's surface. There is no shortage or cost for energy from the sun's power.

There are two technologies that directly convert electromagnetic energy from the sun into useful energy: solar photovoltaic and solar thermal.

Photovoltaic systems convert light into electricity. Silicon cells capture sunlight, including ultraviolet radiation. The sunlight creates a chemical reaction and excites the electrons in a semiconductor, which generates a current of electricity. This photovoltaic reaction is at the core of solar panel systems.

Most people consider photovoltaic a development of the space exploration era; however, a French physicist Alexandre Edmond Becquerel discovered the effect in 1831. About five decades later, the US scientist Willoughby Smith discovered the photovoltaic effect in selenium. He realized that illuminating a junction between selenium and platinum would also create a photovoltaic effect. These two discoveries led to the first selenium solar cell construction in 1877 (PVResources, 2011).

A series of discoveries in the 1950s led to the 1958 launch of Vanguard I, the first satellite to use solar cells for onboard power. The solar cell system ran continuously for eight years. Several other satellites with onboard solar cell-generated power followed: Explorer III and Vanguard II, launched by the Americans, and Sputnik III, launched by the Russians.

A Japanese electronics company, Sharp Corporation, developed the first usable photovoltaic module from silicon solar cells in 1963, setting the stage for the modern solar industry. The big solar companies started forming in the 1970s-1980s, both in the United States and in Japan. ARCO Solar was the first to produce photovoltaic modules, with peak power of over 1 MW per year (PVResources, 2011).

A PV system includes mechanical and electrical connections, mountings, and a means of regulating or modifying the electrical output. Because the voltage of an individual solar cell is low, the cells are wired in series to create a laminate. The laminate is assembled into a protective weatherproof enclosure, thus making a photovoltaic module. Modules are then strung together into an array. Electricity generated by PV systems can be used directly as a standalone power source, stored, fed into a large electricity grid, or linked with many domestic electricity generators to feed into a small grid.

Silicon crystal and thin film are the two main categories of PV technologies. Silicon crystal is used more often because of its higher efficiency and greater abundance. However, the process of refining silicon is expensive. While thin-film technologies have a lower efficiency, they do have the potential to provide solar power at a lower cost per watt than silicon crystals. Now, the third generation of PV designs is even more efficient with lower cost.

Germany made a significant commitment to solar power at the beginning of the twenty-first century by increasing the amounts that the feed-in tariffs (FiT) generate, a move that surprised the industry and kick-started the modern solar power era. Despite being located in a northern European country, not known for an abundance of sunny days, the Germans built several large solar power plants between 2002 and 2003.

In April 2003, the world's largest photovoltaic plant was connected to the public grid in Hemau near Bavaria. The peak power of the "Solarpark Hemau" plant is 4 MW. With the support of Germany's renewable energy law called the EEG, this plant was followed by many larger systems in 2004. Germany's aggressive FiT program led the country into becoming the world leader of installed PV systems. According to the European Photovoltaic Industry Association, Germany installed 3.8 GW by the end of June of 2010 (Gipe, 2014).

Asia's technological tigers—China, Japan, and South Korea—are rapidly overtaking Germany as the world's leader in PV technology and installed capacity. In Japan, a well-coordinated effort to fund research and development of PV technology has led to increased residential use of solar energy.

CHINA'S SOLAR VALLEY CITY

Toward the end of 2010, China overtook Germany as well as Japan to become the world's largest manufacturer of solar panels (Gipe, 2014). However, the Chinese solar industry is primarily an export industry with over 90 percent of PV sales for exports.

Solar industry leaders have lobbied for a more regionally and city-focused active set of government policies to subsidize the domestic use of solar power. Because there is so little domestic use, the potential for growth is strong. Policies intended to jump-start domestic solar power demand are emerging in their next five-year plan (Lo, 2011).

To provide more domestic use of solar, the Chinese Ministry of Finance is pushing an on-site or local Solar-Powered Rooftops Plan. This plan will develop demonstration projects for building solar power systems in medium to large cities that are economically developed and want to be sustainable. The plan calls for rooftop units and PV curtain walls and supports the development of systems in villages and remote areas that are outside of the power grid. As part of this effort to improve domestic use of solar panels, the Ministry of Finance has earmarked a special fund to provide subsidies for PV systems that are at least 50 kilowatts (kW) in size and have 16 percent efficiency. The subsidy will cover the cost of the equipment, or approximately 50 to 60 percent of the total cost of an installed system.

Industry analysts say that much needs to be done to develop a thriving solar industry in China. The country will need to reorient the industry to one that is balanced between domestic consumption and export. To achieve this balance, the Chinese will need to create a new domestic system that matches the industry's export capabilities.

To showcase its commitment to solar energy, the nation built the Solar Valley City in Dezhou, Shandong Province, in 2008. This ambitious project created a new sustainable, environmentally sound center for manufacturing, research and development, education, and tourism focusing on solar energy technologies. Solar Valley City is part of China's efforts to promote green energy technology and grow global market share.

More than 100 solar enterprises, including major solar thermal firms, are based in Solar Valley City. The solar industry in China employs about 800,000 people, and China's solar thermal industry and the accompanying industrial chain are examples for the rest of the world. A leading company, Himin, produces more than twice the annual sales of all solar thermal systems in the United States and is quickly expanding into solar photovoltaic and other technologies.

US SOLAR

The US solar industry is growing quickly, particularly in California and the Southwest. According to the Solar Energy Industries Association, by the end of 2013, there was over 10,250 MW of cumulative solar electric capacity in the United States. This was enough to power 1.7 million American homes. The association points out that falling prices continue to make solar more affordable. Prices, they say, have declined by 60 percent since 2011 (PVResources, 2011).

In the United States and other countries, the owners of central plants typically under government or public regulation have focused on concentrated solar systems. These create energy on solar farms and then transmit it hundreds of miles to the end users, which are typically large customers like shopping malls, office complexes, and government and educational buildings. However, now distributed and on-site power systems using solar in sixteen states are making the residential PV market a growing opportunity with a variety of programs being developed on the local level to encourage PV adoption. For example, Berkeley, California, adopted an innovative program called Berkeley First, which allows homeowners to avoid high upfront costs by amortizing the cost of systems over 20 years and paying for them through property taxes.

Besides PV systems, the sun's energy can be harnessed through solar thermal systems. These systems use sunlight to heat oil or water. Solar thermal systems are available on a small scale for individual consumers. There are many solar thermal systems installed as pool heaters or water heating systems for the residential market.

On a large scale, thermal technology is concentrated in one location so that is it used for utility-scale power generation. These large systems use mirrors or lenses to focus solar energy on the liquid. The heated liquid then drives a turbine that generates electricity. These concentrated solar power (CSP) systems are of four basic types: trough, linear Fresnel reflector, tower, and dish. The most common is the trough system, which uses parabolic mirrors that concentrate solar heat on a fluid-filled receiver that runs the length of the trough (PVResources, 2011).

The operation of a CSP solar plant requires large tracts of land and substantial volumes of water to provide cooling for the steam turbine. Many of the prime locations for these systems are remote, placing a burden on transmission lines and connection activities.

Although both PV and CSP are dependent on solar power input, CSP systems are easily fitted with thermal storage systems such as molten salt.

Adding a storage component allows for operation at night or on cloudy days and turns solar energy into a more consistent power resource.

The United States pioneered CSP technologies, and there are a number of large solar thermal installations in California, Nevada, and now Hawaii. In the mid-2010, the United States produced more than half of all solar thermal power in the world, although Spain is rapidly building solar energy plants.

Use of solar energy is increasing rapidly. In 2000, there was only 170 MW of solar power generated globally. In 2010, the global market reached 20 GW of installed solar power capacity, according to Greentech Media. They anticipate that the global PV market will reach 25 GW by 2013 and come close to 100 GW by 2020. They also anticipate that a number of large solar firms will approach the 1 GW capacity threshold in 2011.

A report prepared by research and publishing firm Clean Edge found that solar power could provide 10 percent of the US power by 2025. The report projected that nearly two percent of the nation's electricity would come from concentrating solar power systems and eight percent would come from solar PV systems. The report noted that as solar power has been rapidly expanding, the cost per kilowatt-hour of PV systems has been dropping. At the same time, electricity generated from fossil fuels is becoming more expensive. As a result, the report projects that solar power will reach cost parity with conventional power sources in many markets by 2015 (PVResources, 2011).

In order to reach their full potential, solar photovoltaic companies must streamline installations and make solar power a "plug-and-play" technology. It must be simple and straightforward to buy the components of the system, connect them together, and connect the system to the power grid. This is the business model that IKEA is following as it rolls out its residential system in the United Kingdom.

Electric utilities need to take advantage of the benefits of solar power, incorporate them into future smart grid technologies, and create new business models for building solar power capacity. The Clean Edge report calls for establishing long-term extensions of today's investment and production tax credits, creating open standards for connecting solar power systems to the grid, and giving utilities the ability to include solar power in their rate base.

PV systems can be integrated into a building during construction. This fast-growing segment of the solar industry includes incorporating PV solar panels in building elements such as roofs, window overhangs, and walls. This reduces the material costs of the building construction and the installation cost of the PV panels. Passive solar building design can also take advantage of solar energy, using windows and interior surfaces to regulate indoor air temperature. As the price of panels and generation drops, solar energy takes on a much bigger role in carbonless energy generation.

New technologies and scientific advancements offer the keys to a new way of life. In this future scenario, most rooftops will have PV systems. The windows of office buildings will be covered with a thin film that, in reality, is another PV system. If the sun is shining in your neighborhood and your PV system is generating more electricity than you are using, you can store it in your car's battery or pass it on to the neighbor a few blocks away that needs it. A whole new era of flexible energy generation is coming and solar is a key element.

WATER AS ENERGY: FROM HYDROELECTRIC TO OCEAN WAVES AND THE RUN OF RIVERS

Throughout the world, water is moving. Gravity is always at work, moving water through tidal action and pulling water from the mountains to the sea. For decades, humans have dammed rivers to generate electricity. Dams like the Grand Coulee Dam across the Columbia River in Washington, the Hoover Dam across the Colorado, and the Three Gorges Dam across the Yangtze River in China are some of the world's most imposing structural engineering accomplishments.

Hydroelectric power has been one of the key features of Western industrialization. While damming a river produces carbonless electricity, it destroys multiple ecosystems that cannot be recovered, and public resentment to dam construction is increasing. There are far better, cheaper, and more environmentally sensitive ways to generate electricity from renewable sources than building large dams.

Excluding dams, water movement offers a huge potential for sustainable and consistent energy generation. Through tidal action, the world's oceans are constantly moving, providing the potential for enormous amounts of renewable energy. For example, the United States receives 2,100 terawatt-hours of wave energy along its coastlines each year. Tapping just a quarter of this potential could produce as much energy as the entire US hydropower system (Wave Tidal Energy, 2014).

There are four basic types of marine water currents: tidal, oceanic, wave action, and river.

Humans have used tidal energy for centuries. There are European tide mills that date to AD 787. Medieval tide mills consisted of a storage pond, filled by the incoming tide through a sluice, and emptied during the outgoing tide through a waterwheel. The tides turned waterwheels, producing mechanical power to mill grain.

Traditional tidal electricity generation involves the construction of a barrage, or barrier, across an estuary to block the incoming and outgoing tide. The dam includes a sluice that is opened to allow the tide to flow into the basin. The sluice is then closed, and as the sea level drops, the elevated water drives turbines to generate electricity.

Tidal power is expensive, and there is only one major tidal energy generating station in operation. This is a 240 MW system at the mouth of the Rance river estuary on the northern coast of France. The Rance generating station has been in operation since 1966 and has been a very reliable source of electricity. The Rance was supposed to be one of many tidal power plants in France, until their nuclear program was expanded in the late 1960s. Elsewhere, there are a 20 MW experimental facility at Annapolis Royal in Nova Scotia and a 0.4 MW tidal power plant near Murmansk in Russia.

There are several other potential tidal power sites worldwide, including constructing a barrage across the Severn River in western England. Similarly, several sites in the Bay of Fundy, Cook Inlet in Alaska, and the White Sea in Russia have the potential to generate large amounts of electricity. Scotland completed an assessment of the narrow waterway between mainland Scotland and the Orkney Islands. Called the Pentland Firth, the study showed that underwater turbines stretched across the strait could generate about half of the nation's electricity. The waterway has exceptionally high tides and an estimated 1.9 GW could be available (Scotland, 2014).

The waters off the US Pacific Northwest are ideal for tapping into ocean power using newly developed undersea turbines. The tides along the Northwest coast fluctuate dramatically, as much as 12 feet a day, and have exceptional energy-producing potential. On the Atlantic seaboard, Maine is also an excellent candidate for tide-generated power.

Tidal turbines that would have less impact on the environment are in development. Because water is denser than air, tidal turbines are smaller than wind turbines and can produce more electricity in a given area. In 2006, Verdant Power installed a pilot-scale tidal project in New York's East River. Verdant received the go ahead to expand the project to 1 MW in 2012 (Verdant, 2012).

Waves are produced by winds blowing across the surface of the ocean. Buoys, turbines, and other technologies can capture the power of the waves and convert it into clean, pollution-free energy. Pelamis Wave Power, a Scottish company, may have developed a breakthrough technology for wave energy. Named after a sea snake, the Pelamis absorbs the energy of the ocean waves and converts it into electricity. About the length of a jumbo jet, but thinner, this extraordinary machine is made up of five large brightly colored tube sections (Pelamis Wave Power, 2014).

The sections are linked by universal joints, which allow flexing in two directions. Facing the direction of the waves, the machine undulates semi-submerged on the surface of the water. As waves pass down the length of the machine and the sections bend, the movement is converted into electricity by using hydraulic power systems inside each joint of the machine tubes. Power is transmitted to shore using standard subsea cables and equipment.

Tethered by cables and an underway structure, the machine sits offshore and operates in depths greater than 50 m. Rated at 750 kW, one machine will provide sufficient power to meet the annual electricity demand of about 500 homes. The company has produced six full-scale Pelamis machines, with major wave farms in development on Scotland's northern coast near the Hebrides islands. The company is working closely with the Scottish government and has extraordinary potential. They say that as manufacturing prices come down, the Pelamis machines may become one of the lowest cost forms of energy generation (Pelamis Wave Power, 2014).

The potential for wave energy is enormous. The estimated wave energy around the British Isles equals about three times the United Kingdom's electricity demand and could become a major factor in the UK's energy mix. Europe's western seaboard offers an enormous number of potential sites off the coast of Ireland, France, Spain, Portugal, and Norway. The Pacific coastlines of North and South America, southern Africa, Australia, and New Zealand are also highly energetic. In fact, most seacoast areas have the potential to generate wave energy at competitive prices.

In several parts of the world, small power turbines are mounted along rivers. As the water flows, the turbines generate electricity. This is being done in the Europe where the water generates a considerable amount of energy without harming the environment or changing the water's natural elements. This could be done in many other locations worldwide. Electricity based on these offshore hydrokinetic resources could provide substantial amounts of power to high-demand electricity markets.

GEOTHERMAL

Geothermal energy creates power by extracting heat that is stored in the Earth. Heat was created and captured inside the Earth when the planet was formed. Heat is also created by the radioactive decay of minerals and solar energy absorbed at the surface. The Earth's heat has been used for space heating and bathing since ancient Roman times, but it is now being used to generate electricity.

Most geothermal energy comes from heated water, or hydrothermal resources, that exists where magma comes close enough to the Earth's surface to transfer heat to groundwater reservoirs. This produces steam or high-pressure hot water. If the reservoir is close enough to the surface, a well can be drilled and the steam or hot water can be used to drive a turbine. The steam or hot water can also be used as a heat source. If the temperature is moderate enough, the hot water can be used directly to heat buildings or for agriculture or industrial processes.

The first geothermal electricity generator was brought online in 1922 at the Geysers in Northern California. Other power plants were added, and this geothermal field is now the largest in the world. In all, the United States has 77 geothermal power plants that provide about four percent of America's renewable electricity. The United States leads the world in geothermal electricity production, with 3,086 MW of installed capacity (GEA, 2010).

In 2010, the International Geothermal Association (IGA) reported that 24 countries have a total of 10,715 MW of geothermal power online, which is expected to generate 67,246 GW of electricity. This represents a 20 percent increase in online capacity since 2005. IGA projects growth to 18,500 MW by 2015, due to the projects presently under consideration (GEA, 2010).

Geothermal electric plants were traditionally built at the edges of tectonic plates, where high-temperature geothermal resources are available near the surface. Recent technological advances have dramatically expanded the scope of the resources that can be tapped and the range for the geothermal energy. This is especially important for applications such as home heating and opens the potential for widespread exploitation.

Geothermal power is very cost-effective, reliable, sustainable, and environmentally friendly. While geothermal wells release greenhouse gases trapped deep within the Earth, these emissions are much lower per energy unit than those produced by fossil fuels. As a result, geothermal power has the potential to help mitigate global warming, if used in place of fossil fuels.

Though the Earth's geothermal resources are vast, only a very small fraction may be profitably exploited. Drilling and exploration for deep resources are expensive, making this resource a limited one for the future. In the last few years, engineers have developed remarkable devices, such as geothermal heat pumps, ground source heat pumps, and geoexchangers, which gather ground heat to provide heating for buildings in cold climates. Through a similar process, they can use ground sources for building cooling in hot climates. More communities with concentrations of buildings, like colleges, government centers, and shopping malls, are trying to adapt geothermal systems to maximize energy efficiency.

BIOMASS, RECYCLED AND REUSABLE GENERATION

Biomass may be the oldest of all sources of renewable energy, dating back to the ancients and their discovery of the secrets of fire. Today, most people think of biomass as garbage because it uses dead trees, yard clippings, or even livestock manure to generate power. Denmark has been a remarkable builder of biomass plants for power from animal and food waste (Lund, 2009).

Solar energy is stored in this organic matter, so under pressure and over time, a chemical process takes place that converts plant sugars into gases. The gas can then be burned as ethanol or used to generate electricity. The process is referred to as digestive, and it's not unlike an animal's digestive system. The appealing feature for energy generation is that it can use abundant and seemingly unusable plant debris—rye grass, wood chips, weeds, grape sludge, and almond hulls, to name a few.

To create usable energy from biomass, materials like waste wood, tree branches, and other scraps are burned to heat water in a boiler. The steam is used to turn turbines or run generators to produce electricity. Biomass can also be tapped right at the landfill or waste treatment plant, by burning waste products.

Biomass can produce energy without the need for burning. Most garbage is organic, so when it decomposes it gives off methane gas, which is similar to natural gas. Pipelines can be put into the landfills to collect the methane gas, which is then used in power plants to make electricity and create hydrogen (Atwood and Clark, 2004).

Animal feedlots can process manure in a similar way using anaerobic digesters. The digesters can create biogas from the manure, which is then burned to produce power. For example, dairy farms can use methane digesters to produce biogas from manure. In turn, the biogas can be burned to produce energy or used like propane. Biogas can also be derived from poultry litter.

Although biomass is a renewable energy source, the combustion process creates pollution. Biomass resources also vary by area and depend on the conversion efficiency to power or heat. For these reasons, many countries see biomass as a transitional renewable energy source, for use as they look for new technologies to either convert the emissions or process waste differently. For example, Denmark creates much of their energy with biomass but they plan to start converting and limiting its use totally in the next decade.

On the positive side, biomass raw materials get their energy from the sun and regrow quickly. Through photosynthesis, plants use chlorophyll to convert carbon dioxide into carbohydrates. When the carbohydrates are burned, they release the energy captured from the sun. Energy crops can be grown on marginal lands and pastures or planted as double crops (Jenkins et al., 2001). While most scientists say that making ethanol from corn is not efficient, it has been produced in the Midwest for years. Converting sugarcane to ethanol is considerably more efficient, and Brazil has adapted sugarcane to ethanol conversion as a major fuel source for transportation.

The Union of Concerned Scientists cites Minnesota's Koda Energy plant as an excellent example of generating energy from biomass. It is a combined heat and power (CHP) plant that uses biomass to generate renewable electricity as well as waste heat from the boiler. In 2009, Koda began generating electricity from oat hulls, wood chips, prairie grasses, and barley malt dust. About 170,000 tons of these agricultural wastes is used a year (Koda, 2014).

Since the beginning of time, humans have captured the energy from biomass by burning it to make heat. In the first industrial revolution, biomass-fired heat produced steam power, and more recently, this biomass-fired steam power has been used to generate electricity. Advances in recent years have shown that there are even more efficient and cleaner ways to use biomass.

FOR THE PLANET'S BENEFIT AND HUMAN HEALTH: RENEWABLES HAVE TO COME FIRST

The shift from a fossil fuel-, nuclear power-based economy to a renewable energy–centered economy has begun. China, Japan, South Korea, and the European Union are well into the process. China, Japan, and South Korea have national energy policies, programs, and financing. The EU has taken a regional approach and is pushing its members toward energy independence through renewable energy generation and the use of storage devices and other technologies.

Around the world, nations are coming to realize that there is a clear need for a consistent, intelligent, and long-term energy policy that will stabilize the energy markets. There is also a whole new era of job creation and business opportunities available to nations with the political foresight to take advantage of these opportunities.

So far, America clearly lags in this effort, while China has embraced it. Other countries, like Scotland and Chile, have taken notice and are working quickly to power their economies with renewable energy.

The new global economy is being led by what scholars characterize as social capitalism. This is apparent in China, which has an enormous opportunity to push the GIR forward. The new economy will be a green one that is focused on sustainable development for the public good. It will focus on renewable energy generation and encompass the social infrastructures of energy, transportation, telecommunications, buildings, and natural resources.

Renewable energy is available in multiple ways—wind, solar, geothermal, marine, and biomass are the most common and most accessible. Wind, sun, and tides are ancient processes that have been used to generate energy for most of human history. They are readily available, highly efficient, and environmentally friendly. Their use produces little or no increase in GHGs, global warming, or acid rain. Most importantly, they are sustainable and not subject to dwindling supplies, extraction problems, price disruptions, or political leverage. Their energy can be produced on utility scale, community scale, village scale, individual facility scale, or even nanoscale.

RENEWABLE ENERGY SYSTEMS ARE PROTECTING THE ENVIRONMENT AND CHANGING LOCAL COMMUNITIES

Converting to 100 or even 50 percent renewable sources for energy power will require a significant financial and political commitment. The United Nations estimates that in 2009, there was $162 billion invested in renewable energy worldwide. About $44 billion was spent in China, India, and Brazil, collectively and $7.5 billion in many poorer countries. Since then, China alone had spent over $100 billion with its 12th Five-Year Plan, primarily on wind turbines.

Although the United Nations and other world bodies have urged developing nations to look first at environmentally friendly renewable energy instead of carbon- and fossil-based fuels as they add capacity, it is extremely difficult to build any system that is not grid-connected. Relatively speaking, it is easier to build $300 million of solar systems in California, than wind farms or industrial-size solar systems in developing countries, even when they feed into a grid for distribution.

Slowly, the GIR is seeping into the nongrid world of the rural poor. Small self-contained solar and biomass power generation systems are being used to create enough power for a house or a hamlet. As small-scale renewable power becomes cheaper and more dependable, it is providing the first access to electricity for people far removed from the power grids that serve the cities of developing countries.

In these rural villages, two modern megatrends—globalization and the GIR—are coming together to change lives. In Africa, small solar power systems are appearing, balanced precariously atop of tin-roofed shacks far away from power grids. In other areas, simple subterranean biogas chambers make fuel and electricity from cow manure. In Nepal, minihydroelectric dams derive enough energy from a local river to power a village. In India, since 2008, Husk Power Systems has installed 80 systems that use rice husks to generate power, providing electricity to over 200,000 people across 300 villages and hamlets (Husk Power, 2014).

Solar systems allow people in remote areas to use cell phones. In Kenyan villages, owning a cell phone is a lifeline. By selling a goat or two, a villager can buy a tiny solar system for $80. Coupled with a $20 cell phone, a villager can make small money transactions, check the price of chickens, or connect with family members who have moved to the city. In areas without electricity, charging a cell phone can require a long walk or bicycle ride to a shop where the phone can be charged. Some shops are so backlogged that it takes days to charge a phone. Under these conditions, the cell phone is a precarious lifeline at best. Now, with the small solar-powered systems, the cell phone can be charged at home, and one villager can develop a small business charging the cell phones of others.

The United Nations estimates that 1.5 billion of the world's 7 billion people live without electricity. As many as 70-80 percent of the population in nations like Kenya—where people still cook and heat with primitive fuels like wood, cow dung, or charcoal—is without electricity.

In poor, rural areas, cheap solar systems are playing an epic, transformative role. These tiny power centers are allowing the rural poor to save money on candles, batteries, kerosene, and wood, which in many parts of the world is becoming less available as governments try to halt the deforestation of rural areas. Infants and children are no longer inhaling the fumes of smoky kerosene lamps. Solar panels coupled to a car battery and a powerful, low-wattage LED lamp can provide lighting. After sundown, older children can study mathematics and do their reading lessons just like their cousins in the city.

In some African countries, villagers use underground tanks in which manure from cows is converted to biogas. Then, the gas is pumped through a rubber tube to a burner for cooking. In other areas, the biogas is used to generate electricity.

These and other tiny, individual, renewable power-generating systems are slowly becoming available in undeveloped areas with the help of non-profit agencies and the United Nations. However, they buck the traditional grid-connected power mold, and investors are reluctant to put money into fragmented markets consisting of poor rural consumers.

The benefits of the GIR are coming slowly, but they are coming to the poor. In Africa, there is evidence that a true market is emerging for home-scale renewable energy and low-consuming appliances. As costs keep coming down, families are willing to sell a goat or borrow money from a relative in the city to buy a tiny solar system. The cell phone is a must-have for African villagers, and this passion is driving the need for tiny, nongrid, and renewable energy systems. It is this nexus of digital electronics and renewable energy that is at the forefront of the GIR.

As the green industrial revolution becomes robust, and more and more nations embrace it, renewable energy generation from sustainable sources will become the norm. A few decades from now, people will look back with contempt at the way the 2IR lifestyle squandered fossil fuels, destroyed ecosystems with coal extraction, and threatened the health and survival of the planet through ruthless disregard of global warming and climate change.

REFERENCES

Atwood, Ted and Woodrow W. Clark II, "Bio Mass and Hydrogen production", ANPA Report, Rome: 2004.
American Wind Energy Association (AWEA) 2011.
Clark, Woodrow W., 2012. Introduction: the green industrial revolution. The Next Economics: Global Cases in Energy, Environment, and Climate Change. Springer Press, Fall, New York.
Clark II, Woodrow W., Isherwood, William, 2010. Creating an Energy Base for Inner Mongolia, China: the leapfrog into the climate neutral future. Utilities Policy Journal.
Clark II, Woodrow W., Sowell, Arnie, 2002. Standard Economic Practices Manual: life cycle analysis for project/program finance. International Journal of Revenue Management. Inderscience Press, London, UK, Nov.
Clark II, Woodrow W. and William Isherwood, Utilities Policy Journal. Special Issue on China: environmental and energy sustainable development. Winter 2010 from Asian Development Bank Report in 2009.
Dodge, Darrel M. 2009. Illustrated History of Wind power: Early History through 1875. Littleton, CO: TelosNet. http://telosnet.com/wind/index.html.2001.
Geothermal Energy Association (GEA) 2010.

Gipe, Paul. Feed-in-Tariff Monthly Reports 2014 at: http://www.wind-works.org/FeedLaws/RenewableTariffs.qpw.

Hong, Lixuan, Developing an analytical approach model for Offshore Wind in China, PhD Thesis, 2013.

Husk Power in India, 2014: http://www.huskpowersystems.com/innerPage.php?pageT=About%20Us&page_id=76

Jenkins, B.M., W.W. Clark, II, V.A. Fung, T. Atwood, V. Tiangco and R. Julian. 2001. Distributed power generation for agro-energy in sustainable development in the Philippines. Final Report, State Environmental Initiative, Council of State Governments, Lexington, Kentucky.

Tribe Koda Energy, 2014: http://www.shakopeedakota.org/enviro/koda.html.

Lo, Vincent, (2011) "China's Role in Global Economic Development", Speech by Chairman of Shui On Land, given at Asian Society, Los Angeles, CA, April 25, 2011.

Lund, Henrik, 2009. Sustainable towns: the case of frederikshavn, Denmark. In: Clark II, Woodrow W. (Ed.), Sustainable Communities. Springer Press, New York, Chapter# 10.

Lund, Henrik, Clark II, Woodrow W., 2008. Sustainable energy and transportation systems introduction and overview. Utilities Policy 16. Elsevier Press, New York, NY, pp. 59-62.

Lund, Henrik, Hvelplund, Frede, 2012. The economic crisis and sustainable development: the design of job creation strategies by use of concrete institutional economics. Energy. Elsevier Press, September.

Lund, Henrik, Østergaard, Poul Alberg, 2010. Climate Change mitigation from a bottom up community approach: A case in Denmark. In: Clark II, Woodrow W. (Ed.), Sustainable Communities Design Handbook. Elsevier Press, Chapter 14.

O'Toole, Sarah, 2010. World's biggest offshore wind farm opens today. Global Energy Magazine. (September 23).

Think Progress, 2012: http://thinkprogress.org/climate/2012/02/22/430441/coal-consumption-in-china/.

PVResources 2011.

Renewable Energy World (REW) http://www.renewableenergyworld.com/rea/news/article/2011/03/project-profile-shanghai-donghai-bridge.

Scotland. 2014 at: http://phys.org/news/2014-01-island-channel-power-scotland.html#jCp.

Stansberry, Mark.A, 2012. America Needs America's Energy: Creating Together the Peoples Energy Plan. Brown Books Publishing Group, USA.

Sun, Xiaolei, Jianping Li, Yongfeng Wang and Woodrow W. Clark, China's Sovereign Wealth Fund Investments in overseas energy: the energy security perspective, Energy Policy, October 2013, http://dx.doi.org/10.1016/j.enpol.2013.09.056.

Wave Tidal Energy, 2014 at: http://www.rnp.org/node/wave-tidal-energy-technology.

U.S. Department of Energy (US DoE, EIA) Energy Information Administration 2011.

U.S. Department of Energy (US DoE, EIA), Energy Information Administration 2013.

US Department of Energy (DoE) International Energy Agency (IEA) 2010.

Verdant, 2012 at: http://verdantpower.com/what-initiative/.

Pelamis Wave Power 2014 at: http://www.pelamiswave.com.

World Wind Energy Association (WWEA), 2011.

CHAPTER 7

Storage Technologies

The Incredible Hulk Roller Coaster ride at the Universal Studios in California features a rapidly accelerating, uphill, full-speed launch, as opposed to the typical gravity drop. Voted the number 1 roller coaster by Discovery Channel viewers in 1999, the cars move up the track pushed by powerful traction motors. To achieve the brief, but very high current required to accelerate a roller coaster train to full speed uphill, the park uses several motor generator sets attached to large flywheels to store the energy. The Hulk uses so much electricity that if it did not have multiple large flywheels to store energy, the company would have to invest in a new substation and risk brownouts on the local grid every time the ride launched.

Flywheels as an energy storage technology have been around since the ancient Greeks. Other common forms of energy storage include pumped hydroelectric, compressed air, batteries, capacitors, Superconducting Magnetic Energy Storage, flywheels, thermal storage, and hydrogen. Today and now into the GIR, issues of energy are critical as they impact not only the increasing demand for energy but also the climate and health reproductions from greenhouse gases, air pollution, and water pollution. China is now the world's leader in atmospheric pollution, which is causing massive and environmental problems due to its repaid modernization over the last two decades (CREIA, 2011).

The biggest criticism of using large-scale renewable energy is that it is intermittent, meaning the power is not always available (e.g., sun and wind sources), and therefore not reliable as fossil fuels. Thus, the search has been for a cheap way to store energy from sun and wind when it is plentiful, so that it can be used when it is not. There are huge potentials in storage technologies and each has benefits. However, some experts consider storage technologies as the "holy grail" since they are not plentiful and can be at a high cost.

Generally, energy storage separates into two broad categories—those that store electricity for either a central or on-site power grid and those that store electricity for transportation. The following table from the US Congressional Research Service is based upon data from the National Renewable Energy Laboratory (NREL), which gives an overview of the uses of energy storage technologies (Parfomak, 2013).

Energy storage applications and technologies

	Electric grid (stationary)	Transportation (vehicular)
High power/rapid discharge	Batteries • Lead–acid • Nickel • Lithium-ion Capacitors Flywheels Superconducting Magnetic Energy Storage (SMES)	Batteries • Nickel Capacitors Flywheels
Energy Management	Batteries • Advanced lead–acid • Flow • High temperature Hydrogen Compressed air Pumped hydroelectric Thermal • Concentrating Solar Power • End Use	Batteries • Lithium-ion • Lithium metal • Metal air Hydrogen

Note: Electric power and transportation applications may elsewhere be referred to as "stationary" and "vehicular," respectively.
Source: P. Denholm, National Renewable Energy Laboratory, 2011

Storage technology can make electric grids more efficient and reliable. It can manage power flows and support the integration of renewable energy generation. In particular, renewable energy, from wind and solar generation, is not constant. Instead, it needs to be integrated with storage devices and shared with a smart grid. If green energy such as wind and solar is to be the core of the GIR, then storage technologies are critical.

Grid storage devices can provide electricity for several hours or more. These devices could shift energy during periods of low demand or high renewable supply to periods of high demand or low renewable supply. Many of them can also provide the same services as high-power or rapid discharge devices.

The main technology in this group is *pumped hydroelectric storage* (PHS). Basically, this stores potential energy by pumping water from a lower elevation reservoir to a higher elevation one. The pumping is done during low-cost off-peak demand periods. Then, during periods of high electric demand, the stored water is released through turbines to produce electric power. While it doesn't save or increase energy, this system increases overall revenue by selling more electricity during periods of peak demand, or when prices are the highest.

Pumped storage is the largest system for grid energy storage, with the United States and China having the largest capacity. This technology has high reliability, high efficiency, and long lifetime but requires suitable geologic conditions and a long development time. Pumped storage requires energy at a cost to move the water into a dam or reservoir for future use. Some communities are getting their energy from wind power so that the entire system is at very little cost and based upon integrated green energy systems.

Flywheels are an old technology that has roots in the Neolithic spindle and the ancient Greek potter's wheel. The potter's wheel features a heavy round stone, connected to a pedal that is pumped by the potter. The flywheel stores the fluctuating pedal movements as inertia and creates a smooth, steady, turn of the wheel. Through the wonders of mechanical engineering, and again the commercial research at the Lawrence Livermore National Laboratory (LLNL), funded by the US Department of Energy, this old technology is being refitted as a modern-day energy-storing device. Flywheel energy storage (FES) works by accelerating a rotor, or flywheel, to a very high speed and maintaining the energy in the system as rotational energy. When energy is extracted from the system, the flywheel's rotational speed is reduced; adding energy to the system correspondingly increases the speed of the flywheel (LLNL Flywheels, 2014).

Most FES systems use electricity to accelerate and decelerate the flywheel, but devices that use mechanical energy are in development. Advanced FES systems have rotors made of high-strength carbon filaments. The filaments are suspended by magnetic bearings and spin at speeds from 20,000 to over 50,000 revolutions per minute (rpm) in a vacuum. Such flywheels can come up to speed in a matter of minutes—much quicker than some other forms of energy storage.

Over the past two decades, scientists have studied flywheels extensively and created ways to use them as power storage devices in vehicles and power plants. Flywheels can be used to produce high-power pulses in situations where drawing the power from the public network would produce unacceptable spikes. A small motor can accelerate the flywheel between pulses. Another advantage of flywheels is that it is possible to know the exact amount of energy stored by simply measuring the rotation speed (Clark and Isherwood, 2010).

Flywheel technology can be used as a replacement for conventional chemical batteries. They have a long life cycle and require little maintenance. Flywheels are also less damaging to the environment, being made mainly from inert materials (Clark and Isherwood, 2007). Some start-up companies have now commercialized flywheels into systems that supply all the power needed for buildings and complexes. The key factors are that these flywheel systems are self-contained, cost-effective, and with a zero environmental impact (Eco-Gen, 2014).

Eco-Gen, a southern California company, markets the JouleBox, which uses a hybrid system of flywheels with solar, wind, and lithium–ion batteries. The flywheel is teamed with conventional renewable energies, solar and wind, to meet the base load needs of customers. Currently, the Eco-Gen systems are being sold all over the world for on-site power using power purchase agreements to finance them (Eco-Gen, 2014).

Flywheels also power transportation systems. In Switzerland, in the 1940s, OC Oerlikon created flywheel-powered buses called gyrobuses. In England, Parry People Movers, Ltd., created a railcar powered by a flywheel for the Stourbridge Town Branch Line. It went operational in 2010 with two units (Stourbridge, 2010).

FES can even help regulate the line voltage for electrified railways. This will improve the acceleration of unmodified electric trains and increase the amount of energy recovered back to the line during regenerative braking, helping to keep costs down. Several large cities including London, New York, Lyon, and Tokyo have pilot FES projects.

Current FES systems have storage capacities comparable with batteries and faster discharge rates. Many are used to provide load leveling for large battery systems, such as uninterruptible power supply systems for data centers, where they save considerable amount of space when compared with battery systems. On average, flywheel maintenance runs about half the cost of traditional battery systems. All that is needed is a basic annual preventive routine that includes replacing the bearings every 5 to 10 years. Newer flywheel systems completely levitate the spinning mass using maintenance-free magnetic bearings, thus eliminating mechanical bearing maintenance and preventing failures.

Thermal energy storage (TES) is often overlooked as an electricity storage technology, because it does not store or discharge electricity directly. However, in some applications, it will provide efficiencies that exceed other storage technologies. TES can store thermal energy from the sun, which is later converted into electricity. Naturally occurring molten-salt formations are the low-cost storage mediums. The main limitation is that TES is usually tied to a specific application, such as *concentrating solar power* (CSP), which has the challenges of high cost and limited locations, mostly in southwest

United States. Current research includes developing storage materials with higher working temperatures that, combined with higher temperature CSP plants, will increase efficiency and decrease costs. Spain has implemented CSP with thermal energy storage systems.

Another application of TES is hot and cold storage in buildings. Cold storage used to reduce peak demand from air-conditioning has been used on a relatively large scale. This is a commercially mature technology that provides system capacity at very high round-trip efficiency, with the capability of providing multiple grid services.

There are two types of *batteries* that have electric grid applications—liquid electrolyte batteries and high-temperature batteries. High-temperature sodium–sulfur batteries are the most mature and commercially available, though manufactured by a single Japanese company. There are over 270 MW in use worldwide, including US installations. They also have the advantage of relying on low-cost and abundant materials, although manufacturing costs have limited larger-scale use.

Energy storage plays a critical part in transportation. Hybrid or all-electric vehicles depend on batteries to store electricity and regulate power flow. The number of plug-in electric cars sold last year almost doubled from the previous year to 96,702 (Electric Drive Transportation Association, 2014). Many of these new sales are the result of Tesla Motors' remarkable Model S sedan. Tesla uses a different lithium-ion battery pack design that is superior to most.

In the spring of 2014, Tesla announced that it would build a giant battery factory in southwest United States to feed a steady stream of lithium-ion batteries for its electric cars and other electric applications (Tesla GigaFactory, 2014). Calling it a "Gigafactory," Tesla said that it would cost about $5 billion and will open in 2017 and employ about 6,500 people. It would reach full capacity by 2020, enough to supply power for 500,000 cars a year (Tesla GigaFactory, 2014).

"The Gigafactory is designed to reduce cell costs much faster than the status quo and, by 2020, produce more lithium-ion batteries annually than were produced worldwide in 2013," Tesla said (Tesla GigaFactory, 2014). The price of the batteries is expected to drop considerably, making a low-cost all-electric midrange sedans accessible to the mass market.

In addition to its own use, the plant could supply batteries to other car-makers and for other uses. Tesla also announced plans to sell battery packs to companies to use for emergency backup power storage. Tesla also said that the Gigafactory will be heavily powered by renewables, wind and solar.

While Panasonic is partnered with Tesla's battery development, a speculation is that Apple Inc., the maker of computers and smartphones, may join Tesla in the project (Tesla Partners, 2014).

While Tesla's Gigafactory may be a game changer in the lithium-ion battery industry, there are other high-temperature battery chemistries under various stages of research, development, and commercialization (Tesla Partners, 2014). Flow batteries are in the early stages of development and commercialization, with a few US demonstration projects of vanadium and zinc–bromine technologies, with several other technologies under development.

The zinc–air fuel cell (ZAFC) technology was developed in the 1990s (Clark et al., 2002). The original research and commercialization project for this battery was never implemented. Later, the original project scientists linked up with international groups that saw value in the technology for transportation and as a replacement for batteries (Economist, 2014). Zinc–air fuel cell in Montana is now commercializing the technology (Annon, 2012). The ZAFC will be a viable cost-effective and environmentally neutral battery for vehicles and buildings before the end of the decade.

V2G POWER STORAGE IN ELECTRIC CARS

In the United States, the University of Delaware is hosting a pilot project that pays electric car owners to store energy for the grid. "Cash-Back Cars" involves seven electric cars that interact with the grid and receive monthly payments. Besides drawing electricity from the grid, the plugged-in cars send electricity stored in their batteries back into the grid as needed, acting like tiny power plants.

Because the grid needs short-, medium-, and longer-term storage to run smoothly, car batteries could meet the short-term need, in a process called frequency regulation. The pilot project actually provides better frequency regulation than traditional methods (Clark and Bradshaw, 2004).

The US Federal Energy Regulatory Commission (FERC) is supporting the project and proposed a mechanism that will pay participating car owners on a national level. Renewable energy, particularly wind energy, is a good match for car battery storage, because with wind power, they mostly charge at night, when people are sleeping, and wind tends to peak at night. Right now, because demand plunges at night, wind generated at night is of lower value or sometimes even dumped.

To work, electric cars need to be retrofitted with a bidirectional power system, so energy can flow both from and to the grid, and software to allow the car to communicate with the grid. FERC's support should encourage automakers to make these features the norm in future models. Because most regional transmission systems have to work with multiple cars, it is expected that auto fleets will be the first to participate. The U.S. Department of Defense, with its fleet of 200,000 autos, is considering early participation.

The concept may become international. NUVVE, a company with offices in San Diego and Copenhagen, has stepped in as an aggregator outside the United States. The company plans to have the program running in Denmark, Hong Kong, and Taiwan by the end of 2014. Soon after, it plans to move into Germany, the Netherlands, Spain, and the United Kingdom—places where wind generation is taking a more important role because more wind energy means a high need for regulation. According to NUVVE, the market could be huge. The company says that by 2020, it's projected to be $12 billion worldwide, with $9.5 billion outside of the United States.

Much of the drawback to mass-scale battery storage is how to scale them up in size and make them cheaper. Scientists from Harvard University announced in March 2014 a technological breakthrough that may make giant, cheap batteries a reality. The scientists, Drs. Huskinson and Marshak, have developed a new, more durable and cheaper chemical mix for a flow battery (Huskinson et al., 2014).

Flow batteries use two liquids, each in contact with an electrode and separated by a membrane that is permeable to hydrogen ions. Jointly, they store the energy put into the battery when it is charged. They do this with chemical reactions that push ions through the membrane without the liquids coming into direct contact. Later, the liquids can release the stored energy by transferring hydrogen back through the membrane. Electrons are generated and flow through an outer circuit to light a lamp, or do other useful work.

Since the media that stores the energy is liquid, they can be pumped into large containers, or even holding tanks. They could in principle harbor huge amounts of electricity. However, they are expensive, needing the salts of expensive metals like vanadium, with coatings from platinum or other rare catalytic metals.

The solution, argued by Huskinson and Marshak in *Nature* (January, 2014: 195-198), is to use quinone molecules, mixed with sulfuric acid on one side of the membrane and a mixture of bromine and hydrobromic acid on the other side. The electrodes, they suggested, will be made out of carbon. The advantage is that this type of battery could be made out of organic molecules instead of rare metals.

While not in production yet, this type of battery has enormous potential. Cost for batteries using these organic molecules would be around $30 per kWh, compared with about $80 for conventional metal molecules, according to the scientists (Forbes Electric Cars, 2011).

HYDROGEN: A BREAKTHROUGH TECHNOLOGY

Hydrogen and electricity-derived fuel cells have an enormous potential when used with renewable energy sources like water, wind, and sun. In a hydrogen fuel cell, electricity is created through reactions between a fuel and an oxidant, triggered in the presence of an electrolyte. Unlike a conventional battery, fuel cells consume reactant from an external source, which must be replaced but lasts considerably longer. They are also more likely to be environmentally sound, in terms of their manufacturing and disposal. The reactants flow into the cell, and the reaction products are separated and flow out of it, while the electrolyte remains.

An electric current is produced that can be directed outside the cell to do work, such as powering an electric motor or illuminating a light bulb. Fuel cells can power an engine or an entire city. Because of the way electricity behaves, the current returns to the fuel cell, completing an electric circuit. There are several kinds of fuel cells, and each operates a bit differently. But, in general terms, hydrogen atoms enter a fuel cell at the anode, where a chemical reaction strips the atoms of their electrons. The hydrogen atoms are now "ionized" and carry a positive electric charge. The negatively charged electrons provide the current through wires to do work (Clark et al., 2002).

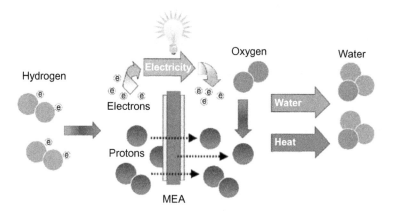

As shown above, oxygen enters the fuel cell at the cathode, where it combines with electrons returning from the electric circuit and hydrogen ions that have traveled through the electrolyte from the anode. Whether they combine at anode or cathode, hydrogen and oxygen combine to form water, which drains from the cell. As long as a fuel cell is supplied with hydrogen and oxygen, it will generate electricity. Even better, since fuel cells create electricity chemically rather than by combustion, they are not subject to the thermodynamic laws that limit a conventional power plant. This makes them more efficient in extracting energy from a fuel. Waste heat from some cells can also be harnessed, boosting system efficiency still further.

The basic workings of a fuel cell are not complicated, but building inexpensive, efficient, reliable fuel cells has proved difficult. Scientists and inventors have designed many different types and sizes of fuel cells in the search for greater efficiency. The choices available to fuel cell developers are constrained by the choice of electrolyte. Today, the main electrolyte types are alkali, molten carbonate, phosphoric acid, proton exchange membrane (PEM), and solid oxide. The first three are liquid electrolytes; the last two are solids.

The type of fuel also depends on the electrolyte. Some cells need pure hydrogen and therefore demand extra equipment such as a "reformer" to purify the fuel. Other cells can tolerate some impurities but need higher temperatures to run efficiently. Liquid electrolytes circulate in some cells, which requires a pump. The type of electrolyte also dictates a cell's operating temperature. For example, molten carbonate cells run hot, as the name implies. Each type of fuel cell has advantages and drawbacks when compared with the others, and none is yet cheap and efficient enough to widely replace traditional ways of generating power.

Some fuel cells depend upon fossil fuels, greatly reducing their value as a GIR technology. In 2010, the US television news show 60 Minutes featured Bloom Energy's fuel cell, which uses natural gas as a fuel source (Bloom, 2010). While Bloom Energy's fuel cell has low-carbon emissions and high efficiencies, it still uses natural gas—a fossil fuel with emissions and particulates. The Bloom Energy makes a 100 kW solid oxide fuel cell that sells for about $700,000. After incentives, Bloom Energy claims its server generates power for 9-11 cents per kWh, a calculation that includes fuel, maintenance, and hardware expenses. Bloom Energy is known for its customer base of high-tech and environmentally sensitive corporate clients like Wal-Mart, Google, and FedEx (Bloom, 2010). Yet, because Bloom Energy's fuel cells use natural gas to generate electricity, they are a fossil fuel-based system.

The real goal in the fuel cell industry is to perfect a hydrogen-based fuel cell to take the place of the gasoline-based internal combustion engine. Hydrogen fuel cells use hydrogen as the fuel and oxygen as the oxidant. For more than three decades, the US Department of Energy (DOE) national research labs have been investigating how to use hydrogen fuel cells for transportation, industry, and homes.

Most hydrogen cars use fuel cells to generate electricity and electric motors to power the car. A few use internal combustion engines modified to accept hydrogen and burn it as fuel, and some use a hydrogen compound to generate hydrogen-on-demand to power the vehicle. Significant funding is being poured into hydrogen fuel cell research, as this is seen as the ultimate in green-car technology.

Every major car company has developed a hydrogen fuel cell vehicle and they are all planning to market them, starting in Europe and Japan. In the United States, at least nine major car manufacturers are introducing hydrogen fuel cell vehicles in 2015, under lease agreements rather than for sale. This strategy is not only partly to control the market but also to monitor and measure performance and focus attention on the need for refueling stations. Within the next decade, these cars will be sold and there will be refueling stations in the homes of the hydrogen fuel car owners.

Hydrogen offers the promise of a zero-emission engine, where the only by-product created is a small amount of environmentally friendly water vapor. Current fossil fuel-burning vehicles emit pollutants such as carbon dioxide, carbon monoxide, nitrous oxide, ozone, and microscopic particulate matter. Hybrids and other green cars address these issues to a large extent, but hydrogen cars are the only ones that don't produce a single pollutant. The US Environmental Protection Agency estimates that fossil fuel automobiles emit 1.5 billion tons of greenhouse gases into the atmosphere each year (HAA, 2014). Switching to hydrogen fuel-based transportation would eliminate this serious cause of climate change.

Roughly speaking, the energy output from one kilogram of hydrogen is equivalent to the energy output of one gallon of gasoline. However, the production cost of the one kilogram of hydrogen is considerably lower. Typically, a gasoline internal combustion engine with a mechanical drive train is 15-20 percent efficient, while one powered with hydrogen is about 25 percent efficient. Hydrogen fuel cell vehicles with electric hybrid drive trains can be up to 55 percent efficient, or about three times better than today's gasoline-fueled engines. Because the production of hydrogen (by steam reformation of natural gas or electrolysis of water) is expected to be about 75-85 percent efficient, the net energy efficiency of hydrogen fuel cell vehicles will still be more than twice that of gasoline vehicles. The delivered price of hydrogen is approximately $3 per kilogram.

Honda released the first commercial hydrogen car in 2005. In 2008, the company created the first production-line hydrogen fuel cell car. The Honda FCX Clarity is powered by a 100 kW V Flow fuel cell stack, a lithium-ion battery pack (50 percent smaller than the one on the previous FCX), a 95 kW electric motor, and 5,000 psi (pounds per square inch) compressed hydrogen gas storage tank that yields a range of 270 miles. In 2011, the Honda FCX Clarity was available for lease only to customers in southern California. Customers need to live near one of the active hydrogen fueling stations in Torrance, Santa Monica, Culver City, or Irvine.

In early 2012, Daimler leased its Mercedes hydrogen fuel cell car in southern California and now needs refueling stations. The California Air Resources Board along with the California Energy Commission has begun a statewide bidding process for these stations, which will be implemented in 2016. The map below shows from the California Hydrogen Refueling Stations Program (CEC, 2014).

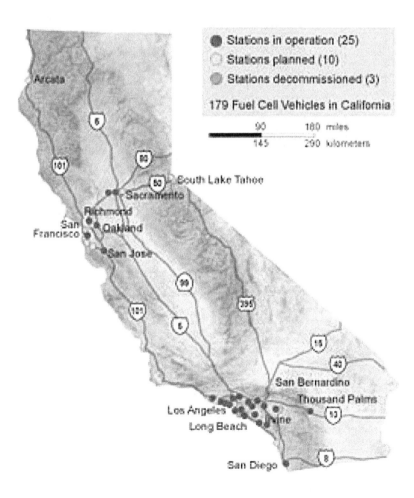

Germany already has a hydrogen highway. Other EU countries have similar plans. With California expecting to get the car manufacturers producing hydrogen fuel cars for its market by 2016, the hydrogen refueling stations are now being partly funded with California state funds from both the California Air Resources Board and the California Energy Commission (California Energy Commission (CEC), 2013). Other states and nations are tracking the California situation, strategies, public policy, and finances (HAA, 2014).

Below is a map of Germany with its hydrogen refueling stations already completed and in operation along with 50 new stations proposed by 2020 (Daimler et al., 2012) called the H2 Mobility plan.

By the end of 2013, estimates of stations grew to 100 with some predictions for 400 by the year 2023 (Plan, 2013). There are dozens of other GIR hydrogen auto prototypes being road-tested throughout the world. In fact, Sweden and Norway are working to construct a hydrogen highway system, complete with fuel cell cars and hydrogen refueling stations. A public–private partnership called Hydrogen Sweden is promoting hydrogen as a green energy carrier for cars and is working to develop a public refueling infrastructure. Hydrogen Sweden, founded in 2007, is a nonprofit organization that currently has 40 members including Honda, BMW, Volvo, Statoil-Hydro, H2 Solution, Air Liquide, and Arise Windpower (Sweden, 2014).

Hydrogen should be produced from renewable resources. Using biomass is one method, but the process emits some carbon dioxide. Hydrogen can also be derived by using wind, hydroelectric, or solar power to electrolyze water. Today, electrolysis is still expensive, but companies in Canada and Norway predict rapidly declining costs. Other companies around the world are developing hydrogen systems for fixed energy generation facilities, such as power plants. Declining production costs through lower-cost electrolyzers and the use of low-cost, off-peak, renewable electricity could dramatically reduce the future cost of electrolytic hydrogen.

Elemental hydrogen has been widely discussed as a possible carrier of energy on an economic scale. Used in transportation, hydrogen would burn relatively cleanly, with some NO_x emissions, but without carbon emissions. The infrastructure costs associated with full conversion to a hydrogen economy would be substantial. However, if refueling was done in the home or workplace using water or other renewable sources, electrolyzers could produce the hydrogen needed for fuel cells.

Today, industrial production of hydrogen is mainly from the steam reforming of natural gas and, less often, from more energy-intensive hydrogen production methods, such as the electrolysis of water. However, hydrogen from electrolysis and other renewable energy sources is gaining momentum. Many GIR countries around the world are making plans to provide it commercially for vehicles in 2015 in line with the global marketing timetable plan developed by the major automakers.

Fuel cell technologies, particularly hydrogen fuel cells, are an integral part of the GIR; they offer an exceptional alternative to oil dependency and fossil fuels. Scientists and manufacturers have a lot of work to do before fuel cells become a practical alternative to current energy production methods, but with worldwide support and cooperation, a viable hydrogen fuel cell-based energy system may be a reality.

REFERENCES

Confidential Information Agreement with Dr. Woodrow Clark, November 2, 2012.

Bloom, 2010: http://www.bloomenergy.com/fuel-cell/solid-oxide/.

California Energy Commission (CEC), 2013. In: Program Opportunity Notice, Alternative and renewable Fuel and Vehicle Technology Program. Subject Area: Hydrogen refueling Infrastructure, PON – 13–607, November 2013 (decisions April 2014). http://www.energy.ca.gov/contracts/index.html.

California Energy Commission (CEC) Hydrogen Refueling, 2014: www.arb.ca.gov/hydrogen.

Clark II, Woodrow W., Bradshaw, Ted, 2004. Agile Energy Systems: Global Solutions to the California Energy Crisis. Elsevier Press.

Clark II, Woodrow W., Isherwood, William, 2007. "Energy Infrastructure for Inner Mongolia Autonomous Region: five nation comparative case studies", Asian Development Bank, Manila, PRC National Government, Beijing.

Clark II, Woodrow W., Isherwood, William, 2010. Creating an Energy Base for Inner Mongolia, China: the leapfrog into the climate neutral future. Utilities Policy Journal.

Clark II, Woodrow W., Paulocci, Emilio, Cooper, John, 2002. Commercial Development of energy–environmentally sound technologies for the auto-industry: the case of fuel cells. J. Clean. Prod. 11, 427–437.

CREIA (Chinese Renewable Energy Industries Association) Spring 2011.

Daimler, Press Release: "50 hydrogen filling stations for Germany: Federal Ministry of Transportation and industrial partners build nationwide network of filling stations", Berlin, June 2012.

Electric Drive Transportation Association, 2014: http://www.electricdrive.org/index.php?ht=d/sp/i/2324/pid/2324.

Eco-Gen Flywheel Company 2014 at: eco-genenergy.com.

Economist, March 8 2014. Going with the flow. The Economist Technology Quarterly, pg. 4.

Forbes Electric Cars 2011 http://www.forbes.com/sites/ericagies/2011/06/22/the-cash-back-car-monetizing-electric-vehicles/.

Huskinson, Brian, Marshak, Michael, Suh, Changwon, Er, Süleyman, Gerhardt, Michael R., Galvin, Cooper J., Chen, Xudong, Aspuru-Guzik, Alán, Gordon, Roy G., Aziz, Michael J., January, 2014. A metal-free organic–inorganic aqueous flow battery". Nature 505, 195–198.

Hydrogen Association of America (HAA), 2014: http://www.hydrogencarsnow.com.

Lawrence Livermore National Laboratory (LLNL) Flywheels. 2014: www.llnl.gov.

Parfomak, Paul W. "Energy Storage for power grids and electric transportation: A technology assessment (Congressional Research Service, 2013) http://www.fas.org/sgp/crs/misc/R42455.pdf.

Mobility Plan, October1, 2013, "15 Stations in place" predicts are for 400 by 2023: www.fuelcelltoday.com.

Stourbridge, 2010: http://www.stourbridgenews.co.uk/news/8362831.Revolutionary_tram_notches_up_500_000_passengers/.

Hydrogen Sweden, 2014: http://www.hydrogensweden.com.

Tesla GigaFactory, 2014 http://blogs.marketwatch.com/energy-ticker/2014/02/26/teslas-gigafactory-what-elon-musk-didnt-say/.

Tesla Partners, 2014: http://www.usatoday.com/story/money/cars/2014/02/22/tesla-gigafactory/5706889/.

CHAPTER 8

Smart, Green Grids*

Every day, all over the globe, people flip on a light switch, turn on a TV, or warm a meal in a microwave. It is an unconscious behavior that requires no forethought, yet it triggers the flow of electricity that overlays modernity and defines conventional life.

To reach a house, electricity must first flow across hundreds of miles of power lines and pass along a series of stations and substations. A technical marvel of the fossil fuel era, this electrical transmission network ties a central power plant through transmission lines to a local distribution grid and then to an end user.

Thomas Edison built the first central power plant built in the United States in New York City. September 4, 1882, Edison flipped the switch that set another standard for the second industrial revolution. Steam engines spun the dynamos and the Pearl Street Station in Manhattan started generating electricity. Initially, it lighted 400 lamps at 85 customers, and 2 years later, it served 508 customers with 10,164 lamps (Josephson, 1959).

The conventional central power plants that emerged and came to dominate the 2IR burned coal, oil, or natural gas or were driven by hydroelectric power from dams on rivers that were hundreds of miles away. The power plants were housed in large concrete structures with massive silos and pipes surrounded by high wire fences and security guards. The New Jersey Turnpike, leading into New York City, is lined with these massive plants, their huge conical exhaust stacks billowing out toxic emissions. Most central power plants were built on a similar concept—fossil fuel is burned to create heat that drives a steam turbine, which generates electrical power.

Electricity is produced in different forms. Large electrical generators produce what is called three-phase alternating current (AC) power, as opposed to direct current (DC) power. Transmitting electricity along the power grid requires additional steps. Electricity leaves the generator and enters a transmission substation. Large transformers then boost the voltage for long-distance travel along transmission lines. There is a line for each phase of the three-phase AC electricity, plus ground lines. The transmission lines

* Drs. A.J. Jin and Wenbo Peng contributed to the section on China's grids.

are usually held aloft by huge steel structures and a typical transmission distance is 300 miles.

Electricity must then be distributed, or shared, with the end users. The electricity produced by these older systems is so powerful that it must be stepped down, or reduced, along the distribution grid to provide the power connected to the household switch. The conversion from power "transmission" to power "distribution" takes place through a power substation. The substation uses transformers to break the electricity down and at the same time uses a "bus" to split the power off in multiple directions.

From the distribution bus, it goes along wires through regulator banks and finally to a tap, bringing electricity to consumers. A key flaw is that the amount of electricity used by the consumer is about 30-40 percent of that generated at the power plant. The transmission over hundreds of miles dramatically reduces the amount that a consumer gets to use. It also degrades the environment and is subject to disruption by storms or grid failures.

The dependency on the central power plant is one of the key reasons for the GIR. In the transition to the GIR, renewable energy must replace fossil fuel to generate electricity at the central power plant, but it can so be used to create on-site power for buildings as well. This distributed power will mitigate global warming and preserve the environment. To get the most out of renewable energy, there needs to be a transition from the old central power grid with its overlapping lines to a local on-site power grid and distribution network that is technologically smart. The need for a smart grid to distribute large amounts of electricity and meet the on-site needs as well is fundamental to developing the modern energy networks. The end result will be a combination of central and on-site power—this combination is called an agile (meaning "integrating") energy system because it combines both energy sources (Clark and Bradshaw, 2004).

The term "smart grid" started to appear at the end of the 20th century during the dot-com era. Basically, it refers to using digital or information technology to control and enhance the electricity networks that form a power grid. Originally, these systems used landlines, such as telephone wires, but today, the systems are primarily wireless.

A smart grid starts with a smart meter, usually installed at the end users building or home. The smart meters enable customers to provide immediate feedback to utilities that are able to use the data to set pricing and smooth fluctuations in consumption. Depending on the size and type, other energy control and monitoring devices, software, networking, and communication systems are installed along the electricity distribution network. Combined, these elements form the grid's nervous system, which allows energy managers and end users to monitor and control consumption in real time. Smart grids use transmission or pipeline systems covering miles from a central power plant to the customer. Or conversely, it can work on a micro level within a building or small community.

The smart grid with its digital communications ability can interact with smart appliances that can turn themselves on and off as part of a sophisticated energy management system. The smart technology would allow the grid to support a fleet of electric cars as well as the buildings where they are parked and being recharged. The smart grid would vastly improve the efficiency of transmission power and lower the cost of electricity.

A smart integrated system merges Internet and grid features with power sources, the data response, and a load center such as a residential home. A smart meter collects power usage data for the utilities and consumers, and it has Internet communication and the capability of using the digital cloud. These technologies are being incorporated in numerous companies including Facebook, Google, Twitter, and Apple. One company has created a cell phone application that can measure and evaluate products for their environmental, health, and production costs when shopping (Earth Accounting, 2014). Real-time data are fed back to a large distribution and transmission power grid that will enable efficient overall load management. Energy storage is extremely important as part of a smart grid because it allows for the load to be leveled, or optimized, between major power activities and a load center. For example, power generation could be from solar or a charge for an electric car battery. Real-time data are critical to predict and hedge power usage.

The smart grid is ideally suited to meet the challenging demands of sharing electricity generated from multiple renewable sources. With the cost of fossil fuels rising from depleting resources along with the need to mitigate GHGs emissions, customers will need green energy. The challenge for engineers and regulators will be how to construct a smart grid system off of a century-old infrastructure, connect it to numerous renewable energy sources, and manage it all in real time for conservation and efficiency.

Solar PV and wind power may have significant power output, but they must be managed with load leveling suited to their output. Moreover, power may be wasted since the traditional power plants do not shut down when the consumer is asleep.

EUROPE'S PARALLEL LINES

Europe, and particularly Germany, is facing a major problem with their grid structures. Germany's decision to close down its aging nuclear reactors leaves areas of the country with insufficient electricity supplies. Bringing power from elsewhere means rebuilding transmission lines, which is not popular. Given Russia's 2014 incursion into the Ukraine, which threatened the natural gas pipelines to Germany, the concern about power and where it comes from is critical.

Amprion and TransnetBW, two German electricity-transmission firms, are experimenting with a system that will run two parallel lines together. One of the cables will carry AC current and the other will carry DC current. At the start, engineers feared that the AC/DC cables would interfere with each other, causing capacitance (that is, accidental storage that would disrupt the flow), induction (e.g., potential changes in voltage), or resistance (e.g., reducing the speed of the flow).

Experiments done in March 2014 suggest that none of these are major problems (Economist, 2014). It seems that the only required changes will be upgraded insulators that control the voltage as it passes through the cables.

This discovery has far greater impact than the solution of a local problem. It opens up the potential to move a lot more electricity around as DC and increase the grid capacity, in what is called the ultranet. This should reduce the need to construct new lines across Europe and help bring expanding wind- and solar-generated power to the cities. These connections would help balance supply and smooth the variability from renewable energy.

In Europe as most other parts of the world, grids are aging and in need of repair and subject to weather disruptions and accidents. Germany's experiment with a parallel cable system points argues that there is a practical way to make the existing transmission lines more efficient. It will go a long way in helping accelerate the transition to electricity generated by renewable energy sources.

Transforming the energy market requires an advanced grid infrastructure with superior energy efficiency and green technology. A smart grid is becoming increasingly important with the growing use of solar, wind, and other renewable energy. A careful upgrade to the old power grid should begin with rebuilding the backbone with a new system.

Many universities are now focusing on the microgrid and as well as nano-technologies. For example, the University of California, Los Angeles (UCLA), has a program called SMERC, which holds quarterly meetings on the smart grid technologies (UCLA, 2014). Fortunately, the enormous opportunities that would be created by the transformation have caught the attention of major investors and businesses. Replacing the world's old grids with new smart grids represents a huge business opportunity. Investors will realize that they can make a substantial social contribution while generating a good return on the investment.

CHINA'S LEADING SMART GRID

China is the world's largest consumer of electricity and demand is expected to double over the next decade and triple by 2035. In 2010, coal generation accounted for 70 percent of China's electricity generation, but now, the Chinese government is investing heavily in renewable energy technologies. Ultimately, China wants to dominate the clean energy technology market worldwide. As of 2012, 17 percent of China's electricity generation came from renewable sources, and their latest goal is to increase renewable energy to 9.5 percent of overall consumption by 2015. To implement China's new clean energy capacity into the national power grid requires upgrades and, ultimately, a national smart grid (Wall Street Journal, 2011).

China's power industry began in 1882 with the birth of the Shanghai Electric Power Company and the completion of the first-generation grid. By 1949, the nation's installed capacity reached a modest level of 1.85 GW. In 1970s, construction of a second-generation grid began with the aim of creating a national grid. The Northwest Power Grid's 750 kV transmission line was put into operation in 2005, and China's first 1000 kV UHV transmission lines were built in 2009. As of July 2010, China's 220 kV or above transmission lines were over 375,000 km in length, the largest in the world.

The Chinese Twelfth (12th) Five-Year Plan enacted in March 2011 includes the development of national renewable energy systems along with a smart grid as a key part of the nation's power system. The plan has support from the nation's large government stimulus package. Rapid development of smart grid systems in China will show significant social and economic benefits (Lo, 2011).

China's smart grid development is extraordinary, according to Bloomberg New Energy Finance. China spent $4.3 billion in 2013 on smart grids, almost a third of the world's total. Conversely, North America's spending declined as much as 33 percent. The Chinese-style power grid is trying to achieve the integration of power, information, and business flow, with a strong and secure grid. For example, China has installed over 250 million smart meters, planned for smart charging stations and networks to serve electric vehicles (with thousands of charging piles completed), and built the world largest wind and solar energy generation and storage system.

China's Zhangbei power station is the world's largest hybrid green power station. Built as a demonstration project for China's ambitious smart grid system, the power plant was commissioned December 2011, in Hebei province. With an initial investment of $500 million, the power station combines 140 MW of renewable wind and solar energy generation with 36 MWh of energy storage and smart power transmission technologies.

The Zhangbei project is a success. It uses a battery storage system to enhance the renewable power generation. The wind turbine production rate has been increased by 5-10 percent, and its whole renewable energy efficiency improved by 5-10 percent. During its first 100 days of safe and stable operation, the power station generated over 100 GWH. Overflow or excess energy that is generated is fed back into the utility grid after the storage has been filled. The Zhangbei project is an excellent example for the renewable energy solution for China and elsewhere.

Another high-profile demonstration project is the China Huaneng Group's microgrid in the Future Science and Technology City, Beijing (Bloomberg, 2014). A smart microgrid with 50 kW of PV was set up in 2012. This project is being hailed as a model for constructing a scalable smart grid. It is the first smart microgrid power system and indicates that the company has begun to enter the field of distributed microgrid power generation. Based on a microgrid controller, the system integrates its PV power, with 300 VAh of energy storage, grid power, and 30 kW load. Under normal circumstances, the load is completely powered by the PV modules.

When PV power decreases, the controller uses battery energy for base load demands. Under the extreme cases that the DC energy is too small to meet the base load, the controller can switch the electricity supply to grid power within 8 milliseconds, to ensure a stable power supply.

These large-scale smart grids are safe, reliable, and stable while accommodating production from renewable energy sources. China has a very uneven distribution of electricity production and demand, which requires that the nation pay more attention to smart transmission grids. So far, the Chinese smart green grid has handled the world's largest wind, solar, and energy storage integrated demonstration project.

The 12th Five-Year development plan for smart grids by the Chinese State Grid has the following goals:

- *Generation link*: Grid to meet the demand of 60 GW wind generation and 5 GW PV generation by 2015 and 100 GW wind generation and 20 GW PV generation by 2020. The capability is over 400 GW.
- *Transmission link*: The next 5-years construction will connect large-scale energy bases and major load centers that will build "3 vertical 3 horizontal" backbone for the EHV grid. This will provide a high-level transmission smart grid and transmission line availability factor of 99.6 percent and a "5 vertical 6 horizontal" backbone EHV grid will be built by 2020.
- *Transformer link functions at high voltages*: 6100 smart substations above 110 (66) kV should be completed by 2015, which will account for about 38 percent of the total number of China's substations. 110(66) kV or above smart substations will account for about 65 percent of the total substations in 2020.

China with its five-year plans does something unique, which is not seen in other countries; it puts financing into the plans to make them work and become viable. For example, the last two five-year plans set the stage for the full-scale construction and improvement of China's smart grid. Now in accordance with the Chinese latest five-year plan, the development from year 2011 to 2015, for the smart grid construction, investment amount will be over $300 billion and the total investment will reach $600 billion by 2020 (Lo, 2011).

Meanwhile, the national smart grid investment funds will be multiplied 10-fold. According to the China Huaneng Group's (2014) analysis for the central government, the national smart grid plan will focus on renewable energy systems. The range will be extensive and new industries will be added so that the market size is extremely attractive.

Moreover, smart grid construction provides a huge benefit. Specifically, by 2020, the benefits will be the following:

- The power generation benefits will be around $5.5 billion, saving the system effective capacity investment and reducing power generation costs by 1-1.5 cents/kWh.
- Grid link benefit will be about $3.2 billion; grid loss will be reduced by 7 billion kWh, and the maximum peak load decreased by 3.8 percent.
- User benefit will be about $5.1 billion, including new offerings of a variety of services, saving 44.5 billion kWh of electricity.
- The environmental benefits will be about $7 billion, conservation of land about 2000 acres per year, emission reduction of SO_2 about

1 million tons, and CO_2 emission reduction of approximately 250 million tons.

- Other social benefits will be about $9.2 billion, increasing employment opportunities of 145,000 per year, and significant electricity cost savings, and the promotion of balanced regional development.

China is committed to being the world's largest smart green grid user in the power industry. It is imperative to have a robust and low-cost smart grid that can accommodate clean renewable energy. In fact, China's total installation reached 1 TW by the end of year 2011 with an annual total electricity consumption of 4.7 trillion kWh. The grid-connected new energy power generation capacity reached 51.6 GW (of which 45.1 GW of wind power, accounting for 4.27 percent of the total installed capacity; grid-connected solar PV capacity of 2.1 GW, accounting for 0.2%; and biomass-installed power capacity of 4.4 GW, accounting for 0.4 percent), geothermal power generation capacity of 24 MW, and ocean energy power generation capacity of 6 MW (China Huaneng Group's, 2014).

The attributes of a Chinese smart green grid or third-generation grid are substantial and include being the following:

- Strong, robust, and flexible.
- *Clean and green*: the smart grid makes the large-scale use of clean energy possible.
- *Transparent*: the information of grid, power, and user is transparently shared, and grid is nondiscriminatory.
- *Efficient*: it improves the transmission efficiency, reduces operating costs, and promotes the efficient use of energy resources and electricity assets.
- *Good interface*: it is compatible with various types of power and users and promotes the generation companies and users to actively participate in the grid regulation.

China's successful integration of renewable energy into smart grid systems is a prime example of what can and should be done (Clark and Isherwood, 2010). Solar, wind, and other on-site distributed energy resources are becoming common and evermore important to our electricity hungry lifestyles. Smart green grid systems will optimize these resources and make the way we share and distribute electricity simple and efficient.

The results are critical. China is now measuring and evaluating the results in which they have found some issues that need to be resolved. A critical one is that the installation of wind and solar farms far from the energy consumers requires massive transmission, monitoring, and smart grid connectivity. That issue needs to be solved. In part, the resolution will be a focus on distributed or on-site power, which does not require massive transmission systems.

REFERENCES

Bloomberg, 2014. http://www.bloomberg.com/news/2014-02-18/china-spends-more-on-energy-efficiency-than-u-s-for-first-time.html.

China Huaneng Group, 2014. http://www.chng.com.cn/eng/.

Clark II, W.W., Bradshaw, T., 2004. Agile energy systems: global solutions to the California energy crisis. Elsevier Press, London, UK.

Clark II, W.W., Isherwood, W., 2010. Creating an energy base for inner Mongolia, China: the leapfrog into the climate neutral future. Utilities Policy Journal.

Earth Accounting, 2014. www.EarthAccounting.com.

Economist, 2014. Can parallel lines meet? The Economist Technology Quarterly (March 8), 6.

Josephson, M.E., 1959. McGraw Hill, New York, OCLC 485621, ISBN 0-07-033046-8, p. 255.

Lo, V., April 25, 2011. China's role in global economic development, Speech by Chairman of Shui on Land, given at Asian Society, Los Angeles, CA.

University of California, Los Angeles (UCLA), 2014. Smart Grid Conferences. SMERC, School of Electrical Engineering.

Wall Street Journal, 2011. China's energy consumption rises the Wall Street Journal. Wall Street Journal.

CHAPTER 9

Emerging Green Industrial Revolution Technologies

The green industrial revolution (GIR) is spawning truly remarkable technologies that are exponentially more stunning, numerous, and revolutionary as the inventions that emerged from the first industrial revolution and second industrial revolution (2IR). From Aalborg to Tokyo, and Beijing to California, amazing technologies—from tiny nanocrystals to 200 mph trains propelled by magnetic force—are being designed by scientists and engineers who are changing human history. A look at a few emerging green technologies will provide a sense of what can be done in the face of truly monumental societal, financial, and policy challenges.

As the twentieth century ended, the GIR started to emerge. The first modern renewable energy systems were wind farms and concentrated solar power installations for central plants, located long distances from the energy buyers and end users. Long transmission lines were a major problem with environmental concerns and high costs of transporting energy over power lines. These systems are inherently inefficient, losing over half the energy generated at the plant before reaching the end users (Lawrence Berkeley National Laboratory (LBNL), 2010). These systems served for decades, but they are an old technology not suited for the twenty-first century.

Today, these long-distance renewable systems are changing to on-site energy generation (Clark and Bradshaw, 2004). Incentives are being offered to consumers who install on-site solar and wind energy systems. Solar panels are being installed on rooftops, and wind farms are being built in large agricultural areas. Small wind turbines are even becoming part of building designs. This distributed power generation requires careful load balancing and creates the need for new storage technologies (NREL, 2014).

Government-led initiatives are encouraging European and Asian nations to research and develop innovative storage devices and to create systems that better integrate and share renewable energy. More pressure is being put on consumers to conserve and on manufacturers to create products that are more efficient. This helps drive the development of new technologies for lights, smart meters, and green grids to maximize energy use and efficiency.

Energy Star from the US Department of Energy took the lead in measuring and setting standards to reduce the use of energy for appliances and other building items (US EIA, 2013). The US Green Building Council's Leadership in Energy and Environmental Design (LEED) created domestic and international standards to measure buildings in the reduction of carbon pollution and greenhouse gases while promoting better energy management. In the initial stages, these programs focused on individual buildings but are slowly expanding to communities.

A REVOLUTION IN LIGHTING TECHNOLOGIES AND PEAK DEMAND RESPONSE

The United States uses about 17 percent of its total electric consumption to light facilities (US EIA, 2013). This includes exterior parking and street lamps (Nularis, 2014). For most of the twentieth century, lighting came from incandescent bulbs and fluorescent tubes.

Incandescent bulbs make light by heating a metal filament wire to a high temperature until it glows. As a legacy technology, incandescent light bulbs use a large amount of electricity and are being replaced by newer technologies that improve the ratio of visible light to heat generation. Europe, Brazil, and the United States are phasing out incandescent light bulbs in favor of more efficient lighting.

After World War II, fluorescent tubes gradually replaced incandescent bulbs in commercial office facilities and high-use residential areas. A fluorescent lamp or fluorescent tube is a gas-discharge lamp that uses electricity to excite mercury vapor. It is more efficient than an incandescent lamp in converting electric power into useful light. The fixture requires a ballast to regulate the current through the lamp. Fluorescent lights were first available to the public at the 1939 New York World's Fair. Improvements since then include better phosphors, longer life, more consistent internal discharge, and easier-to-use shapes. Eventually, more efficient electric ballasts started to replace the older magnetic ballasts, which eliminated the odd humming noise.

The lighting industry is quiet and low-key, so new technological advances come into the market place without great fanfare. However, the GIR's digital communications and Internet connectivity have stimulated a whole new generation of commercial lights that are energy-efficient, dimmable, and, most importantly, able to adjust to exterior daylight or high peak-load demands. Electrical and chemical engineers who are intent on providing the perfect lighting at all times are creating these transformative technologies. Optimal energy-efficient commercial lighting is a combination of building design and task-ambient lighting placement. The essence of this new generation of lighting is the Internet-connected dimmable ballast, working in combination with a high-output, low-wattage light-emitting diode, or LED lamp.

In the energy and utility industries, the goal is to provide just enough electricity to meet the end user's demand. In the 2IR system, the central power grid delivered a constant stream of electricity, whether the end user needs that much power or not. GIR efficiencies come from building end user equipment that can respond to the ebb and flow of consumer and grid demand, rather than providing a constant stream of electricity that may not be needed during nonintensive use periods. For example, why use a constant pulse of electricity to light an office that is flooded with daylight in the morning hours but shadowed in the afternoon? Ideally, the sun would be used for lighting in the morning, and electricity would be used in the afternoon.

The new generation of lighting will respond to these changes by using smart sensors on the window glass. The sensors will determine the light level and transmit the information through an Internet connection to dimmable ballast that will then provide just enough electricity to power the lights at the optimal level for worker comfort and productivity.

The move toward increased efficiency is not so much about workplace comfort as it is about utilities clinging to the remnants of their central power authority. The old central grid systems are nearing capacity in many major metropolitan areas in the developed world. New York City's grid is at capacity, which means more energy is not being generated and ever-increasing demand means greater and greater efficiency.

Throughout the world, major cities as well as emerging nations are struggling for economic growth and growth means more electricity. Additional electricity is needed to power new businesses and an increasing number of electronic devices (iPads, smart phones, computers, televisions, etc.), but the grids and generation systems are at their maximum load. The only way to get quick relief is through conservation and an energy efficiency program that reduces demand at a pace that is faster than the growth of new commerce.

For many reasons, utilities have difficulty responding to large swings in demand. For example, heat, ventilation, and air conditioning (HVAC) systems demand more power on hot summer afternoons. Demand also rises in the evening, when people return home after work. In high-humidity locations like Dallas, Texas, demand is fairly constant during a workday, when many people are in office buildings. Under these conditions, the grid delivers a steady stream of electricity. Then, around 5 p.m., workers begin to flee offices and head home—and in Texas, the homes are large. People walking into a large house at the hottest time of the day immediately turn their air-conditioning units to full blast throughout the house. Several million units surging at the same time sends the grid and power supply into overload. These demand extremes can cause power interruptions or failures. Utility companies go into a panic at this point because power interruptions or failures are expensive and time-consuming and focus consumer attention on their shortcomings.

The easiest way to prevent surges, or overload, is to keep everything in balance. So, when demand spikes, electricity is delivered, but not at such a high level. There comes a point when, to avoid major problems, electricity delivery has to be rationed, which means that some people have to get less. In this world, rationing doesn't work, so the world's utilities have been pushing what is called demand response, or "peak-load management," which offers the ability to reduce electricity use during extreme demand times.

Peak-load management tries to transfer some of the electricity use from high-demand periods to low-demand periods. Originally, peak-load strategies centered on offering utility rate discounts to customers willing to reduce their energy use during surge time periods, referred to as incidents. For example, a utility program might offer a large industrial customer a 5 percent rate discount if it reduces electricity use during days that are declared incident days. Incident days may happen 10-15 times a year in some areas. Once notified of an incident day, the customer would turn off lights or nonessential machinery, reduce the amount of air-conditioning, or take other measures to meet the agreed-upon reduction in electricity use. If the customer fails to meet the reduction, then a penalty is assessed.

Notification of an incident day was initially done by phone and then through an e-mail to the customer. Now, with the GIR's emerging technologies and smart devices, peak-load management is rapidly becoming more flexible and more effective. Utilities are offering deeper discounts to lure customers into letting the utilities connect directly to the customer's energy management system (EMS). The EMS is connected via the Internet to the facility's dimmable ballasts for the lights and the large central HVAC systems for air-conditioning. In this way, the utility can automatically reduce a customer's systems to meet the agreed-upon level.

Eventually, as the new technologies come into the marketplace, simple equipment like thermostats, refrigerators, washing machines, and dryers will have Internet connectivity and smart operating systems. They can then be controlled remotely, and the utilities will notify the machines to reduce demand during peak-load periods. While most people would consider this intrusive, this type of mass-marketplace energy management will go a long way toward optimizing and balancing energy use throughout a smart grid system that integrates renewable energy generation with other technologies.

LEDs are an extraordinary new generation of lighting. A 6 watt LED can provide the same amount of light as a standard 60 watt commercial overhead interior light. LEDs have a longer lifetime (measured in 8–10 years, instead of months), improved robustness, smaller size, faster switching, and greater durability and reliability. As a new technology, they are more expensive than traditional light bulbs, but the price is dropping quickly as new manufacturers come to market and cost-effective ways are developed to finance lighting projects (Nularis, 2014).

Technological advances in lighting may seem mundane compared to some of the other emerging technologies, but lighting impacts every office, home, and room in the modern world. It can even be life-changing. In a mud-walled shack in a rural village in an undeveloped part of the world, solar-powered lighting can offer a child a chance to study his or her lessons and slip the bonds of poverty.

COOL ROOFS WILL OFFSET CARBON

One uncomplicated technology that has a huge potential impact on mitigating climate change and reducing carbon emissions is "cool" roofs. Basically, a cool roof is a roofing coating or outside layer that is white or light in color, which will reflect the sun's rays. A black or dark roofing material absorbs the sun's heat, causing excess air-conditioning in warm climates. Usually, a cool roof is made from some sort of acrylic fluid or membrane material and is applied over a flat roof as new construction or a retrofit.

In hot or warm climates, cool roofs can have an enormous effect on reducing building heat and subsequently building cooling. Not only does it reduce the heat of an individual building, but also it reduces the amount of heat that is carried in the wind. This is particularly important in cities where ambient heat from dark rooftops often causes urban heat island effect.

A 2010 study from researchers at LBNL (2010) calculated that if 80 percent of the roofs in urban areas in the tropical and temperate climate zones were white or cool, it would offset 24 billion metric tons of carbon dioxide emissions. This is about the equivalent of removing the emissions of 300 million autos. The study also pointed out that cool roof technology would reduce energy consumption in over 90 percent of India.

The study also pointed out that the urban heat island effect would be greatly reduced if highway or roadway pavement was changed from a dark to a reflective color. The combination of white, reflective roofs and pavements would have a dramatic impact on mitigating excessive heat in cities, since half the world's population now lives in cities. By 2040, the number of people in cities is expected to reach 70 percent, adding urgency to reducing the urban heat island problem. The LBNL created a new center whose annual reports on "Global Model Confirms: Cool Roofs Can Offset Carbon Dioxide Emissions and Mitigate Global Warming" provide guidance and solutions to climate change from the local and on-site level of buildings (LBNL, 2010).

NANOTECHNOLOGY: "REALLY" SMALL THINGS

Nanotechnology is the study of really small things—substances at the sub-atomic or molecular levels. Like additive manufacturing, it is another marvel of the new age with unlimited potential. To grasp the concept, think of the smallest distance that can be formed between the thumb and finger without actually touching. This is about a millimeter. Divide that space by a million, and that is a nanometer.

A variety of disciplines—biologists, chemists, physicists, and engineers—are converging on this new field and studying substances at the nanoscale. Astoundingly, at the nanoscale, not all things behave normally. For example, a person cannot walk up to a wall and immediately teleport to the other side of it, but at the nanoscale, an electron can—it's called electron tunneling. The nanoworld is one where quantum mechanics plays a major role. The rules of quantum mechanics are very different from classical physics, which means that the behavior of substances at the nanoscale can sometimes contradict common sense by behaving erratically.

For example, substances that are insulators, meaning they cannot carry an electric charge, might become semiconductors when reduced to the nanoscale. Melting points can change due to an increase in surface area (Nanotechnology, 2014).

Scientists working in this field are coming up with major advances in biomedicine, renewable energy, light modification, and carbon capture, among many, many other applications. In Chapter 1, we mentioned the electrochromic nanotechnology from LBNL's Molecular Foundry. They have created electrochromic glass, or "smart windows," which when embedded with a thin coating of nanocrystals can permit visible sunlight into a room while deflecting its near-infrared heat or, conversely, block out visible light while allowing in heat. The change can be done with low-voltage current and a light switch (LBNL, 2014).

Another discovery, also from the Molecular Foundry, is a metal organic framework that is designed to capture carbon dioxide. Sort of like a high-tech sponge held together by metal joints, the system can be fitted inside a smokestack.

While most elements will pass through it, the system's little organic chambers capture and bind carbon dioxide. This technology could result in a practical carbon capture system that could have major implications for the coal industry.

Nanotechnology has invented many new building products. For example, a company called Nano-Architech makes advanced building materials like "nanocement," which is high-tech cement. Nanocement is far superior to normal stucco or concrete as a building skin. The product is lightweight, easy to work with, and strong and can be embedded with photovoltaic, LEDs, or sensors. Of particular importance to a world of increasing climate change is that the product is largely disaster-proof, offering superior building protection to extreme weather patterns.

A major breakthrough in LED technology was announced in spring 2014. Using nanotechnology, an international team working under grants from different nations has developed new thin, flexible LEDs. About three atoms thick, these are the thinnest-known LEDs that be used as a light source. The LED is based off of two-dimensional, flexible semiconductors, making it possible to stack them or use them for applications not allowed by current technology.

Besides light applications, the technology can be used as interconnects to run nanoscale computer chips instead of standard devices that operate off the movement of electrons. This would reduce heat and be far more efficient and allow for the development of highly integrated and energy-efficient devices in areas such as lighting, optical communication, and nano lasers (UW, 2014).

REGENERATION BRAKING: FROM TRAINS TO CARS TO TRAINS AND BACK AGAIN

When a vehicle uses its brakes to slow down, the reduction in speed creates kinetic energy. With conventional braking systems, excess kinetic energy is converted to heat by friction in the brake linings. As a result, the energy is wasted. Regenerative braking systems transform kinetic energy into another form of energy, which can be saved in a storage battery. This energy recovery mechanism is used on hybrid gas and electric automobiles to recoup most of the energy lost during braking. The stored energy is then used to power the motor whenever the car is in the electric mode. The most common form of regenerative braking is used in hybrid cars like the Toyota Prius and involves using an electric motor as an electric generator (Toyota, 2014).

Regenerative braking has emerged as a viable technology for electric railways. For railways, the generated electricity is fed back into the onboard energy supply system, rather than stored in a battery or bank of capacitors, as is done with hybrid electric vehicles (Toyota, 2014). Energy may also be stored using pneumatics, hydraulics, or the kinetic energy of a rotating flywheel (Lawrence Livermore National Laboratory (LLNL), 2014).

Regenerative braking systems have a long and interesting history. Louis Antoine Krieger used them in the late nineteenth and early twentieth centuries as front-wheel drive conversions for horse-drawn cabs. The Krieger electric landaulet had a drive motor in each front wheel, with a second set of parallel windings for regenerative braking. The Raworth system of regenerative control was introduced in England in the early 1900s. It offered tramway operators economic and operational benefits.

Tramcar motors worked as generators and brakes by slowing down the speed of the cars and keeping them under control on descending gradients. The tramcars also had wheel brakes and track slipper brakes, which could stop the tram, should the electric braking systems fail. Following a serious accident, an embargo was placed on this form of traction in 1911. Twenty years later, the regenerative braking system was reintroduced but failed to become the dominant technology because of the political and economic influences of the fossil fuel and internal combustion vehicle industries.

Regenerative braking has been in limited use on railways for many decades. For example, the Baku–Tbilisi–Batumi railway (Transcaucasia railway or Georgian railway) started using regenerative braking in the early 1930s. This technology was especially effective on the steep and dangerous Surami Pass. In Scandinavia, the Kiruna to Narvik railway carries thousands of tons of iron ore from the mines in Kiruna in the north of Sweden down to the port of Narvik in Norway. These trains generate large amounts of electricity with their regenerative braking systems. For example, on the route from Riksgränsen on the national border of Sweden to the port of Narvik, the trains use only a fifth of the power they regenerate. The regenerated energy is sufficient to power the empty trains back up to the national border. Any excess energy from the railway braking is pumped into the power grid and supplied to homes and businesses in the region, making the railway a net generator of electricity (Hellmund, 1917 and retrieved 2014).

The first United States use of regenerative braking was the 1967 AMC Amitron. The American Motor Car Company developed an energy regeneration brake for this concept car. The AMC Amitron was a completely battery-powered urban car, with batteries that were recharged by regenerative braking, which increased the range of the automobile. However, AMC went out of business. So the LLNL (2014), which is part of the US Department of Energy, refined regenerative braking further and offered the intellectual property rights to all three American car companies in the late 1990s. Each one turned it down (Clark and Bradshaw, 2004; Clark, 2009).

The Japanese bought the rights and then commercialized the technology. Later, Ford and Chevrolet licensed it back from Toyota for their hybrid cars. At the time in the late 1990s, the American big three car companies could care less about saving fuel and protecting the environment. They misread the general public because the hybrid Prius made Toyota the number one car company in the world, while General Motors went into bankruptcy, Chrysler was bought by Fiat, and Ford downsized. Meanwhile, regenerative braking is also used in the Vectrix electric maxi-scooter. The future is in technologies that are good for the environment.

COMBINED HEAT AND POWER

Engineers hate to let a good idea go to waste, or so it seems with combined heat and power (CHP), systems. In the 1880s, reciprocating steam engines powered the first electric generators. Because these plants were inefficient, a large amount of waste steam was available for process use or building heat. Early developers provided electricity to customers and sent the waste heat through steam pipes for space heating. New York City's Manhattan still uses steam from decades-old cogeneration plants to heat over 100,000 buildings.

Modern engineers have brought this idea back and renamed it as CHP system. CHP produces heat as a by-product of producing power, or electricity. With a fuel source that can be fossil fuel, like natural gas, or a renewable energy like hydrogen or biomass, a turbine is driven by an engine to produce power. As the turbine generates power, heat or thermal energy is also generated. This heat is captured and used to heat buildings (CEERE, 2014).

New technologies are allowing CHP to enter new markets, including small commercial buildings, food service operations, and even for heating Olympic-size swimming pools. Cooling seems the opposite of heating with the hot thermal outputs from CHP plants. However, by using another marvel from mechanical engineers, an absorption cycle chiller, the hot thermal output of the CHP plant can be converted to a chilled water supply for use in the summer for space cooling.

CHP systems are most efficient in colder climates where heat can be used on-site or very close to it. Northern Europe is a large user of cogeneration, particularly Denmark where biowaste and biomass are used to power the systems. As a whole, the European Union generates 11 percent of its electricity using cogeneration. Denmark, the Netherlands, and Finland have the world's most intensive cogeneration economies (Finish Energia, 2014).

Other European countries are also making great efforts to increase efficiency. Germany reported that over 50 percent of its total electricity demand could be provided through cogeneration and plans to double electricity cogeneration to 25 percent of the country's electricity by 2020. The United Kingdom is also actively supporting CHP and encourages CHP growth with financial incentives, grant support, a greater regulatory framework, and government leadership and partnership.

According to the IEA 2008 modeling of cogeneration expansion for the G8 countries, the expansion of cogeneration in France, Germany, Italy, and the United Kingdom alone would effectively double the existing primary fuel savings by 2030 (IEA, 2008).

Under the EU, a public-private partnership called the Fuel Cells and Hydrogen Joint Undertaking Seventh Framework Programme project was established. The partnership intends to do 1000 residential fuel cell CHP installations in 12 states by 2017 (EU HFC, 2014).

HEAT PUMPS AND SEAWATER HEAT PUMPS

Simply, a heat pump is a mechanical device that uses a small amount of energy to move heat from one location to another. Typically, heat pumps are used to pull heat out of the ground to heat a home or office building, but they can be reversed to cool a building (Heat Pumps, 2014).

Advances in heat pump technology have made these systems extremely efficient, and unlike a typical HVAC unit, there is no need to install separate systems for heating and cooling. A geothermal heat pump or ground source heat pump (GSHP) is a central heating and/or cooling system that transfers heat to or from the ground. In winter, it uses the Earth as a heat source and inversely uses it to cool in summer. The system uses the moderate temperatures in the ground to boost efficiency and reduce heating and cooling costs. Combining it with solar heating to form a geosolar system gives even greater efficiency. GSHPs are usually called "geothermal heat pumps" although the heat comes from the sun, not the planet's center. GSHPs harvest heat absorbed at the Earth's surface from solar energy. Like CHP systems, heat pumps are popular in Northern Europe. In Finland, a geothermal heat pump is now the most common heating system for new homes (Heat Pumps, 2014).

One remarkable adaptation of the heat pump is the seawater source heat pump, like the ones being tried in Alaska and cities in China like Dalian and Rizhao (Kwan, 2009, pp. 215-222). A seawater heat pump is a water-to-water system that harnesses the energy released by seawater temperature differences to provide buildings with air-conditioning and heating, bringing huge potential energy savings and environmental benefits.

Electric compressors are used with a refrigerant, which is an evaporating and condensing fluid. Latent heat from raw seawater is "lifted" and transferred to building heat. The Alaskan system uses a dual high-efficiency rotor screw to help lift the heat from the cold temperatures. A stainless steel and titanium-coated plate-and-frame heat exchanger is also required to prevent corrosion during the process of removing heat from the raw seawater flow in advance of the heat pump.

While this technology has been used in Europe, this innovative process of removing latent heat from seawater and using it to heat buildings is an emerging technology in Alaska (Seawater Heat Pump, 2014).

HIGH-SPEED RAIL AND MAGLEV TRAINS HAVE BECOME REALITIES

Magnetic levitation, or maglev, may do for the GIR in the twenty-first century what airplanes did for the 2IR in the twentieth century. Most Asian and European countries have developed and implemented high-speed rail train systems. Like regenerative breaking, this technology was perfected at the LLNLs in the 1990s and offered to American companies, which turned it down (Clark and Bradshaw, 2004). Now, the GIR countries like China are developing maglev or floating high-speed train systems. Maglev train systems use powerful electromagnets to float the trains over a guideway, instead of the old steel wheel and track system. A system called electromagnetic suspension suspends, guides, and propels the trains. A large number of magnets provide controlled tension for lift and propulsion along a track.

Maglev trains do not need an engine and, therefore, produce no emissions. They are faster, quieter, and smoother than conventional systems. The power needed for levitation is usually not a large percentage of the overall consumption. In fact, most of the power is used to overcome air drag, which is a factor with any high-speed train.

Maglev technology is based on a 1934 patent and was pioneered by German Transrapid International after World War II. Transrapid completed the first commercial implementation for the Chinese in 2004 with the Shanghai Maglev Train, which connects the city subway network to the Pudong International Airport. This system transports people more than 19 miles in just over 7 minutes. In 2010, the Chinese started a new project to extend the line to Hangzhou, about 105 miles away, and construction should be completed in 2014. The proposed speed is over 200 miles per hour, which would allow the train to travel the distance in 27 minutes. The line will become the first intercity maglev rail line in commercial service in the world and also the fastest intercity train. The Chinese have plans for similar maglev trains throughout the country. They reason that going from one city to another via a maglev train is far easier, is more efficient, uses less fuel, and is better for the environment than any other form of transportation except the bicycle.

BIOFUEL: A TRANSITIONAL ENERGY POWER

Although renewable energy is the goal, transitional energy sources are an interim step. Ironically, Henry Ford, since he was a farmer in Michigan, used biofuels from his farm corps to fuel his cars for several decades until 1923 when the oil and gas industries pushed him to use their fuel sources.

This approach to technologies from basic products and sources has been called "from the bottom up" and seen to work well in Denmark where recycled products from waste have been used to create environmentally friendly energy supplies as well as the reuse of waste (Alberg Østergaard and Henrik, 2010, p. 247).

Biofuels, then and today, are fuels made from living or recently living organisms (i.e., algae) that can be burned and used in ways similar to fossil fuels, but they are not carbon-based. Ethanol from corn or sugarcane is an example of a biofuel. Unfortunately, it takes about the same amount of fossil fuel energy to make corn-based ethanol, so there is no real benefit to its use as an alternative to gasoline. Sugarcane is more efficient, and it is widely used in Brazil.

Two promising sources for biofuels are algae and a process called metabolic engineering. Though both must be burned to create energy, they are significantly cleaner than fossil fuels as substitutes for gasoline and diesel, and they can be sustainably produced. In one of history's most delightful ironies, metabolic engineering produces a clean fuel from switchgrass. This is the plant that the great herds of prairie bison fed on for centuries before America's Great Plains became Nebraska, Iowa, and Kansas and were crisscrossed by the highways and cornfields used to support the fossil fuel industry.

ALGAE AS A BIOFUEL SOURCE

Algae are a group of simple organisms that are among the world's most ancient creatures. They have a fossil record that goes back three billion years to the Precambrian era. The US Algal Collection lists almost 300,000 specimens, ranging from one-cell organisms to large plants like the giant ocean kelp that grows to 150 feet in length. The glory of algae is that they are photosynthetic (able to use sunlight to convert carbon dioxide and produce oxygen) and "simple," because their tissues are not organized into the many distinct organs found in land plants.

Scientists and researchers are particularly excited about algae's rapid growth cycle—up to 30 times faster than corn—and the ease with which they can be turned into lipids (a green, goopy vegetable oil). This oil, much like any vegetable oil, can be burned and used as a substitute for carbon-based diesel oil or corn-based ethanol. In a strict sense, burning algae or other biofuels does not reduce atmospheric carbon dioxide, because any CO_2 taken out of the atmosphere by the algae is returned when the biofuels are burned. However, it does reduce the introduction of new CO_2 by reducing the use of fossil hydrocarbon fuels.

Algal oils have many attractive features. They can be farmed on land that is not suitable for agriculture. They do not affect fresh water resources and can be produced using ocean and wastewater. They are biodegradable and relatively harmless to the environment if spilled.

At current production costs, oil made from algae is more expensive than other biofuel crops such as corn but could theoretically yield between 10 and 100 times more energy per unit area. One biofuel company claims that algae can produce more oil in an area the size of a two-car garage than soybeans can produce in an area the size of a football field, because almost the entire algal organism can use sunlight to produce oil. The US Department of

Energy estimates that if algae fuel replaced all the petroleum fuel in the United States, it would require just 15,000 square miles of farming area, which is only 0.42 percent of the nation's land mass. This is less than one-seventh the area that is currently used to grow corn in the United States. The US Algal Biomass Organization claims that algae fuel can reach price parity with oil in 2018, if granted production tax credits (Feldman, 2010).

While much of the research on algae is focused on creating oils, either for food or as a transitional fuel for vehicles, one Canadian cement company has discovered a unique application. Ontario's St. Marys Cement plant is using algae from the nearby Thames River, which runs through Ontario to absorb carbon dioxide. The plant started a pilot project in 2010, using algae's photosynthesis to absorb the carbon dioxide produced during cement manufacturing. Martin Vroegh, the plant's environment manager, says the algae project is believed to be the first in the world to demonstrate the capture of CO_2 from a cement plant (Hamilton, 2010).

Through this process, the St. Marys plant is turning CO_2 into a commodity rather than treating it as a liability. The CO_2-consuming algae will be continually harvested, dried using waste heat from the plant, and then burned as a fuel inside the plant's cement kilns. Additionally, the goopy oil can be used as a biofuel for the company's truck fleet.

The company is preparing for a carbon-constrained future that will not treat cement makers and other energy-intensive industries kindly. "The amount of exposure to carbon pricing we face as an industry is very high," says Vroegh. "If we want to be around tomorrow we have to be sustainable. This project helps us achieve that" (Hamilton, 2010).

Algae can be used to make vegetable oil, biodiesel, ethanol, biogasoline, biomethanol, biobutanol, and other biofuels. Algae-based oil can be used for a variety of products, ranging from jet fuel to skin care and food supplements. The potential for large-scale production of biofuel made from algae holds great promise, because algae can produce more biomass per unit area in a year than any other raw material. The breakeven point for algae-based biofuels should be within reach in about 10-15 years.

GIR FUEL FROM PLANTS

Some of the most advanced scientific minds in biology, chemistry, and now metabolic engineering are working to develop useful microbes that will break down simple plants into starches and sugars and eventually into clean fuel.

For over a century, scientists have made fuels and chemicals from the fatty acids in plant and animal oils. The hope is that a synthetic microbe can cost-effectively break down tough plant materials like wood chips and plant stalks and extract the simple sugars so they can easily be converted to fuel. In the United States, the University of Illinois, Urbana, and the University of California, Berkeley, have a 10-year, $500 million grant program to develop algae and other biofuels.

While scientists are engineering fuel-producing microbes, farmers and agriculture experts are developing the inexpensive plants needed to produce biofuels.

Wood waste can also be a source of ethanol. Two US companies, Ineos Bio of Florida and KiOR Inc., are using gasified wood waste to create biofuels (Ineos, 2014). KiOR (2014) built the first commercial-scale cellulosic fuel facility in Columbus, Mississippi, and says it will make one-two million gallons of diesel fuel from its plant.

A major breakthrough for biofuels may have been made in February 2014 by chemists with the Davis campus of the University of California. While it is relatively easy to create diesel from biofuels, until now, creating gasoline from farm and forestry waste has eluded researchers. Led by Mark Mascal, professor of chemistry, the researchers invented a new process for making a gasoline-like fuel from cellulosic material.

Gasoline requires branched hydrocarbons with a lot of volatility, and Mascal and his researchers used levulinic acid as a feedstock for the process. Levulinic acid can be produced by almost any cellulosic material like straw, corn stalks, or municipal waste. While biodiesel fuels from plant-based oils are commercially available for modified diesel engines, a plant-based gasoline replacement would open up a much bigger market for renewable fuels (UC Davis, 2014).

WASTE TO ENERGY

California's Sierra Energy is selling the US Army a breakthrough technology that turns waste—basically, any kind of trash—into clean energy. Called the FastOx Pathfinder, the technology uses a waste gasifier to heat trash to extreme temperatures without combustion. The output includes hydrogen and carbon monoxide, which together forms synthetic gas, or syngas. This syngas can be burned to generate electricity or made into ethanol or diesel fuel.

About the size of a shower stall, the FastOx is basically a small, portable blast furnace that uses a chemical reaction to heat waste. The company says the system can use organic or inorganic materials—banana peels, old iPods, and raw sewage. The concept comes from the old steel blast furnaces and, in fact, was patented by two retired engineers from Kaiser Steel, a long-closed US steel company.

Rugged, simple to use, and modular, the FastOx would be used by the Army to reduce its oil consumption. It can also be used to supply the front lines with fuel for its vehicles and generators (Sierra Energy Group, 2014).

COMMERCIALIZING EMERGING TECHNOLOGIES

Truly innovative technologies are not transformative solely by themselves. Today's mass market is far too vast and complex to support these remarkable technologies by traditional means. To be commercialized, these technologies need government support through a consistent and long-term incentive process from the research and development stage to public financing. Similarly, today's electricity companies were made possible because the costs were supported by local governments and the electricity could be made available at reasonable prices. Since the end of World War II, the industrialized nations have all used government research and development monies to support the commercialization of everything from diesel fuels to the Internet.

Government incentives, tax breaks, and even procurement are critical to the commercialization of new technologies. Governments can also assist in the introduction of new technologies through regulations and standards. When such government actions are geared to climate change issues, environment, health care, and other societal concerns, they have become known as social capitalism similar to some Nordic countries and Chinese national policies, plans, and funding (Clark and Li, 2004, 2012). Today, the advancements of technology to speed communications and to slow climate change are all linked to government regulations and oversight. Nations all over the world should consider their own unique ways to foster and support these emerging industries.

A key component to commercializing these wondrous technologies is updating the patents and intellectual property rights derived from original research. Since much of the advanced research and development comes from government support and funding, there needs to be a new intellectual property model that crosses national borders. This would encourage advance collaborative work between international research laboratories and universities. Aside from creating and moving new green and cleantech ideas into commercial use, there will be less time and money wasted on legal and patent conflicts. Instead of the legal profession making money on intellectual property rights, there would be a model for collaboration that takes the GIR into environmental, sustainable communities and international implementation.

Joint business ventures between public and private organizations can work together to create new industries and jobs. For example, California used such a partnership for the zero-emission vehicle regulations introduced in the early 1990s with a focus on electric battery-powered vehicles. Today, the state is trying to continue that tradition with the implementation of a hydrogen highway, started in 2002 as part of the solution to the state's energy crisis that resulted from energy deregulation. The state was to partially fund the refueling stations for automakers that had hydrogen fuel cell cars ready in 2015. Now, under Governor Brown, the program is moving forward so that by 2016, both the hydrogen highway and every automaker in the world will be able to have its hydrogen fuel cell cars in California.

No matter how impressive these technological advances may be, dependence on fossil fuels hinders the ability to bring them into the mainstream. This makes introducing game-changing technologies and transforming markets that much harder. Social and political support is required to help these transformative innovations reach their full potential to change the way we power our world.

In addition, it will require complete vertical integration. For example, creating broad market acceptance for using algae to reduce CO_2 in the cement industry will require training thousands of support people plus incentives to make the change. Yet, the various technologies emerging from the GIR offer hope and healing for a planet that is getting more polluted and hotter each year.

REFERENCES

Alberg Østergaard, P., Henrik, L., 2010. Climate change mitigation from a bottom-up community approach. In: Clark II, W.W. (Ed.), Sustainable Communities Design Handbook. Elsevier Press, London, UK, p. 247.

CEERE, 2014. http://www.ceere.org/iac/iac_combined.html.

Clark II, W.W., 2009. Sustainable Communities. Springer Press.

Clark II, W.W., Bradshaw, T., 2004. Agile Energy Systems: Global Solutions to the California Energy Crisis. Elsevier Press.

Clark II, W.W., Li, X., 2004. Social capitalism: transfer of technology for developing nations. International Journal of Technology Transfer and Commercialization 3 (1).

Clark II, W.W., Li, X., 2012. Social capitalism: China's economic rise. In: Clark II, W.W. (Ed.), The Next Economics. Springer Press.

EU Hydrogen Fuel Cells (EU HFC), 2014. http://www.h2fc-fair.com/hm13/images/ppt/10we/1420-1.pdf.

Feldman, S., 2010. Algae fuel inches toward price parity with oil. Reuters (November 22). http://www.reuters.com/article/2010/11/22/idUS108599411820101122?pageNumber=.

Finish Energia, 2014. http://energia.fi/en.

Hamilton, T., 2010. CO_2-eating algae turns cement maker green. The Toronto Start (March 18). http://www.thestar.com/business/2010/03/18/co2eating_algae_turns_cement_maker_green.html.

Heat Pumps, 2014. http://home.howstuffworks.com/home-improvement/heating-and-cooling/heat-pump.htm.

Hellmund, 1917. Discussion on the 'regenerative braking of electric vehicles' Pittsburg, PA. Trans. Am. Inst. Elec. Eng. 36, 68, Retrieved March 11, 2014.

IEA, 2008.

Ineos, 2014. http://www.ineos.com.

KiOR, 2014. http://www.kior.com.

Kwan, C.L., 2009. Rizhao: China's Beacon for Sustainable Chinese Cities. Sustainable Communities. Springer Press, pp. 215-222.

Lawrence Berkeley National Laboratory (LBNL), 2010. Global model confirms: cool roofs can offset carbon dioxide emissions and mitigate global warming (July 19), http://newscenter.lbl.gov/news-releases/2010/07/19/cool-roofs-offset-carbon-dioxide-emissions/.

Lawrence Berkeley National Laboratory (LBNL), 2014. http://foundry.lbl.gov.

Lawrence Livermore National Laboratory (LLNL), 2014. www.llnl.gov/energy/regenerativebreaking.

Nanotechnology, 2014. http://science.howstuffworks.com/nanotechnology1.htm.

National Renewable Energy Laboratory (NREL), 2014. www.nrel.gov.

Nularis, 2014. LED bulbs. www.nularis.com.

Seawater Heat Pump, 2014. http://energy-alaska.wikidot.com/seawater-heat-pump-demonstration-project.

Sierra Energy Group, 2014. http://www.sierraenergycorp.com.

Toyota, 2014. www.toyota.com/regenerativebreaking.

UC Davis, 2014. http://news.ucdavis.edu/search/news_detail.lasso?id=10823.

University of Washington (UW) LEDs, 2014. http://www.washington.edu/news/2014/03/10/scientists-build-thinnest-possible-leds-to-be-stronger-more-energy-efficient/.

US Energy Information Administration (EIA), 2013. Residential lighting consumption. http://www.eia.gov/tools/faqs/faq.cfm?id=99&t=3.

CHAPTER 10

China: The Twenty-First-Century Green Powerhouse

While the green industrial revolution surprised America, it did not catch China napping. Starting in the 1990s (as the United States was thinking that cheap fossil fuel would go on forever and 10 mile per gallon SUVs were everyone's natural right), China was thinking ahead with its five-year plans. Since the 1949 revolution, Chinese leaders have consciously set in motion five-year plans and policies with strong economic backing in terms of funds, resources, and long-term relations. The latest is the 12th Five-Year Plan that began in March 2011. In the first decade of the twenty-first century, the Chinese leaders saw the need to use their latest 5-year plan to "leapfrog" the infrastructure and environmental mistakes made by Western developed nations to gain social, economic advantages to mitigate climate change. The result is that China has become a global financial leader with its technological, economic, and commercial applications of the GIR through state-owned or involved companies.

The People's Republic of China went beyond the Western market-driven market economics and created its own economic form of social and government-led capitalism (Clark and Li, 2012). The results were extraordinary. For the last decade, China has led the world with an average seven percent GDP growth. Initially, China followed the United States in its fossil fuel-driven second industrial revolution. But as the twenty-first century began, China switched to the social economic approach taken by Northern European (Germany and the Nordic countries) and other Asian nations (Japan and South Korea). As part of China's "social capitalism" economic model, the Chinese leadership merged their interest in making money with their concern for protecting their society and the environment (Clark and Cooke, 2011). Now, with the new Chinese national government of Mr. Hou Li in place through 2023, there will be significant changes in the 13th Five-Year Plan because of China's strong concerns with urban growth and the subsequent demands for energy, water, waste, and transportation. Under the 13th Five-Year Plan, market economics will be more controlled, regulated, and evaluated in terms of environmental and societal impacts throughout China.

Over the past two decades, China converted many of its government-built and government-operated infrastructure industries into state-controlled companies that included foreign investors and public shareholders (Clark and Li, 2004). This "joint venture" business model is similar to the one used in Germany and the Nordic countries when they converted their infrastructure industries during the 1990s. As the world saw with the spectacularly successful 2008 Beijing Olympics, China has successfully leaped into the GIR.

China used the 2008 Olympics to show that it had arrived as a world leader. While criticized for air pollution and violations of individual human rights, China set the stage and demonstrated its leadership for the twenty-first century with extraordinary displays, performances, and buildings using solar energy to generate power. The Chinese central government is fully aware of the nation's environmental and social issues and has taken aggressive steps to rectify them with its 13th Five-Year Plan.

Since the Olympics, China has continued to invest in clean energy technology. In 2009, China led the United States and the other G-20 nations in annual clean energy investments and finance, according to a 2010 study by the Pew Charitable Trusts (2010). In May 2010, it was reported that China used 34 percent of its stimulus funds ($586 billion) for clean technology energy generation. It will have the capacity for more than 100 GW of renewable energy installed and operating by 2020.

In early 2011, China replaced Japan as the second largest economy in the world (based on gross domestic product) behind the United States. China has shown how the GIR can spur economic growth through industrial and manufacturing expansion and by building high-speed trains, magnetic levitation train systems, subways, housing, renewable energy systems, and onsite power for heating and cooling that are more and more environmentally sound.

By the end of 2011, the United States lost the "distinction" of being the world's worse air polluter. China is now number one. However, if the data were calculated on a per capita basis, the United States would still rank first. Given its focus on sustainable energy sources, China will certainly reverse that trend and be well below the United States and other Western nations in a few years. China's next five-year plan will begin to move the entire country into renewable energy, including electric and hybrid vehicles. The Chinese have the money and resources to make these goals not only achievable but also sustainable. The new Chinese central government leadership will make such decisions as they move further into the 13th and 14th Five-Year Plans with planning under way.

The benefits of the GIR go beyond mitigating climate change and halting environmental degradation. They include social benefits such as more jobs, increased entrepreneurship, and new business ventures, all of which have been, and continue to be, enjoyed in Europe (by Germany and the Nordic countries, in particular). In the United States, California is seeing small-scale growth as it tries to move into the GIR with new green jobs.

The shift to renewable energy requires a more educated workforce, upgraded labor skills, and businesses that can be certified as environmentally responsible for the short and long terms. Along with environmentally sound technologies comes a new green workforce that must learn new GIR technologies that range from nanotechnology to chemical engineering systems. China knows this and understands that the shift to renewable energy will require extensive workforce retraining. Most of China's senior leaders have degrees in engineering or science, unlike other nations whose elected leaders tend to have backgrounds in economics or law. China's dominance in the wind turbine and solar panel manufacturing sectors is a good example why having leaders with scientific and engineering knowledge can be beneficial to national leadership.

China's social capitalist economic model has also played an important role (Clark and Li, 2004). This model requires that foreign businesses located in China have a 50.1 percent or higher Chinese ownership. The government or newly created government-owned companies then become the majority owners of any new venture. The central government sets a plan, and the foreign company enacts it with projects and financing. Additionally, China requires that the profits made by the new ventures be kept in China as reinvestments.

Vestas, the large Danish wind turbine manufacturer, saw early on that China and Asia were large emerging markets. In the early 1990s, Vestas agreed to China's social capitalist business model and established a joint venture. The result was that the wind industry and all the ancillary businesses needed to support it (mechanics, software, plumbing, and electrical work), and those needed to install, repair, and maintain it, grew. Today, China is a world leader in wind energy production and manufacturing. The world-dominating solar industry in China followed a similar business path.

China intends to learn from the West's mistakes as it moves its energy infrastructures into the GIR. The nation plans to do this through a centralized economic policy that is mainly shaped by the Communist Party of China (CPC) through plenary sessions of the People's Republic of China Central Committee. The committee plays the leading role in establishing the foundations and principles of Chinese policy by mapping strategies for economic development, setting growth targets, and launching reforms.

Long-term planning is a key characteristic of centralized social economies, as the overall plan normally contains detailed economic development guidelines for the various regions. There have been 12 such five-year plans; the name of the 11th plan was changed to "guideline" to reflect China's transition from a Soviet-style communist economy to a social capitalism economy.

In October 2010, China announced the 12th Five-Year Plan with final approval and implementation plans for early 2011. The 12th plan ends in 2015. The guideline addresses rising inequality and sustainable development. It establishes priorities for more equitable wealth distribution, increased domestic consumption, and improved social infrastructure and social safety nets. The plan represents China's efforts to rebalance its economy, shifting emphasis from investment to consumption and from urban and coastal growth to rural and inland development. The plan also continues to advocate objectives set out in the 11th Five-Year Plan to enhance environmental protection, which called for a 10 percent reduction of the total discharge of major pollutants in five years.

The 12th Five-Year Plan focuses the nation on reducing its carbon footprint and will address climate change and global warming. Not only will it be well financed, with the equivalent of $1 trillion U.S. dollars, but also it will set in motion the possibility that China will be able to surpass Western nations in the technologies and industries that support and make up the GIR. One key element is that the Chinese will change their central power plants into agile, sustainable, and distributed-energy infrastructures with local onsite power systems that use renewable energy to power the facilities.

EMERGING WORLD LEADER IN ENVIRONMENTAL SUSTAINABILITY

There has been a dramatic transformation in Beijing and several other major cities in the world's most populous nation. In the 1990s, Beijing was ranked third among world cities with the highest levels of air pollution. Today, air pollution has been reduced dramatically.

Sustainable development is now official government policy in China, and it has been implemented at a remarkable pace in some regions. The

Inner Mongolia Autonomous Region (IMAR) north of Beijing was one of the first (Clark and Isherwood, 2007; Clark et al., 2010). At the same time, in the early part of the first decade of the twenty-first century, Shanghai built the first commercial high-speed magnetic levitation, or maglev, train line, which was based on cutting-edge German train manufactures, who licensed it from the US Department of Energy in the late 1990s. The Shanghai Maglev Train connects their new Pudong International Airport to the city's rapidly growing subway system, making the 30 km trip in just seven minutes. The city wanted the train in operation since it was hosting the World Expo in the summer and fall of 2010. Furthermore, Shanghai, Beijing, Shenzhen, Qingdao, and Chengdu are all building or planning new underground railway lines and, in some cases, light rail lines at ground level. High-speed rail now connects many Chinese cities, providing shorter travel times than can be achieved by airplane.

The transformation of Beijing, Qingdao, and Nanjing into modern GIR cities was driven in large part by China's goal of making their 2008 Olympics "green." For example, PV panels and concentrated solar systems were constructed on many of the Olympic buildings. The determination to achieve sustainable development in other cities like Shanghai, which was not an Olympic city, can be explained in somewhat the same way. For example, Shanghai did host the World Expo in 2010, which provided it an opportunity to showcase its implementation of significant GIR technologies (such as the maglev train from their new airport) and the city's plans for the future.

Shanghai's urban planning and sustainable exhibition areas attract thousands of visitors and business people each day. Within the exhibit, the city says that it wants to become one of the world's leading commercial cities in the twenty-first century. To do this, Shanghai cleaned up its air and water and waste drainage systems. As new buildings, communities, and areas were added, the city installed subways and roads that allowed for bikes and walkways.

The Chinese recognize that sustainable development of an area or region is good for business and tourism, as well as its citizens. Numerous cities and regions are now labeled as sustainable and have established benchmarks and criteria for official certification. Hundreds of conventions and conferences are held in these cities, primarily not only to show visitors the positive environmental impacts but also to create business opportunities.

To encourage sustainable development, the Chinese national government continues to strengthen environmental legislation and make huge investments in green technology and sustainable infrastructure improvements. The nation's environmental protection sector is projected to grow at a 15 percent annual rate. Cities are closing their most polluting factories and moving others to locations far from residential and commercial areas. China encourages industries to modernize, which has improved energy efficiency over the past decade. However, the growing demand for personal cars and new buildings and homes has increased national energy consumption. To meet this demand, the government is building traditional fossil fuel facilities and nuclear power plants, as well as large wind and solar farms, which it will combine with renewable energy onsite and distributed power systems.

CHINA'S ENERGY NEEDS ARE MASSIVE

As the world's most populous country, and one with a rapidly growing economy, China has huge energy needs that will continue to grow in the future. China's GDP grew at an average of 10 percent between 2000 and 2008. In 2011, China was ranked number 2 globally in GDP. China is also the second largest oil consumer behind the United States. A net oil exporter in the early 1990s, China is now the world's second largest net importer of oil. In 2009, China's oil consumption growth accounted for about a third of the world's oil consumption growth.

Natural gas usage in China has also increased rapidly in recent years, and China has looked to raise natural gas imports via pipeline and liquefied natural gas. China continues to buy and invest in energy-producing companies from around the world.

China is also the world's largest producer and consumer of coal, an important factor in world energy markets, but one that also creates significant water and emissions impacts from drilling, shipping, and burning coal. With 70 percent of its energy derived from coal, another 20 percent from oil, and less than one percent from renewable sources, China has a ways to go to reach energy sustainability. Nevertheless, China is intent on reducing its fossil fuel dependence.

Unlike the United States, China has removed most of the subsidies for the production and use of fossil fuels. By 2035, China plans to reduce coal use to 62 percent through increased efficiencies and reduce its carbon emissions by at least 40 percent from 2005 levels by 2020. With its focus on reaping the benefits of the GIR, China plans to increase non-fossil fuel energy consumption to 15 percent of the energy mix in the same time period.

China's dramatic growth and increased energy demands put added pressure on global supplies of fossil fuels and have motivated China to purchase large fossil fuel companies and contracts around the world. In addition, energy demand will have an enormous impact on China's already burgeoning solar and wind energy generation industries. Spurred by the government's social capitalist economic policies, which have resulted in rapid business and job growth in the green industry, China is now the largest manufacturer of solar panels in the world. China is poised to lead in advanced batteries, high-speed rail, hybrid and electric vehicles, nuclear, and advanced coal technology.

CHINA'S SOLAR VALLEY CITY

China has a strong commitment to solar energy generation. The Chinese have built Solar Valley City in Dezhou, Shandong Province. This ambitious project will create a new sustainable, environmentally sound center for manufacturing, research and development, education, and tourism focusing on solar energy technologies. Solar Valley City is part of China's efforts to promote green energy technology and grow global market share.

More than 100 solar enterprises, including major solar thermal firms, are based in Solar Valley City. The solar industry in China employs about 800,000 people, and China's solar thermal industry and the accompanying industrial chain are examples for the rest of the world. A leading company, Himin, produces more than twice the annual sales of all solar thermal systems in the United States and is quickly expanding into solar photovoltaic and other technologies.

The Chinese solar industry is an export industry. Toward the end of 2010, China became the world's largest producer of PV cells, but because approximately 98 percent of sales of PV products were exports, the industry was hard hit by the worldwide financial crisis. Solar industry leaders have lobbied for a more active set of government policies (similar to those in the wind industry) to subsidize the domestic use of solar power. Because there is so little domestic use of solar power, the potential for growth is strong. Policies intended to jumpstart domestic solar power demand and turn around China's overly export-oriented PV industry are emerging and the Chinese Ministry of Finance is pushing an onsite or local Solar-Powered Rooftops Plan.

The Solar-Powered Rooftops Plan will develop demonstration projects for building integrated solar power (including solar power rooftop units and PV curtain walls) in large- and mid-sized cities that are economically developed and want to be sustainable. The plan also supports the development of

PV systems in villages and remote areas that are outside the reach of the power grid. As part of this effort to improve domestic use of solar panels, the Ministry of Finance has earmarked a special fund to provide subsidies for PV systems that are at least 50 kilowatts (kW) in size and have 16 percent efficiency. The subsidy will cover the cost of the equipment, or approximately 50 to 60 percent of the total cost of an installed system.

Industry analysts say that much needs to be done to develop a thriving solar industry in China. The country will need to reorient the solar industries from one that relies on foreign trade to one that is balanced between domestic consumption and export. To achieve this balance, the Chinese will need to create a new domestic system that matches the industry's export capabilities.

WIND POWER

China's most economically competitive new energy source is wind power. The nation's wind industry emerged in 2005, after a decade of joint ventures and collaborations with Northern European companies. Favorable government policies were key to doubling the country's wind power capacity each year. According to the Chinese Renewable Energy Industries Association, China has the world's largest installed wind turbine capacity. In 2010, China had a total wind power capacity of 41.8 GW, an increase of 16 GW, or 62 percent, from a year earlier.

The wind industry is essential to achieving China's goals of secure and diversified energy production. The industry also contributes to economic growth, environmental and pollution control, and GHG reductions. The Chinese Renewable Energy Industries Association (CREIA, 2011) estimated that in 2009, the wind turbine industry provided an output value of 150 billion renminbi (RMB), which with taxes and fees was valued at more than 30 billion RMB.

China also plans to reduce emissions through the use of wind-generated power. If the Chinese wind power industry installs 200 GW by 2020 (with a power generation output of 440,000 gigawatt hours (GWh)), it will reduce GHG emissions by 440 million tons. China will also limit air pollution by reducing coal consumption, and CREIA predicts that at the same time the country will generate more than 400 billion RMB in added value and create 500,000 jobs.

While work is needed to integrate renewable energy into China's electricity grid, the country is intent on a massive increase in wind generation. China is rich in wind energy resources, with a long coastline and a large landmass. Wind energy resources are particularly abundant in the southeast coastal regions, the islands off the coast, and the northern part of the country. The western inland regions are also rich in wind energy potential.

Offshore wind energy resources are also plentiful, and in 2010, the first large offshore project was completed at Shanghai's Donghai Bridge. Thirty-four large 3 MW turbines, producing 100 MW, were installed. Analysts estimate that as much as 32,800 MW could be installed by 2020.

Wind energy has enormous potential in China and could easily become a major part of the country's energy supply. Some scientists estimate that the total capacity for land-based and offshore wind energy could be as high as 2,500 gW.

China's wind turbine equipment manufacturing industry has developed rapidly by reaping the benefits of the green industrial revolution. The GIR has provided substantial new business and job growth through the development of new green technologies. Domestic wind turbine manufacturers now account for about 70 percent of China's supply market and are beginning to export their products. The largest manufacturers are Sinovel, Goldwind, and Dongfang Electric. China now leads the world and accounts for roughly a third of the global total, both in installed wind turbine capacity and in equipment manufacturing capability.

The state-owned power supply companies have developed the largest wind farms. These companies are pushed by national law to steadily increase their proportion of renewable energy. The CREIA reported that by the end of 2009, a total of 24 provinces and autonomous regions in China had their own wind farms and more than nine provinces had a cumulative installed capacity of more than 1,000 MW, including four provinces exceeding 2,000 MW. The Inner Mongolia Autonomous Region (IMAR) was the lead region, with newly installed capacity of 5,545 MW and a cumulative installed capacity of 9,196 MW. At the 2009 Copenhagen Conference on climate change, China committed that by 2020, it would meet 15 percent of the nation's energy demand with non-fossil fuels. Achieving this goal will require a huge increase in green energy development, including a much greater concentration on wind power.

Through the Renewable Energy Law and other policies, China has made a major commitment to wind energy. A major part of future efforts involves the creation of seven major-scale wind power bases. Each wind base has potential for at least 10 GW of installed capacity.

The National Energy Bureau is developing these bases. They plan to create a total installed capacity of 138 GW by 2020, but only if the supporting grid network is established. A significant problem is that many of these bases are located in remote areas with a weak transmission grid and a long distance from China's main electricity load centers. There are also concerns about how to integrate large quantities of variable wind power into a grid built for coal-burning power stations.

Pricing is another important element. China's support mechanism for wind power has evolved from a price based on return on capital to a feed-in tariff, with variations based on differences in wind energy resources.

The FiT system was introduced to China in 2009. The system divides the country into four categories of wind energy areas. This regional FiT policy seems to be a positive step in the development of wind power and is stimulating stronger economic growth, increasing manufacturing output, and adding jobs. Additionally, the Chinese see the need for trained workers for building, operating, and maintaining these new systems, so they have created engineering and science programs to train people to work in wind and other renewable technology industries.

China faces several challenges when it comes to integrating large-scale wind-generated energy into its local and regional grid networks and infrastructures. Wind farms in China are located mainly in areas far from load centers and where the grid network is relatively weak. This causes a loss in efficiency, so the present design of the infrastructure grids places constraints on the development and use of wind power. This has become the biggest problem for the future development of wind power throughout the country. However, the Global Wind Energy Council projects exceptional growth for China's wind power capacity. They predict it could reach 129 GW by 2015, 253 GW by 2020, and 509 GW by 2030. Wind power would account for 10 percent of total national electricity supply by 2020 and reach 16.7 percent in 2030. These figures do not take into account more local and regional wind farm systems or smaller systems that are integrated into buildings.

THE GREEN TECHNOLOGY CULTURE

In a pattern reminiscent of the post-World War II rise of Asia's manufacturing sectors, China, Japan, and South Korea are leading the way in developing green energy technologies. As a result, these countries have been dubbed the "green technology tigers." The three nations have already passed the United States in the production of renewable energy technologies. According to the Breakthrough Institute and the Information Technology and Innovation

Foundation 2009 report *Rising Tigers, Sleeping Giants*, these nations will out-invest the United States three-to-one in renewable energy technologies as confirmed by the latest studies in 2011. This will attract a significant share of private sector investments in green energy technology, perhaps trillions of dollars over the next decade. Asia's green technology tigers will, therefore, receive the benefits of new jobs and increased tax revenues at US expense.

Government policies and investments are the keys to helping China, Japan, and South Korea gain a competitive advantage over the United States, and even Europe, in the green energy sectors. These Asian nations are making a large direct public investment in GIR technologies. Government investments in research and development, green energy manufacturing capacity, the deployment of green energy technologies, and the establishment of enabling infrastructure will allow them to capture economies of scale, learning-by-doing, and innovation advantages.

While the Asian governments are making large-scale investments, the United States relies on modest tax incentives that are indirect, create more risk for private market investors, and do less to overcome the many barriers to green energy adoption. Tax breaks make sense when the economy is strong and they work well for large companies. However, during a long global economic recession, they pose a barrier to building and operating green energy systems in the United States (especially for smaller companies).

Companies that can establish economies of scale and create learning-by-doing opportunities ahead of competitors can achieve lower production costs and manufacture higher-quality products. This will make it harder for new entrants to break into the market. Direct government investments will help Asia's green technology tigers form industry clusters, similar to California's Silicon Valley, where inventors, investors, manufacturers, suppliers, universities, and others can establish a dense network of relationships. Even in an era of increasingly globalized commerce, the structure of these regional economies can provide enduring competitive advantages.

In China, national, regional, and local governments are offering green energy companies generous subsidies—including free land, funding, low-cost financing, tax incentives, and money for research and development—to establish operations in their localities. It took just three years for the Chinese city of Baoding to transform from an automobile and textile town to the fastest-growing hub for wind and solar energy equipment makers in China. The city is home to "Electricity Valley," an industrial cluster modeled after Silicon Valley, that is composed of nearly 200 renewable energy companies that focus on wind power, solar PVs, solar thermal, biomass, and energy efficiency technologies. Baoding is the center of green energy development in China and operates as a platform that links China's green energy manufacturing industry with policy support, research institutions, and social systems.

In Jiangsu, a province on the eastern coast of China, local government officials have provided large subsidies for solar energy with a goal of reaching 260 MW of installed capacity by 2011. Jiangsu already houses many of China's major solar PV manufacturers, and the new policy is targeted to create substantial market demand and attract a cluster of polysilicon suppliers and solar technology manufacturers.

Another Chinese city, Tianjin, is now home to the Danish company Vestas, the largest wind energy equipment production company in the world. This base not only enhances the company's production capacity but also increases the number of locally installed wind turbines and helps component suppliers develop expertise with the company's advanced wind power technology while providing a learning laboratory for students and researchers.

Regionally based programs provide cost and innovation advantages, including access to specialized labor, materials, and equipment at lower operating costs, as well as lower search costs, economies of scale, and price competition. A regional focus provides member organizations with preferred access to market, technical, and competitive information while creating knowledge spillovers that can accelerate the pace of innovation. Relationships between companies are leveraged and integrated so they can help each other learn about evolving technologies as well as new market opportunities. Workforce mobility enhances the rate of innovation for the whole region, making it both sustainable and part of the GIR. These regional areas provide an attractive business environment for particular industries; if one or two companies fail or move out of the area, others can quickly replace them.

The United States created some of the first examples of these kinds of economic regions. Detroit became a center for automobile production, and its early leadership in auto technology made it a world industry leader for most of the twentieth century. Later, Silicon Valley became a center for information technology. Developing regions to promote science, innovation, venture capital, and relationships among organizations provided strong competitive advantages that made it costly for other nations to catch up. But the difference with the GIR is that the US government rarely provides outright financial support other than tax breaks and limited research funds.

Conversely, China and other GIR nations have provided proactive and long-term governmental commitments through planning, financial investment, and strong international networks for marketing, sales, and support.

Establishing industrial regions does not guarantee continued market dominance. In the case of the automotive industry, US firms eventually lost market dominance after East Asian nations spent years implementing an industrial policy that sheltered their nascent auto industry from competition. At the same time, these nations invested billions in direct subsidies to support the industry's growth and technological progress. Above all, the Asian nations held a high regard and value for the environment and moved aggressively into hybrid, electrical, and hydrogen technologies that combine renewable energy and infrastructures like highways, rail, and air systems.

China's extraordinary emergence as a twenty-first-century powerhouse is a tribute to the vision and practicality of its leaders. While too many Western nations rested on their twentieth-century successes, China surged ahead, adopting the opportunities of the green industrial revolution. By deftly using its system of planning and government-supported development, China leapfrogged the mistakes made by other nations that were bogged down in a second industrial revolution mind-set. Knowing that fossil fuels are a declining resource that is destroying a fragile global ecosystem, China has acknowledged the need to protect the environment through the expanded use of renewable energy and sustainable communities.

REFERENCES

Chinese Renewable Energy Industries Association, 2011.

Clark II, Woodrow W., Cooke, Grant, 2011. Global Energy Innovation. Praeger Press, U.S.

Clark II, Woodrow W., Li, Xing, 2012. Social Capitalism: China's Economic Rise. In: The Next EconomicsSpringer Press, Chapter #7.

Clark, I.I., Woodrow, W., Li, Xing, 2004. Social Capitalism: transfer of technology for developing nations. International Journal of Technology Transfer and Commercialization, vol. 3 Inderscience, London, UK, No.1.

Clark, I.I., Woodrow, W., Isherwood, William, 2010. Report on Energy Strategies for the Inner Mongolia Autonomous Region. Utilities Policy 18 (1), 3–10.http://www.sciencedirect.com/science/article/pii/S0957178709000290.

Pew Charitable Trusts. 2010. "Pew Charitable Trusts. 2010. "Who's Winning the Clean Energy Race? Growth, Competition and Opportunity in the World's Largest Economies." G-20 Clean Energy Fact Book. http://www.pewtrusts.org/uploadedFiles/wwwpewtrustsorg/Reports/Global_warming/G-20%20Report.pdf.

CHAPTER 11

The Green Industrial Revolution Is Spreading Around the World

ASIA LEADING

Although the world is just now starting to embrace the GIR, its origins go back to medieval Japan's political and social no waste philosophies. As an island nation with a large population, Japan used its own natural resources for energy and development for centuries. The Arab oil embargoes of the 1970s pushed Japan and South Korea toward social policies that eventually led to their development of the GIR. These nations wanted more energy security, and they developed national policies and programs to reduce their growing dependency on foreign fuels.

By the late 1980s, Japan was headed into the GIR with its historical and cultural awareness of the need to protect the environment. Many leading Japanese companies had begun environmental programs that would enable their products to reduce water use and recycle waste. At the same time, the government supported global searches for technologies that would mitigate and reverse vehicle pollution. Toyota then developed the hybrid Prius and went on to become the leading environment-sensitive vehicle manufacturer, surpassing General Motors in 2010 as the world's largest automaker (Clark, 2008-2014).

Today, Japan is responsible for only 4% of the global CO_2 emissions, which is the lowest percentage of all major industrialized nations. Nonetheless, Japan intends to reduce its emissions by 60-80% by 2050. Japan intends to increase solar power generation by more than ten times today's levels by 2020 and 40 times by 2030 (Funaki and Adams, 2009).

In the summer of 2012, Japan started a national feed-in-tariff program like Germany's, which provides strong and stable economic support for renewable energy systems. Because some solar power generators are large, companies are encouraged to concentrate in communities that can support large systems. In Japan, the future for sustainable communities can also be seen in the growth of zero-emission homes, which are models built by the Ministry of Economy, Trade, and Industry (METI) to advertise Japan's energy-efficient and environmentally friendly technologies.

Since the 1970s, Japan has been a leader in the GIR. The Japanese readily accept green technology and their culture and traditions embrace sustainability. For them, it's a matter of survival. The Japanese pioneered the GIR, not because they were smarter or more responsible, but simply because the nation needed to sustain a large population in a small space. Japan's GIR is not about big things, but about small and practical solutions (Funaki and Adams, 2009).

While the Japanese moved toward the GIR, they had to rely heavily on nuclear power as a base energy generator since they lacked other domestic energy sources. Unfortunately, during the 1970s, nuclear power plants were promoted as the ultimate solution to cheap, continuous energy.

In March 2011, disaster struck Japan when a massive earthquake of 9.0 magnitude followed by a devastating tsunami overran northeastern Japan. It destroyed communities and killed thousands and washed out Fukushima I Nuclear Power Plant. The reactor's cooling system failed, leading to a shutdown. Since then, most of Japan's nuclear plants have been closed, or their operation has been suspended for safety inspections.

This tragic earthquake presented many challenges to Japan's leadership. The current leadership in Japan, however, does not seem able do without nuclear power plants. By early 2014, many nuclear power plants were being planned on restarting despite growing reports of nuclear water pollution in the Pacific Ocean, now coming to the Northern Hemisphere (Gresser, 2013).

The Fukushima disaster has forced other nations to rethink their nuclear reactor development plans. For example, China has stopped building more nuclear power plants. Chile has suspended its plans. Germany has taken the most dramatic and best strategy by shutting down its existing plants and moving ahead with renewable energy sources. Siemens, a German company, has decided to shut down all its global nuclear power construction programs (Siemens, 2013). Sweden has shut down its nuclear power plants.

SOUTH KOREA

Like Japan, South Korea has no oil or natural gas resources and in the 1970s turned to nuclear power. Needing a major source of energy for South Korea's heavy industry economy, the nation increased its nuclear reactors to the point where now 30% of its electricity comes from nuclear. As the rush to build nuclear reactors faded, South Korea increased its oil- and natural gas-driven generation. Today, it generates about 65% of its electricity by thermal, which uses mostly natural gas and some coal, which are imported.

In fact, the nation is currently the world's number five crude importer and number two importer of liquefied natural gas (Kim, 2010).

However, also like Japan, South Korea has been working toward the Green Industrial Revolution. To deal with its carbon emissions, a Green Growth Task Force was established in 2009 with the goal to reverse greenhouse gas emissions to early 1990s levels. As part of the effort, South Korea targeted 79% of its stimulus money to green technologies. South Korea also plans to spend 40 trillion won by 2015 in a combined push by the public and private sectors to boost renewable energy resources (South Korea Energy Cap, 2011).

South Korea has considerable offshore wind resources and ambitious plans to become the world's third largest offshore wind power generator. The largest project is a 2.5 GW wind farm, to be constructed off the southwest coast and comprising 500 5 MW turbines.

Offshore tidal power is also being developed. The nation's first tidal power plant began full operations in 2011 at the artificial seawater Lake Shihwa. With a total power output capacity of 254 MW, it is now the world's largest tidal power installation, surpassing the 240 MW Rance Tidal Power Station in France. The southern and western coasts of South Korea are well known for high tides and strong tidal currents. Long-term feasibility studies have been completed on even larger tidal power plants at two other sites—Garolim Bay, with a planned 480 MW capacity, and Incheon Bay with a proposed 1 GW capacity (South Korea Carbon Trading, 2012).

In spring of 2011, South Korea approved a national emissions trading program. The program is aimed at reducing the country's GHG pollution and will start in 2015. It caps carbon pollution across the economy, from steelmakers and shipbuilders to power generators and even large universities, and encourages them to become more energy-efficient. To meet the mandatory cap, firms buy or trade emissions permits or carbon offsets credits.

With the emissions cap program, South Korea joins a growing number of nations putting a price on carbon. Europe has the world's largest emissions trading program, Australia and New Zealand have programs, and the Australian government said it was exploring carbon-trading links with South Korea.

In implementing the program, South Korea's government said that it was crucial to rein in emissions from Asia's fourth largest economy, which have doubled since 1990. The government says the scheme is needed to meet a pledged goal of reducing emissions by 30% from projected levels by 2020 (South Korea Carbon Trading, 2012; China was detailed in Chapter 10.).

EUROPE

During the 1960s, Europe's historic miasma started to clear. Europeans realized that small bickering nations without a common currency, lacking the ability to transport themselves and commercial goods easily between trading partners, and without the ability to agree on even the simplest mutual political and legal concerns—like copyrights and patents—were going to end up as underdeveloped nations subject to the whims of the world's superpowers. Much to the surprise of everyone, including the Europeans, the European Union came out of this new consciousness, crafted by some of the most brilliant and artful politicians in Europe's history.

The end of the Cold War and the reformation of Eastern and Central Europe fostered European cooperation, replacing nationalistic jingoism. Additionally, the Arab oil embargoes of 1970s were worldwide wakeup calls. However, it seemed that the Europeans, who were suddenly paying $5 a liter for gasoline, were the only ones really listening.

The embargoes spurred renewed oil exploration, and fortunately, this led to the discovery of the North Sea's nominal but critical supplies. The North Sea oil deposits—which are rapidly depleting—were enough to prevent social disarray while the EU was coming together and provided a buffer as the search for alternative energy generation began.

This new regional cooperation, coupled with severe energy dislocation and the realization that pollutants were overrunning the continent's fragile environment, triggered the so-called green or environment-first political movement. With roots that go back to the student radicalism of the late 1960s and early 1970s, the green movement was a uniquely European phenomenon that found support from college students and young activists. Eventually, it led to the creation of the European Green Party, and environmental issues morphed into legitimate political concerns and were subsequently followed by greenhouse gas reduction directives and incentives for renewable energy.

As the EU gained political and economic purpose, it helped birth the GIR. The ingredients for this game-changing megatrend came together. Regional cooperation, the shock of spiking oil prices, resource depletion, and environmental degradation were the drivers—the midwives of the GIR. Politicians were slowly rounding into action, Germany eventually passed the renewable energy laws, and soon, major corporations were investing and innovating, happy to make a profit on new markets and new demand. Today, major European companies like Holland's Vesta,

Germany's BESCO, and France's Somfy are international in scope, getting positioned for decades of expansion and profits.

By the 1990s, the EU nations became aware that there were serious social, political, and environmental issues connected to an overreliance on fossil fuels and the internal combustion engine. Policy leaders began to understand that dependency on Middle Eastern and North African oil and gas was a problem. In 2007, the EU really moved ahead with new policies and plans, backed with over 3 billion Euro (about $4.5 billion) in funding for technologies and support (Clark and Cooke, 2011).

Germany kick-started its efforts with the FiT, formally called the Act of Granting Priority to Renewable Energy Sources. By creating variable cost-based pricing, the Germans were able to encourage the use of new energy technologies, such as wind power, biomass-generated power, hydropower, geothermal power, and solar PV power.

Other European countries have similar programs. Germany, Finland, France, the United Kingdom, Luxembourg, Norway, Portugal, Denmark, Spain, and Sweden are on track to achieve their renewable energy generation goals. The Danes are seeking 100% renewable energy power generation by 2050, and they will have 50% renewable energy generation by 2015.

Driven to meet its carbon emissions reduction goals, the United Kingdom has increased its commitment to renewable energy through commercial incentives such as the Renewable Obligation Certificates (ROCs) scheme and feed-in tariffs and the promotion of renewable heat through the Renewable Heat Incentive. The United Kingdom has potentially huge wind resources and in 2012 had 3,400 wind turbines, generating 3% of the nation's electricity. A further 4,000 turbines are planned to be built by 2020 (UK Wind Farms, 2012). Scotland has huge tidal and wave resources and remarkable new renewable technologies that could have an enormous global impact.

The various EU countries have widely different resource availability and energy policy stipulations. France and Finland, for example, are heavy backers of nuclear energy. The United Kingdom and the Netherlands have gas deposits, although with reduced output predictions. In a major breakthrough, the EU developed the world's first Emissions Trading System, and trading started in January 2005.

In addition to the trade in emissions allowances, the Kyoto Protocol provides for Joint Implementation and a Clean Development Mechanism as further means to meet commitments. Joint Implementation enables an offset for emissions reductions obtained by projects between two industrialized nations. The Clean Development Mechanism concerns projects in which investors from industrialized countries lower GHG emissions in developing countries. Suitable underlying conditions permitting, both mechanisms can offer a cost-effective alternative to expensive emissions reduction measures taken in the home country.

An aging grid structure and the need to increase capacity complicate the EU's policy decisions and mean the nations must crank up investment in new energy generation. Estimates indicate that to meet demand in the next 25 years, they will need to generate half again as much electricity as they are now. This could result in profound increases in renewable energy generation.

ASM, a public utility located in Settimo, near the city of Torino, in the Piedmont region of northern Italy is helping to create an ecotown called Laguna Verde that includes a Green Technology Park. The project also has economic value and is based on the concept that energy production should begin with a "dialogue with the environment." Initially, ASM did a detailed analysis of how much energy was being consumed in buildings, districts, cities, and rural territories. Then, it used that knowledge to develop sustainable building and infrastructure designs (Asola and Riolfo, 2009).

A similar project, called City Life, is taking place in Milan in the Lombardy region just east of Piedmont. The regions collaborated and shared information because they needed the help of expert designers and scientists and wanted to include historic areas. For Milan, the goal was to regain a primary role in design experimentation, formal research, and technology innovation. The environmental quality in Milan needs to be improved, which requires a remarkable effort. They focused on an old steel milling area that was located near Settimo so that ASM could provide support and guidance for both regions.

The need for new designs for buildings and infrastructures provided an opportunity for alternative energy products, such as electric vehicles, biomass energy through biogas, and hydrogen fuel cells. Companies were created to develop new environmentally sound technologies like hydrogen fuel cells for homes and motorbikes. Other new companies started to produce products, systems, and technologies at the Green Technology Park in Torino. This effort was led by ASM in Settimo, where they created "isles" of good practices for energy efficiency and innovative technology integration. This included PV systems; thermal solar systems; cogeneration with microturbines, natural gas endothermic engines, and hydrogen fuel cells; mini/micro wind systems; ground and water heat pumps; small hydroelectric systems; geothermal systems for electricity; waste and biomass gasifiers; fuel cell backup systems; and stationary hydrogen plants (Asola and Riolfo, 2009).

The result was that the Piedmont region and ASM led the GIR in Italy and set standards for the rest of Europe. In particular, the region and cities involved in these projects aimed at enhancing good, ecological, energy-sensible practices. They created a regulatory document that outlined the requirements for certifying buildings as "ecoenergetic" or "bioenergetic." They also established incentives for good practices, which have become standards.

As the Europeans and the Asians are discovering, the GIR must start in the home. Energy efficiency and conservation must become part of everyday life. The home is also the place to start with other elements of the GIR, such as renewable energy generation, energy storage devices, and new fuels for transportation. Understanding the GIR on a personal level leads directly to the larger community in which people live and work. People in Europe are beginning to connect sustainability with what they do on a personal level and how they work, go to market, or seek entertainment. When it comes to conservation and efficient use of energy, as well as renewable power generation, storage, and telecommunications, the issues are the same for the wider community as within a person's home.

AMERICA

The United States, unlike China, Japan, South Korea, and Europe, has been slow to embrace the enormous potential of the Green Industrial Revolution. Sadly, the American public has grown to think that it is entitled to cheap energy, particularly oil and coal. Though Americans once had a deep appreciation for the beauty and wonder of their natural heritage, over the years, this has given way to greed-driven exploitation. The appalling lack of concern for the environment makes it hard for the Unites States to develop sound environment-sensitive policies. This is a cultural and political issue as much as a resource issue, and it impacts far too much of the US social fabric.

This is in sharp contrast to most regions in the world, where oil and fossil fuels are expensive and the environment is treasured and protected. In the end, it is a leadership issue, and since two out of the last four US Presidents had tight connections to the oil industry, environmental sensitivity and pollution are not prominent issues in the political discourse. In fact, there are regular attempts by some members of Congress to roll back or dilute an existing policy or regulation established by the US government's own Environmental Protection Agency.

However, led by California, the United States is slowly waking to the benefits of the Green Industrial Revolution. California has some unique geographic features—a long agricultural valley in the state's center rimmed with high mountains and a southern urban region hemmed by mountains. While the southern area, which includes metropolitan Los Angeles, is widely known for its smog, the great central valley has some of the nation's worst air pollution. For decades, California has been aware of its declining air quality and the related health problems and environmental degradation.

The growing concern about air quality, and a massive electricity crisis at the end of the twentieth century that almost derailed the state's high-tech economy, pushed policy makers to action. The result was the California Energy Efficiency Strategic Plan, which launched its major three-year cycles in 2006-2008. With funding from the utility ratepayers, the state has poured over $6 billion into energy efficiency with the goal of reducing electricity consumption by 20%. The program will enter its next three-year cycle in 2015 with several billion more dollars for energy efficiency (CPUC, 2010). Since the late 1970s when California launched its energy efficiency program, the state's per capita energy use has remained flat, while the rest of the United States has increased by about 33% (CPUC, 2010).

California has put hundreds of thousands of workers into jobs in the green and cleantech industries, ranging from light retrofits to electric autos, to developing hydrogen fuel cells.

Honda refueling station in California. Note solar panels in back of covering.

In 2016, California will launch the world's longest hydrogen highway, eclipsing Germany's. Honda, Toyota, and Daimler followed by most other carmakers are getting ready to introduce hydrogen fuel cell autos into California's market.

California also adopted a goal to generate 33% of its electricity from renewable sources by 2020. To move this effort forward, California's Public Utility Commission approved a version of a FiT, called a Renewable Auction Mechanism (RAM), which has resulted in economic gains (Grose, 2012). As mentioned, California has extensive solar energy generation, including the Ivanpah thermal park.

In ambition and in results, California's Energy Efficiency Program is the world's largest, and it has triggered similar plans in several states. Started in 2008, New York has the nation's second largest energy efficiency program. New York's program was prompted after realizing that Manhattan, which is one of the world's major financial centers, had reached the limit of its electricity capacity. Something had to give or New York City was going to suffer electricity outages that would damage the financial and tourist industries.

Subsequently, Texas, New Jersey, and many other states have adopted some form of an energy efficiency program. As part of its stimulus package in 2009, the US government adopted plans detailing energy efficiency upgrades to federal facilities. While the United States acknowledges the benefits of energy efficiency, it still does not have a national energy policy.

This lack of a national consensus and direction pushes the issues of energy renewable generation and carbon emissions control back to the states. Left to the states, the results are a confused mix, greatly slowing down the potential benefits—healthier lifestyles and jobs—of the Green Industrial Revolution.

The United States, particularly in the Great Plains states and along the coastal shores, has enormous wind resources. However, without incentives or government support that clearly provides long-term financial clarity, the private investment sector is reluctant to take the risk.

Fortunately, the western states like California, Nevada, and Arizona are committed to supporting and expanding solar. While generating electricity by wind is cheaper than solar, wind works best in large clusters, or farms, which require larger capital investment. Solar, as the German's demonstrated a decade ago, is easily adapted to individual buildings and facilities. The US western regions have an abundance of sunshine coupled with high air-conditioning costs so the cost benefits for small distributive systems for residential, businesses, and schools are clearer. As the cost of PV declines and reaches parity, installations will expand exponentially.

In 1997 the US Senate rejected the United Nations Kyoto Protocol, which formalized the developed nation's efforts to reduce carbon emissions below the 1990 levels. The rejection marked a low in US international environmental cooperation. However, California with its history of liberalism and support for environmental concerns passed Assembly Bill 32, California Global Warming Solution Act.

AB32 is a truly groundbreaking piece of state legislation and far and away the most significant step taken by the United States or a US state to reduce GHG. The law requires the state to reduce total carbon emissions to 1990 levels (427 million metric tons) by 2020. The act covers cities and public identities and almost all business sectors including the high polluting electric power, oil and fuel refiners, and heavy industrial sectors.

The act forces the companies and cities to report their annual carbon emissions. Those with emissions greater than 25,000 metric tons are subject to caps. Companies are then responsible to reduce their carbon emissions or purchase carbon allowances. Companies that fail to obtain sufficient offset will be punished.

Theoretically, the carbon allowances or offsets are supposed to be equivalent to reductions in actual carbon emissions. However, since the carbon offsets are not part of a larger carbon emissions market scheme, the reality is that the companies are actually paying the state a form of a carbon tax. Policy makers will address many of the problems making AB32 effective in the coming years; however, the intent to reduce carbon emissions is clear.

One promising area for the United States is the work being done by the ENERGY STAR program and the Green Building Council. The ENERGY STAR program is a joint program of the US Environmental Protection Agency and the US Department of Energy that identifies energy-efficient products or practices. The program tests and verifies appliances or provides standards for buildings that are energy-efficient. An ENERGY STAR label on a product helps consumers choose energy-efficient items. The program has been highly successful, particularly in the consumer product sector. A 2010 report said that with the help of ENERGY STAR, the United States saved enough energy to avoid greenhouse gas emissions equivalent to those from 33 million cars and saved nearly $18 billion on utility bills (Energy Star, 2010).

The US Green Building Council set best-practice building standards in the 1990 that have now been accepted worldwide (USGBC, 2014). Though sometimes confused with a government agency, the is a nonprofit trade organization that was started in 1993 with the goal of promoting sustainability in how buildings are designed, built, and operated. The organization is best known for the development of the Leadership in Energy and Environmental Design (LEED) green building rating systems and its work to promote the green building practices, including environmentally responsible materials, sustainable architecture techniques, and public policy. The program has spread to eight national councils, which helped found the World Green Building Council.

USGBC works to promote buildings that are environmentally responsible, profitable, and healthy places to live and work. The organization is having a major impact on new facility construction, particularly with colleges and universities, and its sustainability theories and methods are being taught at architecture and construction departments in all US and many international colleges. There are regional LEED certification councils in most parts of the world, and there are now community neighborhoods or clusters of buildings using LEED standards. This set of criteria reflects broader concerns for clusters of buildings with integrated designs for basic infrastructure needs, including schools, universities, and cities.

Despite some isolated efforts, the United States must move faster into the GIR for its own sake and the sake of the global environment. As a beginning, the nation needs to establish a national energy policy that is consistent and free of the heavy influence of the fossil fuel interests. The move to the GIR is being led by China and others, yet as one of the world's top polluters and heaviest user of fossil fuels, it is very important that the United States join the effort to reduce GHGs and mitigate climate change.

CANADA

Overall, Canada's main energy resource is hydroelectric power, some of which is exported to the United States. Canada does have a national goal of generating 90% of electricity through nonemitting sources by 2020, reducing GHG emissions to 17% below 2005 levels by 2025. However, much like the US states, the Canadian provinces are leading the move into the GIR (All-Energy, 2014).

Ontario is the largest province and the most aggressive in developing renewable energy through an aggressive feed-in tariff. The tariff, enacted in October 2009, has propelled Ontario to trailing only California in North

American solar energy market activity. The province sees wind develop-ment and solar power as a way to create green jobs and now has 18 solar panel and 15 inverter manufacturers (All-Energy, 2014).

In addition, the province has approved 40 new large-scale renewable energy projects including solar, wind, and water that will attract $3 billion in private sector investment, according to the Ontario Ministry of Energy. These projects represent more than 872 MW of renewable power, including 35 solar projects totaling 357 MW, four wind projects totaling 615 MW, and one 500 kW water project. The government said these projects will result in at least 240 more wind turbines and at least one million more solar panels in Ontario (Canada Market, 2011).

Wind energy has increased tenfold over the last six years in Canada. In 2010, Canada had 3,549 MW of installed wind capacity. One-third of that was in Ontario with another one-third installed in Quebec and Alberta. Geo-thermal has a large potential to generate electricity. The country shares the same continental shelf and geology as the United States, Mexico, and Latin America and counts some 200 hot springs that have yet to be developed.

CENTRAL AND SOUTH AMERICA

Latin America, which extends south of the United States to Antarctica, is a huge geographic area with a mix of 23 poor and emerging nations. Brazil with one of the world's largest economies is the acknowledge leader. Ven-ezuela with huge oil resources and Argentina with tremendous agricultural potential are also prominent in influence and politics. Nations like Guate-mala, Belize, and the Caribbean island nations like Haiti suffer with grinding poverty and little resources.

Overall, the region is growing fast and has major development potential. However, energy is expensive and demand is increasing rapidly in all the nations, especially in Brazil, Argentina, and Chile. With limited fossil fuels, many of these nations are trying to develop sustainability and renewable energy generation. The potential for these countries to leverage the Green Industrial Revolution benefits is enormous; however, this region has a his-tory of ineffectual governments that make policy decisions and implementa-tions extremely difficult.

As a region, about 85% of energy generation comes from conventional fossil fuels—oil from Venezuela, and some natural gas from Bolivia and Argentina, and coal in Brazil. About 10% of the region's energy comes from hydroelectric and less than 5% from renewables.

There is enormous potential for renewables in this region and Brazil is starting to take advantage, particularly with biomass and biofuel. In fact, Brazil is close to 95% energy-independent, mostly through a combination of domestic coal and oil supplies and sugarcane ethanol. Brazil also has huge solar and wind resources and is slowly developing them.

To the west of Brazil, Argentina and Chile are slowly trying to develop renewable energy generation. Both countries have excellent solar and wind potential and are trying to develop policies that will secure the needed international investment to make use of the resources. Chile believes it has the potential to develop a sophisticated economy, and its electricity demand is doubling every decade.

Like Chile, Argentina and Uruguay are in the early phases of wind development. Both nations have outstanding wind resources and have held site auctions with the hope of taking advantage of the current slump in the world's wind industry. Uruguay has very few carbon resources and renewable energy generation; wind as well as biomass would be important for a viable energy mix (Uruguay, 2011).

Argentina has natural gas and some oil deposits and a great potential for solar and wind energy development. The nation used to export natural gas to Chile but has almost eliminated it, putting added pressure on Chile to develop renewable energy resources. Argentina, like Uruguay, is in the process of moving forward with wind.

Other parts of Latin America are looking at renewable energy generation to facilitate their development. Like Venezuela, Mexico has large oil reserves, but these Latin American oil reserves are declining steadily. Mexico is starting to develop solar energy in the northern region and the nation has plans to develop as much as 20 GW of wind power by 2020. Panama and Honduras are also looking at wind as a resource, as those nations develop.

Central America's Republic of Costa Rica is leading the GIR in the Southern Hemisphere. Wedged between Nicaragua to the north and Panama to the south, Costa Rica is a stable democracy of about 4.3 million people. A beautiful tropical country, with pristine rain forests, sparkling beaches, and freshwater rivers and lakes, Costa Rica has put over one fourth of the country's 19,653 square miles in public reserve land areas. In 2007, the Costa Rican government announced plans for Costa Rica to become the first carbon-neutral country by 2021.

With the lead from Central America, South America has tremendous potential to maximize the benefits of the GIR. The region has high electricity rates; for example, kWh rates in northern Mexico are twice what they are in nearby Texas. As a whole, with a tropical climate that generates large quantities of biomass, plus abundant solar and wind resources, this region has the potential to maximize the benefits of the GIR (South America, 2013).

INDIA

A huge nation with over a billion in population, India is the fourth largest energy consumer in the world behind the United States, China, and Russia, according to the US EIA (India 2013, U.S. EIA). Coal is the major source of energy generation, and though much of it is mined in Jharia, India is heavily dependent on importing coal and crude oil.

A nation of extreme contrasts—with extraordinarily modern-looking high-tech campuses in Bangalore, to crushing population density in Delhi, and abject poverty in much of the rural areas—India is struggling to meet its rapidly growing demand for electricity. With huge energy needs, a large carbon footprint and limited fossil fuel resources, India is adding renewable energy to its energy mix. In fact, India was the first of the world's nation to set up a ministry of nonconventional energy resources in the 1980s.

In 2013, India had about 30 GW of renewable energy installed with about 70% coming from wind. Solar represents about 5% but is growing quickly. India has high solar insolation and is ideal for solar power. Much of the country does not have an electric grid, and solar power is being used to replace diesel-powered water pumps. India has some 4-5 million of these pumps and replacing them with solar is a priority.

Some large solar projects are being proposed for the Thar Desert, a vast desert region in the interior. In 2003, India launched a solar loan program that was supported by the UN Environment Programme. The loan program won an Energy Globe World Sustainability award and has helped finance thousands of solar home systems, many of them in the rural south, a region without grids.

The nation is also developing waste to energy and biomass waste systems.

SOUTHEAST ASIA

The Republic of Singapore is a Southeast Asian city-state off the southern tip of the Malay Peninsula. An island country with 63 islands and about 19 thousand square miles, Singapore has about 5.3 million people most of them living in the City of Singapore. An exceptionally prosperous nation, Singapore has the highest trade to GDP ratio in the world. Over a decade ago, Singapore agreed to share programs and strategies about sustainability with other Asian cities. It has now become the model for an Asian ecocity (Yun and Zhang, 2013).

While Singapore is exceptionally wealthy, Thailand's Bangkok is at the other end of the economic spectrum. Yet much of what other countries and cities like Singapore have done to become sustainable can apply and be adjusted to fit the needs of poorer cities. Bangkok, the capital of Thailand with a population of over 8 million people, is threatened by climate change year round.

Bangkok and its suburbs are still regularly affected by flooding with monsoon rainstorms that are getting progressively heavier. The resulting heavy downpours overwhelm drainage systems, and runoff discharge is a major danger. Severe flooding hit much of the city in 2011, and most of Bangkok's northern, eastern, and western districts became inundated, in some places for over two months. Coastal erosion is also an issue in the gulf coastal area, a small length of which lies within Bangkok's Bang Khun Thian District. Global warming poses further serious risks, and a study by the OECD has estimated that 5.138 million people may be exposed to coastal flooding by 2070, the seventh highest among the world's port cities.

Bangkok has turned to public transportation in an attempt to reduce carbon emissions and make the city more healthy and sustainable. Four rapid transit lines are now in operation, with more systems under construction or planned by the national government and the Bangkok Metropolitan Administration.

AUSTRALIA AND NEW ZEALAND

Australia is known for its vast coal, natural gas, and mineral resources; yet, it knows that the production and use of energy comes with a major environmental challenge. Despite its size and low-density population, Australia has a sensitive environment that is subject to drought. Over the last few years, there has been a growing awareness that the nation needs to reduce its global emissions of GHG and manage its resources for the long term.

The nation has significant wind and solar resources and understands that renewable energy is an essential part of Australia's low emissions energy mix. In July 2012, the government announced a major program to support the development and deployment of clean energy technologies through Clean Energy Future package.

This program includes a carbon price and complementary measures—including the Australian Renewable Energy Agency, the Renewable

Energy Target (RET), the Clean Energy Finance Corporation, and the Clean Technology Innovation Program

The RET scheme is designed to deliver on the Australian Government's commitment to ensure that the equivalent of at least 20 percent of Australia's electricity comes from renewable sources by 2020. The RET expands on the previous Mandatory Renewable Energy Target (MRET), which began in 2001.

Australia has ample wind and solar resources. In fact, it has the highest average solar radiation per square meter of any continent in the world. Hundreds of thousands of Australian households now have solar hot water systems or solar PV systems on their roofs. Megawatt-scale solar electricity generation systems, though in the early stages, are coming (Australia, 2013).

New Zealand depends on hydropower and geothermal for most of its electricity generation. However, in 2007, New Zealand broadened its portfolio of energy generation. In 2007, a national target of 90% renewable electricity by 2025 was set, with wind energy to make up much of that increase.

New Zealand has large ocean energy resources but does not yet generate any power from them. The greater Cook Strait and Kaipara Harbour seem to offer the most promising sites for using underwater turbines. Two permits have been granted for pilot projects in Cook Strait itself and in the Tory Channel, and consent has been granted for up to 200 tidal turbines at the Kaipara Tidal Power Station. Other potential locations include the Manukau and Hokianga Harbours and French Pass. The harbors produce currents up to 6 knots with tidal flows up to 100,000 cubic meters a second. These tidal volumes are 12 times greater than the flows in the largest New Zealand rivers.

New Zealand has an emissions trading program in which New Zealand Units (NZUs) are traded. Effectively, one NZU is the right to emit one ton of carbon dioxide or the equivalent amount of certain other greenhouse gases (New Zealand, 2014).

MIDDLE EAST AND NORTH AFRICA (MENA)

About 25 nations make up the Middle East and North Africa. They range from Algeria and Egypt to Israel and Iran to Tunisia and Yemen. Occasionally, Turkey and Somalia are included. While the nations are primarily Muslim in religion (except for Israel), they differ significantly in politically and in their amount of natural resources. Collectively, the region is becoming referred to as MENA, for Middle East/North Africa.

Since this region has the world's largest oil and gas reserves, it would seem unlikely to be included in a discussion of the Green Industrial Revolution. However, the reality is that MENA just like the rest of the world is evolving into this new era. Actually, a case can be made that from 2010-2013, MENA has felt the impact of the GIR more than any other region in the world, if not with the transition from fossil fuel to renewable energy and sustainability, then with the incredible power of the GIR's digital communication channels.

Beginning in December of 2010, this region has undergone extraordinary political change and revolt by citizens against authoritarian governments and tyrannical leaders who had controlled their nations for decades. Tunisia, Egypt, Libya, and Yemen forced out rulers, and Syria is the middle of horribly violent revolution against its despot. Powering the demonstrations, protests, rallies, and wars has been the ability of citizens, dissidents, and revolutionaries to organize and communicate digitally via the Internet and various local and international social networks.

While the parts of the region are still involved in political turmoil and social conflicts, other areas are experiencing early signs that renewable energy offers significant potential. There are excellent solar and wind conditions across much of the region. This and a tightening gas market plus a growing recognition that hydrocarbon reserves may have peaked and the remainder of the reserves can be put to better use than generating electricity have the region starting several renewable energy projects.

Morocco is building one of the world's largest solar thermal plant. Saudi Arabia is investing $109 billion into solar energy, with the goal of building a solar industry that can provide a third of its electricity by 2032. The first solar farm is expected to begin operations in 2015. Eventually, the nation is targeting around 40,000 MW of solar capacity within two decades. About 16,000 MW will come from photovoltaic panels, and the other 25,000, from solar thermal technology. The country currently has only around 3 MW of solar installations.

Saudis say that solar energy makes sense from an economic standpoint. Generating electricity from solar will free up larger quantities of its reserves for international sales (Saudi Arabia, 2012).

Following Morocco and Saudi Arabia, other nations have renewable energy projects in development, including Qatar, Bahrain, Kuwait, Egypt, Algeria, Libya, Jordan, Iraq, and the United Arab Emirates. Recently, Dubai announced the Mohammed bin Rashid Al Maktoum Solar Park, a huge

$3.5 billion project that will be funded by a combination of government and private investment.

Very much part of the GIR is the remarkable effort being made by the United Arab Emirates to build its showcase city of Dubai to LEED standards. Sheikh Mohammed, UAE's prime minister and ruler of Dubai, has created a wonderland of architectural marvels including buildings, facilities, structures, and transportation systems. These extraordinary examples of design genius include the Burj Al Arab hotel, located on its own island, and Burj Khalifa, the tallest structure in the world. At the same time, these buildings are built in the most sustainable manner possible with LEED silver-certification at a minimum. The reduction of water and energy usage is the major driver for this amazing testament to the imaginations of the world's best architects and builders. By 2011, over 40 major skyscrapers, office buildings, and major facilities were LEED-certified (Saudi Arabia, 2012).

Besides the benefit to Dubai, the LEED mandate is having a major impact among the world's architects, contractors, and engineers, since so many large American and European firms are involved in designing and building in Dubai and Abu Dhabi, key cities in the UAE. The valuable design and construction lessons learned are being exported to the rest of the Middle East and around the world.

As a small-island state, the Republic of Mauritius faces many sustainable challenges. Mauritius includes three main islands in the Indian Ocean off the southeast coast of Africa. Originally settled by Portuguese explorers, this tropical paradise is one of the world's top luxury tourist destinations. It possesses a wide range of natural and man-made attractions and enjoys a subtropical climate with clear warm seawaters, attractive beaches, tropical fauna, and flora complemented by a multiethnic and cultural population that is friendly and welcoming.

Mauritius's natural beauty hides the fact that the nation has fragile ecosystems, which are vulnerable to climate change. Without fossil fuel deposits, the nation relies on imported coal from Australia for energy generation. In 2009, the government introduced a long-term energy strategy that would bring renewable energy generation to the island. The plan calls for an increase in renewable energy generation to 35% of the nation's total by 2025. Mauritius uses bagasse or sugarcane as a supplement to its coal burning. This sustainable biofuel process contributes about 17% to the island's power needs. Solar, wind, and waste-to-energy are areas that the nation is pursuing to increase the renewable energy generation (Elahee, 2014).

REFERENCES

All-Energy Conference, 2014. http://www.all-energy.ca.

Asola, T., Riolfo, A., 2009. Sustainable communities: the piedmont region, Settimo. In: Clark II., W. (Ed.), Sustainable Communities. Springer, New York, pp. 169–192.

Australia, 2013. http://windenergy.org.nz/documents/Windenergybasics.pdfLike Australia.

http://www.renewableenergyworld.com/rea/news/article/2011/04/canada-market-overview-o-canada.

Clark II., W.W., Cooke, G., 2011. Global Energy Innovations: Why America Must Lead. Praeger, Santa Barbara.

http://www.cpuc.ca.gov/NR/rdonlyres/5D0472D1-0D21-46D5-8A00-B223B8C70340/0/StrategicPlanProgressReportOct2011.pdf.

http://www.energystar.gov/index.cfm?c=about.ab_index.

Elahee, K., 2014. Energy management in a small island developing economy. In: Clark II., W. (Ed.), Global Sustainable Communities Handbook: Green Design Technologies and Economics. Elsevier, New York, pp. 293–305.

Funaki and Adams, 2009.

http://cleantechnica.com/2012/11/25/saudi-arabia-investing-109-billion-into-solar-energy-wants-13-of-electricity-from-solar-by-2032/#32KXIIHiyLHRvWs1.99.

http://www.energici.com/energy-profiles/by-country/central-a-south-america-a-l.

http://www.huffingtonpost.com/2012/05/02/south-korea-carbon-trading-scheme_n_1470297.html.

New Zealand Climate Change Information, 2014. https://www.climatechange.govt.nz/emissions-trading-scheme/.

http://www.renewableenergyworld.com/rea/news/article/2011/12/renewable-energy-recap-south-korea.

India, 2013. U.S. Energy Information Administration. http://www.eia.gov/countries/cab.cfm?fips=in.

http://www.bloomberg.com/news/print/2011-09-08/uruguay-plans-wind-farms-worth-1-3-billion-to-cut-power-costs.html.

Siemens, 2013. http://www.siemens.com/press/en/index.php?content=cc.

http://new.usgbc.org.

Kim, J., 2010. Korean Economy: Past and the Future. Unpublished paper.

Yun, Feng, Zhang, Mingzhuo, 2013. A Path to Future Urban Planning and Sustainable Development: A Case Study of Sino-Singapore Tianjin Eco-city. In: UCLA, Cross-disciplinary Scholars in Science and Technology Program, paper to be published in 2013.

CHAPTER 12

Economics of the Green Industrial Revolution

More than 100 million people will die and the international economy will lose out on over 3 percent of gross GDP by 2030, if the world fails to tackle climate change. This is the dire prediction of *A Guide to the Cold Calculus of a Hot Planet*, published in the fall of 2012 by Climate Vulnerable Forum (Monitor Climate, 2012). The report's authors calculate that 5 million deaths occur each year from air pollution, hunger, and disease as a result of climate change caused by carbon-intensive economies. Further, that would likely rise to 6 million a year by 2030 if current patterns of fossil fuel use continue (Monitor Climate, 2012).

Other recent reports are identifying similar problems. The United Nations Environment Programme, the United Nations Intergovernmental Panel on Climate Change (UN IPCC, 2014) and earlier reports, and the United Nations Framework Convention on Climate Change (Clark, 1997–1999) have long histories of tracking the data on climate change, as do global agencies like GRID-Arendal in Norway. The US Department of Defense (US DOD) for over three decades has been concerned about the issues (such as quantities, countries, and security) of energy and fuel for the American economy. The DOD's 2014 report stated clearly that climate change is costing the world resources, money, and above all lives—including those of American troops who will be seeing more and more conflicts, costs, and wars over energy supplies around the world (US DOD, 2014).

As the Earth's average temperature rises because of greenhouse gases, the effects on the planet, such as melting ice caps, extreme weather, drought, and rising sea levels, will threaten populations and livelihoods, said the report conducted by the humanitarian organization DARA, which is a partnership of 20 developing countries threatened by climate change (DARA, 2012). The report says that these costs are going to continue and get far more intense as regions and nations grow economically, needing more and more energy from other areas of the world.

Tragically, the DARA report concluded that more than 90 percent of those deaths will occur in developing countries, since the world's poorest nations are the most vulnerable to increased risk of drought, water shortages, crop failure, poverty, and disease. Economically, the world's poorest nations could see an average 11 percent loss in GDP by 2030 from climate change. Agriculture and fisheries, two areas that most poor nations depend upon, could face losses of more than $500 billion per year by 2030 (DARA, 2012).

The effects of climate change already cost the global economy a potential 1.6 percent of annual output or about $1.2 trillion a year, and this could double to 3.2 percent by 2030 if global temperatures are allowed to rise, noted the authors. These current reports go way beyond what Al Gore predicted in his film "An Inconvenient Truth" and the UN IPCC Fourth Assessment Report, both released in 2007 (UN IPCC, 2007).

Commenting on the DARA report, Bangladeshi Prime Minister Sheikh Hasina told the news agency Reuters that "One degree Celsius rise in temperature is associated with a 10 percent productivity loss in farming. For us, it means losing about 4 million metric tons of food grain, amounting to about $2.5 billion. That is about 2 percent of our GDP. Adding up the damages to property and other losses, we are faced with a total loss of about 3-4 percent of GDP" (Huffington Post. Reuters, 2014). Even the biggest and most rapidly developing economies will not escape unscathed, noted the report. The United States and China could see a 2.1 percent reduction in their potential GDPs by 2030, while India could experience more than 5 percent loss of potential output.

British economist Nicholas Stern, in a 2006 report on the economics of climate change, calculated that without any action, the overall costs and risks of climate change would be equivalent to a cut in *per capita* consumption of about 20 percent (Stern, 2006). In short, the prosperity of every nation is at risk with climate change.

Global temperatures have been steadily rising since preindustrial times and have already risen about 0.8 degrees Celsius. In 2010, 200 nations agreed to limit the rise in global average temperature to below 2 °C (3.6 Fahrenheit) to avoid dangerous impacts from climate change. However, climate scientists have warned that the chance of limiting the rise to below 2 °C is getting smaller as global greenhouse gas emissions rise because of the impacts from burning fossil fuels.

In its assessment, the International Energy Agency (IEA) estimated that before 2020, for every $1.00 not invested in cleaner technology in the power sector, an additional $4.30 would be needed to compensate for the increased emissions (IEA, 2014).

The IEA made these calculations before Hurricane Sandy devastated the US northeastern seaboard, Typhoon Haiyan savaged the Philippines, or the megadrought in California threatened agriculture. In a dense, metropolitan area like New York and New Jersey, there can be staggering costs associated with the extreme weather created by a warming planet. In early 2013, the United States estimated that repairing the damage done by Hurricane Sandy would cost over $100 billion. This is in addition to the $146 billion spent to repair the damage done by Hurricane Katrina when it shattered New Orleans in 2005 (NYT, 2012).

In California, a "megadrought" that has been growing since 2008 and that could last decades threatens the state's vast agricultural industry. Agriculture is a $45 billion-a-year business for California and the state produces about half of the US fruits and vegetables, along with a significant amount of dairy and wine. Americans all over the country count on California for our grapes, walnuts, almonds, rice, tomatoes, oranges, and strawberries, to name just a few crucial crops. Even guacamole, the wonderful avocado mix, and new fruits like kiwi that have an international market for California seem to be threatened.

Chipotle, the largest chain of Mexican restaurants in the United States, warned in March 2014 that "Increasing weather volatility or other long-term changes in global weather patterns, including any changes associated with global climate change, could have a significant impact on the price or availability of some of our ingredients" and "we may choose to temporarily suspend serving menu items, such as guacamole or one or more of our salsas, rather than paying the increased cost for the ingredients" (Chipotle, 2014).

The state's reservoirs and groundwater are being depleted on an unprecedented scale, and the odds that California will get enough rain to replenish them are slim. Scientists and researchers say long-term trends point to the likelihood of more frequent and severe droughts that could render this region unsuited to grow these crops reliably in the future.

Given the immense and increasing cost of global warming and extensive carbon emission pollution, how is the world going to pay for or much less recover from centuries of using the environment as a garbage can?

Clearly, the world needs to rethink the cost of delaying to move into the green industrial revolution. Since energy is the lifeblood of a modern society, societies need to fully calculate the direct and the indirect and residual costs of using fossil fuels for energy generation.

Particularly in the United States, the real cost of energy has never been reflected in the market because of government financial support for oil, gas, coal, and nuclear power. This support has taken the form of land grants to changes in environmental laws to actual subsidies and grants. For example, *Forbes* magazine reported in April 2010 that, according to its 2009 report to the Security and Exchange Commission, ExxonMobil Corporation (the largest grossing company in the world) paid no US taxes, while reporting a record profit of $45.2 billion (Forbes, 2010). ExxonMobil minimizes the taxes it pays by using 20 wholly owned subsidiaries in the Bahamas, Bermuda, and the Cayman Islands to legally shelter cash from its operations in Angola, Azerbaijan, and Abu Dhabi. ExxonMobil did pay $17 billion in taxes to other countries but paid nothing to the United States (Forbes, 2010). While ExxonMobil does not contribute anything to the US federal government, it spends millions on lobbying for the continuation of oil subsidies. According to the Center for Responsive Politics, in 2009 alone, ExxonMobil spent over $27 million on political lobbying (NYT, 2011).

In 2011, as gasoline prices reached $4 per gallon in the United States and an angry public prodded the Congress, the Senate Democrats tried to pass a bill to repeal federal tax breaks and subsidies for five major oil companies. Following several days of hearings during which the Senate criticized "Big Oil" executives, the bill failed 52-48 when three Senate Democrats from oil states voted against it. The bill would have stripped tax breaks from five major oil companies—ExxonMobil, ConocoPhillips, BP America, Shell Oil, and Chevron. Ending their tax breaks would have cost them about $21 billion over 10 years, according to Congress' Joint Economic Committee (NYT, 2011).

In early 2014, a scandal erupted in North Carolina on the US eastern seaboard. The state's governor and administration were reported to have stripped the state's Environmental Protection Agency of resources in an attempt to have the agency relax its efforts to force the nation's largest utility, Duke Energy, to clean up 39,000 tons of coal ash that was spilled into the Dan River in February. The spill ruined the river, coating the river bottom for 70 miles, and threatened drinking water and aquatic life. As events unfolded, it turned out that Duke Energy had 32 ash ponds that were leaking (NYT, 2014).

The Duke Energy ash coal spill came a month after 10,000 gallons of 4-methylcyclohexane methanol, or MCHM, spilled into West Virginia's Elk River, ruining the water supply of Charleston, the state's capital. A second chemical, a mix of polyglycol ethers, known as PPH, was part of the leak, the company involved, Freedom Industries, told Federal regulators. The company uses the chemicals to wash coal prior to shipping to use by coal-powered utilities. Over 300,000 West Virginians were impacted and several hundred residents went to the hospitals with various symptoms (CNN, 2014).

In 2011, Dr. Paul Epstein of the Center for Health and the Global Environment at Harvard Medical School tried to quantify the actual costs of coal in the United States (Reuters, 2011). Their comprehensive review of total economically quantifiable costs found that it costs the US economy annually some $345.3 billion or close to an additional 17.8 cents per kWh. These externalities or hidden costs are not borne by miners or utilities, but by the American taxpayer in multiple ways. Coal companies benefit from the following:

- *Tax breaks.* Just like the oil and gas companies, coal companies receive preferential treatment from the IRS. The Department of Treasury estimates that eliminating just three tax preferences for coal would save $2.6 billion between through 2022. These include expensing exploration and development costs—coal companies can expense costs incurred by locating coal ore deposits. The tax codes allow coal companies to claim a tax deduction to cover the costs of investments in mines. Coal royalties for private owners are treated as long-term capital gains, not current revenue.
- *Public land loopholes.* According to the Energy Information Administration, 43.2 percent of the US coal comes from public lands. The coal industry benefits from various loopholes that make obtaining leases easier and cheaper. For example, the nation's largest coal-producing region, the Powder River Basin in Wyoming, is not legally classified as a "coal-producing region." Therefore, the coal tracts are not competitively leased, which shortchanges taxpayers for the value of the land and the coal beneath it.
- *Subsidized railroads.* Coal is the most important commodity transported on railroads. US railroads get loans and loan guarantees from the governmental agencies and have received numerous tax incentives for investments. This relationship is important when considering coal export, which has surged to all-time highs. A large portion of the exports going to Asia include subsidies from the US taxpayer.

TRUE COST OF OIL

The externalities and true costs of oil are a subject that governments and the international oil industry would rather not talk about. While the oil industry is a powerful wealth builder, there are enormous long-term risks and impacts. Some costs are obvious. Oil spills and environmental disasters like the 1989 Exxon Valdez in Alaska and 2010 Deepwater Horizon oil spill in the Gulf of Mexico are never fully calculated when economists reckon the costs of fossil fuel-based energy generation. The actual costs of the destruction to the environment and the damage to the atmosphere are rarely recovered in full from the offending companies.

In fact, 25 years after the Alaska spill, the coastal ecosystem has still not fully recovered, and toxic oil remains in shoreline sediment. Several of the world's oil-producing areas, such as the Niger Delta, the Caspian, and Siberia, have repeatedly suffered oil spills.

But the true costs of oil go far beyond the obvious damage from spills, writes Richard Steiner, professor and conservation biologist in *The True Cost of Our Oil Addiction* (Steiner, 2014). The additional or external costs, Steiner points out, are far less visible than oil spills. Little considered costs include ecological habitat degradation from exploration, production, and pipelines; health costs from breathing air polluted with fossil fuel emissions; urban sprawl and traffic congestion around all major cities of the world; and seemingly endless wars fought to secure oil supplies, like Iraq and Sudan, costing thousands of lives and trillions of dollars.

Steiner notes that climate change from carbon emissions is incurring enormous present and future costs. Costs for storm damage, drought, wildfires, lost agricultural productivity, infrastructure damage, climate refugees, disease, forest decline, marine ecosystem collapse, species extinctions, and lost ecosystem services need to be considered. Global climate change costs already exceed $1 trillion a year, with 5 million human deaths, and both will continue to rise.

Because of the wealth and political influence it generates, there is a "sociopolitical toxicity" connect to oil. From the earliest of times, oil has significantly distorted economic, social, and political systems. Rather than the prosperity promised, oil discoveries around the world often become more curse than blessing, causing social dysfunction, assimilation of indigenous cultures, runaway inflation, a decline in traditional exports, overconsumption, abuse of power, overextended government spending, and unsustainable growth. Former Venezuelan oil minister Juan Pablo Perez Alfonzo, a founder of OPEC and once a true believer in the promise of oil, thought differently after he saw the corruption, greed, waste, and debt it caused, calling oil "the devil's excrement."

The world continues to use more oil, hitting a historic high of 91 million barrels a day in 2013. To date, the world has pumped and burned about 1 trillion barrels of oil (Steiner, 2014).

Oil-producing governments including the United States are heavily influenced by oil interests that dictate policies to limit regulation, to lower taxation, and to favor production and demand for oil over development of low-carbon alternatives. The 2010 US Supreme Court Citizens United ruling now allows oil companies to pour unlimited funds into oil-friendly candidates and issues, without public disclosure. US oil production has steadily increased since 2008, largely from oil shale, prompting calls from some to lift the export ban.

The International Monetary Fund reports that worldwide, governments provide annual subsidies of some $1.9 trillion, including $480 billion per year in direct subsidies to the oil industry. Such subsidies artificially depress prices and encourage overconsumption; detract from government spending on health care, education, and social services; and keep alternative energy uncompetitive.

Milton Copulos, president of the US National Defense Council Foundation, testified before the Senate Committee on Foreign Relations in 2007 on the true cost of oil. He estimated that the true cost for a gallon of gas made from imported oil to the US consumer was $26 dollars. This amounts to about $15,400 per year for the average car in the United States or 70 percent of the median household income given 2.3 cars per household. Copulos' estimate was based on $60 per barrel oil. Oil in 2014 is on average around $105 per barrel, so the true costs may actually be twice Copulos' figures.

The true cost of importing oil to the United States includes the following aspects:

- The cost of the oil ($2.50)
- Oil-related defense expenditures ($3.79)
- The loss of domestic employment and related economic activity due to cash outflow for oil ($3.23)
- The reduction in investment capital ($10.85)
- The loss of local, state, and federal tax revenues ($1.18)
- The economic toll periodic oil supply disruptions impose on the domestic economy ($3.65)
- Federal subsidies for oil and gas industry ($0.69)
- The market cost of carbon ($0.18)
- Pricing shown per US taxpayer; assumes $60 per barrel of oil and $20 per ton of CO_2

Source: Copulos (2007).

These costs are ignored and the US taxpayers are paying for some through income taxes but deferring most to future generations. With the use of cheap oil over the past century, human population has quadrupled and resource consumption has increased many times more. Without access to fossil fuels, humanity would almost certainly have evolved on a more sustainable path. But by not accounting for its true cost, oil has allowed us to dig ourselves deeper into an unsustainable hole. The environmental debt we are accruing is far larger and more consequential than our national financial debt, Steiner points out (Steiner, 2014).

The US government has estimated that the full social cost of carbon is about $50-100 per ton of CO_2. Global emissions now exceed 39 billion tons per year, which amounts to $2-$4 trillion annually. Including these very real costs, sustainable alternatives become competitive.

FREE-MARKET ECONOMICS HAS FAILED

In light of the October 2008 world financial meltdown, it seems silly to think that the supply-side, deregulated, free-market economics, so passionately espoused by President Ronald Reagan and Prime Minister Margaret Thatcher in the 1980s, would work for a 21st-century world threatened by irreversible environmental degradation (Clark and Fast, 2008a,b). Even the bastion of supply-side economics, *The Economist*, ran a special issue in July 2009, with a Bible melting on the cover, almost a year after the global economic collapse that discussed the failures of modern economic theory (Economist, 2009). That started a debate that continues to this day about the value, since economists did not predict the global economic collapse in October 2008, nor do economists 5 or 6 years later. They are able to conduct research that sets rules and formulas and provide data that predict economic trends in the future, like science does (Clark, 2012).

The 2008 global economic implosion from trillions of dollars in credit swaps, hedge funds, subprime mortgages, and related marginal derivatives (which nearly pushed the world's financial structure into the abyss) underlined what happens when governments ignore their responsibility to govern. In the end, the Great Recession was the worst financial disaster since the 1930s, and a testament to the venal side of free-market capitalism—greed, stupidity, carelessness, and total disregard for risk management (Li and Clark, 2009).

Market economists and others had argued that there was no need for regulation. Government would act as "the invisible hand." Adam Smith was just as wrong centuries ago as his proponents are today. The neoclassical economics are not standards, let alone viable rules that can be repeated. New economic processes must be developed if the planet is going to survive climate change and its impact on the Earth and its inhabitants (Clark and Fast, 2008a,b; Fast and Clark, 2012).

Unlike the first and second industrial revolutions, the green industrial revolution must develop an economy that fits its social and political structures. The first industrial revolution replaced an agrarian, draft animal-powered economy with one powered by steam engines and combustion machine-driven manufacturing, an evolution that was accelerated by colonial expansion. Then, the second industrial revolution created a fossil fuel-powered economy that extracted natural resources in an unregulated, free-market capitalist manner without regard to people, health, or environmental costs.

Toward the end of the 20th century, there were extraordinary breakthroughs in information and digital technology driven by innovation, science, and investment capital. California's Silicon Valley was, and to some extent continues to be, the center of this capitalist market-driven universe that was propelled by cheap fossil fuel, ingenuity, and intelligent and productive venture capital. American culture spread as the world's regions and nations became more interconnected.

It continued unabated until September 11, 2001, when 19 al-Qaeda terrorists hijacked four commercial passenger jets, crashing 2 into New York City's World Trade Center and 1 into the Pentagon in Arlington, Virginia. Passengers in the fourth jet managed to overtake the terrorists, but then, the plane crashed in a field near Shanks Ville, Pennsylvania.

Since then, free-market economics have been shaken by a series of events originating in the Middle East: al-Qaeda and Taliban insurgences in Afghanistan, the US invasion of Iraq, terrorist attacks on civilians in Yemen, and Islamic terrorist bombings in England and Spain. A shift occurred and the oil-rich Middle East has gone from the world's gas station to its battleground. The Middle East has become a volatile, transitional region driven by demands for modernity and economic participation, contrasted with the fundamentalists' desire for self-identity and community. Many developed nations are embracing GIR economics just to reduce the leverage this region has on politics and economics.

Globally, the economics of the 2IR resulted in a widely disproportionate amount of money in the hands of the industries involved with fossil fuels and related products and the manufacturers who were able to prosper by supplying cheap energy. It was wealth built at a significant cost to the health of the general population, which had to live with polluted air and water, climate change, acid rain, and greenhouse gases that impacted health. The public, particularly in the United States, has been forced historically to pay undeserved taxes to subsidize the oil, coal, and now the natural gas industries, all of which continue to be the wealthiest industries on Earth.

If the United States developed a culture based on cheap fossil fuels, Europe and Asia offer a telling contrast (Clark and Cooke, 2011). In those regions of the world, consumers pay three to four times more than Americans pay for fossil fuels. This creates an economic barrier against overuse and provides an incentive to conserve fuel, seek other options, and use alternative modes of energy.

As the GIR emerges, the world becomes much more interdependent. What happens in one part of the world, be it weather, pollution, politics, or economics, impacts other regions. For example, the dramatic change in the Egyptian government in early 2010 has affected the rest of the Middle East and will result in global changes of oil and gas supplies. Further, the 2014 Russian aggression in Ukraine and Russia's subsequent threats to reduce natural gas shipments to European nations are particularly unsettling. The result might well be the forced end of the 2IR so that other nations can become energy-independent as they transition to the GIR.

There is historical precedence for a forced transition from the 2IR to the GIR (Clark and Cooke, 2014). The Arab oil embargoes in the early 1970s pushed Europe and Asia toward socioeconomic policies that eventually led to the development of the GIR. As early as the mid-1970s, Japan and South Korea along with the Scandinavian countries decided that they needed to get off coal and other fossil fuels to help the environment and their own economies. Energy independence, climate change, and environmental protection became serious political issues.

Social and environmental factors—sustainable communities, climate change mitigation, and environmental protection—are growing in importance and will soon demand far greater international cooperation and agreement (Clark, 2009, 2010, 2014). Rampant economic growth and individual accumulation of wealth is being replaced by social and environmental values that benefit the larger community.

Without a national policy, countries cannot address their basic infrastructures, and there can be no action, no improvement, no resources, and certainly no response to environmental degradation. For example, the United States' inability to develop a national energy policy that addresses climate change or renewable energy generation is often cited as a monumental failure of its free market and deregulation economic model. Energy and infrastructures, the argument goes, are extraordinarily important national issues. To address them for the greater good, a nation needs to have plans, which are outlined and offered by the central government, to address these basic systems that interact with citizens and the environment.

The key is to have each of the major infrastructure components—energy, water, waste, telecommunications, and transportation—linked and integrated. That way, these components overlap, and costs for construction, operations, and maintenance can be contained and reduced. If the basic infrastructure components can be constructed, operated, and maintained on the local level and meet regional, state, and national goals such as carbon reduction, they take on a different perspective, format, and cost structure. While the United States has no national energy policy, states such as California have created energy and environmental policies of their own. These days, America's leaders on energy policy are not in Washington, DC, but at the local level.

CHINA'S CENTRAL PLANNING MODEL

The People's Republic of China, not the United States, is showing real global leadership as the world heads into the GIR. More than anything, China demonstrates how important a role the central government plays in overseeing, directing, and supporting the economics of technologies and creation of employment. Nonetheless, the actual implementation of the GIR in China is at the regional and city levels where new communities and their infrastructures are being built. The market and capitalism have nothing to do with this remarkable sustainable growth. Instead, it has more to do with the needs of the people to help improve their lives, work, and education, as well as rest or retirement areas.

China's economic system is the prototype of social capitalism. Since the 1949 revolution, the Chinese have moved toward economic development through a series of five-year plans. The central party plays the leading role in establishing the foundations and principles of Chinese policy by mapping strategies for economic development, setting growth targets, and launching reforms. Long-term planning and financing are the key characteristics of centralized social economies, as one overall plan normally contains detailed economic development guidelines for the various regions and communities. Each plan comes with considerable funding to implement the plans based on measureable results.

China's 12th five-year plan addresses rising inequality and sustainable development. It establishes priorities for more equitable wealth distribution, increased domestic consumption, and improved social infrastructure and social safety nets. The plan represents China's efforts to rebalance its economy, shifting emphasis from investment to consumption and from urban and coastal growth to rural and inland development. The plan also continues to advocate objectives set out in the 11th five-year plan to enhance environmental protection, which called for a 10 percent reduction in the total discharge of major pollutants in 5 years.

The current plan will focus the nation on reducing its carbon footprint and addressing climate change and global warming. Not only is it well financed, with the equivalent of $1 trillion US dollars, but it also sets in motion the possibility that China will be able to surpass Western nations in addressing environmental concerns, creating sustainable communities, and reaping the benefits of the GIR. A case can be made that China has shifted to renewable energy as its key area for political stability, both within the country and for its international policies and programs (Clark and Li, 2012).

The energy economics for China are considerably different than those of other countries. For example, one strategy for the Chinese has been to focus on emerging nations and their gas and oil supplies. Chinese companies have created joint ventures and merged with or acquired companies in these regions, causing concern for economic and climate change impacts. The current push back by Western developed nations needs to be watched to see where the alliances and partnerships can be created since the energy generation of power and its impact on the environment have global results in terms of economics, weather, and security.

Over the years, Chinese economics have changed from a state-controlled communist system to a more socialistic one. In the post-Mao era, China moved aggressively into a market-capitalism system, but one where state institutions were owned in part by the Chinese government and shared in joint ventures with foreign companies. Companies wanting to do business in China had to keep their profits there for reinvestment and have at least 49 percent of the company owned by the Chinese government. By the end of the late 1990s, China embarked on more of a social-capitalist approach to economic development. Based on the Chinese five-year plans, these finance and capital systems worked extremely well. Today, China has the largest wind and solar companies in the world. The goal was to create sustainable communities while continuing to grow and expand responsibly to meet environmental needs and concerns. The environment became an economic asset for the entire nation.

The significant change in China was its economic growth, which required more secure supplies of fuel and energy. This meant that the Chinese government had to be far more proactive than market economic philosophy would support. While China's economic growth was made possible by an increasing involvement in the capitalist world system, that was also its weakest point. China's economic growth is inseparable from its increasing dependence on global markets, with some estimates suggesting that more than 40 percent of its GNP is derived from international trade (Clark and Li, 2012). An area of increasing concern is China's energy demand and supply. China's escalating economic consumption puts pressure on the global energy market supply, affects prices, and causes political and social conflict, especially since China can pay above-market prices for energy.

The question is, can China maintain economic growth in an environmentally sound and responsible manner according to its pledges and commitments? It is apparent that China has identified energy security as one of its vital national interests (Clark and Li, 2012). It has instituted plans and funds to provide for more energy from renewable sources, instead of relying on its own limited fossil fuels or importing fuel from other nations. Analysts predict that China will meet or exceed its 2020 renewable energy targets. China's rise today is in large part due to its rapid emergence as a major force in the energy geopolitics (Sun et al., 2013).

THE FEED-IN-TARIFF (FiT) MODEL

Europeans adjusted their economies to meet the requirements of the GIR early on. Both the Scandinavians and the Germans realized that the move away from fossil fuels to renewable energy distribution would require more than what the neoclassical free-market economics could deliver (Gipe, 2010–2014). While the Danes and other Scandinavians shifted national resources toward renewable energy power by national consensus, the Germans developed the innovative FiT financial process (Morris, 2014).

Germany's FiT was part of their 2000 Energy Renewable Sources Act, formally called the Act of Granting Priority to Renewable Energy Sources. This remarkable policy, called Energiewende, was designed to create a total energy transformation by encouraging the adoption of renewable energy sources and to help accelerate the move toward grid parity, making renewable energy for the same price as existing power from the grid. Under a FiT, those generating eligible renewable energy, either homeowners or businesses, would be paid a premium price for the renewable electricity that they produced. Different tariff rates were set for different renewable energy technologies, based on the development costs for each resource. By creating variable-cost-based pricing, the Germans were able to encourage the use of new energy technologies such as wind power, biomass, hydropower, geothermal power, and solar photovoltaic and to support the development of new technologies.

The most significant result of the German FiT was that it stabilized the renewable energy market and reduced the financial risk for energy investment. By guaranteeing investors' compensatory payments, the FiT program created a secure climate for investment. The program covered up to 20 years per plant, with the exception of hydroelectricity installations, which required longer amortization periods. The law also offered a means for altering the compensation rates for future installations, if necessary. The executive summary of the original document says:

> This remuneration system does not mean the abandonment of market principles, but only creates the security needed for investment under present market conditions. There is adequate provision to safeguard the future existence of all the plants already in operation. The new act has abolished the regulation contained in the Electricity Feed Act, which limits the uptake at preferential rates of electricity from renewable energy sources to a maximum share of five percent of overall output.
>
> Instead, we have introduced a nation-wide cost-sharing arrangement. The act should put an end to any fears of excessive financial burdens. The contribution resulting from the new cost-sharing mechanism amounts to a mere 0.1 Pf per kWh. Even if, as we hope, there were powerful growth in renewable energy sources, this would still only rise to 0.2 Pf per kWh in a few years time. That, indeed, is a small price to pay for the development of this key sector.
>
> **GR ME (2000)**

The designers of this policy had exceptional foresight and GIR intuition. The result of the 2000 Energy Renewable Sources Act was the creation of Germany's renewable energy industry. The policy triggered the creation of wind turbine farms and launched the German solar miracle. Despite having an Alaskan-latitude climate, Germany was—for almost a decade—the number one world leader in solar power manufacturing and installation.

The German FiT sets in place a GIR economic model, which other nations can follow. By 2005, 10 percent of Germany's electricity came from renewable sources, when only 5 years earlier that portion was less than 1 percent. Germany estimated that the total level of subsidy was about 3 percent of household electricity costs. The FiT rates are lowered each year to encourage more efficient production of renewable energy. By 2008, the annual reductions were 1.5 percent for electricity from wind, 5 percent for electricity from photovoltaic, and 1 percent for electricity from biomass. In 2010, Germany met their goal of 12.5 percent of electricity consumption, thus avoiding the creation of more than 52 million tons of carbon dioxide. They are on track to reach their goal of 20 percent renewable power generation by 202 (Gipe, 2010–2014). By 2050, they hope to power almost entirely by renewable sources.

The German GIR economic model is being adopted by other nations that are developing renewable energy sources. According to the *Renewables Global Status Report: 2009 Update*, FiT policies have been enacted in 63 jurisdictions around the world, including Australia, Austria, Belgium, Brazil, Canada, China, Cyprus, the Czech Republic, Denmark, Estonia, France, Germany, Greece, Hungary, Iran, Republic of Ireland, Israel, Italy, the Republic of Korea, Lithuania, Luxembourg, the Netherlands, Portugal, South Africa, Spain, Sweden, Switzerland, Thailand, and Turkey (Renewable, 2009). Despite the lack of a national policy in the United States, several states are considering some form of a FiT, and the concept seems to be gaining momentum in China, India, and Mongolia.

Recently, the European Commission and the International Energy Agency, among other groups, had completed various analyses of the FiT policy. Their conclusion was that well-adapted FiT policies are the most efficient and effective support systems for promoting renewable electricity.

The German FiT model continues to be highly successful and certainly moves the country beyond conventional 2IR economic theory. The idea that ratepayers can get funds from higher rates to purchase renewable energy systems, which then generate power for their own buildings with the excess sold back to the central power company, is not part of the neoclassic 2IR economic model. This is because neoclassic economic theory does not consider infrastructure calculations (Borden and Stonington, 2014).

FiT: CALIFORNIA STYLE

California is tugging the United States toward the GIR. The state was the first to launch a major energy efficiency effort after two decades of being the most conservation-oriented. The initial 3-year funding cycle (2006-2008) has evolved into a multibillion-dollar effort that is now in its third cycle. As part of the state's commitment, California's goal is to generate 33 percent of its electricity from renewable sources by 2020. In 2010, to move this effort forward, California's Public Utility Commission approved a version of a FiT called the Renewable Auction Mechanism (RAM). While not a classic FiT, the program is intended to drive small to medium-sized renewable energy development. It will require investor-owned utilities to purchase electricity from solar and other renewable energy systems of 1.5-20 MW.

While it is too early to tell if the RAM program will be successful, the renewable energy industry—particularly the solar industry—is optimistic. Several industry leaders say that RAM improves the traditional FiT programs because it allows for market-based pricing, while still providing a long-term, stable power agreement for project developers. Other endorsers think that, because it sets an outcome instead of fixing a price, it will help eliminate speculators and keep high-quality developers involved.

However, the RAM program may not be successful. Critics say that California's RAM does too little in terms of both financial support and payback to consumers. The renewable energy industry is skeptical, too, saying that the RAM does not go far enough. The results appear to be similar to the 2IR economic model, since direct government rules and standards will likely not be very effective. The point is that RAM like California's coming cap-and-trade scheme fails to support the GIR since its basic economic argument is that market forces will bring renewable energy systems down in cost. This may not be the case. As noted above, the 2IR was based upon government financial and regulatory support. There were never any market forces at work. When at the turn of the 21st-century market forces were tried in California, under deregulation, they failed in all aspects. Instead, corruption and market manipulation prevailed.

PAYING TO MITIGATE CLIMATE CHANGE

Climate change is the greatest challenge of modern times; yet, addressing it offers the potential for tremendous economic growth. As the world embraces the GIR, efforts to mitigate climate change will unleash a wave of new economic development, generating jobs and revitalizing local, regional, and national economies. However, while the GIR nations may have the technologies to jump-start a green energy economy, developing the mechanism to curb climate-changing emissions is another matter.

Climate change stems from a single fact: human beings treat the atmosphere as a free dumping ground. No one has to pay to pollute the shared air. The result has been increasing concentrations of climate-warming gases, a blanket of carbon that is keeping heat in the Earth's atmosphere.

In order to transition to a green economy, a price must be put on climate-changing emissions (Nijaki, 2012). Mechanisms have to be developed that make the polluters pay, while guaranteeing that emission-reduction goals are met. Further, a workable system has to be built on three principles: efficiency, effectiveness, and fairness. Increasing taxes in areas or for issues that the society targets have resulted in a more aware public and subsequent behavior changes. The best example is the tobacco smoking taxes in California that reduced smoking, kept secondhand smoke under control, and labeled California as a smoke-free state. Now, around the world, other countries are following that same lead with taxes on tobacco products and prohibitions to smokers in public places.

In theory, it seems straightforward. However, in practice, and in the political world, it is extremely hard to develop international agreements, especially on economic issues. While several schemes are being discussed, the one being promoted the most by Western countries is cap and trade. Proponents say that a fair cap-and-trade system must be comprehensive, operate upstream, allow energy to be auctioned, limit the use of offsets, and have built-in protections for consumers. According to a 2009 report from the Sightline Institute (a Seattle-based think tank), even if the system meets those criteria, it must require nations to set responsible limits on climate change emissions and gradually ratchet down those limits over time. It must also harness the power of the marketplace to reduce emissions as smoothly, efficiently, and cost-effectively as possible, allowing the economy to adjust and thrive (Cap and Trade, 2012).

So, what does cap and trade mean? A "cap" is a legal limit on the quantity of greenhouse gases a nation's economy can emit each year. Over time, the legal limit goes down—the cap gets tighter—until the country hits its targets and achieves a clean-energy economy. The cap serves as a guarantee that a nation reaches its goal. Countries would use energy efficiency standards for vehicles and appliances, smart-growth plans, building codes, transit investments, tax credits for renewable energy, public investment in energy research and development, and utility regulatory reforms to ensure that the goals are met.

"Trade" refers to a legal system that allows companies to swap the ability to emit greenhouse gases among themselves, thus creating a market for pollution permits or allowances. The point of a trading system is to place a price on pollution that is dispersed through the economy, motivating businesses and consumers to find ways to reduce greenhouse gases. By turning the permission to pollute into a commodity that can be bought and sold, everyone up and down the economic ladder gets new opportunities to make and save money. Trade leverages the flexible power of the marketplace—the mobilized ingenuity of millions of diverse, dispersed, innovative, self-interested people—to help meet climate goals.

Several cap-and-trade climate policy schemes have already been created; one is being partially deployed in Europe, and another is being initiated for California. The European cap-and-trade system began in 2005, with a first-phase, 3-year learning period. Despite US criticism that it was a failure, the EU's Emission Trading System (ETS) is stable. According to a report from the World Resources Institute, the system has improved through the lessons learned when it was first implemented, and European companies are confident that carbon pricing is here to stay (GHG, 2012).

Critics claim that the trading system has not changed behavior, but there is evidence that it has. For example, in 2006, only 15 percent of the companies covered by the ETS were taking the future cost of carbon into account. Point Carbon and other researchers found that a year later, about 65 percent of companies in the trading system were making their future investment decisions based on having a carbon price, which is the system's goal (Price of Climate Change, 2012).

The trading system, as an economic model, tries to achieve a balance between supply and demand, which is the core of the elements of the neo-classic economic model used in the 2IR. The problem is that this economic model does not change anything—it allows companies to continue to produce carbon emissions, rather than stopping them. Companies that agree to eliminate their carbon emissions at some point in the future can continue to pollute, postponing their commitment for decades.

A better approach to reducing carbon and stopping greenhouse gases are programs like the Carbon Tax Center. This New York nonprofit organization argues for a carbon tax. They regard a carbon tax as superior to a carbon cap-and-trade system, for five fundamental reasons:

1. Carbon taxes will lend predictability to energy prices, whereas cap-and-trade systems exacerbate the price volatility that historically has discouraged investments in less carbon-intensive electricity generation, carbon-reducing energy efficiency, and carbon-replacing renewable energy.

2. Carbon taxes can be implemented much sooner than complex cap-and-trade systems. Because of the urgency of the climate crisis, we do not have the luxury of waiting while the myriad details of a cap-and-trade system are resolved through lengthy negotiations.
3. Carbon taxes are transparent and easily understandable, making them more likely to elicit the necessary public support than an opaque and difficult to understand cap-and-trade system.
4. Carbon taxes can be implemented with far less opportunity for manipulation by special interests, while a cap-and-trade system's complexity opens it to exploitation by special interests and perverse incentives that can undermine public confidence and undercut its effectiveness.
5. Carbon tax revenues can be rebated to the public through dividends or tax shifting, while the costs of cap-and-trade systems are likely to become a hidden tax as dollars flow to market participants, lawyers, and consultants. (Carbon Tax, 2011).

Arguments and criticism abound over cap-and-trade systems, economics, and their efficacy. Critics point to the big American companies that are creating trading desks to facilitate "carbon credit" trading (Dole, 2012). The market could be worth trillions of dollars as emission reduction becomes an international priority, and critics say that this much money will lead to corruption and cheating. Most of these arguments are being pushed aside as Europe and California adopt cap-and-trade systems. Momentum is building for some way to put a price on pollution. Most likely, an internationally regulated carbon credit trading process will be implemented.

The question is how can climate change be stopped and who will pay for it? Most argue that the voters and politicians need to decide; however, few governments are focused on environmental issues and societal concerns. The GIR economics combines social capitalism with standard economic mechanisms that use externalities. GIR economics is neither the form of economics used in the 2IR nor a totally government-controlled form of finance, regulations, and markets. In short, there is a need for a new economics, which includes externalities, life cycle analyses, and effective monetary policies that will reduce carbon emissions and pollution and can be measured and controlled.

GREEN JOBS: THE GIR'S RESULTS

The Great Recession of 2008 has started to fade away amid signs that the world's financial structure has stabilized and the world economy is in recovery. Regionally, Asia appears the strongest, though China's growth slowed in 2010-2011 but has now picked back up. Some EU nations and the United States are lagging for different reasons. Europe is trying to maintain a unified structure while balancing the wealth and stability of the more prosperous nations such as Germany and France and the Nordic countries, against the near-bankrupt smaller nations such as Greece, Ireland, Spain, and the potentially 25 new central and eastern EU member nations that are rapidly trying to catch up.

The United States, unlike China, Japan, South Korea, and Europe, has been slow to embrace the enormous potential for job creation in the green industrial revolution. While the current presidential administration identifies green job growth as the way to reduce the unemployment level, the US Congress seems locked into the fossil fuel-driven economics and its associated public policies, backed by large corporations that profit from their continuance.

The Chinese have leapfrogged the rest of the world into GIR economics, technology, and job creation (Clark and Isherwood, 2007, 2010). In fact, China is now leading the world with economic and career innovations. Analyzing China's phenomenal economic growth, it is clear how much of that growth can be attributed to the development of green industries. Not only is China creating massive systems to generate renewable energy for its own use, but also it has quickly become the world leader in exporting these technologies. If anyone around the world wants to buy solar panels or wind turbines, China is able to provide the best pricing and quality.

What the Chinese clearly understand, and other nations do not, is that once committed to the GIR, a nation creates new economic development and business opportunities that lead directly to job generation, new career paths, and the revitalization of local economies in a sustainable way. This results in further research and development supporting the GIR. Because of the extraordinary interconnectedness of a modern nation's economy, once local economies start to come back, they revitalize the service industries and corporations that support them—the markets, retail stores, and small businesses, as well as schools, city governments, and all the other public agencies dependent on tax revenues.

So, what exactly is a "green" job? The label is much like the "knowledge-based" job label, a generic term that describes an industry or service rather than a specific type of activity. Raquel Pinderhughes, a professor of urban studies at San Francisco State University, defines green jobs as a generic term for people doing any kind of work, whether mental or manual, that in some way relates to improvements in environmental quality (Pinderhughes, 2006).

The United Nations Environment Programme's *2008 Green Jobs Report: Towards Decent Work in a Sustainable Low-Carbon World* added a subset called "green-collar" jobs. The UN tried for a more rigorous definition of green jobs, saying, "This includes jobs that help to protect ecosystems and biodiversity; reduce energy, materials, and water consumption through high efficiency strategies; de-carbonize the economy; and minimize or altogether avoid generation of all forms of waste and pollution" (UN EP, 2008). Subsequent reports provide data and details for green jobs.

It is difficult to put a number on how many jobs would be created if the US economy focused on climate change mitigation. Economists struggle when they analyze green job data and try to interpret the results. Robert Pollin, codirector of the Political Economy Research Institute at the University of Massachusetts, wrote a report in 2008 that calculated that the United States could generate 2 million new jobs over 2 years with a $100 billion investment in a green recovery (Pollin et al., 2008). President Obama talked about creating 5 million new green jobs in his 2011 State of the Union Address. As with any emerging industry, the actual number of jobs that will be created is hard to predict, although a 2014 report on the solar industry indicated a 400 percent growth in the industry with large numbers of new jobs (Cusick, 2014).

While economists can provide calculations that show on average how many jobs would be created, based on the number of dollars invested, they cannot measure the corollary impact or predict the number of related jobs that will be created. However, one precedent for green job creation is Germany's FiT. Germany attributes strong growth in the renewable energy sector to blunting the recession. According to Deputy Environment Minister Astrid Klug, there were 250,000 jobs in Germany's renewable energies sector and an overall total of 1.8 million in environmental protection. The number of jobs in the renewable energy sector will triple by 2020 and hit 900,000 by 2030.

Another example is California's world-leading energy efficiency program, which has put thousands of workers back to work retrofitting commercial buildings. At the same time, the program is driving the lighting industry to develop extraordinary new products like cost-effective LED lights, dimmable ballasts, and smart networks that make peak load management simple and effective. This new generation of lighting products is transforming the market and along the way creating a $1 billion industry for

California. If LED and dimmable ballast changeovers were made a national priority, it would be a multibillion-dollar industry in the United States alone.

PRIVATE INVESTMENT IS NEEDED

Reducing carbon emissions and mitigating climate change will require a radical transformation in the ways the global economy currently functions. It will require a rapid increase in renewable energy: sharp falls in fossil fuel use or massive deployment of carbon capture schemes, mitigation of industrial emissions, and elimination of deforestation.

Given the growing awareness of the costs associated with global warming, business-as-usual is not an option. The World Resources Institute estimates that to make this transition to the GIR will require massive amounts of capital by the world's nations, about $300 billion annually by 2020, growing to $500 billion by 2030. This compares against the $100 billion annual funding committed by industrialized nations in the UNFCCC's Green Climate Fund.

Yet, this is modest considering the alternative of doing nothing. "The longer we delay the higher will be the cost," said the UN IPCC chairman Rajendra Pachauri in April 2014 (UN IPCC, 2014). The cost is relatively modest now, he added, but only if the world acts quickly to reverse the buildup of heat-trapping gases in the atmosphere. His comments came as the IPCC released a report projecting that shifting the world's energy system from fossil fuels to zero- or low-carbon renewable sources would reduce consumption growth by a minimal 0.06 percentage points per year.

The UN IPCC said large changes in investment would be required. Fossil fuel investments in the power sector would drop by about $30 billion annually, while investments in low-carbon sources would grow by $147 billion. Meanwhile, annual investments in energy efficiency in transport, building, and industry sectors would grow by $336 billion (UN IPCC, 2014).

Given the large amounts of money needed to address climate change, public funds will be a small portion. Unlocking private sector investment will be the key, Rachel Kyte, World Bank vice president for Sustainable Development, told in the September 2012's United Nations General Assembly meeting. Private investors—equity firms, venture capitalists, pension funds, insurance companies, and sovereign wealth funds—currently control several trillions of dollars' worth of assets that can provide climate-smart infrastructure development. They must be more actively engaged, to identify the risks that they perceive in green investment and to develop new policy and financing mechanisms that make for attractive returns (Climate Finance, 2012).

In 2011, $257 billion was invested in renewable energy, according to the Global Trends in Renewable Energy Investment 2012 report. This United Nations Environment Programme-backed study has tracked the finance flowing into green energy since 2004. China attracted more money than any other country with $52.2 billion. Even though 2011 represented the largest investment in renewable energy, renewables still only represented about 6 percent of the world's energy requirements. So, there's a huge upside potential for investment capital (Renewable Finance, 2012).

Goldman Sachs, the world's premier investment bank, noted on its website that harnessing an important natural resource like wind requires a substantial amount of capital. That is why they helped renewable energy companies find the capital they needed to embrace this innovative technology. They go on to note that they have invested $3 billion in clean energy since 2006 and raised more than $10 billion in financing for clean-energy clients around the world since 2006 (Sachs, 2009).

GOOGLE INVESTS OVER $1 BILLION IN GREEN TECH

Google, Inc., perhaps the most successful and innovative technology company in the world, is a gigantic user of energy. The company has farms of computer servers spread throughout the world to power it search engine function and related information technology (Jin, 2010). In Q1 2014, the company said that they spent $2.25 billion on data center and infrastructure spending, a major cost for the company. Google is also one of the world's most aggressive major companies when it comes to advancing a clean-energy agenda, investing over $1 billion in renewable energy (Google, Inc., 2012).

The company says that they are getting about 34 percent of their power with renewable energy, either directly from solar panels or indirectly by buying green power near their data centers. Google says that renewable energy projects must make good business sense and have potential for long-term significant impact. In 2007, they installed the largest corporate solar panel installation of its kind—a 1.9 MW system that provides 30 percent of the peak load for its massive Mountain View headquarters.

While the company says that they are striving to power their operation completely with renewable energy, they are also investing heavily in large-scale projects (Jensen and Schoenberg, 2010). As of April 2014, they had put money into 15 solar and wind power generation projects, totaling over 2 GW of power. While some of the projects are in South Africa and Germany, most are in the United States like the huge Ivanpah project in the Southwest desert. The project is one of the largest solar thermal farms in the world and uses 347,000 sun-facing mirrors to help produce 392 MW. Ivanpah's green energy powers more than 140,000 California homes.

Google says that they are investing in green energy because doing so makes it more accessible. Rick Needham, Google's director of energy and sustainability, says that their efforts in procuring and investing in green energy all make business sense. "They make sense for us as a company to do," he said. "We rely on power for our business" (Google, Inc., 2012).

If Goldman Sachs, one of the icons of 2IR economy, and Google, one of the leaders in the GIR, are prominently promoting GIR investments, it must mean that venture capital investors are starting to pay close attention. Venture Business Research (VB/Research), a leading data provider, tracks the international financial activity in the green industry. They reported in the second quarter of 2010 that venture capital and private equity investment in clean technology and renewable energy exceeded $5 billion worldwide, despite a 30 percent decline in early-stage venture capital activity. They also reported that a record number of merger and acquisition deals valued at over $14.5 billion were transacted during that period.

They noted that the initial public offering sector was dominated by Chinese companies, which accounted for 75 percent of new issues, and said, "The market lacks pre-credit crunch exuberance but has recovered significantly from the moribund levels of 2009" (Venture Business research, 2010). Then, they noted that the $5 billion invested during the second quarter of 2010 in global venture capital and private equity funds was slightly ahead of average quarterly investment of $4.9 billion since the second quarter of 2008 and 45 percent above the low 2009 level.

The bulk of the money being invested in green and clean technology is going to China, according to the accounting firm Ernst & Young. The firm does a quarterly assessment of the most attractive countries for renewable energy investment. In their September 2010 report, they noted that China was now the world's biggest energy consumer and the world's most attractive country for renewable energy investment. China has been encouraging investment in its clean-energy companies as part of its goal of generating

15 percent of its electricity from renewable sources by 2020 (Energy Assets, 2012).

Ernst and Young also compared regulations, access to capital, land availability, planning barriers, subsidies, and access to the power grid. The report ranked investments in onshore and offshore wind, solar, biomass, and geothermal energy projects. After China, the next most attractive countries for renewable energy investment were the United States, Germany, India, and Italy. Ernst and Young noted that government support in China gives it a huge advantage over other countries in pursuing clean-energy projects. In the second half of 2009, China almost doubled consumer subsidies for generating renewable power, bringing the amount to $545 million. The fact that China has both five-year plans and substantial financial support for the green technology sector gives it a huge advantage. Even more significantly, the Chinese require government participation or partial ownership of many of the new firms, in keeping with their tradition of government control of infrastructures.

The Ernst and Young report underscores the extraordinary progress China has been making. Clearly, China recognizes the enormous potential for economic growth that can occur through a concentrated national policy to mitigate climate change. Over the last decade, they have pushed past the United States and Europe in renewable technologies, business development, green job generation, and economic revitalization. Not only has China taken the lead in renewable energy technologies, they have been far advanced in developing a national planning process that sets environmental protection as a priority. Now, as Ernst and Young's report makes clear, they are capturing the world's investment capital.

There are huge amounts of money at stake, as the world moves toward a carbonless economy. The big losers will be the entrenched fossil fuel and centralized utility industries. Both industries have business plans dependent on increasing demand, and neither is capable of adjusting to a declining demand environment. Without a doubt, some of the world's most powerful and wealthiest industries have much to lose. It is no wonder that the fossil fuel interests are going to such lengths to argue against climate change science and putting enormous pressure on governments and geopolitics to maintain their privileged status.

However, the green industrial revolution is not all about money. In fact, the world is realizing that there are other tremendous costs—in human lives and reduced GDP—by not reducing our dependency on carbon-generated energy. The GIR is about climate change mitigation, renewable energy, smart grids, and environmental sensitivity. But achieving the benefits of the GIR—a wave of new technologies, business enterprises, and green jobs—will require substantial public and private financing. The green GIR economy will be needed to accelerate the necessary changes and stop climate change.

REFERENCES

Monitor Climate, 2012. http://daraint.org/climate-vulnerability-monitor/climate-vulnerability-monitor-2012/.

UN IPCC, April 2014. Fourth report 2007 and WGII AR5, phase 1 report launch.

Clark II, W.W., 1997–1999. First Research Director, Framework Convention on Climate Change (FCCC) Bonn, Germany and UN EPA Paris, France.

US Department of Defense (U.S. DOD, Quadrennial Defense Review 2014, "Executive Summary" Secretary of Defense, Washington, DC). http://www.defense.gov/pubs/2014_Quadrennial_Defense_Review.pdf.

DARA, 2012. http://daraint.org/climate-vulnerability-monitor/climate-vulnerability-monitor-2012/.

Huffington Post. Reuters, 2014. http://www.huffingtonpost.com/2012/09/26/climate-change-deaths_n_1915365.html.

Stern, N., 2006. What is the economics of climate change? World Econ. 7 (2).

International Energy Agency (IEA), 2014. http://www.worldenergyoutlook.org.

NY Times Hurricane Sandy (NYT), 2012. http://www.nytimes.com/2012/11/28/opinion/hurricane-sandys-rising-costs.html?_r=0.

Chipotle, 2014. http://www.huffingtonpost.com/2014/03/04/chipotle-guacamole-climate-change_n_4899340.html.

Forbes, 2010. http://www.forbes.com/2010/04/01/ge-exxon-walmart-business-washington-corporate-taxes.html.

NY Times (NYT), 2011. http://www.nytimes.com/2011/05/13/business/13oil.html?_r=0.

NY Times (NYT), Coal Ash, 2014. http://www.nytimes.com/2014/03/01/us/coal-ash-spill-reveals-transformation-of-north-carolina-agency.html?hpw&rref=us&_r=0.

CNN West Virginia, 2014. http://www.cnn.com/2014/02/19/health/west-virginia-water/.

Reuters, February 16, 2011. Coal's Hidden Costs Top $345 Billion in U.S. Study. Boston. http://www.reuters.com/article/2011/02/16/usa-coal-study-idUSN1628366220110216.

Clark II, W.W., Fast, M., 2008a. Qualitative Economics: Toward a Science of Economics. Coxmoor Press, London, UK.

Steiner, R., 2014. The True Cost of Our Oil Addiction. Huffington Post. http://www.huffingtonpost.com/richard-steiner/true-cost-of-our-oil-addiction_b_4591323.html.

Copulos, M.R., January 8, 2007. Testimony of Copulos, President, National Defense Council Foundation, before the Senate Foreign Relations Committee, March 30, 2006; IMF, 2008; EIA; CleanTech Group, 2007; US Census Bureau; Experian Automotive; Paper presented to Congressional staff members by NDCF President Milt Copulos. http://www.sapphireenergy.com/learn-more/59518-the-true-cost-of-oil-to.

Economist, July 16, 2009. Collapse of Modern Economic Theory, Cover page and Special Section.

Clark II, W.W., 2012. The Next Economics: Global Cases in Energy, Environment, and Climate Change. Springer Press, New York, NY, December 2014.

Li, X., Clark, W.W., 2009. Crises, opportunities and alternatives globalization and the next economy: a theoretical and critical review. Chapter #4, In: Xing, L., Winther, G. (Eds.), Globalization and Transnational Capitalism. Aalborg University Press, Denmark.

Fast, M., Clark II, W., 2012. Qualitative Economics: A perspective on Organization and Economic Science, Theoretical Economic Letters, Scientific Research, Online January. http://www.scirp.org/journal/teldoi:10.42362011.

Clark II, W.W., Fast, M., 2008b. Qualitative Economics: Toward a Science of Economics. Coxmoor Press, London, UK.

Clark II, W.W., Cooke, G., 2011. Global Energy Innovation: Why America Must Lead. Praeger Press, Santa Barbara.

Clark II, W.W., Cooke, G., 2014. The green industrial revolution. Chapter #2, Global Sustainable Communities Design Handbook: Green Design, Engineering, Health, Technologies, Education, Economics, Contracts, Policy, Law and Entrepreneurship. Elsevier Press, Amsterdam.

Clark II, W.W., 2009. Sustainable Communities. Editor and Author, Springer Press, New York, NY.

Clark II, W.W., 2010. Sustainable Communities Design Handbook. Editor and Author, Elsevier Press, New York, NY.

Clark II, W.W., 2014. Global Sustainable Communities Design Handbook. Editor/Author, Elsevier Press, New York, NY.

Clark II, W.W., Li, X., 2012. Social Capitalism: China's Economic Rise. The Next Economics. Springer Press, New York, NY, pp. 143–164.

Sun, X., Li, J., Wang, Y., Clark, W.W., 2013. China's sovereign wealth fund investments in overseas energy: the energy security perspective. Energy Policy 65, 654–661.http://dx.doi.org/10.1016/j.enpol.2013.09.056.

Gipe, P., 2010b–2014. Feed-in-tariff monthly reports. http://www.wind-works.org/FeedLaws/RenewableTariffs.qpw.

Morris, C., 2014. Energiewende—Germany's community-driven since the 1970s. Chapter #7, In: Global Sustainable Communities Design Handbook: Green Design, Engineering, Health, Technologies, Education, Economics, Contracts, Policy, Law and Entrepreneurship. Elsevier Press.

Federal Ministry for the Environment (GR ME), 2000. Nature Conservation and Nuclear Safety, Germany.

Gipe, P., 2010–2014. Feed-in-tariff monthly reports. http://www.wind-works.org/FeedLaws/RenewableTariffs.qpw.

Renewable Energy Policy Network for the 21st Century, 2009.

Borden, E., Stonington, J., 2014. Germany's Energiewende. In: Global Sustainable Communities Design Handbook: Green Design, Engineering, Health, Technologies, Education, Economics, Contracts, Policy, Law and Entrepreneurship. Elsevier Press.

Nijaki, L.K., 2012. The green economy as sustainable economic development strategy. In: Clark II, W.W. (Ed.), The Next Economics: Global Cases in Energy, Environment, and Climate Change. Springer Press, New York, NY, pp. 251–286, Fall.

Cap and Trade, 2012. http://www.sightline.org/wp-content/uploads/downloads/2012/02/Cap-Trade_online.pdf.

Greenhouse Gas (GHG) EU Mitigation, 2012. http://www.wri.org/publication/greenhouse-gas-mitigation-european-union.

Price of Climate Change, 2012. http://www.huffingtonpost.com/thomas-kerr/paying-price-climate-change_b_2206791.html.

Carbon Tax Center, 2011 http://www.carbontax.org/

Dole Jr., M., 2012. Market solutions for climate change. In: Clark II, W.W. (Ed.), The Next Economics: Global Cases in Energy, Environment, and Climate Change. Springer Press, New York, NY, pp. 43–70, Fall.

Clark II, W.W., Isherwood, W., 2010. Creating an Energy Base for Inner Mongolia, China: the leapfrog into the climate neutral future, (Utilities Policy Journal, 2010).

Clark II, W.W., Isherwood, W., 2007. Energy Infrastructure for Inner Mongolia Autonomous Region: five nation comparative case studies, Asian Development Bank, Manila, PI and PRC National Government, Beijing, PRC.

Pinderhughes, R., 2006. Study of Green Jobs: Small businesses. http://www.sfsu.edu/~news/2008/spring/15.html.

UN Environment Programme (UN EP), 2008. Green Jobs Report: Towards Decent Work in a Sustainable Low-Carbon World. http://www.unep.org/documents.multilingual/default.asp?documentid=545&articleid=5929&l=en.

Pollin, R., Heidi G.-P., Heintz, J., Scharber, H., 2008. Green Recovery: A Program to Create Good Jobs and Start Building a Low-Carbon Economy, Center for American Progress and Political Economy Research Institute, University of Massachusetts.

Cusick, D., 2014. Solar Power Grows 400 percent in only Four years, Scientific American.

Climate Finance, 2012. www.insights.wri.org/news/2012/10/wri-launches-project-climate-finance-and-private-sector.

Renewable Finance, 2012. http://www.cnn.com/2012/06/12/world/renewables-finance-unep/.

Sachs, G., 2009. Goldman Sachs makes over $10 billion on profits. www.washingtonpost.com.

Jin, A.J., 2010. Transformational relationship of renewable energies and the smart grid. In: Clark II, W.W. (Ed.), Sustainable Communities. Springer Press, New York, NY, pp. 217–232, Editor and Author.

Google Inc., 2012. http://www.cnbc.com/id/101417698.

Jensen, T., Schoenberg, D., 2010. Google's Clean Energy 2030 plan: why it matters. In: Clark II, W.W. (Ed.), Sustainable Communities. Springer Press, New York, NY, pp. 125–134, Editor and Author.

Venture Business Research Limited, 2010. www.Bloomberg.com/news/venturebusinessresearch.

Energy Assets, 2012. http://www.ey.com/Publication/vwLUAssets/Renewable_energy_country_attractiveness_indices_-_Issue_27/$FILE/EY_RECAI_issue_27.pdf.

CHAPTER 13

Smart Sustainable Communities: The Way We Will Live in the Future

The icon for modern civilization is the city. Cities such as Beijing, New York, Paris, Buenos Aires, Tokyo, Dubai, and London have extraordinary, glittering monuments to civilization that have attracted residents for centuries. The lure of cities are why people in developing nations give up subsistence farming and where they go to find work and get a paycheck, social services, and, most importantly, schools for their children so they can escape the poverty cycle. To do all this, cities require large infrastructures of which water, waste, transportation, and energy are key. It is the green industrial revolution that will change cities into sustainable communities.

Today, half the world's population of 7 billion people lives in cities. Assuming these trends continue, by midcentury, 75 percent, or about 7 billion people, will live in a city. China already has several cities with more than 10 million residents and regions such as Shanghai with 30-40 million. In 1975, there were five megacities with populations of more than 10 million around the world. By 1995, there were 14 megacities mostly in China, and, according to the United Nations, by 2015, there will be 26, with at least two-thirds of them in China (UN DESA, 2011), which appears now (2014) to be the case in less than a year.

Humanity has been trending toward ever-larger social organizations since humans gave up the hunter-gatherer lifestyle. The more humans evolve, the less willing they are to tolerate a rural or agrarian existence. The lures of bright lights, easy money, and the rest of modernity are too much to ignore, and so the steady march to the cities continues.

While planners, architects, and politicians have watched and profited from this phenomenon, they have allowed most cities, particularly those in America, to grow without regard to available resources or ecological impacts. Why else would a city like Las Vegas be allowed to sprawl unchecked across a fragile desert without apparent concern? There are about 2 million people living in this water-constricted environment. It's just a question of time until the water runs out.

LOS ANGELES: THE CAR-CENTRIC 2IR CITY

California's Los Angeles is the United States' second largest city behind New York. It is the classic second industrial revolution city. Like many cities built during the post-World War II period, it was shaped without regard for ecology or the environment. Towns, neighborhoods, and communities line the freeways that were built in Los Angeles in the 1950s, systematically replacing train tracks that were built and used since the 1920s (Los Angeles, 2014).

The freeways replaced early railroad systems and ran into one another, without regard for geography or logic. Today, the Greater Los Angeles Area has a population of about 15 million people. It is a region locked into a fossil-fuel addiction, a car-centric culture dependent on a network of interlaced freeways and highways. The Los Angeles 2020 Commission describes in its report, "A Time for Truth" that "Los Angeles is barely treading water while the rest of the world is moving forward. We risk falling further behind in adapting to the realities of the 21st century and becoming a city in decline" (Los Angeles, 2013).

Not only does Los Angeles have the largest high-speed roadway network in the world, but also it has the highest *per capita* car population in the world. While the United States may be the country with the most registered vehicles (around 232 million), California and in particular Los Angeles are the places most committed to and dependent on the automobile—the backbone of the 2IR. The Greater Los Angeles Area holds the highest concentration of cars in the world, with more than 26 million, or about 1.8 cars per person. Los Angeles is the world's most car-populated area of urban sprawl in the world.

The Los Angeles freeway system handles more than 12 million cars daily. This helps explain why Los Angeles holds the number one spot on the list of North America's most congested and polluted roadways, according to the 2014 Congestion Index. Drivers in Los Angeles spend an average 34 percent more time on the road than during the odd times when traffic is flowing freely (Los Angeles, 2013: 1).

More automobiles and more driving equal more air pollution. The Los Angeles Basin, which is susceptible to atmospheric inversion (that is, a meteorologic anomaly that traps pollution and keeps it close to the ground), contains exhaust from road vehicles, airplanes, locomotives, shipping, and manufacturing (Los Angeles Highway Congestion, 2013). Millions of tons of toxins are released daily into the atmosphere. With a semidesert climate, Los Angeles does not get much rain, so pollution accumulates as a dense cover of smog, threatening the health and well-being of the residents (Los Angeles Highway Congestion, 2013). This heavy layer of carbon-intensive pollution finally triggered enough concern that lawmakers passed environmental legislation, including the Clean Air Act.

As a result of this and other efforts, pollution levels in California have dropped in recent decades. Despite improvement, the 2014 annual report of the American Lung Association of California (ALAC) ranked Los Angeles as the most polluted city in the country (ALAC, 2014). In addition, Los Angeles groundwater is increasingly threatened by methyl tertiary butyl ether (MTBE) from gas stations and perchlorate from rocket fuel (SCAQMD, 2010).

The Greater Los Angeles Area is the uncontested monument to the carbon-heavy, fossil fuel-driven second industrial revolution. It represents everything that was once glorious about cheap fossil fuel but now threatens the world with global warming and environmental degradation (SCAQMD, 2010). Neither Los Angeles nor its lifestyle will easily adapt to a world with less carbon emissions. However, for the sake of the planet, cities like Los Angeles will have to make significant changes, and the sooner the better, if we are to avoid a catastrophic increase in global temperatures.

SUSTAINABLE COMMUNITIES ARE THE ANSWER

The answer for Los Angeles and all cities and communities is to become green and sustainable. Cities must become more walkable, bike friendly, and livable. They need to focus on environmental sustainability as well as economic sustainability. The quality of urban space must improve. The architecture should be inventive with sensitive urban design and a dynamic atmosphere. Sustainable living and sustainable business development must be promoted, along with infrastructure needs of water, recycling, transportation, waste, and materials. Above all else, a green sustainable city needs to generate renewable energy and use energy storage with a smart integrated grid system to balance and share energy (Clark, 2009, 2010, 2014).

Sustainable communities started in Europe and Japan as a reaction to the Arab oil embargo in the mid-1970s. They gained a toehold in Germany, then Denmark, and the Netherlands. These countries had historically used wind power as a source of renewable energy, and so they quickly embraced sustainability. Already experts with transmission and grid connections, they created on-site energy systems for farmers, communities, and towns.

Norway, for example, was one of the largest oil-producing nations in the world with its offshore drilling but realized that there was a limit to oil so planned in the 1990s to convert its economy into other areas, including the creation of hydrogen from renewable energy sources.

Other European nations followed. Germany led the way in solving the problem of paying for these renewable energy systems with its feed-in tariff, which made it one of the world's largest manufacturers and installers of solar panels. Scotland and Britain have major wind farms and biomass generation plants. Pushed by the EU, Spain, Italy, and the other European countries have adopted policies, governmental programs, and economic plans to move sustainable communities and the GIR forward, reaping environmental and economic benefits along the way.

With a long tradition of conservation and sustainability, Japan led the movement in Asia (Funaki and Adams, 2009). After their blitz modernization during the first half of the 20th century ended with defeat in World War II, the Japanese regained their historic sense of sustainability. Of course, their entrance into the GIR was also triggered by a glaring lack of natural energy resources. Unfortunately, and despite being the target for US atomic bombs in WWII, the Japanese started building nuclear power plants a few decades later. In the 1970s, many nuclear plants were built to provide power to the growing Japanese economy. The tragic result was Fukushima, which is now a lesson for all nations leaning toward nuclear power generation.

Today, the rest of Asia with Chinese leadership has leaped into the GIR. Granted, the region has horrendous pollution and environmental issues to overcome, but it is rapidly developing renewable energy technologies. China now leads the world in solar and wind technology production.

The common theme in all these nations, and a theme that is lacking in America, is the notion that sustainability starts at home, in the behaviors and values of families. Communities are usually described as a group of interacting people living in a common location. Psychologists describe a community as one that includes a sense of membership, influence, integration, and fulfillment of needs, as well as shared emotional connections. Generally, communities are organized around common values or beliefs. They share resources; organize around a political structure; agree on preferences, needs, and risks; and agree to tax themselves for the benefit of the whole.

A smart green community has these elements plus a core value of conservation, a respect for natural resources, and an appreciation for the environment (Clark, January 2014). The concept of "achieving more with less" is broadly endorsed. Sustainability is a community-centric activity: the more focused and integrated the community, the more it has a chance of achieving sustainable development through environmental policies that reduce, recycle, and reuse. For power, add energy that comes from renewable resources. The payoff from sustainable development is a cleaner environment and a healthier lifestyle.

Besides these core values, sustainable communities combine common social activities with education, business development, and job creation. This concept was first defined as *sustainable development* more than a quarter of a century ago. The Brundtland Commission, convened by the United Nations in 1983, introduced the term in its *Our Common Future* report in 1987. In that same report, named for its chairperson Ms. Gro Harlem Brundtland, Prime Minister of Norway in the 1980s, the UN commission addressed emerging concern about "the accelerating deterioration of the human environment and natural resources and the consequences of that deterioration for economic and social development" (Brundtland Commission, 1987).

Even in the 1980s, the UN General Assembly recognized that environmental problems were global in nature. Subsequently, it was in the common interest of all nations to establish policies for sustainable development. The Brundtland Commission defined sustainable development as meeting the "needs of the present without compromising the ability of future generations to meet their own needs" (Brundtland Commission, 1987). Today, the term describes how a community's economic concerns interact with its natural resources. Addressing large global problems locally can generate new creative ventures and opportunities, which then provide strong business reasons to pursue sustainable development.

All sustainable communities must address the essential infrastructure elements of energy, transportation, water, waste, and telecommunications. The critical component is renewable energy power generation. Renewable energy provides power generation in harmony with the environment and economic development. Europe and Japan have developed numerous communities that are sustainable and secure through the use of their own renewable energy sources, augmented by storage devices and emerging technologies.

DENMARK

As a nation, Denmark is close to a classic sustainable community model. With gasoline costing $10 a gallon, Danes prefer to take public transportation. In the mid-1990s, Denmark established a CO_2 tax to promote energy efficiency, despite discovering offshore oil. Instead of risking environmental issues with offshore drilling, the Danes installed offshore wind farms. Since 1981, Denmark's economy has grown 70 percent and energy consumption is almost flat (Lund and Clark, 2002).

Denmark's early focus on solar and wind power has paid off, as those sources now provide more than 16 percent of its energy. One-third of the world's terrestrial wind turbines now come from Denmark, and Danish companies Danisco and Novozymes are two of the world's most innovative manufacturers of enzymes for the conversion of biomass to fuel (Lund and Østergaard, 2010).

As an additional benefit, these industries have provided green jobs. In 1970s, people from the northern and western regions were flooding Denmark's capital city of Copenhagen, an island in the nation's east. The government moved to reverse that massive movement to one city by establishing universities, businesses, and economic development in Northern Jutland in the west. People moved back with even more coming from other nations and cities to live in vast regions that were once being abandoned. The move into the Northern Jutland has paid off tremendously, attracting companies that are large, profitable, and sustainable (Clark and Jensen, 2002).

On a local or regional level, a sustainable smart community must have three components (Lund and Østergaard, 2010):

- First, it must have a government-accepted strategic master plan for infrastructure that includes renewable energy, transportation, water, waste, and telecommunications and the technologies to implement and integrate these systems.

- Second, facility planning and financing must be addressed from a green perspective. The planning must include standards that can be measured and evaluated. There are an array of issues pertaining to the design, architecture, and sitting of buildings, complex, and residents that affect sustainability. The community needs to consider efficient conservation and generation of resources that apply to multiple-use design. For example, communities should be dense and walkable and integrate transportation choices powered by renewable sources, which reduce energy consumption as well as eliminate greenhouse gases, carbon emissions, and pollution.

- Third, a sustainable community is a vibrant, "experiential" applied model that should innovate, catalyze, and stimulate entrepreneurial activities, education, and creative learning, along with research, jobs, and new businesses.

As the sustainable movement has developed, communities have sought out policies that direct facilities and infrastructures to be smart and green. Originally, there were several certification processes; however, ENERGY STAR and the US Green Building Council's Leadership in Energy and Environmental Design (LEED) certifications are becoming universal. There are regional LEED certification councils in most parts of the world, and there are now community neighborhoods using LEED standards (Clark and Eisenberg, 2008).

Buildings need to be designed so that they are environmentally sound. The design and construction of buildings need to consider the shift toward on-site power through renewable energy production. Small, relatively self-contained communities within larger cities are more easily made sustainable. It is simpler on a local level to reduce the dependency on central grid energy (Clark, 2004).

Local on-site power is more efficient and can use the region's renewable energy resources. For example, Denmark's many sustainable communities are generating energy with wind and biomass to provide base load power (Lund and Clark, 2008). Denmark has a goal of 50 percent renewable energy generation by 2020 (Lund and Ostergaard, 2010). The country is well on its way to meeting and perhaps exceeding that national goal.

Europe's combined heat and power systems were developed to meet local needs, reduce the use of fossil fuels, and help communities become energy-independent and more self-sufficient. Some US communities are now developing similar systems focused on renewable energy and using cogeneration or combined heat and power systems (Andersen and Lund, 2007).

AGILE SYSTEMS

The second industrial revolution's energy model generated power from a central plant dependent on fossil fuels or nuclear energy. Historically, governments wanted to centralize power for the public good with control, oversight, and regulatory measures. The power was transmitted or piped great distances over a rigid, one-way grid to the user. The old central grids required long transmission lines, pipelines, or ships to deliver the raw fossil fuel for processing. Then it needed to be transmitted and distributed to the end user. The standard approach was for municipalities to manage the capital costs for the central plant with its processing of raw materials but for rate-payers to absorb the transmission costs.

As we transition to the GIR, a hybrid or integrated model has been developed. This new model is agile, that is, flexible, because it can accommodate both green on-site power generation and grid-connected power (Clark and Bradshaw, 2004). Agile systems combine electricity from on-site renewable energy with electricity from traditional central grids hundreds of miles away and manage them both to meet demand.

The agile system is efficient, smart, and rooted in renewable energy power generation. Although there is usually a central grid that depends heavily on fossil fuels to generate power, agile systems allow and even encourage on-site renewable energy sources and then disperse the electricity accordingly. These distributed energy systems can be formed and operated on the local level to serve targeted communities and consumers (Clark and Bradshaw, 2004).

College campuses are examples where a central grid can be combined with green buildings that have rooftop solar panels or in some cases wind or biomass for local renewable energy generation. Solar or wind power can generate significant amounts of power for clusters of buildings, but on days that are not sunny or windy, these buildings must rely on the central energy grid (Eisenberg, 2009).

Regional- and community-level solutions are needed to address the challenges of global warming and climate change. Rather than having centralized power plants that use fossil fuels or nuclear power to generate energy and then transmit it over power lines, local on-site generation of power from renewable sources is better for the environment, far less expensive, and much healthier for the planet.

The energy needs for communities around the world are growing more complex as population increases, cities expand, and power demands climb. Air pollution and water pollution are causing serious and costly health problems for young and old. Regulators are now implementing carbon dioxide regulations to stop pollution. Meeting the challenges of supplying energy for increasing demand while reducing carbon emissions calls for more complex and creative solutions. It requires energy efficiency, renewable energy generation, and new systems to change the way people live and think about using electricity.

Despite its car-centric lifestyle, California is trying to be on the forefront of this transition. Several small California communities like Santa Monica, Berkeley, Beverly Hills, and San Diego are using their own on-site energy from renewable sources (McEneaney, 2009). They are sustainable models for other communities that want to generate their own power from renewable sources. While some communities may be better suited for solar, or PV, systems, others (e.g., mountain communities) may be able to use run-of-the-river systems to generate power. California has considerable amounts of geothermal power as well as ocean and wave power along with significant numbers of wind farms and solar installations. Hydrogen can be produced from renewable or green sources and then stored close to the needs and demands of communities.

Energy systems are evolving and agile energy systems are emerging. In the future, the central grid will be used for redundancy and backup purposes or act as a battery for energy storage when the sun is not shining and the wind is not blowing. These agile systems are not only technologies or market mechanisms but also a new model that is part of the GIR.

SUSTAINABILITY STARTS AT HOME

Researchers and political decision makers around the world are slowly recognizing that they need to do something about climate change and global warming. Sustainability is achievable. It can be done, and must be done, at the community level. Block by block, communities can change how they live.

Mobility is an essential characteristic to the modern lifestyle and it, along with animal protein, is the most sought-after requirement of people emerging from poverty. This behavior and the desire to experience the American dream as personal freedom will not change easily. WardsAuto reported that in 2010, the global number of cars exceeded 1.015 billion, jumping from 980 million the year before, despite the global recession in October 2008.

However, it is critical to reduce, eliminate, and replace fossil-fuel transportation. Pressured to increase gas mileage and reduce carbon emissions, the vehicle industry is making huge, almost revolutionary, changes. Automakers are upgrading gasoline engines, using more efficient turbochargers with computer-assisted transmissions. Ford is substituting aluminum for steel to shave 500 pounds off their F-150 trucks, creating profound gas mileage savings.

Diesel, as opposed to gasoline, is now the main fuel used to transport goods and people worldwide. Diesel is heavier and oiler than gasoline; however, it emits lower amounts of carbon monoxide, hydrocarbons, and carbon dioxide than gasoline.

Researchers, particularly in the EU, have made diesel more environmentally friendly by reducing particulate emissions per mile. The great advantage of diesel over gasoline is that diesel engines are far more efficient and durable, so a diesel-powered auto gets much better mileage and burns cleaner than a comparable gasoline engine. In some areas, diesel is being replaced by natural gas to power city buses and fleet cars and achieve dramatic fuel savings.

Nonetheless, diesel and natural gas are both fossil fuels with substantial emissions. They may be cleaner, but they are not green and the costs are dramatically high for the environment and human health. More and more research shows that the emissions and pollution from fossil fuels are particularly harmful to the older adults and young children, costing uncountable amounts of money.

Research with this focus is being done by health agencies in California as well as the American Lung Association of California. Reports and standards are showing how cities rank throughout California and the nation on lung disease, with a new focus on emissions and greenhouse gases. To support these fossil fuels with tax breaks and changes in environmental rules will only postpone the eventual use of clean hydrogen for vehicular transportation.

Beginning in 2016, most, if not all, vehicles will be fueled by hydrogen from cells that are charged by renewable resources. Lead by state governments such as New York and California, supply, demand, and economics will force this transformation. Every automaker is planning on having thousands of hydrogen fuel cell cars in California after the state awards fueling for refueling stations in 2016.

Until then, the auto industry is using a mix of more efficient natural gas and diesel engines and hybrid and electric-assisted engines to save gasoline. Meanwhile, the price for oil and natural gas rises despite new sources from shale oil, and hydraulic fracking threatens water tables and the environment (Radow, 2011). It begs the question, why go with toxic fossil fuels when clean hydrogen is around the corner?

There is nothing new about hydrogen, electric, or hybrid cars. Germany's Lohner-Porsche Carriage originally developed hybrid electric cars in 1903. Henry Ford ran his cars originally on biofuels from his agricultural products. These early hybrids disappeared as gasoline-powered engines came to dominate the industry before the Great Depression. It was not until Toyota introduced the Prius in 1997 that hybrid vehicles began their extraordinary success story.

While introduced first in Japan, Toyota took the bold step in launching Prius in 2001 into the California market with 2000 vehicles. The cars underwent several upgrades making them bigger, faster, and even more efficient.

In 2012, Toyota introduced a plug-in Prius hybrid. A common sight as taxis in world capitals, the Prius with its 4.4 kWh lithium-ion battery has proved to be remarkably durable and appealing. Toyota says that in June 2013, total Prius sales worldwide reached 3.8 million cars, making it the leading car manufacturer in the world, replacing GM.

Most of the major automakers are now offering hybrid versions of their cars, and hybrids are gaining steadily in popularity. Manufactures like Chevy and even Porsche are offering plug-ins that are gaining market share. However, Tesla Motors out of California's Silicon Valley has proved to be the true game changer.

Named after Nikola Tesla, a late 19th-century electrical engineer and physicist, the car uses an AC motor directly descended from Tesla's original 1882 design. Barely 10 years old, Tesla Motors was founded by Elon Musk, a successful Silicon Valley high-tech entrepreneur. Musk set out to prove that an all-electric car could compete directly with the world's high-end luxury automobiles. In 2006, the company released the Tesla Roadster, which was the first highway-capable all-electric vehicle in serial production in the United States. The car was the first BEV (all-electric) car to travel more than 200 miles per charge and could do over 125 miles per hour. Selling for over $110,000, it was featured as Time Magazine's "Best Invention 2006—Transportation Invention."

While the Roadster sold in limited quantities, Tesla's next auto, the Model S, has seen strong demand. Released in 2012, the Model S is a sleek four-door sedan that has become the "go-to" luxury car of Silicon Valley and Northern California techies. They can be seen plugged into electric fuel stations at tech campuses like Google and Intel or cruising through San Francisco. Parking structures and airports now have recharging stations, but the demand is so high, and electric car owners are all having trouble finding places to recharge their cars.

Beautifully designed with high performance and a range of over 260 miles, the Model S sells for around $80,000. In Europe, the Model S was Norway's top-selling car in September and December 2013. Also in 2013, the Model S was the top-selling luxury car in the United States, outselling Mercedes-Benz, BMW, Lexus, Audi, and Porsche. Globally, the Model S sold over 25,000 units in 2013 and sales are even stronger in 2014.

Some cities are considering banning autos all together. Hamburg, Germany's second largest city, has laid out an initial concept that would eliminate cars by 2034. Named the Green Network Plan, it would expand public transportation and add more routes for pedestrians and bicyclists.

A steady phaseout of automobiles in the center of the city would take place over the next two decades.

London and Florence have adopted green rings that discourage autos and encourage pedestrians. Paris and other cities faced with congestion are examining similar bans and restrictions. Even New York City has made parts of Manhattan, including blocks near the theater district, into pedestrian zones. The idea of banning or reducing automobiles in city centers has become a hot topic among urban planners struggling to deal with issues like congestion and smog.

Banning autos or using hydrogen or electric cars is not enough to achieve sustainability. More is needed, and throughout the world, communities are developing plans for smart, sustainable futures. The use of fossil-fuel energy sources are losing political and community support as advanced GIR technologies are developed. For example, Britain, a country that for years maintained its prosperous lifestyle as a financial go-between for the Middle Eastern oil trade, recently hit a milestone of 1 GW of installed offshore wind turbine capacity with the completion of the Gunfleet Sands and Robin Rigg wind farms. Future plans call for the development of 25 GW from offshore wind farms, with more than 7000 wind turbines. Scotland, the Nordic countries, and Germany have been taking similar actions in political and economic decisions based on their national plans, technologies, and business leadership.

In South America, Brazil is 95 percent energy-independent through a combination of sugar cane ethanol and domestic oil supplies. To the west, Chile is developing renewable energy as a power source, after numerous public demonstrations against additional hydroelectric dams. Chile is one of the world's most beautiful countries, with wild, pristine regions interlaced with extraordinary free-flowing rivers. The Chilean public is adamant about keeping the rivers free from additional hydroelectric dams, and the government is responding by opening the door to renewable energy. In Mejillones in the Atacama Desert region of northern Chile, Algae Fuels S.A. consortium is using microalgae in second-generation biodiesel production. Wind and solar energy development is also headed to the region.

In other parts of the globe, China and Spain are developing sustainability through public policies that support renewable energy power generation, such as FiTs, which fix rates to provide rebates to consumers. Communities in Japan have been sustainable for many years, since Japan must either import all of its energy or generate it within the island nation. Increasingly, Japan is using renewable energy, and until 2008, it was the world leader in solar manufacturing companies. Now, Italy has been active regionally, for different reasons but primarily due to the national historical regionalization and its city-focused policies and programs. In the Baltic Sea region that was part of the Soviet Union, Lithuania has been active and begun a national focus on sustainability. While Ukraine, a former member of Soviet Union, has gotten much "attention" in 2014, other former members of the Soviet Union are rapidly moving into the GIR, such as Serbia (Paunkovic, 2014).

The United States has not been quick to join this movement, but several American city governments, such as San Francisco and Santa Monica in California, are supporting LEED-designed buildings and encouraging local renewable energy generation including solar, wind, and ocean powers, along with electric vehicles and hydrogen-fueled cars. Other states are also working to combine the demand for transportation with electric and hydrogen fuel cell cars, along with the need to have local recharging and refueling stations. In April 2014, the California Energy Commission awarded almost $50 million for these stations. Unfortunately, only six stations (out of 28) will use solely renewable energy to electrolyze into hydrogen (CEC, 2014).

CREATING SUSTAINABLE COMMUNITIES

Creating sustainable communities is an extraordinarily complex task. It begins with addressing key infrastructure elements—energy, transportation, water, waste, and telecommunications—and extends through incorporating the belief systems, values, and behaviors of residents. Codes and standards are required to guide how buildings are designed, sited, and constructed. Certification programs like LEED provide guidelines, expertise, and political influence on how to construct and retrofit environmentally sensitive facilities that allow for the maximum use with fewer resources. Public policies establishing goals are needed to reduce GHG emissions and set thresholds and benchmarks for renewable energy power generation.

Emerging technologies are providing additional tools for achieving smart green sustainable communities. This is apparent not only in the development of new renewable energy technologies but also in the discovery of innovative ways to conserve valuable resources, particularly water. There are devices just coming to market that will help minimize the water used in large HVAC systems. These retrofit systems are gaining credibility in water-constrained locations.

These systems conserve the amount of water required to run the equipment and greatly reduce the outflows. A city like San Francisco not only has high water costs, but the sewer costs are equally high. One recently installed device on a large San Francisco hotel is projected to save $50,000 a year in sewer costs alone. This saving, combined with the reduction in the original water use, means that the cost of the device is recovered in just a few months. Ozone laundry conversions for commercial laundry systems provide similar benefits and a comparable short-term payback.

THREE SMART GREEN SUSTAINABLE COMMUNITIES

Here are three examples of communities that have come or will come very close to achieving sustainability and the benefits it brings.

Frederikshavn

Denmark set a national goal to be sustainable and to use 100 percent renewable energy by 2025. Combined heat and power (CHP) systems are a core approach in local energy systems. In 2006, Denmark chose Frederikshavn a relatively small town on the northeast coast of the Jutland peninsula to demonstrate how it was going to achieve this goal. The city government along with local industry and Aalborg University were included in the project (Lund and Ostergaard, 2010).

Frederikshavn, or in English, Fredrick's Harbor, with about 25,000 residents, is a busy transport hub. As an international ferry terminal, over 3 million people move through its port each year. The city has become Denmark's model city for sustainability. Dubbed the Energy Town Frederikshavn project, the city intends to be 100 percent energy-independent by using renewable energy systems like wind and biomass by the end of year 2015. In 2014, the city is over 45 percent of its goal. Chart A illustrates its progress (Lund and Ostergaard, 2010):

Frederikshavn held community meetings on sustainability to educate residents. Plans were developed that could be publically reviewed, and websites created that allowed residents to keep tabs on which programs were being developed and how much progress was being made. To encourage participation, they conducted campaigns to create awareness and encourage sustainable practices by the residents.

To achieve its goal, the city identified key components, which are now being planned and implemented. A new highly efficient waste incineration CHP plant with a capacity of burning 185 GWh per year was built. The project includes a biogas CHP plant of 15 MW. The rest of the heat production will be supplied from a biomass boiler burning straw.

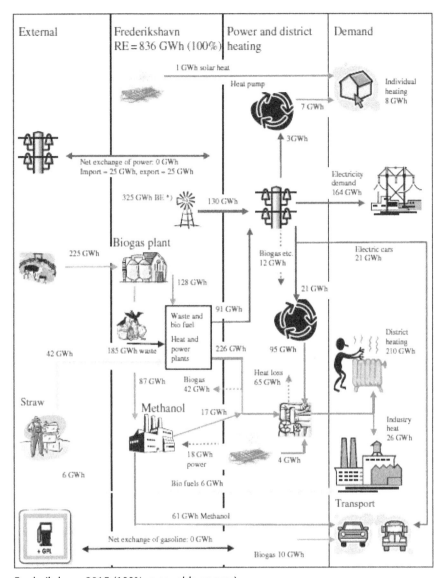

External	Frederikshavn RE = 836 GWh (100%)	Power and district heating	Demand

Frederikshavn 2015 (100% renewable energy)

Expansion of District Heating Grid

The Frederikshavn project also includes an expansion of the existing district heating grid. This will replace about 70 percent of the heat demand in industry and individual houses. The rest will be supplied from biomass boilers, and the individual house heating will be converted to a mixture of solar thermal and electric heat pumps.

Transportation

For transportation, Frederikshavn is using vehicles converted into the use of biogas in combustion engines (biofuel cars), electric cars, and plug-in hybrid cars. To implement electric vehicles or hybrids or combine the use of batteries with fuel cells based on either methanol or hydrogen, the city is establishing refueling stations. Furthermore, motor bikes and vans and buses will be converted into biogas, hydrogen, or methanol.

Biogas Plant and Methanol Production

Frederikshavn includes a biogas plant using 34 million tons of manure per year for the production of biogas to produce methanol for transportation and to replace natural gas for electricity and heat production. The plant itself is being converted into methanol, which will provide heat for district heating.

There are concerns with methanol due to its waste and impact on the environment. Eventually, methanol will be fully or partly produced by electrolysis from renewable energy sources. Moreover, efforts are under way to convert cars to hydrogen. In that case, some of the biogas will be replaced by wind power.

Geothermal and Heat Pumps

The city is located on top of potential geothermal resources, which may be included. The resources can supply hot water with a temperature of about 40 °C. However, the temperature can be increased to district heating level by the use of an absorption heat pump, which can be supplied with steam from the waste incineration CHP plant.

The plan calls for additional compression heat pumps to use the exhaust gases from the CHP plants. The boilers will be supplemented by other sources like wastewater.

Wind Power

Finally, Frederikshavn includes wind turbines to cover the rest of the electricity supply, for about 40 MW.

University of California Davis West Village

While the United States lags behind most European and Asian countries in greening its communities, the City of Davis in California is an exception. Davis is a town of about 65,000 residents wrapped around a campus in the University of California system with 35,000 students. Together, the university and the city probably lead the United States in sustainable concepts and practices.

A campus office of environmental stewardship and sustainability leads the university as well as most of the town in developing policies and regulations that promote sustainability. Multiple university departments like agricultural intertwine concepts of sustainability, renewable energy, and energy-efficient design into the curriculum.

In 2011, the university opened UC Davis West Village (Davis, California, 2013). Built by a partnership that included private builders and state agencies, the community is set on 130 acres next to the campus. Mark Berman, president of Davis Energy Group, which is one of the project's advisors, says "West Village is an example of what is sustainably possible, and it will be the US's largest planned zero net energy community—meaning West Village is designed to generate as much energy as it consumes each year" (Davis, California, 2013). Eventually, West Village will be home to 3000 students and campus employees.

When completed, it will include 662 apartments and 343 single-family homes plus commercial space, a recreation center, and village square. West Village will also have a community college center and what is being called a "uHub." Located in the commercial space surrounding the village square, the uHub will be home to the campus' energy research centers. Promoting sustainable development, the uHub will help commercialize the universities' scientific innovations and patents.

West Village apartments are built on aggressive energy-efficient measures and uses on-site renewable energy. The apartments feature two to four bedrooms, walk-in closets, a full-size washing machine and dryer, stainless steel kitchen appliances, unlimited high-speed Internet service, and air-conditioning.

The energy-efficient measures include solar-reflective roofing, radiant barrier roof sheathing, and extra insulation. Energy-efficient interior and exterior lighting fixtures, indoor occupancy sensors, and daylighting techniques help the community use about 60 percent less energy than if standard lighting had been used.

The first phase is powered by a 4 MW solar array on building rooftops and on parking lot canopies. A biodigester converts campus table scraps; turning animal and plant waste into energy is planned for future build-out.

The community is sustainable in ways beyond energy. It offers an extensive bike network and is served by natural gas-powered buses. Drought-friendly landscaping, water-saving toilets, recycled building materials, and paints low in volatile organic compounds (VOCs) are just some of the green features incorporated into the design.

West Village demonstrates a revolution in California's residential construction, showing how construction practices can be sustainable. In a remarkable leap into the GIR, the state's public utilities commission has called for shifting all new residential construction to zero net energy by 2020 and all new commercial construction by 2030.

Sino-Singapore Tianjin Eco-City

In a landmark example of cooperation, Singapore and China agreed in 2007 to jointly develop Sino-Singapore Tianjin Eco-city (SSTEC, 2013). With the intention of developing a sustainable eco-city that can be a model for future cities, the Chinese picked the site that was on nonarable land with water shortages. The Tianjin area 150 km away from Beijing was chosen. Before the work began, the site has mainly salt pans, barren land, and polluted water bodies, including a 2.6 sq. km large wastewater pond.

The goal for Sino-Singapore Tianjin Eco-city's is to be "A thriving city which is socially harmonious, environmentally friendly and resource-efficient—a model for sustainable development." This vision is underpinned by the concepts of "Three Harmonies" and "Three Abilities" (SSTEC, 2013).

"Three Harmonies" refers to
- people living in harmony with other people, i.e., social harmony;
- people living in harmony with economic activities, i.e., economic vibrancy;
- people living in harmony with the environment, i.e., environmental sustainability.

"Three Abilities" refers to the eco-city being
- practicable (the technologies adopted in the eco-city must be affordable and commercially viable),
- replicable (the principles and models of the eco-city could be applied to other cities in China and even in other countries),

- scalable (the principles and models could be adapted for another project or development of a different scale).

With a completion date of 2020, Sino-Singapore Tianjin Eco-city is a remarkable and fascinating project. It is the largest of its kind in the world, the first attempt to build a smart, green city from the ground up. With a total land area of 30 sq. km., the city will have an eventual population of 350,000 residents. The start-up area is scheduled for completion by the end of 2013, and the site will be transformed in phases over 10-15 years.

The master plan for the Sino-Singapore Tianjin Eco-city was developed by teams of experts from Singapore and China and incorporates the best ideas from the two countries as well as the international community. The city will showcase the new green technologies and use benchmarks to ensure that development will be environmentally friendly, resource-efficient, and economically sustainable. The two governments have established a high-level joint steering council to monitor and support the progress of the Sino-Singapore Tianjin Eco-city.

Planners say the Sino-Singapore Tianjin Eco-city will create a green community where nature is integrated into daily living and adopt affordable technologies and practices to create a strong foundation for sustainable development. To reduce its carbon footprint, the Sino-Singapore Tianjin Eco-city will be a walkable city that promotes public and nonmotored transportation. Hybrid cars and buses will form the bulk of the general transportation. Residential and commercial developments are planned to be located close to either the main city center or subcenter so that residents can live close to work and amenities (Fang and Zhang, 2012).

All buildings in the Sino-Singapore Tianjin Eco-city will be green and comply with the Green Building Evaluation Standards, a unique benchmark that Singapore and Chinese expert teams have developed for the Sino-Singapore Tianjin Eco-city. Eco-solutions will be integrated to enhance sustainability and commercial viability so that homes will be affordable and well designed.

A sound water management system will allow residents to drink directly from their taps, a rare benefit in China, especially in the more arid regions of northern China. Sewage water will be treated to provide a supplementary supply. Clean and renewable energy sources such as solar water heaters and geothermal heating systems will be used to supplement traditional energy supplies.

A collective system of waste management and recycling will be introduced and integrated with waste disposal and incineration processes to regenerate energy as well as minimize the strain on landfills. A pneumatic waste collection system will be used in the business park.

The city will have an advanced light rail transit system and varied eco-landscapes ranging from a sun-powered solarscape to a greenery-clad earthscape for its estimated 350,000 residents to enjoy. A vast, beautiful Eco-Valley, which serves as an ecological green spine, will run through the city linking major transit nodes, residential developments, and commercial centers. Natural habitats will be restored and rivers, water bodies, and wetlands will be cleaned up (SSTEC, 2013).

Developing sustainable communities requires a plan or a set of concepts that will lead to that goal. Frederikshavn set the ball rolling by establishing a set of principles to guide the city toward sustainability. In 2009, the UN IPPC met in Copenhagen, Denmark, and reviewed some of the Frederikshavn programs that were stated in 2008. The plan consisted of statements on how cities can become more sustainable. They are designed to be read by decision makers and provide a starting point on the journey toward sustainability.

Frederikshavn, UC Davis West Village, and Sino-Singapore Tianjin Eco-city are examples of what can be done. They provide a conceptual framework for moving today's sprawling, ever-growing cities toward sustainability. Turning energy, waste, water, transportation, and telecommunications into sustainable processes is critical. Besides the plans, green cities must develop the financial resources to implement the processes. These infrastructure elements are integrated; transportation and energy are connected because transportation systems should use renewable energy. The same is true for water pumping and surface transportation, particularly in places like California that require enormous amounts of energy to move large quantities of water across miles of surface area.

One good example that provides a global perspective is the C40 program of megacities first created by the Clinton Foundation at the beginning of the 21st century. The program is now part of former New York City Mayor Michael R. Bloomberg's programs, which is independent from the city government. C40 works both locally and collaboratively with other cities around the world to reduce both greenhouse gases and mitigate climate risks. Over 4700 actions have been taken around the world based on global criteria that cities need to meet to be considered sustainable. These actions range from planning, conservation and efficiency for energy, mass transportation systems, food supplies, to waste and recycling.

Hurricane Sandy, which devastated New York City and the surrounding coastal areas, is a solid evidence of the needs for communities to become sustainable. The rebuilding of the entire region is estimated at about $100 billion. While this rebuilding will most likely be done in a sustainable plan and program, it needs to be monitored and evaluated for greenhouse gas reductions and pollution elimination.

Each sustainable community must retrofit the traditional central power plant into one using renewable energy generation and smart grid distribution. Further, the sustainable infrastructure systems must provide for recycling, waste control, water and land use, and green building standards that require energy-efficient and compact housing. Today, green sustainable cities and communities are necessary to reduce environmental pollution and to provide a healthier world for tomorrow. The solutions to global warming and climate change exist now; we need to design, finance, and implement them.

REFERENCES

American Lung Association, California, 2014. http://www.lung.org/associations/states/california/advocacy/fight-for-air-quality/sota-2014/state-of-the-air-2014.html.

Andersen, A.N., Lund, H., 2007. New CHP (combined heat and power) partnerships offering balancing of fluctuating renewable electricity productions. J. Cleaner Prod. 15, 288–293, Elsevier Press.

Los Angeles Highway Congestion, 2013. http://www.stateoftheair.org/2013/city-rankings/most-polluted-cities.html.

Los Angeles, 2014. http://www.forbes.com/pictures/ehmk45jhdg/1-los-angeles/.

Brundtland Commission (1983) Report, 1987. Our common future. UN Commission General Assembly 828 Resolution #38/161 for Process of Preparation of the Environmental Perspective to the Year 829 2000 and Beyond. Oxford University Press, Oxford.

California energy Commission (CEC), 2014. http://www.energy.ca.gov/drive/.

Clark II, W.W., 2004. Distributed generation: renewable energy in local communities. Energy Policy, Elsevier, London, UK.

Clark II, W.W., 2009. Sustainable Communities. Springer Press, New York, NY.

Clark II, W.W., 2010. Sustainable Communities Design Handbook. Elsevier Press, New York, NY.

Clark II, W.W., 2014a. Global Sustainable Communities Design Handbook: Green Design, Engineering, Health, Technologies, Education, Economics, Contracts, Policy, Law and Entrepreneurship. Elsevier Press, New York, NY.

Clark II, W.W., 2014b. Founder and Chair, Beverly Hills Centennial, Report on the next 100 years, Beverly Hills, CA.

Clark II, W.W., Bradshaw, T., 2004. Agile Energy Systems: Global Solutions to the California Energy Crisis. Elsevier Press, Amsterdam.

Clark II, W.W., Dan Jensen, J., 2002. Capitalization of environmental technologies in companies: economic schemes in a business perspective. Int. J. Energy Technol. Policy 1 (1/2), Inderscience, London, UK.

Clark II, W.W., Eisenberg, L., 2008. Agile sustainable communities: on-site renewable energy generation. Utility Policies J. 16 (4), 262–274, Elsevier Press.

Davis, California, 2013a. http://www.ucdaviswestvillage.com.

Davis, California, 2013b. www.davisenergy.com.

Eisenberg, L., 2009. Los Angeles Community College District (LACCD). In: Clark II, W.W. (Ed.), Sustainable Communities. Springer Press, New York, NY, pp. 29–44.

Fang, Y., Zhang, M., 2012. A Path to Future Urban Planning and Sustainable Development: A Case Study of Sino-Singapore Tianjin Eco-city. Cross-disciplinary Scholars in Science and Technology Program, UCLA.

Funaki, K., Adams, L., 2009. Japanese experience with efforts at the community level towards a sustainable economy. In: Clark II, W.W. (Ed.), Sustainable Communities. Springer Press, New York, NY, pp. 243–262, Chapter #15.

A Time for Truth, Los Angeles Commission on 2020, December 2013.

Lund, H., 2009. Sustainable towns: the case of Frederikshavn, Denmark. In: Clark II, W.W. (Ed.), Sustainable Communities. Springer Press, New York, NY, Chapter 10.

Lund, H., Clark II, W.W., 2002. Management of fluctuations in wind power and CHP: comparing two possible danish strategies. Energy Policy 27 (5), 471–483, Elsevier Press, November.

Lund, H., Clark II, W.W., 2008. Sustainable energy and transportation systems introduction and overview. Utilities Policies J. 16 (2), 59–62, Elsevier Press, June.

Lund, H., Østergaard, P.A., 2010. Climate change mitigation from a bottom up community approach: a case in Denmark. In: Clark II, W.W. (Ed.), Sustainable Communities Design Handbook. Elsevier Press, Amsterdam, Chapter 14.

McEneaney, B., 2009. Santa Monica sustainable city plan: sustainability in action. In: Clark II, W.W. (Ed.), Sustainable Communities. Springer Press, New York, NY, pp. 77–94.

Paunkovic, J., 2014. Educational programs for sustainable societies using cross-cultural management method: a case study from Serbia. In: Clark II, W.W. (Ed.), Global Sustainable Communities Design Handbook: Green Design, Engineering, Health, Technologies, Education, Economics, Contracts, Policy, Law and Entrepreneurship. Elsevier Press, Amsterdam, pp. 387–405.

Radow, E.N., 2011. Homeowners and gas leases: boon or bust? N. Y. State Bar Assoc. J. 83 (9), 9–21, November/December.

South Coast Air Quality Management District (SCAQMD), 2010.

SSTEC Eco-City Singapore, 2013. http://sstec.dashilan.cn/en/SinglePage.aspx?column_id=10304.

UN Department of Economic and Social Affairs (UN DESA), 2011.

CHAPTER 14

A Smart Green Future

The Green Industrial Revolution is emerging as the next significant political, social, and economic era in world history. As it takes hold, it will result in a complete restructuring of the way energy is generated, supplied, and used. It will be a revolutionary paradigm of extraordinary potential and opportunity, with the already remarkable innovations in science and energy that will lead to new ones in sustainable, smart, and carbonless economies powered by nonpolluting technologies like wind, geothermal, wave, river, and solar with their advanced technologies like flywheels, regenerative and maglev systems, and hydrogen fuel cells (Clark and Rifkin, 2006).

Community-based and on-site renewable energy generation is replacing massive fossil fuel and nuclear-powered central plant utilities. New advances in recyclable efficient batteries and fuel cells will store energy for when it is needed most. Smart green grids will share electricity effortlessly. Additive manufacturing will minimize wasted resources, and new sciences like nanotechnology will have a profound impact on business, careers, human health, and the global economy.

This new era encompasses changes in technology, economics, businesses, manufacturing, jobs, and consumer lifestyles (see Case no. 4 in "Appendix" on these changes). The transition will be as complete as when the steam-driven First Industrial Revolution gave way to the fossil fuel-driven Second Industrial Revolution. This monumental shift is already underway and spreading rapidly around the world.

Industrial revolutions occur when a new energy source intersects with a new form of communication. In the First Industrial Revolution, steam was the energy source and the printing press provided the means to rapidly disseminate new ideas that accelerated scientific breakthroughs and the adoption of inventions. In the Second Industrial Revolution, the fossil fuel-driven internal combustion engine was the power source and analog communication provided the channel for new ideas and technologies.

Today, the world is entering a new industrial revolution. The digital age, with its Internet access to almost all scientific knowledge and Facebook- and Twitter-led social media, has intersected with renewable energy generation, hydrogen storage, and smart grids. While vast fortunes were made by extracting natural resources in the fossil fuel era and despoiling the environment, wealth in this new green era will come from digital and information technology breakthroughs, intelligent machines, and a host of environmentally sensitive inventions.

Despite resistance from the fossil fuel industries and political interests, renewable energy generation is becoming cheaper, more efficient, and mainstream. Solar power is gaining so much momentum that it competes directly in price with oil, diesel, and liquefied natural gas in much of Asia (Evans-Pritchard, 2014). In the United States, almost 30 percent of last year's added electricity capacity came from solar. In Vermont and Massachusetts, almost 100 percent added capacity came from solar (Evans-Pritchard, 2014). According to the US Solar Energy Industries Association, more solar was installed in the United States in the past 18 months than in the last 30 years.

Deutsche Bank, a leading German investment bank, reported in January 2014 that there were 19 regions around the world where unsubsidized solar power costs were competitive with other forms of generation. This equality of costs with fossil fuel and natural gas is creating a solar boom in 2014-2015. Deutsche predicts that a huge 46 gigawatts of solar will be installed in 2014, followed by an additional 56 gigawatts in 2015 (Parkinson, 2014).

The bank notes that in the world's three largest economies—the United States, China, and Japan—solar power is booming. India, Australia, South Africa, Mexico, and the regions in the Middle East, South America, and South East Asia are rapidly installing solar power. Besides solar reaching grid parity with other forms of energy generation, the bank cites other reasons for this worldwide growth:

- The United States-distributed generation business models will stimulate growth in European markets.
- Financing costs and availability for the solar sector are set to improve from 2014.
- It expects downstream solar companies to participate in the "gold rush" to acquire solar customers at an accelerated pace.
- While the past 5 years were about module cost reduction, the next 3 years would be about reductions in the balance of systems costs. This includes the cost of inverters, hardware, customer acquisition, and financing costs.

Deutsche says that its base case demand estimate assumes that the Japanese market increases from around 7 gigawatts in 2013 to around 8 gigawatts in 2014, the US market increases from 6 gigawatts in 2013 to 8 gigawatts in 2014, and the Chinese market increases from 8 gigawatts in 2013 to around 12 gigawatts in 2014. Europe will account for 7-8 gigawatts of installations and international markets to account for around 12-17 gigawatts of demand (Parkinson, 2014).

TO THE END OF THE OIL AGE

The Stone Age came to an end, not because we had a lack of stones, and the oil age will come to an end not because we have a lack of oil.

This was said by Sheikh Ahmed Zaki Yamani, the former oil minister of Saudi Arabia and arguably the world's foremost expert on the oil industry, in 2000. He introduced this extraordinary observation with an even more prescient one, to wit, "Thirty years from now there will be a huge amount of oil—and no buyers. Oil will be left in the ground," he told the United Kingdom's The *Telegraph* (Evans-Pritchard, 2014).

A decade and half later, the world is arriving at the end of the oil age and the domination of the world's geopolitics and economy by the fossil fuel interests for the past century. Correspondingly, the carbon- and nuclear-powered centralized utility industry that was started by Thomas Edison in 1882 when he flipped the switch at the Pearl Street Station in Manhattan has begun its decline.

Over the years, Big Oil and its related industries have wrecked havoc on our environmentally fragile planet and disrupted the way humans manage their affairs. Today, the loss of a major section of the West Antarctic Ice Sheet from global warming caused by excessive carbon-generated heat appears "unstoppable." It is not hard to conclude that if humans had not developed fossil fuels as a major source of energy, the planet would have a much more sustainable environment.

Nor is it hard to conclude that human affairs would have turned out much differently if the oil industry had not gained so much political and economic power. For example, the US oil oligarchs, Charles and David Koch, have made a mockery of American democracy by pouring hundreds of millions of dollars into smear campaigns against scientists, environmentalists, and liberal politicians. The Koch brothers have managed to replace census and compromise with vitriol and dysfunction in US politics.

In the spring of 2014, Vladimir Putin, channeling the ghost of Joseph Stalin, swept up a large portion of Ukraine and threatened astonished European nations that if they opposed him, the result would be a shutdown or price hike of the Russian natural gas that is vital to the EU's economic recovery. And the world seems to have grown accustom to the mayhem in the Middle East, where the biggest transfer of wealth in world history—from the oil users to the oil suppliers—has led to social and political chaos, repression, and bloodshed.

Even after a century of tax payer support, the US Federal Government grants the oil industry, the world richest, with about $4 billion a year in tax subsidies, and Exxon Mobil Corporation (the highest grossing company in the world) minimizes the taxes it pays by using 20 wholly owned subsidiaries in the Bahamas, Bermuda, and the Cayman Islands to legally shelter cash from its operations in Angola, Azerbaijan, and Abu Dhabi.

As pointed out earlier, the coal industry also enjoys US tax breaks, public land loopholes, and subsidized railroads. Few, if any, of the fossil fuel industries' external costs, or externalities, are covered by product costs; instead, most are borne by the world's tax payers.

Fortunately, the carbon-based industries, which include coal, oil, natural gas, and the related industries like centralized utilities and transmission line companies, are coming to the end of their socially useful cycle. Their resources are aging beyond economic justification and their business models are too inflexible to adapt to a new industrial era with a different energy model.

Many factors are coming together to hasten the Green Industrial Revolution. Putin's march on Ukraine has shocked Europe and stirred the region's efforts to generate more renewable energy and cut the ties to fossil fuel. Forty percent of Scotland's domestic electricity generation comes from renewable sources, mostly tidal and wind. Denmark and other Nordic nations intend to generate 100 percent of their energy by midcentury. Germany's Energiewende (energy transition), which aims to power the country almost entirely on renewables by 2050, is accelerating.

Almost daily, scientists in university and national research laboratories are making breakthroughs in developing noncarbon energy sources. Advancements in nanotechnology are making electricity usage much more efficient.

Several nations, as well as California, are creating hydrogen highways. Norway, Sweden, and Germany have them, and California will open its hydrogen highway in 2016. China is considering a ban on fossil-fueled new cars, and major cities across the globe have limited the use of autos in downtown areas. Daimler, Honda, Chevrolet, and most other major automobile manufacturers have hydrogen-powered fuel cell cars ready to go.

Tesla's Model S, a plug-in electric car, is outselling the high-end German autos in the United States and Norway. Tesla's success comes from a superior lithium-ion battery pack that will soon come to market. Superior batteries and hydrogen fuel cells will revolutionize the way renewable energy can be stored and shared across smart Internet-like grids.

Finally, there are key factors that will put to rest the fossil fuel industry and make the wise Sheikh Yamani's prediction come true.

The first is that the carbon emitters will be held accountable and made to pay for using the atmosphere as a garbage can. While still struggling to price the cost of pollution, most nations, as well as California, have come to realize that the heavy carbon emitters need to pay for the damage that they have done. A cap-and-trade process has risen from the bureaucracies as the first method to hold the emitters accountable. While imperfect and not nearly as effective as a straight carbon tax, this system is growing throughout the world. The EU's program, which started several years ago and was descried by the fossil fuel interests as failing, is now deemed a success. It has become an established part of European culture and corporate practice. Various nations such as Australia, New Zealand, Canada, Korea, and China have developed cap-and-trade programs.

In the United States, California's program continues to grow and the carbon offsets are tradable in parts of Canada as well. Other US states are watching California's program as it gains momentum and thinking about adopting their own. Impoverished state governments see cap-and-trade programs as a help to their environment and a way to garner vital tax revenues. Since increases in personal income taxes are so unpopular, this is a way to bring new money into the state treasuries without risking voter rebellions.

The pressure to make the major carbon emitters pay for their pollution is coming from the agreements made at the 2012 UN Climate Change Conference in Doha, Qatar. At this conference, world governments consolidated the gains of the last 3 years of international climate change negotiations and opened a gateway to greater ambition and action. Among the decisions was to concentrate on a universal climate agreement by 2015, which will come into effect in 2020. The 2015 conference will be held in Paris, and world governments are expecting much greater cooperation and agreement for carbon-reduction policies from the United States and other major emitters.

The world is slowly accepting the reality that the mitigation of climate change is a massive problem. Climate Vulnerable Forum estimated that more than 100 million people will die and the international economy will lose out on over 3 percent of gross domestic productivity ($1.2 trillion) by 2030, if the world fails to tackle climate change (Climate Vulnerability Model, 2012). Governments don't want to use their funds for environmental cleanup and climate change mitigation, so it will be the heavy emitters who will pay.

This cost for carbon cleanup added to the increasing costs of extracting hard-to-get fossil fuel resources will hit the oil industry hard.

GRID PARITY

The second major factor hastening the end of today's megalithic fossil fuel industries is grid parity. Grid parity is a technical term meaning that the cost to a consumer for electricity from a renewable source (without subsidies) is about equal to the cost from a traditional source—be it fossil fuel or nuclear. The Germans used grid parity to price their feed-in tariff program that launched Energiewende.

Simply put, a California utility's rate to a resident or small commercial consumer is about 20 (19.9) cents per kilowatt hour for electricity from traditional sources. If that same kilowatt hour came from a renewable source and cost the consumer an equal 20 cents, then the renewable source would be at "parity" or equal to the cost of the traditional source.

However, the cost of conventional energy is rising, driven by higher extraction costs, increased maintenance costs for natural gas pipelines, and increased operational cost at nuclear power plants. At the same time, the costs for renewable energy—wind, solar photovoltaic, and biowaste fuels—are declining. The costs for wind generation have been and still are the lowest. However, the costs for solar are declining rapidly. Solar PV technology, which has been helped by the US military, is improving so fast that it has achieved a virtuous circle.

ENERGY DEFLATION

Sanford C. Bernstein & Co., the highly respected New York investment bank, released a report in the spring of 2014 that introduced the term "global energy deflation." They argued that the fossil fuel-dominated energy market will experience a major decline in costs over the next decade. The market is entering a new order that will erode the viability of oil, gas, and fossil fuel continuum over time (Greentechsolar, 2014).

According to Sanford C. Bernstein & Co., this global energy deflation will be created as the cost of PV (and other renewable energy sources) achieves grid and price parity with fossil fuels. They say that solar is now cheaper than oil and Asian liquid natural gas and will get cheaper over time, while fossil fuel extraction costs will keep rising.

The report argued that the adoption of solar in developing markets will translate into less demand for kerosene and diesel oil. The adoption of oil in the Middle East means less oil demand and the adoption of solar in China and developing Asia means less LNG demand. Further, distributed solar in the United States, Europe, and Australia will likely reduce demand for natural gas.

They reason that while solar has a fractional share of the current market, within one decade, solar PV and related battery storage may have such a large market share that it becomes a trigger for energy price deflation, with huge consequences for the massive fossil fuel industry that is dependent on continued growth. They conclude that it is inevitable that renewable technology and battery storage will turn into behemoths and lead to energy price deflation.

In April 2014, the highly respected Paris-based financial company Kepler Cheuvreux released a research report that has rippled through the fossil fuel industries. The report is in part a response to an Exxon Mobil Corporation report on how it was managing carbon risk. Exxon Mobil Corporation essentially said that it was business as usual and was dismissive of significant change in the energy/power continuum (Parkinson, 2014).

Kepler Cheuvreux is highly critical of the oil industry's clinging to business-as-usual forecasts and describes what is at stake for the fossil fuel industry as the world governments' push for cleaner fuels and reduced greenhouse gases emissions gathers momentum. The firm argues that the global oil, gas, and coal industries are set to lose a combined $28 trillion in revenues over the next two decades as governments take action to address climate change, clean up pollution, and move to decarbonize the global energy system. The report helps to explain the enormous pressure that the industries are exerting on governments not to regulate carbon emissions or greenhouse gases.

The firm used International Energy Agency forecasts for global energy trends to 2035 as a basis for its research. As carbonless energy becomes more available and government policies make steep cuts in carbon emissions, demand for the fossil fuels oil, natural gas, and coal will fall, which will lower prices.

The report said that oil industry revenues could fall by $19.3 trillion over the period 2013-2035, coal industry revenues could fall by $4.9 trillion, and gas revenues could be $4 trillion lower. High production cost extraction such as deepwater wells, oil sands, and shale oil will be most affected. Even under business-as-usual conditions, however, the oil industry will still face risks from increasing costs and more capital-intensive projects, fewer exports, political risks, and the declining costs of renewable energy.

The report continues, "The oil industry's increasingly unsustainable dynamics. . .mean that stranded asset risk exists even under business-as-usual conditions. High oil prices will encourage the shift away from oil towards renewables (whose costs are falling) while also incentivizing greater energy efficiency" (Parkinson, 2014).

As far as renewables are concerned, Kepler Cheuvreux says that tremendous cost reductions are occurring and will continue as the upward trajectory of oil costs becomes steeper.

Kepler Cheuvreux's report is consistent with other reports released in 2014. One report from the United States' Citigroup was titled "Age of Renewables is Beginning—A Levelized Cost of Energy (LCOE)." Released in March 2014, the report argued that there will be significant price decreases in solar and wind power, which will add to the renewable energy generation boom. Citigroup projects price declines based on Moore's law, the same dynamic that drove the boom in information technology.

In brief, Citigroup is looking for cost reductions of as much as 11 percent per year in all phases of PV development and installation. At the same time, they said that the cost of producing wind energy would also significantly decline. As the same time, Citigroup says that natural gas price will continue to go up and the cost of running coal and nuclear plants will slowly become prohibitive (Wile, 2014).

When the world's major financial institutions start to do serious research and quantify the declining costs of renewable energy versus the rising costs of fossil fuels, it becomes easier to understand the monumental impact of the Green Industrial Revolution.

Even the Saudis are betting on solar, investing more than $100 billion in 41 gigawatts of capacity, enough to cover 30 percent of their power needs by 2030. Most of the other Gulf States have similar plans.

ZERO MARGINAL COST

The third element is "zero marginal cost." Marginal cost, to an economist or businessperson, is the cost of producing one more unit of a good or service after fixed costs have been paid. For example, consider a shovel manufacturer. It costs the shovel company $10,000 to create the process and buy the equipment to make a shovel that sells for $10. So, after 1000 shovels or so are sold, the company has recovered its fixed or original costs. Thereafter, each shovel has a marginal of cost of $3, consisting mostly of supplies, labor, and distribution.

Companies have used technology to increase the productivity and reduce marginal costs and return profits. However, as Jeremy Rifkin points out, we have entered an era where technology has unleashed "extreme productivity," driving marginal costs on some items and services to near zero (Rifkin, 2014). File sharing technology and subsequent zero marginal cost almost ruined the record business and shook the movie business. The newspaper and magazine industries have been pushed to the wall and are being replaced by the blogosphere and YouTube. The book industry struggles with the e-book phenomenon.

A growing number of pioneers with pioneering technologies are breaking new ground and going beyond YouTube and Amazon. The example of the Explorers' Wheel from Julian Gresser's latest book is given as an overview in "Appendix" in Case no. 18.

An equally revolutionary change will soon overtake the higher education industry. For the first time in world history, knowledge is becoming free. At last count, the free Massive Open Online Courses had enrolled about six million students. The courses, many of which are for credit and taught by distinguished faculty, operate at almost zero marginal cost. Why pay $10k at a private university for the same course that is free over the Internet? The traditional brick and mortar, football-driven, ivy-covered universities will be scrambling for a new business model.

Airbnb, a room-sharing Internet operation with close to zero marginal cost, is a threat to change the hotel industry in the same way that file sharing changed the record business, especially in the world's expensive cities. For example, young out-of-town high-tech workers coming to San Francisco from Europe use Airbnb to rent a condo or an empty room in a house instead of staying in a hotel. They do this because they cannot find room with the location they need or because their expense reimbursement cap won't cover one of the city's high-end hotel rooms. Industry analysts estimate that Airbnb and similar operations took away over a million rooms from New York City's hotels last year (Rifkin, 2014).

A powerful technology revolution is evolving that will change all aspects of our lives, including how we will access renewable energy. The Internet is becoming sophisticated enough, and soon, it will seamlessly tie together how we share and interact with electricity. It will greatly increase productivity and drive marginal cost of producing and distributing electricity down, possibly to nothing beyond fixed costs.

This is almost the case with the early adopters of solar and wind energy. As they pay off these systems and their fixed costs are covered, additional units of energy are basically free (Clark and Greenfield, 2013). Eventually, we'll be able to buy a home solar system at Ikea, Costco, or Home Depot, have it installed, and recover our costs in under 2 years.

All three of these elements—carbon mitigation costs, grid parity, and zero marginal costs—(and others like additive manufacturing and nanotechnology) are part of the coming Green Industrial Revolution. It will be an era of momentous change in the way we live our lives. It will shake many familiar and accepted processes like twentieth-century capitalism and free-market economics, reductive manufacturing, higher education, and health care.

More to the point, it will see the passing of the carbon-intensive industries. Like the centralized utility industry, the fossil fuel industries have a business model predicated on continued growth in consumption. The Second Industrial Revolution's carbon energy model was about Big Banks financing Big Energy to build Big Power Plants or refineries. This created a centralized, command-and-control-oriented, secretive, and extractive architecture in specific areas that built building, communities, and financial institutions.

The Green Industrial Revolution's energy architecture is completely different. Unlike coal or oil, solar and wind are free and the new financing is about financing smaller, distributed power plants everywhere. The new energy model is distributed, mobile, intelligent, and participatory and will rapidly replace the old energy model with abundant, cheap energy. As the nexus of declining prices for renewables and rising costs of extraction is crossed—and we are there in several regions of the world—demand will rapidly shift and propel us into global energy deflation.

The era of sustainability and renewable energy has begun. The Green Industrial Revolution is more significant and life-changing than either the First or Second Industrial Revolution. It has to be, for there is so much more at stake. Climate change is real, and we are all savaged by it. It is killing five million people a year and costing the world's economy annually about $1.2 trillion. Greenhouse gases from carbon emissions cover our planet, forcing drastic changes in weather patterns that create destructive superstorms. Decades of failing to curb the world's dependency on fossil fuels have made the planet hotter and more polluted. It has killed people and stolen their livelihoods. The world's poorest nations are the most vulnerable as they face increased risk of drought, water shortages, crop failure, poverty, and disease.

The world is in a dangerous time—a point at which global warming and environmental degradation may become irreversible. Critical decisions must be made on a global level for the good of the planet. It is also an age of opportunity, and the Green Industrial Revolution will provide those opportunities.

This new era of sustainability and carbonless energy generation is at our doorstep. The push for renewable energy and a carbonless lifestyle will become history's largest social and economic megatrend, with the potential of extraordinary benefits in the form of economic revival, innovation, emerging technologies, and significant job growth for those nations capable of fast entry.

Humanity is at a crossroad and time is running out. By not acting soon, the cost of inaction goes up and the benefits of the alternative, environmentally sound path are reduced. Doing nothing destroys our environment and holds back economic growth, but strong action has the potential to reap monetary and environmental benefits while safeguarding our planet.

The Green Industrial Revolution is here now, and for the sake of our children and our grandchildren, the world must embrace it.

REFERENCES

Clark II, W.W., Rifkin, J., 2006. A green hydrogen economy. Energy Policy 34 (17), 2630–2639. Special issue on Hydrogen.

Evans-Pritchard, A., 2014. Global solar dominance in sight as science trumps fossil fuels. Telegraph. (April 9). http://www.telegraph.co.uk/finance/comment/ambroseevans_pritchard/10755598/Global-solar-dominance-in-sight-as-science-trumps-fossil-fuels.html.

Forbes, 2014. Solar costs rapid decline. http://www.forbes.com/sites/stevebanker/2014/04/21/falling-solar-energy-costs-are-poisedto-reshape-the-worlds-economy/.

Greenfield, J., Clark II, W.W., 2013. Re-make 'made in America': community real estate construction and sustainable development. Los Angeles Review of Books. (October 28).

Greentechsolar, 2014. Solar's dramatic price plunge could trigger energy price deflation. April 14. http://www.greentechmedia.com/articles/read/Solars-Dramatic-Price-Plunge-Could-Trigger-Energy-Price-Deflation.

Climate Vulnerability Model, 2012. http://daraint.org/climate-vulnerability-monitor/climate-vulnerability-monitor-2012.

Parkinson, G., 2014. Deutsche Bank predicts second solar 'gold rush'. Renew Economy. (January 7). http://reneweconomy.com.au/2014/deutsche-bank-predicts-second-solar-gold-rush-40084.

Rifkin, J., 2014. Say goodbye to capitalism as we know it. Marketwatch. (May 15). http://www.marketwatch.com/story/say-goodbye-to-capitalism-as-we-know-it-2014-05-15.

Wile, R., 2014. CITI: 'the age of renewables is beginning'. Business Insider. (March 29). http://www.businessinsider.com/citi-the-age-of-renewables-is-beginning-2014-3.

The Economic Implications of Green Industrial Revolution (GIR) in Central and Eastern Europe: The Case of Poland

Robert Ruminski[1]

[1]Department of Banking and Comparative Finance, Faculty of Management and Economics of Services, University of Szczecin, Szczecin, Poland.

OVERVIEW

The critical problems faced by humanity include, inter alia, climate change; access to water, food, energy, and raw and secondary materials; and waste management. In view of progressing globalization, those problems assume particular significance. Through the process of globalization, not only markets and business development opportunities become common, but also do the challenges of ecological nature—on a global scale. Without a doubt, the green industrial revolution (GIR) has positive overtones and is highly inspiring.

Nevertheless, will green technologies prove to be capable of stimulating a new wave of growth comparable to the industrial revolutions of the nineteenth and twentieth centuries?[1] The lack of access of 2 billion people to potable water and the impact of climate change on the lives of several hundred million others demonstrate that the challenges that humanity faces require the will to change and the courage of all decision makers: political, social, and corporate ones.

No organization or institution is capable of solving those issues on its own. Every one of us ought to feel equally responsible for the future of our planet. The initiatives undertaken with the aim of energy conservation, taking environmental issues into consideration in production or changing the entire paradigm of profit-based economics indifferent to environmental costs, are fundamental matters.

---

[1] http://www.greeneuropeanjournal.eu/green-industrial-revolution/.

If clean renewable energy with a limitless supply is available, how come the world is still addicted to fossil fuels with their destructive side effects? The answer to this question is far more complex, and it is linked with a group of interests represented by parties financially involved in the sectors crucial to many economies, deriving huge, long-term profits, placing private interest above the people's and planet's needs.

The solutions proposed—due to the global nature of risks—should be of systemic character. Public administration, dedicated to resolving environmental problems, but guided by particular interest of countries or regions it represents and bound by a short-term perspective of the incoming election, has a scant chance of arriving at solutions enabling a truly sustainable development. Supranational organizations (e.g., the UN) try to take action in response to mounting ecological problems.

However, their impact is insufficient to provide solutions adequate to the challenges faced by humanity. Enforcing the implementation of such solutions is of little effect, an example of which could be Rio+20 conference.[2] There is a need for joint actions. The role of business is growing, while effective measures cannot be implemented without the cooperation of enterprises, in particular SMEs (Grzybek, 2011).

In the enterprise sector, investments in environment-friendly business have been growing rapidly over the last decade. From power engineering through construction to automotive industry, ecology has become one of the major directions of innovation. Presently, in developed countries, "green" business is a large, quickly developing branch of significant potential for innovation and further growth. The largest economies of the world and corporate leaders fight for a share in the market and future leadership position.

EU REGULATIONS FAVORING GIR

From the late 1970s, numerous European "green" parties aimed to oppose the negative effects caused by industrialization to people and the environment. They fought with factory pollution, while many political opponents treated them as a threat to the economy and jobs—chiefly in the industrialized parts of Europe. Currently, three decades later, the situation could not be more different. There is a demand for a different type of industry, in keeping with the challenges of ecology and ensuring a new quality and the possibility of sustainable development in Europe. Resistance towards

[2] http://sustainabledevelopment.un.org/rio20.html.

"green" initiatives seems to have been growing in recent years, which is why "green" parties ought to consistently champion the ecological transformation of the European industry.[3]

Closures of toxic plants in Europe in the recent decades and the global recession have contributed to the reduction of greenhouse gas (GHG) emission. Nevertheless, deindustrialization is not a solution that will enable accomplishing environmental objectives, and there are a number of arguments supporting that view. Firstly, there is no concept of how to finance developed countries of Europe without the revenue coming from taxes paid by the industry. Another argument is a perspective of transitioning to more efficient resource management in industry and thereby creating new, green jobs and the fact that the implementation of permanent development model requires the existence of expert knowledge and products of European industry. Furthermore, import of industrial goods from other parts of the world, where lower standards apply, would be environmentally counterproductive. Last but not least, the industry sector is a leader in resource productivity improvement, an example of which may be the reduction of CO_2 emissions over the last two decades in production and construction by approx. 25% and by approx. 12% in housing, services, and agriculture.

The condition requisite for transitioning into green economy is the implementation of fiscal, social, and cultural changes. That process will not be exclusively limited to technological changes, such as the development of renewable power engineering, but the main challenge will entail a change in the present concept of industrial society. Social progress in society has so far been based on indefinite increase of energy of mineral origin. Therefore, the fundamental challenge is to build a new relationship between man and nature.[4]

The European Union invests billions of Euros in green technologies realizing a part of investments within the scope of public-private partnership. Clean technologies, biotechnologies, and nanotechnologies are currently as important to the economy as telecommunication technologies were in the 1990s. The sector of green technologies faces a period of growth, even despite ever-decreasing state support.[5] The significance of social and environmental factors to investment-related decision making is increasingly

---

[3] http://ziemianarozdrozu.pl/artykul/2645/od-zielonej-rewolucji-przemyslowej-do-rewolucji-ekologicznej.

[4] http://ziemianarozdrozu.pl/artykul/2645/od-zielonej-rewolucji-przemyslowej-do-rewolucji-ekologicznej.

[5] http://odpowiedzialnybiznes.pl/aktualno%c5%9bci/zielona-gospodarka-rozwija-sie/.

higher. "Smart innovations" are required to maintain future competitiveness of European enterprises. It does not merely come down to protecting the economy and its potential growth from resource deficit and environment degradation, but it is about a new wave of growth that will stir the industrialized countries out of the present situation of weak growth.[6]

One of the manifestations of GIR is the EU energy and climate package (ECPack), constituting a set of legal acts with which international agreements on the reduction of GHGs emission are realized, including chiefly the reduction of carbon dioxide. According to the climate package, the countries exceeding the limits of carbon dioxide emissions face multibillion penalties. Green energy provides a way of avoiding them.[7]

The ECPack regulations are devised for the implementation of long-term objectives of emission reduction with the use of market instruments (trade in emission allowances) and regulatory activities. The most important assumptions of the EU climate policy by 2020 are[8]

- a reduction in EU GHG emissions of at least 20% below 1990 levels;
- a 20% reduction in primary energy use to be achieved by improving energy efficiency, relative to the reference scenario, BAU (business as usual);
- achieving 20% of EU energy consumption from renewable resources.

This is an initial phase of the implementation of the EU CEP policy, given the complexity of the problem and the apparent long-run horizon. Further climate policy targets and specific measures beyond 2020 were put forward by the European Commission as "Roadmap 2050" and are debated among the EU countries (Hagemejer and Zolkiewski, 2013).

An obligatory emissions trading system was introduced in order to reduce GHG emission. Producers included in the system (enterprises classified within the scope of power engineering, industrial processing, and air transport) are obliged to cover their GHG emissions with suitable allowances.

The emissions trading system affects approx. 12,000 installations in power engineering and in other branches of industry in the territory of the EU, corresponding to over 50% of carbon dioxide and approx. 40% of all GHGs. In 2013, the next stage of ETS development entered into force, in accordance with which the basic principle will involve an auction system

---

[6] http://zielonewiadomosci.pl/zw/zielona-rewolucja-przemyslowa-w-perspektywie-historycznej/.

[7] http://forsal.pl/artykuly/502198,polske_czeka_zielona_rewolucja_w_energetyce.html.

[8] http://ec.europa.eu/clima/policies/package/index_en.htm.

of trading in permits to GHG emissions. Industry will be gradually shifting to the auction system, starting with the introduction of an auction sale of 20% of permits in 2013, to 70% in 2020, to 100% in 2027. In the industry, the allocation of free emission allowances will depend on benchmarks based on 10% of the most efficient installations in a given sector (Hagemejer, 2013).

The conclusions reached at the EU summit on the 2030 package are a testimony to partial modification in EU's line of thinking about climate policy, placing a greater emphasis on the availability of inexpensive energy and independence from fuel imports. The political situation in Ukraine (2012-2014 and continuing) has to a certain degree changed the European perspective on the security of mineral fuel supplies to the EU. The experience of introducing renewable energy sources to the European energy mix brings price-related factors into consideration in determining the framework of climate policy.

The new geopolitical situation after the events in the Crimean Peninsula and the economic effects of a dynamic introduction of RES (renewable energy sources) into European energy mixes call for reconsideration of the main instruments of the EU climate policy in this respect.[9]

THE POTENTIAL OF CENTRAL AND EASTERN EUROPE (CEE)

CEE, due to regional, historic, and cultural differences between individual states of the region, is a difficult market to analyze in terms of GIR. Its countries were shaped by various historical phenomena, including the war in the Balkan Peninsula, accession to the EU, and communism. Those phenomena continue to have significant impact on the rate of green or low-emission economy implementation.

The CEE region seems to be highly promising for the green economy, owing to the need of building operational waterworks infrastructure, sewage treatment, and waste disposal installations. Large foreign companies build plants, changing the entire existing infrastructure with the use of the most modern technology. They also cooperate with local enterprises on minimizing energy consumption in construction.

The European objectives in renewable power engineering, feed-in tariffs, and insecurity of energy supplies cause the countries and governments of

[9] http://energetyka.wnp.pl/ue-zmienia-podejscie-do-polityki-klimatyczno-energetycznej,221714_1_0_1.html.

the region to make large-scale investments into renewable power engineering. However, a majority of them encounter certain difficulties in the accomplishments of the EU objectives.

The countries that acceded the EU in 2004 (the Czech Republic, Hungary, Poland, Slovakia, and Slovenia) have already reached a stable economic situation in terms of their GDP and productivity. In this group of countries, the priorities are to achieve the goals in renewable power engineering, to invest in energy-efficient solutions, and to modernize waste management and recycling systems.

Great development opportunities stem from the fact that numerous countries of this part of Europe committed to observe the EU Renewable Energy Sources Directive and from huge backlog of work that these countries need to catch up with. At the same time, they have access to attractive financing sources (e.g., the EU or the EBRD European Bank for Reconstruction and Development) to achieve those ambitious goals.

The global economic crisis had a negative impact on investors' willingness to support initiatives in the region. Moreover, potential investors are discouraged by complicated legal systems in many countries of that region. Currently, major investments concern
- potable water supply and wastewater treatment,
- increasing the share of renewable energy sources,
- implementation of measures aimed at increasing energy efficiency,
- streamlining waste management systems and introducing innovative transport solutions.

Depending on the country, there are distinct differences in energy efficiency, the structure of waste management market and renewable energy share in total production. Poland, which features the strongest economy in the region, in 2007, invested EUR 902 M in water economy and waste management. The general goal for 2020 assumes that renewable energy is to constitute a 20% share of the total energy production. However, these goals differ for individual countries depending on their relative degree of economic development. For instance, the share of renewable energy in Poland is to eventually grow nearly twofold, from 7.2% in 2015 to 15% in 2020. Wind energy is considered to be the most promising branch of power engineering. Solar energy also offers huge potential. National power companies—similarly as in other regions—demonstrate a rising interest in entering the sector of renewable energy engineering. In CEE countries, large companies are frequently owners of former state power engineering companies, as is the case in the Czech Republic and in Poland, where national companies (CEZ and ENEA) hold a strong position and operate independently.

The degree of energy consumption of the economies in individual CEE countries is highly varied. Overall, the entire region fares rather well in comparison with Western Europe. The energy consumption of CEE countries is eight times higher than the average for 28 EU member states. National priorities include

- energy-efficient heating,
- construction of passive houses,
- the use of renewable energy for house and water heating.

In the recent years, mandatory energy audits have been introduced, while energy certificates gain increasing significance. The EU plays an important role in financing the initiatives that are to improve energy efficiency by providing investment subsidies in the amount of up to 15% of an investment value. The EBRD grants loans to states and takes actions aimed at encouraging private enterprises to engage in undertakings related to energy saving. National subsidies, motivating building owners to their modernization, have also enjoyed popularity.

In the course of the accession of CEE countries to the EU, many of them committed to reduce the amount of waste deposited in landfills and to increase its recycling. Nearly all countries of the region strive to increase recycling.

A significant proportion of EU funds designated to finance regional development is allocated to investments in waste management. The programs financed by EU funds chiefly focus on small communities living in rural areas without access to controlled landfills or other waste management installations. Much has already been done to ensure the same waste management standards enjoyed by the citizens of West European countries.

Many of the largest companies involved in waste management simultaneously provide the necessary technologies. The German company of Remondis is a case in point. The company manages approx. 500 waste management plants all over the world. It is able use its own network of technologies used in other countries to ensure access to the most modern technologies to its plants in CEE. Companies based in Germany and France, including Remondis, Sulo, Alba, Suez, Veolia, CNM, and EDE, hold a particularly strong position in the CEE region.

In terms of water and air purity and waste management, the EU objectives require investments of EUR 115 b. A comparison of energy consumption of economies in CEE countries with the average for 28 EU states suggests that there is a potential for savings as high as over 50% through energy-efficient solutions. A substantial portion of electrical energy comes from coal power plants, which are highly damaging to the environment. The tardiness with which local energy suppliers invest in renewable energy sources provides foreign companies with plentiful opportunities to enter the market. The very scale and numerous modernization opportunities of many obsolete installations are a sufficient basis to recognize CEE countries as attractive targets for foreign investors (Roland Berger Strategy Consultants GmbH, 2011a).

FINANCING GREEN TECHNOLOGIES

From the perspective of economics, the GIR green economy takes several crucial issues into account (Burchard-Dziubinska, 2013):
* The substitutability of various forms of capital
* The value of natural environment and services rendered by ecosystems
* The efficiency of resource use
* The scale of production and consumption
* Fair access to resources
* The question of measuring business activity and the degree of prosperity
The initiatives undertaken within the scope of the abovementioned issues require suitable financing, depending on the nature (type) of the entity involved in green investments. In the sector of green technologies, capital is typically acquired in two ways:
* Bank financing (debt financing)
* Sale of company shares to public or private investors (capital funds)
The rules of financing in the sector of green technologies are similar to those operating in other branches of industry; however, in this case, investment decisions are also affected by regulations of the law and government policy. The latter have also a direct impact on the level of risk and return on investment (ROI). Poland is considered to be one of the most promising markets in terms of risk assessment by the financing institutions.

Banks typically grant loans for financing current business operations with their allocation for new investments or refinancing of assets from previous investments. Loans are sometimes secured with a bank guarantee. They may also provide project financing regarding specific projects, comprising particular goals, actions, and time frames. This type of financing usually entails a higher risk to the bank. In such a case, banks require collateral in the form of assets used in a given project, for example, solar panels and wind turbines. The entities applying for this form of financing are required to present a detailed description of a loan use in the process of project implementation.

Another type of financing is mezzanine financing, which features indirect characteristics of a classic loan with a collateral and financing provided by shareholders or investors. If a project or a company fails, assets will be liquidated, and the funds obtained through their sale will be designated for the loan repayment. However, the risk involved in mezzanine financing is lower than in the case of equity financing.

In financing green technologies, there is no place for insider loans or operating capital financing. Some investments are financed by business angels, VC Venture Capital funds, and private equity, through an initial public offering (IPO) or from innovativeness-supporting funds. Business angels and private equity usually appear at the stage of initial operations, whereas VC funds invest at later stages of a company development. In order to gain access to VC financing, a company needs to demonstrate market successes, history of operations, and a tangible (finished) product. A decision to invest is taken only when the perspectives of high growth are sufficiently promising.[10] Private equity funds tend to get involved at a later stage of an enterprise development, when it has already become an attractive target of acquisition for big players, and the time frame for such a type of investment typically ranges from 3 to 5 years. The required internal rate of return amounts to approx. 25%, while the preferred form of exit from an investment is an IPO (Roland Berger Strategy Consultants GmbH, 2011b).

The minimum investment amount required by these types of funds (VC and PE) constitutes a barrier for private investors, which is why they usually invest in the funds concentrating their operations on consumer goods markets. The companies operating in the sector of green technologies face huge costs incurred on R&D. Independently of those costs, there are other financial needs, that is, investments into fixed and current assets and operating expenses.

VC funds support the development of technologies on many markets, enabling the launch of a product into the market and a quick company's growth. Nevertheless, VC representatives and PE fund managers have so far not demonstrated any particular interest in green technologies, despite the fact that the enterprises of the sector may yield high ROI and

---

[10] http://greenevo.gov.pl/images/zielonetechnologieraportypdf/raport_greenevo_20131022.pdf, p. 219.

demonstrate significant growth. One of the main reasons concerns the burden of significant long-term assets and sales channels requiring significant outlays (Kenney, 2009).

Using the example of financing wind power generators, one may differentiate three sources of capital:

- Commercial banks
- International financial institutions
- EU funds

The basic funding criteria used by the commercial banks and international financial institutions are the following:

- Location—high wind recourses potential confirmed by the quality measurements performed over a year (minimum)
- Risk—project located in a country with stable legal and economic system
- Developer—should be present on the local market and should have global experience
- Own contribution—20-30% of the project (the rate increases with the cost of debt but may be lower for reputable investors)
- Administrative permits
- Equipment—technologies supplied by reputable manufacturers
- Power purchase agreement—for a minimum of 4-6 years

Banks prefer giving loans in EUR, rather than in local currency, for 12 years on average.

As far as the international financial institutions are concerned, three institutions are particularly active in CEE: European Investment Bank, EBRD, and the International Finance Corporation. They provide medium-term financing and long-term financing in forms of debt, equity, leasing, and guarantees.

Last but not least are the EU funds: *Infrastructure* and the *Environment and Smart Growth*, as well as 16 regional programs,[11] with support for renewable energy sources differentiated depending on region. The national programs will be used to finance larger projects in terms of value or volume (Wind Energy in Poland, 2013).

THE SIGNS OF THE GIR IN POLAND

Poland is one of the largest and most dynamically developing countries in the EU. It is also a key representative of CEE, part of the continent with

---

[11] The EU budget perspective for 2014-2020.

great potential for development (Ruminski, 2012). According to Ernst & Young's 2013 European attractiveness survey, Poland is the most attractive country for investment in CEE. It boasted the Europe's strongest increase in the number of investment projects[12] (by 22%) and jobs (67%), whereas the average in Europe is (8%) created by the FDIs. With these results, Poland became one of the European leaders of growth.[13] The UNCTAD's 2013 World Investment Report[14] confirms Poland's strong position on the international investment scene: in the next 2 years, it will be the 4th in Europe and the 14th world's most attractive economy.

Despite the difficult situation in the global investment arena, Poland and Spain were the only EU countries to have seen a growth in the number of greenfield FDI projects. The number of projects in Poland increased in 2012 by 5%—with 237 new FDI projects launched. Worldwide, the number of new projects fell by 16%, while in the EU, it fell by 23%.[15] Poland is an attractive business destination in comparison with other countries of the region. According to the business climate survey for CEE countries conducted by the Polish-German Chamber of Industry and Commerce, 94% of the investors would choose Poland again.[16]

In 2013, FDI into Europe declined and the region received $137.26 billion worth of investments, representing a 12.08% decrease in comparison with 2012. The top 10 countries accounted for 72.64% of foreign capital investment into Europe (Table 1).

Poland, thanks to its developed industry, offers the right conditions to develop green technologies and to profit from them. According to the Ministry of Environment, Poland is now home to 510 green technology companies with approx. 25,000 employees. They work at wind farms, biomass plants, and solar collectors and on energy efficiency solutions, geothermal energy development, and a raft of water and gas treatment technologies (Kureth, 2013). Among the successes so far, one needs to note the use of solar energy and the production of specialist ships for various types of offshore installations, particularly wind turbines, by one of the shipyards. New RES regulations, according to which the greatest extent of support is to be provided to, inter alia, photovoltaics, can help in the development

---

[12] Compared to 2011.

[13] http://www.economist.com/news/special-report/21604684-first-time-half-millennium-poland-thriving-says-vendeline-von-bredow.

[14] http://unctad.org/en/publicationslibrary/wir2013_en.pdf, p. 7.

[15] http://ftbsitessvr01.ft.com/forms/fdi/report2013/files/the_fdi_report_2013.pdf.

[16] http://www.paiz.gov.pl/files/?id_plik=21375, p. 6.

Table 1 Foreign Direct Investments into Europe by Market Share in 2013

Country	Market share
The United Kingdom	19.31%
Spain	8.74%
Russia	7.72%
Turkey	6.69%
Germany	6.69%
Romania	6.13%
The Netherlands	5.01%
France	4.78%
Poland	4.49%
Ireland	3.08%
Other	27.36%
Total	100%

Source: http://ftbsitessvr01.ft.com/forms/fDi/report2014/files/The_fDi_Report_2014.pdf

of those branches of industry. The use of natural renewable energy sources is gaining more and more significance in terms of Poland's energy independence. A departure from Polish coal sources would entail shifting to gas sources, thereby reducing the degree of energy independence (Chart 1).

In Poland, the most rapid development of "green" initiatives is observed in those sectors in which investments result from regulations and in which mechanisms exist that guarantee a return or partial financing of the investment. The quickest developing areas of "green" economy include renewable power engineering, in particular wind and biomass power engineering and construction. The 2011 report by Ernst & Young ranked Poland 10th in

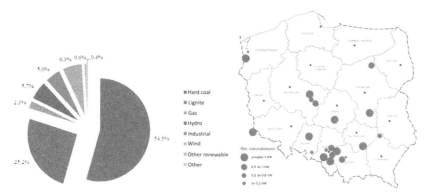

Chart 1 Structure of electricity generation capacity in Poland and the main electricity production centers in 2012. *Source: Energy Market Agency, http://www.paiz.gov.pl/files/? id_plik=21681.*

the world with regard to its wind energy potential.[17] Growth areas also feature biofuels, which attracted farmers' attention, or the construction of modern wastewater treatment plants.

Renewable Energy

According to Directive 2009/28/EC, all EU member states should gradually increase the share of energy from renewable sources in total energy consumption and the transportation sector. Therefore, the development of the renewable energy sector is one of the priorities for the Polish government. One of the main objectives of the Polish energy policy is to increase the proportion of energy from renewable sources in final energy consumption up to 15.5% in 2020 (19.3% for electricity, 17% for heating and cooling, and 10.2% for transportation fuels). These limits should be regarded as the beginning, because the EU is moving towards further increase of the share of energy from renewable sources. It requires investments in new generation capacities. The most dynamically developing renewables are currently wind energy and the use of biomass for energy production. The largest foreign investors in Poland are Vortex, EDP, RWE, E.ON, CEZ, GDF SUEZ, Mitsui, J-POWER, ACCIONA (wind farms), Dalkia (biomass combustion), Poldanor, and AXZON Group (biogas plants). The Polish companies are also investing in renewables, for example, TAURON, Enea, Energa, and PGE. The factors stimulating investments into green energy include, inter alia, the following:

- Reducing connection costs to renewable source networks by half.
- Operator of electrical power engineering system will first transmit green energy.
- Companies generating green energy of the capacity higher than 5 MW will be released from license charges and charges related to acquiring and registering origin certificates.

In the course of the last several years, wind power engineering has witnessed dynamic development, which is the consequence of the regulations obliging big players to sell specific volumes of energy from renewable sources. What is more, the system of green certificates plays a stimulating role. In 2013, with a result of 892.8 MW, Poland occurred to rank as the third EU market in terms of new power increase in wind power engineering, which constitutes a 28.1% rise in relation to the preceding year. The installed capacity of

[17] http://www.eyeim.com/pdf/11eda220_europe_attractiveness_2011_web_resolution.pdf.

Table 2 Renewable Energy Installations in Poland (Electricity)

Type of installation	Quantity	Power (MW)
Biogas power stations	207	136.319
Biomass power stations	29	876.108
Photovoltaic power stations	9	1.289
Wind power stations	743	2644.898
Hydroelectric power stations	771	966.236
Cofired technology	41	n/a

Source: Energy Regulatory Office (as of 31 March 2013)

wind farms grew from 288 MW in 2007 to 3.4 GW in 2013 and wind farms currently generate approx. 6.6 TWh of energy. According to forecasts, by 2015, joint capacity may rise up to 5 GW.[18] Poland has the right geographic conditions with favorable wind zones along the Baltic Sea coastline, in the east and north east and in the mountainous regions of the southern regions in the Lower Silesia and in the lower Carpathian ranges.[19]

Alongside a quickly growing segment of wind power engineering market, companies cooperating with that sector are established. These are farm developers (e.g., EPA Wind[20] and Polish Energy Partners[21]), specialist companies measuring wind power, or enterprises manufacturing wind installations (e.g., GSG Towers[22]—a company belonging to Gdańsk Shipyard, building wind turbine towers). Main components are usually supplied by global leaders. Local companies concentrate on services and supply of construction components (Table 2).

Poland is gradually becoming an attractive destination for investments in manufacturing of devices used in energy generation. According to the Institute for Renewable Energy, there are more than 200 production companies working for the renewable energy sector.[23] According to EurObserv'ER,[24] Poland is ranked 5th in the EU in terms of production of primary energy from solid biomass, and it is a leader among the new EU member states in terms of total installed capacity of wind farms. Poland is also the largest

---

[18] http://www.pse.pl/index.php?dzid=115&did=581.
[19] http://www.reo.pl/o-355-gw-wzrosly-moce-zainstalowane-w-energetyce-wiatrowej-na-swiecie-w-2013-r.
[20] http://en.epawind.pl/.
[21] http://pepsa.com.pl/en/page/wind-energy.
[22] http://www.gdanskshipyard.pl/gsg-towers-offer.html.
[23] http://www.ieo.pl/pl/ekspertyzy.html.
[24] http://www.eurobserv-er.org/.

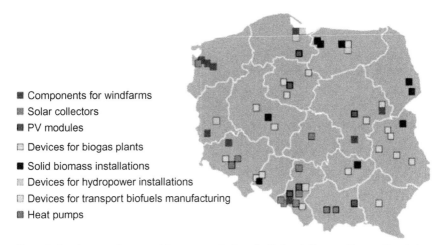

- ■ Components for windfarms
- ▧ Solar collectors
- ■ PV modules
- ▢ Devices for biogas plants
- ■ Solid biomass installations
- ▧ Devices for hydropower installations
- ▢ Devices for transport biofuels manufacturing
- ■ Heat pumps

Chart 2 Producers of renewable energy devices in Poland. *Source: Energy Regulatory Office.*

beneficiary of support from EU funds, which are allocated to, among others, innovative investments, research and development, infrastructural projects, environment protection, renewable sources of energy, and employee trainings (Energy Sector in Poland, 2013).

There are numerous key drivers of renewable energy sector development in Poland. One of them is the country's dynamic economic growth in recent years and large domestic market—approx. 38 million consumers. There is a demand for green energy, and it will grow due to the energy policy objectives (the increase of the proportion of energy from renewable sources in energy consumption up to 15.5% in 2020). Moreover, there are favorable natural wind conditions and a large potential for obtaining biomass and biogas. Last but not least, there are the investment incentives (government support) for the production of energy from renewable sources (Chart 2).[25]

The Use of Biomass and Waste for Energy Production

The interest in building wind farms is extensive; however, in case of biogas plants, it is limited. According to plans, one such installation is to be built in each commune. Today, 13 biogas plants of total power equal to 9.3 MW operate in Poland, while another 9 are under construction. In order for the plans to be fulfilled, over 2.4 thousand biogas plants would need to

[25] http://www.paiz.gov.pl/sectors/renewable_energy#.

be built within 7 years' time and a startup of one such installation requires from 12 to 18 months.

With regard to waste management, it is worth noting that on average, in the entire EU, less than half of waste is deposited in landfills. In Sweden, it is merely several percent. Swedes incinerate nearly half of their rubbish, and consequently, they obtain 40% of heat from waste, not only domestic waste but also wood-origin, agricultural, and industrial waste. Moreover, their incineration plants also generate electricity. They are clean and meet high eco-standards.

In Poland, 86% out of 12 million tons of annual waste is deposited in landfills. So far, there is only one waste incineration plant—in Warsaw. Next ones, partly owing to the EU funds, are to be built in six cities.[26] Out of the projects of municipal waste incineration plants currently under construction, only one (in Poznan) is being realized through public-private partnership. It is an attractive model of investment implementation, which allows for building an installation that over the subsequent 3 years will be operated by an experienced contractor.[27] What makes the project additionally distinctive is the combination of private investor's capital, a bank debt, and an EU subsidy.

The Poznan project features some pioneer components on a European scale. The debt service accounts for availability payment, which is incurred by the municipality. Banks agreed to a long investment financing period, and one of them (BGK[28]) provided a tranche with a repayment date of 22 years. The involvement of PPP in the project is treated by BGK as a mission element that is as important as the investment profitability.[29]

Construction

It is not the industry or transportation that is responsible for the highest energy consumption, but buildings. They account for approx. 40% of total final energy consumption in the EU and for the highest volume of CO_2

---

[26] http://wyborcza.pl/piatekekstra/1,133150,14223289,nadchodzi_zielona_rewolucja_na_smiecie.html.

[27] Founded by Sita Polska and its French partner—Marguerite Waste Polska.

[28] Bank Gospodarstwa Krajowego (BGK) is Poland's only state-owned bank. The primary business objective of BGK is to provide banking services for the public finance sector, in particular through the support of the government's economic programs and local government and regional development programs implemented with the use of public funds, including those from the EU.

[29] http://www.forbes.pl/sa-pieniadze-na-nowe-spalarnie,artykuly,157263,1,1.html.

emissions into the atmosphere, which is why the EU set a target of reducing energy consumption in buildings. The parliament and European Commission directive[30] obliges the member states to construct buildings with nearly zero energy consumption after 31 December 2020. Some of the member states are already prepared for the revolution of "green" construction.

In Poland, decision makers are already aware of the idea of energy-efficient construction. The Energy Efficiency Act passed in March 2011 provided for a model role of the public sector in raising energy efficiency, which contributed to awakening the interest in energy efficiency construction, particularly among public utility companies.[31]

Two trends can be detected in Polish construction: investing in thermal isolation of buildings and an increase of interest in installing solar panels for water heating. The factor motivating investors to insulate buildings is a system of subsidies and loans for repairs and investments granted through one of Polish "green" banks (Wierzbicka, 2012), that is, Bank for Environmental Protection[32] and the Voivodeship Fund for Environmental Protection and Water Management.[33] The most important factor contributing to the rise in solar collector sales is a system of subsidies, in line with which six banks offer preferential loans with a subsidy from the National Fund for Environmental Protection and Water Management.[34]

Automotive Industry

As regards the automotive industry, Solaris Bus & Coach,[35] which has been producing hybrid and electric drive buses, is a good example of a successfully developing Polish enterprise. The company is a major European producer of city, intercity, and special-purpose buses and low-floor trams. Since the start of production in 1996, over 10,000 vehicles have left the factory near the city of Poznan. They are running in 28 countries, including Germany, France, and Switzerland. Solaris is still rapidly developing, introducing advanced hybrid engines.[36] The development of new Solaris technologies minimizes the impact of public transport vehicles on the environment,

---

[30] 2010/31/EU.

[31] http://www.biura.pl/pl/artykul/1609/gornoslaski-park-przemyslowy-dolacza-do-zielonej-rewolucji.

[32] http://www.bosbank.pl/index.php?page=root_en.

[33] http://www.wfosigw.pl/.

[34] http://www.nfosigw.gov.pl/en/financing-environmental-protection/.

[35] http://www.solarisbus.com/vehicles/.

[36] http://www.profinfo.pl/img/401/pdf40160330_3.pdf.

which is one of the key business objectives.[37] The offer is continually enhanced not only in terms of quality but also in terms of reduced environmental impact. Solaris is active in bringing down the vehicles' overall weight and energy consumption.[38]

Furthermore, the international success of Polish students from the Students' Club of Vehicle Aerodynamics at the Warsaw University of Technology deserves to be mentioned. They won the second place at an international competition of Shell Eco-marathon 2014 with a result of covering 376.46 km on one liter of petrol. The competition, in which each year hundreds of student teams from all over the world compete, were held in Rotterdam this year. The students competed with a vehicle of their own construction called PAKS in the category of a petrol-driven urban concept. The goal of each team was to design and construct a vehicle featuring the lowest possible fuel consumption, and the competition involves covering a distance of 16.2 km in simulated urban conditions at a minimum average speed of 25 km/h.[39]

Students of the same university also won an international competition of Smartmoto Challenge 2014 held in Barcelona. Their *light electrical motorcycle* defeated all teams and won the general classification of the competition in which 10 teams from Spain, Russia, and Ecuador participated. The motorcycle created by the members of the Vehicles and Mobile Robots Science Club also received an award for design and best business plan.[40]

Biotechnology

Biotechnology in business and science is becoming more and more important for the economic development of Poland. It is used in medicine, pharmacy, plant growing, and animal breeding. Biotechnology is still an emerging sector in Poland. Nevertheless, it has been developing very fast. Further dynamic growth of the domestic biotechnology is expected thanks to innovative research projects undertaken by the Polish biotech

---

[37] http://solarisbus.com/firm/#goto|firm_scene10.
[38] http://solarisbus.com/firm/#goto|firm_scene13.
[39] http://www.reo.pl/miedzynarodowy-sukces-studentow-politechniki-warszawskiej.
[40] http://www.reo.pl/elektryczny-motocykl-wroclawskich-studentow-najlepszy-w-barcelonie.

companies and academic institutions and by the inflow of foreign investment.

Poland attracts investors by offering broad access to highly qualified employees and researchers[41] and competitive labor costs. There is a network of more than 110 scientific institutions employing over 2800 scientists who are active in the field of biotechnology and molecular biology. Moreover, there are six mature biotech clusters (Warsaw, Lodz, Tri-City, Krakow, Wroclaw, and Poznan) where biotech companies and research institutes collaborate. Business spending on R&D in Poland has increased over the last year by 800% (from 61 to 500 million EUR).[42]

The following factors facilitate the development of the biotechnology sector in Poland (Chart 3)[43]:

- Favorable government policy supporting investments in new technologies
- Development of biotechnology-related sectors
- Government grants for biotechnology projects
- Numerous facilities with high research capabilities
- Wide range of financing programs (both national and the EU)
- Competitive labor costs and increasing labor productivity

The Innovative Polish "Green Technology" Companies

Entrepreneurs have plenty of ideas about how to respond to global environmental problems. They undertake actions intended to improve eco-efficiency of their own processes, an example of which can be minimization of generated

---

[41] The top Polish researches (500) have had a great opportunity to take part in the unprecedented program called *Top 500 Innovators, science-management commercialization*, created by the Ministry of Science and Higher Education of the Republic of Poland to help bridge the gap between academia and business. In the course of the program, selected individuals complete a 9-week academic program taught at partnering universities from the academic ranking of world universities—Stanford University and the University of California, Berkeley. The courses provide the participants from Polish universities, research institutes, and technology transfer offices with skills needed to manage cooperative research projects and bring high-technology products to the market. Participants also have the opportunity to visit key technology companies from a variety of areas (including energy, biotechnology, information technology, and technology transfer) in order to obtain a broad understanding of how companies have taken intellectual property and changed the landscape of various industries. For more information on this program, see http://www.top500innovators.org/english.

[42] http://www.paiz.gov.pl/sectors/biotechnology#.

[43] http://www.paiz.gov.pl/sectors/biotechnology#.

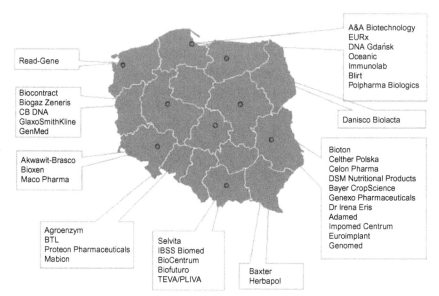

A&A Biotechnology
EURx
DNA Gdańsk
Oceanic
Immunolab
Blirt
Polpharma Biologics

Read-Gene

Biocontract
Biogaz Zeneris
CB DNA
GlaxoSmithKline
GenMed

Danisco Biolacta

Akwawit-Brasco
Bioxen
Maco Pharma

Bioton
Celther Polska
Celon Pharma
DSM Nutritional Products
Bayer CropScience
Genexo Pharmaceuticals
Dr Irena Eris
Adamed
Impomed Centrum
Euroimplant
Genomed

Agroenzym
BTL
Proteon Pharmaceuticals
Mabion

Selvita
IBSS Biomed
BioCentrum
Biofuturo
TEVA/PLIVA

Baxter
Herbapol

Chart 3 Map of the Polish biotechnology sector. *Source: Polish Information and Foreign Investment Agency, http://www.paiz.gov.pl/sectors/biotechnology#.*

waste. The main goal of such an approach is implementing a closed cycle, assuming minimizing waste to zero thorough materials reuse, recycling, and composting. Addressing the consumers and persuading them to join in the activities for environment protection are a highly important trend emerging among proenvironment companies. More and more companies (also global corporations) announce that they have managed to reduce non-recyclable waste to zero. Procter & Gamble can serve as an apt example, as it communicated that 45 of the concern plants globally (including 3 plants in Poland) no longer send any production waste to landfills. Over the last 5 years, the company's activities related to waste reuse and recycling generated more than $1 billion in savings. Sainsbury's comes to mind as yet another example of a company that managed to reach the "zero waste" target within barely 3 years of having set the target.[44]

At this juncture, enterprises offering environmental technologies deserve to be mentioned. Among those, the winners of GreenEvo competition

---

[44] http://odpowiedzialnybiznes.pl/wp-content/uploads/2014/03/forum_ odpowiedzialnego_biznesu_ekologia-przyszlosci.pdf, p. 5.

merit particular attention.[45] In 2014, winning solutions included, inter alia, domestic wastewater treatment plant and high-efficiency heat recuperation technology in ventilation systems. GreenEvo[46] is a project of the Ministry of Environment, which aims to support the international transfer of Polish green environmental technologies and popularize the technical thought contributing to climate protection, support of sustainable development, and building of green economy. The competition award committee, composed of, among others, representatives of business and government institutions, evaluates the authors of innovative environmental technologies. The GreenEvo technology leaders and their field of expertise are presented in Table 3.

As it transpires from the information and examples presented above, Poland undertakes numerous "green" initiatives contributing to sustainable economic development and being the testimony of the existence of GIR. A definite majority of the "green" investments planned and completed so far are a direct or indirect consequence of the policy conducted by the EU and the directives that member states need to implement.

In most of the cases, the realization of such investments requires substantial financial outlays, including also on research and development. Apart from private financial institutions and capital funds, public institutions are also financially involved. In certain cases, projects are realized through public-private partnership, providing measurable benefits for the parties engaged in an investment.

The financial dimension related to new green technologies, green business ventures and green professions is of huge importance, however—as it was already noted in this monograph—it is not all about money. There are other significant costs incurred, like human lives and reduced GDP, by not reducing the world's dependency on carbon-generated energy. The green economy is needed to accelerate the necessary changes and stop climate change.

---

[45] http://www.greenevo.gov.pl.

[46] GreenEvo—green technology accelerator is a project devised with the purpose of international transfer of Polish, green solutions to foreign markets. In the course of four editions held so far, 48 winners have been selected. They were promoting their solutions nearly all over the world including, inter alia, China, Indies, Thailand, United Arab Emirates, and Russia.

Table 3 Polish Leaders in Green Technologies

	Company name	Technology offered
1.	ASKET Roman Długi	RES technologies (BIOMASSER—technology of briquetting moist agrobiomass)
2.	BIOGRADEX-Holding Sp. z o.o.	Water and sewage technologies (BIOGRADEX Technology—treatment of municipal and industrial sewage using the low-loaded activated sludge method)
3.	Ecotech Polska Sp. z o.o. (EnviroMix technology)	Technologies supporting waste management (EnviroMix technology—neutralization and recycling of substances harmful to the environment through immobilization of pollutants as thermodynamically stable phases of minerals and microencapsulation of organic pollutants into mineral structures)
4.	Energoinstal S.A.	Energy-saving technologies (the use of waste heat and chemical energy of excess gas from the process of coke dry cooling)
5.	LEDIKO Walendowski i Wilanowski Sp. j.	Energy-saving technologies (intelligent streetlight LED CLEVEO)
6.	Neon Wojciech Norberciak	RES technologies (solar collector Neosol 250)
7.	Petroster Sp. j.	Low-emission technologies (SAFE-Tank package—equipping and securing storage tanks for liquid fuels)
8.	PP-EKO Sp. z o.o.	Water and sewage technologies (ROVAPO—technology and installation for high-purity water recovery from industrial sewage (together with software for automation of the technological process))
9.	PPHU MARBET WIL Sp. z o.o.	Technologies supporting waste management (Sultech—technology of recovery of hazardous waste through stabilization and solidification in sulfur concrete)
10.	PROMAR Sp. z o.o.	Organizational or software solutions supporting energy saving (PROM@R MONITORING SYSTEM (PMS)—system of optimizing energy consumption in the building, combining sensors, controllers, and software, offered to customers as a specialist service managed by the PROMAR company)
11.	SUNEX Sp. z o.o.	RES technologies (collector with a stainless steel frame)
12.	Watt Sp. z o.o.	RES technologies (flat solar collectors and vacuum solar collectors)
13.	WOFIL Robert Muszański	Water and sewage technologies (ozoning technology—used for water treatment)

14.	ekotop Roman Sobczyk	Water and sewage technologies (dryers of sewage sludges)
15.	Jacek Habryło Firma NPF	RES technologies (technology of pellet combustion using the boiler LESTER 20)
16.	Dagas Ltd	Looking for eco-friendly solutions for rubber, plastic, municipal waste, and solid biomass treatment, Dagas Ltd offers pyrolysis plants. Unlike competitive solutions, the use of Dagas technology produces heat power, which can be processed into electricity or thermal energy
17.	Energo Natura Ltd	For institutions managing water reservoirs and industrial plants using processing water, which experience problems with biological and chemical (including oil) pollution, they offer a comprehensive reagent-based technology for water treatment and counteracting the effects of eutrophication. In contrast to other, hazardous chemical methods of a narrow use and expensive and logistically complex physical methods, they offer secure, quick, and simple removal of impurities and the ability to restore ecological balance
18.	Energetyka Solarna ENSOL Ltd	For all those who want to significantly reduce the growing cost of maintaining household, production hall, or public building, they offer solar panels. In contrast to traditional products, the ENSOL hybrid collector produces electricity and heats utility water, which reduces investment costs and allows you to take maximum advantage of the available roof surface
19.	Far Data Ltd Sp. k.	For all those who are struggling with the problem of traffic noise, they offer an automatic monitoring station. Unlike other noise measurement systems, they deliver a solution for long-term monitoring with flexible expansion capabilities, saving time and resources
20.	Frapol Ltd	For all owners of single-family houses, apartments, and small public buildings, seeking high-quality parameters of the air in the room and their convenient control and checks, they offer the Onyx recuperator with an integrated control system. Unlike conventional ventilation systems based on regenerators, Frapol Ltd offers a comprehensive control system for easy and precise control of air parameters in the room and controlling all devices working with the air treatment system

Continued

Table 3 Polish Leaders in Green Technologies—cont'd

	Company name	Technology offered
21.	M3System Ltd	For investors, developers, and construction companies seeking high profits, they offer a unique energy-saving construction technology. Unlike other available technologies, their solution helps to achieve a passive housing standard (zero or plus energy) with costs similar to traditional construction. An average building takes only 6–8 weeks to erect. Buildings can be erected on land with very low bearing capacity, and no heavy equipment or highly qualified employees are needed.
22.	Mielec Diesel GAZ Ltd	For companies extracting or processing crude oil, which consume significant amounts of energy or operate at a large distance from energy networks, they offer gas and combined heat and power generators to produce technology electricity and heat from waste gas. Unlike the gas flares and boilers, pump drives available on the market, the Vapor CHP technology effectively and economically uses waste gases for simultaneous, distributed generation of electricity for one's own business or for sale
23.	NIKOL Jan Nikołajuk	For all those building or renovating buildings who want to significantly reduce the cost of heating or air conditioning and get rid of problems resulting from the absence or ill-functioning of ventilation, they offer NIKOL, a heat recovery air-handling unit. As opposed to natural or other simple ventilation systems, their device allows to reduce the cost of heating or air conditioning and ensure efficient and user-friendly ventilation

Source: http://greenevo.gov.pl/en/greenevo–participants

REFERENCES

Burchard-Dziubinska, M., 2013. Green economy as a new area of interest in economics. In: 9th Congress of Polish Economists. Department of Development Economics, Lodz University, Lodz, Poland, pp. 3.

Clark II, W.W., Cooke, G., 2014. The Green Industrial Revolution, Draft. Elsevier.

Energy sector in Poland, Sector profile, USA 2013.

Energy sector in Poland, Sector profile, Polish Information and Foreign Investment Agency, Poland 2013.

Grzybek, M., 2011. Business and Ecology—Common Future? Ecology of the Future. Responsible Business Forum, Warsaw.

Hagemejer, J. (Ed.), 2013. Short-term Effects of the Macroeconomic Climate and Energy Package in the Polish Economy. Implications for Monetary Policy. Institute for Economics, National Bank of Poland, Warsaw, pp. 3.

Hagemejer, J., Zolkiewski, Z., 2013. Short-run impact of the implementation of EU climate and energy package for Poland, computable general equilibrium model simulations. Bank Credit 44 (3), pp. 238.

Kenney, M., 2009. Venture Capital Investment in the Greentech Industries: A Provocative Essay. Department of Human and Community Development, University of California, Davis.

Kureth, A. (Ed.), 2013. Investing in Poland 2013. Valkea Media SA, Warsaw, p. 104.

Roland Berger Strategy Consultants GmbH, 2011a. Green Growth, Green Profit. Wolters Kluwer Poland Ltd, Warsaw, pp. 186–201.

Roland Berger Strategy Consultants GmbH, 2011b. Green Growth, Green Profit. Wolters Kluwer Poland Ltd, Warsaw, pp. 281–282.

Ruminski, R., 2012. Small business financing in CEE—the case of Poland. In: 57th ICBS World Conference, Wellington, June 2012, pp. 1–2.

Wierzbicka, E.M., 2012. Environment Protection and the European Integration. The Polish Experiences. Difin Publishing, Warsaw, Poland, pp. 219.

Wind Energy in Poland—report, 2013. The Polish Information and Foreign Investment Agency, November, pp. 53–54.

APPENDIX 4

The Green Revolution Applied in Everyday Life

Bruce Hector[1,3], Woodrow W. Clark, II[2,3]

[1]Dr. Hector is a family physician, inventor of the Universal Ecolabel and founder of Earth Accounting. He can be reached at: bphmd43@gmail.com

[2]Woodrow W. Clark II, MA[3], PhD is an expert in the solutions on climate change with seven published books and now producing documentary, educational and feature films. He can be reached at: www.clarkstrategicpartners.net

[3]The authors want to thank Josh Schenk for his work on this case.

OVERVIEW

Today, most environmental initiatives have been the result of government regulations, but for the last few years now corporations have been pressured by shareholders to provide information on the impact of their products and services on the environment. This is due in part to the fear of legal actions that could reduce share value as well as genuine environmental impact concerns and recognition that in the long run production, supplies and services must be sustainable to all companies in order for them to survive as well as to protect the earth for future generations.

Shareholder and now pension, investment, and other fund investors' demand for this information has led to the specialized field of "sustainability auditors" who are hired by corporations to produce annual reports documenting progress in defined sustainability areas. In many instances, these audits ask the appropriate questions for assessing the environmental impact of a company's products, but tend to do so only at the corporate level, rather than at the basic "footprint" of each item sold, shipped, and produced. In other instances where products are produced at only one location, the audit also assesses the facility's environmental impact.

However, even these reports overlook several important elements including environmental impact of use of the product and post-use environmental impact. Further, certain "externalities" (Clark, 2012) like quality of discharge water, use of recycled materials, type of energy used in production may be overlooked in these sustainability reports making everyone suspicious that even the data provided may not be complete, valid, and accurate.

CONVENTIONAL ECONOMICS: ON PRODUCT CODES AND LABELS

Most economists (Clark and Fast, 2008) note that consumer purchases account statistically in data and percentages for about 70% of a nation's GDP. "What is needed for economics and for both product and sustainability performance are qualitative" data such is the norm in all sciences. That means the understanding of data with definitions and meaning of numbers, data, equations, and actual language with words and concepts must be obtained. Thus, conventional economist usually concludes that consumer potentially has an extremely powerful role to play in accelerating the sustainability movement, if the consumer can make purchases based as much on environmental, product content, costs, and their social impact. What is lacking are both the definitions and meanings of the data, product content and manufacturing (including shipping and handling) as a method for consumers to easily learn the complete product footprint preferably at the point of purchase, so as to facilitate selection of products that most promote sustainability. In short, qualitative economics does that through the analysis of definitions and meanings for what products are made of, produced, shipped, and sold need to be defined, monitored, and evaluated for all consumers.

THE PRODUCT, SERVICE, AND DISTRIBUTION CHANGE: ECOLABELS

Ecolabels are the proven case in point and a significant aspect for all consumers in the Green Industrial Revolution (GIR). They were introduced in Germany in the early 1990s as an attempt to provide consumers with assurance that the producer had considered the environmental impact of production and sought to reduce adverse impact. Over the last 3 decades, more than 490 ecolabels have come into existence. In some cases the manufacturers produced the label to distinguish their product from others. In other cases, an industry as a whole has developed environmental and social production standards and offers third party certification usually requested and financed by the company. The entire process is called "social capitalism" as both government and business work together.

In other cases, government regulation has required companies to provide data on some resource utilization (US EPA Energy Star rating). In a recent paper (Hector et al., 2014), a detailed new tool for the Universal Ecolabel (UE) was presented that seeks to be the first comprehensive and generalized method for assessing the full environmental and social impact of product production, use and post-use fate.

Generic Universal Ecolabel—that is, the "green bar code."

The UE is designed for use as a smart phone application that allows the consumer, at the point of purchase to know the full impact of the product, determine if it meets a predetermined consumer preference set, identify alternate products more compatible with the consumer's environmental and social preferences and even redirect the consumer to a preferred product at the same time notifying the first producer that his product was not chosen because of a less desirable footprint.

Woman consumer scanning ecolabels with an iPhone app.

The dizzying array of ecolabels now have lack of specific information on the label's meaning; uncertainty of what is left off the label and general consumer suspicion that such efforts are just a new marketing technique have resulted in only limited influence of ecolabels by consumers thus far. In the USA, an ongoing political rejection of environmental concerns by many has greatly reduced the likelihood that the federal government will create universal environmental and social production standards. In Europe, the role of the European Union has been much more aggressive in this endeavor though the degree of success has not been fully determined and there is still no single consistent environmental and social impact presentation method.

The majority of people today in the industrialized world now acknowledge that industrialization has resulted in global warming, species loss, environmental depredation, and pollution that threatens continued progress especially during the last half of the twenty-first century when world population is expected to reach 9-10 billion. People particularly realize that the adverse impact of fossil fuel dependency noted elsewhere throughout the GIR book. A Cone Communication survey done in April 2013 noted that more than 70% of shoppers would like more environmental information and almost 30% uses environmental information to influence purchases (http://www.conecomm.com/stuff/contentmgr/files/0/a70891b83b6f1056074156e8b4646f42/files/2013_cone_communications_green_gap_trend_tracker_press_release_and_fact_sheet.pdf).

This 30+% of shoppers therefore represent a powerful economic force that producers are likely to respond to if provided with adequate feedback and remediation options are available at reasonable cost. Companies obtaining electricity from fossil fuels probably make products with a larger greenhouse gas (GHG) footprint. These companies would be encouraged by market forces to install or switch to renewable energy sources especially if they learned that their main competition had done so.

Efforts are underway in the corporate world to better document impact. The largest effort is the Walmart funded Sustainability Consortium (Wal-Mart—Thinking outside the big box, Ethical Corporation, 7 September 2009) that has been addressing the problem for more than 4 years. Recent assessment of progress notes that the process of full life cycle assessment (LCA) is hampered by inability to document the impact of materials received from the supply chain. While in principle, no one is asking the company to change production methodology, LCA does require measuring elements of production, not commonly measured.

For example, a proper LCA of any material extraction process would require documenting the amount of waste (unused but extracted material) and noting its composition and disposition just as importantly as noting the concentration of material extracted and energy used to extract and concentrate the material. This documentation will require companies to measure production elements formerly not acknowledged and felt to be of no consequence. These companies may be reticent to do so especially if the "footprint" of the materials is less than optimal and makes their competitors look more "green."

Likewise, identifying dependencies on other industries for production elements, like fossil fuel-generated electricity, could cause the end line producer to select an alternate supplier less adversely impactful to the environment (Clark and Cooke, 2011). Hence, there arises a challenge of both incentivizing suppliers to gather new data germane to environmental impact and potentially expose some current less than optimal production practices, something few companies are willing to chance for fear of loss of market share. Other incentives have to be created for both companies to become honest concerning their production footprint and to encourage them to learn the impact of their supply chain.

In response to this challenge the designers of the UE have developed another new concept, an Information Cooperative. In corporations, owners are shareholders and value is determined by the number of shares owned. In a cooperative, reward is determined according to contribution. Cooperatives are quite common in agricultural settings where all producers bring their product to a central and usually shared facility for processing, packaging, and distribution. Proceeds from sale go to the cooperative and then are distributed to participant according to a predetermined formula.

GLOBAL TRANSITION ECONOMY: SOCIAL CAPITALISM

An information cooperative operates in a similar manner but the product, information, has varying levels of importance from simply data on one element of production to meta-analysis of large data sets to determine trends applicable on an industry, national or global level. A predetermined value once established would reward participants according to the quality and quantity of data provided to the cooperative with dispensation according to the level of participation.

Clearly, collecting and distributing large volumes of production information as well as knowledge of consumer preferences should be of value to most producers. Information on new production methods, tools for reducing footprint, and sources of subassemblies that are most environmentally friendly would also likely be valued by producers and purchased from an information resource. Further, consumer redirection to a more sustainable product should have value to the producer receiving the redirected customer. Thus, there are several economic models that should result in sales by the UE founding company, Earth Accounting. In return, the cooperative would be an independent organization composed of a wide variety of information contributors including corporations, Ecolabel producers, not-for-profit environmental organizations and social justice ones, governments collecting environmental impact data and even individuals with special useful information. A portion of Earth Accounting revenues would be provided to the cooperative on a regular basis using the predetermined formula. Participants in the cooperative would be rewarded based on their contribution of information to the cooperative and use of that information by Earth Accounting.

Earth Accounting's Universal Ecolabel

	INPUT	USE	POST USE
MATTER	Adverse Effect on Humans ___ Plants ___ Animals ___ Air ___ Water ___ Ecosystems ___ Recycled matter (% wt.) ___ Petroleum based (% wt.) ___ Waste Water (gal.) ___ Prod. Waste (%wt.) Upcycle ___ Down ___	Type(s) _____ Quantity: _____ Single use: _____ Lifetime: _____ Waste water (gal/use). _____	Biodegradable Short Term ___ Long ___ Technical (%wt) Reuse ___ Recycle ___ Return ___ Atmos. Emissions ___ Waste Water (gal) ___ Landfill or Burn (%wt.) ___
ENERGY	Total Kwh or gallons: ___ % Fossil Fuel ___ % Nuclear ___ % Renewable ___ Greenhouse Gas Emissions (Gg) ___ (tons)	Kwh or gallons fuel Per Use ___ Lifetime ___ Energy Source % Fossil Fuel ___ Nucl ___ Renew ___ Gg emissions Single ___ Life ___	Energy Produced (Kwh) ___ Gg Emissions ___ Energy Use (Kwh) Fossil Fuel ___ Nuc ___ Renew ___
HUMAN	FL ___ CL ___ LU ___ Dsc ___ WS ___ LW ___ WH ___ WB ___	(Determined by Consumer Choice)	FL ___ CL ___ LU ___ Dsc ___ WS ___ LW ___ WH ___ WB ___ Community Exposure Type ___ Quantity ___
PACKAGING	Matter(% Wt) Petro ___ Org ___ Inorg ___ Energy (%) FF ___ Nuc ___ Renew ___ Human FL ___ CL ___ LU ___ Dsc ___ WS ___ LW ___ WH ___ WB ___	Barcode Space	(% Weight) Biodegradable short term ___ long ___ Reuse ___ Recycle ___ Return: ___ Landfill or Burn ___
DISTRIBUTION	(map image)	KEY Human Codes: FL = Forced Labor WH = Working Hours LU = Labor Unions CL = Child Labor WS = Worker Safety Dsc = Discrimination WB = Whistle Blower LW = Living Wage Y: Standard Met N. Standard Not Met	Matter Adverse Effect Scale: A = Strong Evidence B = Good Evidence C = Possible Correlation D = Low Probability E = No Evidence F = Effect Unknown [Refers to any element of product or production methods.] ENERGY Units: 1 gal gas = 35.5 Kwh

Close up barcode input coding system variables.

The cooperative would also have a component similar to Wikipedia where anyone is free to contribute information; all information collected is source identified, checked, and validated with opposing voices provided the opportunity to respond. For example, if a company said its wastewater was free of adverse impactful discharge but a nearby resident had observed and documented hazardous discharge, the "whistleblower" could receive financial reward for providing data proving the allegation.

Such a format could alleviate the need for governmental regulations and monitoring of the system and provide the necessary transparency for a market-based application to be successful. No amount of advertising could deceive the facts available by smart phone scanning a barcode. It would also encourage producers to advertise more their product's positive environmental and social attributes but not prevent hiding of negative ones. Using private enterprise and market-based solutions should appeal to political conservatives and avoid need for political consensus at a government level. No laws would need to be passed; unwilling producers could simply decline to provide data to Earth Accounting but could face rejection of their product by consumers for such refusal.

SUCCESSFUL CASES: ECOLABLES AS PART OF CORPORATIONS

Over the past decade (turn of the twenty-first century) several large companies have undertaken the study of their full impact and sought to improve their footprint noting that limitations in resources, tools, or methods may prevent them at present from providing a fully sustainable product. The idea was well put by the late Ray Anderson (1999), owner of Interface-Flor, a carpet tile company that in the early 2000s decided to examine and alter its footprint after realizing the amount of carpet that ends up in landfills. Mr. Anderson viewed sustainability as a hill to climb, starting ignorant and wasteful but seeking over time to become fully sustainable. He knew he had not made it at the time of his passing but had instilled the spirit of the endeavor in his employees who are committed to continuing the effort.

A visit to their production facility in La Grange, GA demonstrated their commitment with endeavors to reuse old carpet in new production, obtain energy from a nearby waste dump, and develop a cogeneration system at the plant. Other even larger companies have made pledges of zero waste including some auto manufacturers (Zero Waste, 2008).

Companies that seek to exist at the end of the century know that sustainable production is a must and is slowly changing of their own volition. The presence of the numerous ecolabels indicates that companies are both aware of the need to move toward sustainability and the potential market advantage to be had by doing so. Provision of easily accessible data on the social and environmental impact of products may accelerate greater consumer participation in the sustainability movement avoiding the need for punitive legislation or political consensus.

Consider how this knowledge could impact consumer behavior and promote a new green economics as we transition to a sustainable economy (Clark, 2009). Anyone who has used any of the available websites to calculate their GHG footprint quickly learns that automobile driving, home heating and air conditioning, and especially airplane travelling are highly energy consumptive. Driving footprint can be altered by purchase of a more fuel-efficient vehicle or even electric vehicle. Instillation of home solar or wind generation can reduce the home footprint but it is likely to be a long time before we have zero footprint airlines. Ironically, some of the best-known advocates of sustainability who lecture around the world encouraging others to reduce their footprint likely have one of the largest footprints. "Do as I say, not as I do" will have to be their motto. If we were incentivized to minimize our footprint how would behavior change?

Since transportation is one of our most polluting of activities, low footprint people would prefer to work at home, would ensure that they live in an efficient sealed-envelope home and build or retrofit with renewable energy sources. Information technology has so advanced that virtual meetings are common and tools for even greater interaction at a distance rapidly developing. Food production and distribution also carries a heavy footprint of both GHGs and water especially meat production.

A vegetarian diet leaves a smaller footprint (Food, 2014; Vegetarian, 2014) with the production of food at home would be much more sustainable and one can envision the redistribution of population to more rural areas for agricultural production. Establishing the impact of industrial food production especially the associated fossil fuel use, the cost of which would be ultimately attached to the purchasing consumer could lead to more and more family farms producing for themselves in a more sustainable manner, albeit less efficient from the perspective of tons of crop per acre. The current agricultural economic system allows producers to not account for the impact of fertilizers and pesticides except financially passing the environmental cost onto the public not to mention the social cost of using immigrant low wage labor that is often abused.

Failure to account for these "externalities" has resulted in a false economic model that does not account for the planetary impact of production. In order for our communities and the world to become sustainable, basic goods and services must be held to high measurable standards that require evaluations and oversight (Clark, 2010).

BEHAVIOR CHANGE LIKELY

Consumers in this transition economy would likely be more interested in products that last, can be reused by others when personal use is done, and avoiding of products that have their constituent components not biodegradable or reusable. The longer a product is kept theoretically the less its footprint. However, as one has seen with the evolution of information technology devices like televisions and computers, each new generation is more powerful and more efficient than the last and usually requires fewer resources. If the old product can be easily reused by others or easily broken down to reusable components, consumers may well demonstrate a preference for leasing products like electronics, autos, and appliances, trading them in every few years for a more efficient, smaller footprint model, rather than continuing to use the old energy hog model.

Consumers furthermore may also prefer hand-produced items over industrially produced ones as the footprint could be significantly smaller. It is likely items produced closer to home would have a smaller footprint and therefore more economically beneficial to the local economy. Anyone who has inspected his weekly trash realizes the amount of packaging that has the shortest home life span. Packaging has changed dramatically with the need to catch customers' eyes and occupy larger shelf space.

Packaging is produced as much for these factors as product protection. No sooner does the consumer get the product home than he deposits the packaging in the trash never to be used again if not composed of recyclable materials. Why can't the trucks that deliver new products to the store return with used packaging to be reused around another new item? Probably the answer is that it is currently not economical as judged by consumers who have to return packaging, receivers who have to sort and possibly clean the packaging, and packaging facilities who have to install new technology to reuse/recycle packaging and be responsible for inadvertent contamination that could be hazardous to someone's health. It is quite possible that some states even prohibit reuse because of this fear. If so, this and similar changes could require legislative action.

In the 1950s and 1960s, the concept of planned obsolescence was in vogue with consumers alleging they "just don't make things like they used to" (Packard, 1957). This was coincident with a single-use, throw-away production ideal insuring customers would purchase hopefully the same product over and over. We were all made to feel this thing was made "just for me, just for now." This seemed to be an alluring aesthetic that was enhanced by claims of greater cleanliness, more appealing appearance and longer shelf life, elements also attractive to the retailer. Again, we failed to realize the full impact of these choices on the environment and their long-term consequences. It wasn't necessarily planned that way. The other priorities just seemed more important and environmental impact may not have even been considered. However, such practices are not sustainable, especially with 9 billion people all needing relatively similar resources.

The field of medicine is one of the most dramatic (Sastri, xxxx) 50 years ago physician's offices and hospitals had metal instruments that were packaged and autoclaved by staff each evening for the next day. Even syringes were still made of glass for reuse but with a new sterile needle attached at the time of use. Sheets and linens were washed and reused. Now virtually all materials are packaged at a factory, designed for single use and disposal thereafter often requiring special discarding processes because of tainting of the product with human blood or waste. The transition next economy will have to achieve the proper balance between safety and resource conservation to achieve sustainability.

Consumers in the current global transition economy will also likely prefer refurbished items to new ones presuming functionality is unchanged and the footprint of repair is smaller than replacement. Heirlooms may again be popular with families treasuring old but continued functional household items. Craft items are often less resource intense than industrially produced ones making this another element of the transition economy. Luxury items may dramatically change if environmental and social impact costs are counted. Art treasures may replace yachts, small, energy-efficient homes replace mansions, private jets sold to reduce footprint or a social movement like "Occupy Wall Street" that identifies those with large footprints and challenges them to use their fame to be role models for a sustainable future rather than demonstration of opulence. More, bigger, and better "bling" may cease to be in vogue during this transition phase.

If forecasts about human-induced climate change prove accurate but underestimate the impact, governments could be compelled to develop methods to tax resource use, especially products that have large footprints.

Theoretically, each individual's interaction with the planet has a "footprint," some bigger, some smaller. Indeed, one can imagine an ideal or average footprint calculated based on an average person's resource utilization. If this was assigned an annual value, individuals whose footprint is less would be rewarded, those with a footprint exceeding the average, taxed extra for their additional impact on the planet and future sustainability. One can envision consumers scanning into a home computer, each product receipt with its embedded footprint information, compiling, monitoring, and submitting the data like their taxes to collect a "refund" for maintaining a minimal footprint. In the sustainability society, it may be that what resources one uses is a more important element to tax than the income. Certainly from the planet's perspective, income itself is not relevant but how the money is spent (impact).

These ideas are of course speculative but logical if society were to change priorities from consumption to maximize corporate profit and consumer self-indulgence to development of a sustainable economy for future centuries. Next presented is another speculative discussion about what life would be like in a world where there are still finite resources but essentially unlimited clean energy.

UNLIMITED ENERGY

Over 4 decades ago, there was a gasoline shortage in the 1970s, when the US energy future then looked bleak with discovery of new reserves not keeping pace with growing use (Stobaugh and Yergin, 1980). In 1950 the US was the world's chief oil producer, but the US percentage of global output gradually declined over the next 20 years and by 1970, it was pronounced that the nation had reached "Peak Oil," meaning that reserves had been fully discovered and Americans were now consuming the last 50% of that available within its continental boundaries and off shore.

A future dependency on oil from Arab nations seemed eminent and undoubtedly influenced national leaders (Perron, 1988). Solar energy was in its infancy and mostly confined to water heating though wind energy had been operative on the American plains for more than a century but not to produce electricity. We realized the dependency of our economy on fossil fuels and feared inability to maintain our living standard in the presence of high-priced fuel or exhaustion of its supply (Perron, 1988; Stobaugh and Yergin, 1980).

A look to the sky, however, would have assured the viewer that there was no shortage of energy in Universe. At night everywhere we look are stars, each a powerhouse of nuclear energy and each day the Earth reflects more sunlight energy than man uses in a year. We just needed to learn how to harness it to perform the functions currently done by fossil fuel energy. Environmentalists were somewhat relieved by this pessimistic prospect since we would soon be forced to develop clean fuels but in the years since 2000 the picture has changed.

Not only have new oil reserves been discovered globally, but new methods now allow the removal of oil previously unobtainable and coal reserves are now projected to be sufficiently abundant to provide energy for 500 years. Warming of the arctic has led to new exploration there with promises of still more. Oil removal from old wells by hydraulic fracturing (commonly called "fracking") has allowed oil firms to extract more petroleum resource from previously spent wells. All this has led environmentalists to realize that now humanity has enough fossil fuel reserves to accelerate global warming from carbon dioxide release by fuel use to the point of threatening human existence. These fuels represent the solar energy storage of the planet's last 200 million years. Considered reflection on the issue would have caused one to realize that if in 150 years we had used about 50% of available resources, humanity must find other sources if the human race is to continue successfully in the future centuries even if global warming presented no threat.

Literally speaking almost all energy is ultimately derived from star power including nuclear energy. Fossil fuels were all the result of living matter's conversion of sunlight to carbon and hydrocarbons. Nuclear fuels were created under the massive gravity generated by star explosions and collapse; tides and wind motion are created by solar and lunar influenced gravitational forces and since the Earth itself was once stardust, even geothermal energy is in some way solar related. Consequently a more appropriate distinction is between renewable and nonrenewable energy sources. This would exclude all fossil fuels as well as nuclear sources that once spent for energy production, usually by boiling water in generators, create waste toxic to humans for hundreds of thousands of years.

Over this same period technological advances in solar photovoltaic cells have dramatically improved in efficiency and there is promise of continued progress. The improved efficiency from each solar cell combined with improved production techniques suggests that the cost of solar cell production will soon be competitive with fossil fuels without even counting the unpaid externalities of pollution and CO_2 and current oil producer subsidies. Thus, a world of unlimited clean renewable energy is no longer a fantasy. This section will address some speculative impacts of such a scenario. How would life change if we had unlimited renewable fuel available to all who desired it? How would the transition economy discussed above be altered?

First, unlimited clean electrical energy would resolve ground transportation problems currently dependent on fossil fuels. Electric cars could be driven by everyone, as much as desired without creation of GHGs. Alternately, we may find fuels' cells using hydrogen derived from hydrolysis of water rather than hydrogen atom stripping from fossil fuels to be easier to implement and not requiring major battery innovation. Thus, the environmental impact of land transport would be greatly reduced.

The same could not be said for air transport unless airplane fuel cells or more lightweight batteries are developed. High-speed railway systems instead might then compete more effectively with airplanes due to smaller GHG footprint. Reduction of GHG will still be a major social priority to curtail global warming and desire to return to preindustrial revolution atmospheric CO_2 levels. Most climate scientists believe that even with immediate cessation of additional CO_2 release, it will take at least a century to return to preindustrial revolution levels. However, this should clearly be a goal of the new GIR.

The impact of unlimited renewable energy on industrialization is somewhat more uncertain. Most of the prior industrial revolution processes may be characterized as "heat, beat, or treat." Materials are either heated to break them down, crushed into subcomponents or chemically treated to produce new molecules or free atoms. Each of these processes is dependent on introduction of energy to move the reaction; energy heretofore derived from fossil fuels. The same processes could operate using clean, renewable energy without dramatic change in industrial production methods.

However, these processes have also had other adverse environmental consequences that would not be eliminated by switching to renewable energy including toxic waste products, habitat destruction to obtain raw materials, and incomplete breakdown to products not able to enter the natural reuse cycle. With the availability of unlimited clean energy these processes could be modified to capture and neutralize toxic waste, restore exploited resource habitats, and make products whose components can be separated into biodegradable and reusable components. In other words, many of the old processes could continue but with additional energy modified to be more environmentally safe.

In recent years other options have developed that may make more sense and some were always before our eyes but ignored. Nature has been evolving life on planet Earth for 4 billion years. During this time many species have come and gone but nature has developed certain processes that endure and provide the underpinning of all life processes. This idea was best espoused by Janine Benyus in a book "Biomimicry" published first in 1998. She notes that Nature uses only sunlight for energy, has no waste energy, recycles everything, exploits diversity over specialization, utilizes adaptability to meet changing environments, and relies on species local expertise (Benyus and Janine, 1997). Heat, beat, and treat have no role in Nature yet spiders produce fibers stronger than steel, biological tissue in birds and butterflies that is sensitive to gravitational fields and organs that generate an electrical current all made by living organisms whose precursor substrates are mostly carbohydrates, fats, and proteins all done at standard temperature and pressure.

As part of the retooling from the GIR, greater emphasis on learning and reproducing Nature's problem-solving designs to address human needs would be more inline with an Earth where human behavior compliments Nature rather than fights it. This pathway would not be inhibited by unlimited renewable energy but implementation could be slowed by those who prefer to continue to use outdated first industrial revolution practices out of convenience, habit or lack of added expense.

Global warming and melting of glaciers and annual snow pack reduction as well as draining of aquifers means that traditional fresh water supplies are likely to continue to be depleted even with access to unlimited clean energy. However, energy cost has largely precluded desalination plants from becoming more abundant. With free, clean energy with ocean side fresh water produced by desalination plants, shipped throughout the planet via clean energy vehicles or pipes fresh water supply problems could be addressed without compelling residents to relocate due to water shortage.

Unlimited renewable energy will also not clean polluted streams or reforest eroded hillsides. It will not bring species near extinction back any quicker or prevent loss of more species, at least in the near term. These are all prices from the first industrial revolution for which we must continue to pay probably for more than a century or two. Abundant renewable energy will not completely remove our need for petroleum as so many of the goods we have come to rely upon are petroleum derived.

These objects, however, unlike fossil fuels which are converted with use to CO_2 and water could be designed with recycling or reuse in mind reducing the need for continuous extraction greatly. Alternately, products made from plastic could be phased out for ones more environmentally friendly but more consumptive of energy during production. Having unlimited clean energy will allow mankind to rethink industrial processes with more in mind than economics. The impact on economics with the introduction of unlimited clean energy is also worthy of at least brief exploration.

If the new energy designs are developed and controlled by the robber-baron types who built the railroads, developed mining and fossil fuel extraction in the nineteenth century industrial revolution, and the clean energy is delivered by centralized monopolies, then little could change except for reduction in GHG production. These parties would control the cost of energy and like the current fossil fuel producers limit the production of energy to maintain prices that insure an ample return on investment. While new clean energy is certainly laudable, it need not be revolutionary.

The few controlling the energy sources could hold most of rest of the population hostage to the whims of its controlled market. However, if particularly the cost of photovoltaic-produced energy continues to fall, decentralized energy production is much more likely and the ability of it to be monopolized greatly reduced unless home producers could be prohibited from feeding into the existing system and forced to go "off-grid" if they used personally installed solar or wind energy devices. Further, among the greatest opponents of abundant clean, renewable energy are those billionaires who own fossil fuel energy-producing assets. More likely, they will fight tooth and nail to slow renewable energy growth, push politicians to reject any economic incentives promoting clean energy and allege abundant, clean energy is a myth.

This seems to be their current posture with many conservative think tanks opining that clean energy options will never be more than a small percentage of our energy mix. These are the same folks who deny climate change and assert "Drill Baby Drill" is our nation's best strategy. What is ironic about supporters of this position is that they allege they are promoting US energy independence without acknowledging that prices for coal and petroleum are based on international markets and subject to supply and demand by all nations of the planet. Therefore, the only way that the expanded extraction of national resources makes sense is if there is enough to supply all the planet's needs.

Even more salient is a look at the past. As stated earlier in 1950, the US was the world's largest exporter of energy, both oil and coal. At that time the price of gasoline was less than 30 cents a gallon. So during the first half of the twentieth century the US sold almost 50% of its known reserves for what today is a pittance. Finite resources in most instances are expected to increase in value as they are depleted and therefore a prudent owner of such resources should prefer to hold onto them as protection against rising international market prices and to sell at a greater value later. To the astute observer this would seem to be the prudent conservative position rather than squandering all our resources now to fuel inefficient vehicles and production methods becoming dependent on others in the long term.

How would abundant clean energy affect jobs? Most jobs in the US are derived from small businesses. For most of these employers energy costs are not their most significant expense perhaps with the exception of the transportation sector. Hence, clean cheap renewable energy would be more important as a stabilizing factor of an uncontrollable employer variable rather than a rapid job promoter. Nonetheless, such secure cost control for energy would eliminate one variable business owners are confronted with on a regular basis and likely therefore to promote increased hiring if product demand grows.

For corporate employers the same stabilization of a cost variable would apply again removing this barrier to growth. According to the Bureau of Labor Statistics, the fossil fuel industry employs slightly over 1/2 million people in drilling, extraction, and support industries (http://www.eia.gov/todayinenergy/detail.cfm?id=12451).

These jobs would likely suffer if fossil fuel extraction was to be greatly reduced. Half million out of roughly 140 million jobs in the US in 2014 is quite small and those affected would be reasonably expected to be absorbed into the new green economy. Due to increased hiring for reasons noted above, service sector jobs would be essentially unaffected by abundant clean energy as their demand is determined by social need rather than energy limits.

With an aging population, growth in healthcare and related social service jobs is expected to increase steadily for most of the twenty-first century until population stabilizes. New technologic innovations using the abundant clean energy could allow greater autonomy and mobility for the aging population. Devices to monitor vital functions, quickly report illness, and promote rapid response could reduce need for more intensive care services but even then the need for emergent services would only be delayed for many. Overall abundant clean energy would not be expected to greatly alter service sector jobs.

While footprint monitoring and reduction as noted above could promote more family farms and local crop production to reduce use of fossil fuels for food transport, fertilizers, and pesticides during the transition period, with abundant clean energy transport concerns would be alleviated but not the potential environmental hazards associated with fertilizer and pesticide use. The cost of these products and their resultant increased acreage production may contribute significantly to farmers' choice of farm monoculture rather than synergistically mixing some crops that mutually fertilize and reduce pests.

Abundant, clean, cheap energy could promote greater efforts to grow organic reducing the liabilities of ecosystem contamination from farming. Use of irrigation farming would be expected to increase as a primary cost of irrigation is the expense of moving water from one place to another. However, this would not overcome reduced supply due to climate change though, when coupled with drip irrigation, the overall impact could be minimized.

If clean energy was everywhere available and humanity agreed on cooperatively supporting the basic human needs of all the world's population, the need for an accurate accounting system to know what resources are being utilized where would become an extremely important issue. The physical resources would still remain finite, matter cannot be created or destroyed but can be converted to more or less usable forms by human intervention. Only energy would become practically infinite. With a growing population, shared basic needs by all and finite physical resources to meet those needs, a system is needed to not only account for but also monitor the impact and distribution of resources.

Such a system should be adaptable for use by government and industry planners but also by consumers to monitor personal environmental impact that will continue to have planetary significance even with abundant clean energy. The previously described UE provides one model capable of meeting this need and certainly others could be developed. Indeed, usual money based economics structured on the ability of translating all production impacts into monetary terms seems to lack the discrimination necessary to assess environmental impacts as well as social ones (Clark, 2012).

CONCLUSION: THE NEXT STEPS

If energy was viewed as abundant as air, its value would be greatly reduced and the need to monitor it unnecessary. It is the production matter and changes it undergoes that become more important than the monetary value of assets in the larger economic management scheme. Knowing the environmental impact of all products as well as constituent components will become more important than monetary value and a more likely source of rate limiting production. Therefore, tools that distinguish the full impact of production and production methods should be of increased value in the GIR.

Quite possibly, the need for tools to assess success or failure of chosen GIR interventions will demand such an accounting system, to date nonoperative nor even considered by most economists as was general environmental impact until the information was demanded by shareholders who were valuing sustainability issues in their investing and by corporate attorneys fearful of legal battles over corporate environmental depredation. Thus, one can anticipate in the GIR the need for a true and accurate accounting system that looks at impact from the natural world's perspective.

As has been discussed earlier, economic principles are not the same as natural ones. Economists wish they were but it is not true. The former are subject to "market deviations" and twists of thinking by economists, the latter responding only to the nonchanging immutable laws of nature, ever and always behaving the same. Indeed, monetary economic theory has no relevance to nature except as humans choose to award value for some things and not others.

Nature does respond to every pound of pollution, ton of runoff soil and its contamination added to the water. It responds to decimated species often resulting in different species often less desirable to man. For most of the past centuries the human population and its planetary impact has been sufficiently small that the effects, though occurring were not readily noticed and responded to by man. With an anticipated population of 9 billion by 2050, the impact of man's activities on the planet can only be expected to increase. This will not change in the presence of abundant clean energy but the potential for greater adverse impact will probably increase once the energy constraint on production is removed. Alternately, removing the key constraint of production energy could accelerate the transition to more sustainable but energy intensive options.

The foregoing discussion is not meant to be fully comprehensive and is highly speculative subject to the variations of human behavior but hopefully raises several pertinent issues and discusses the societal impact of the GIR. It also is intended to promote a realization that a new economics, based on the behavior of nature and natural systems as well as man will be of critical importance in the GIR and without which humans will find themselves not able to determine if real progress toward sustainability is being achieved nor able to anticipate the environmental impact of interventions alleged to promote sustainability.

REFERENCES

Anderson, R., 1999. Mid-course correction: toward a sustainable enterprise: the interface model.

Benyus, Janine M., 1997. Biomimicry Innovations Inspired by Nature. Harper Perennial.

Clark II, W.W., 2009. Sustainable Communities. Springer Press.

Clark II, W.W., 2010. Sustainable Communities Design Handbook. Elsevier Press.

Clark II, W.W., 2012. The Next Economics: Global Cases in Energy, Environment and Climate Change. Springer Press.

Clark II, W.W., Cooke, G., 2011. Global Energy Innovation: Why America Must Lead. Preager Press.

Clark II, W.W., Cooke, G., 2015. The Green Industrial Revolution. Elsevier Press, to be published in early 2015.

Clark II, W.W., Fast, M., 2008. Qualitative Economics: Toward A Science of Economics. Coxmoor Press.

Clark, W.W., Xing, L.I., 2012. Social capitalism: China's Economic Rise. In: Clark, W.W. (Ed.), The Next Economics: Global Cases in Energy, Environment, and Climate Change. Springer Press, pp. 143–164.

Food Documentary, 2014. http://www.nationofchange.org/reader/42590P.

Hector, Bruce, Schenk, J., Clark II., W.W., Saavedra, A., 2014. The Universal Ecolabel. Int. J. Appl. Sci. Tech. 4.

Packard, Vance, 1957. The Hidden Persuaders.

Perron, P., 1998. The Great Crash, the Oil Price Shock and the Unit Root Hypothesis. University, Princeton; Program, Econometric Research. Retrieved February 3, 2012.

Sastri, V.R. Plastics in Medical Devices, xxxx.

Stobaugh, R., Yergin, D., 1980. Energy future.

Universal Ecolable (EarthAccounting.com). Earth Accounting.

Vegetarian, 2014. http://www.nationofchange.org/food-documentary-revealing-look-sourcing-our-modern-food-supply-1389970149.

Zero Waste, 2008. http://www.edmunds.com/car-buying/an-auto-factory-as-green-as-its-cars.html.

Reforming the Energy Vision in New York State

CLEAN COALITION COMMENTS ON MATTER 14-00581 / CASE 14-M-0101 BEFORE THE NEW YORK STATE PUBLIC SERVICE COMMISSION (July 18, 2014)

Stephanie Wang, David Miller*
*Stephanie Wang, Esq., Policy Director, Clean Coalition and David Miller, PhD and LEED AP, Policy & Technical Advisor, Clean Coalition

INTRODUCTION

The Clean Coalition applauds the Public Service Commission for laying out a visionary proposal for transforming the relationships between customers, utilities, and system operators to guide New York State toward a clean energy future. The Clean Coalition offers the following recommendations for implementing the Reforming the Energy Vision (REV) staff report's proposals to increase reliance on distributed energy resources (DERs). These comments are submitted in accordance with Administrative Law Judge Eleanor Stein's Ruling Issuing Track 2 Questions and Establishing a Response Schedule, issued on May 1, 2014, in response to the questions regarding outcomes-based ratemaking. Specifically, these comments address questions 2 (new outcomes/metrics) and 8 (removing bias toward increasing capital expenditures).

The Clean Coalition is a nonprofit organization whose mission is to accelerate the transition to renewable energy and a modern grid through technical, policy, and project development expertise. The Clean Coalition drives policy innovation to remove barriers to procurement, interconnection, and realizing the full potential of integrated DERs, such as distributed generation, advanced inverters, demand response, and energy storage. The Clean Coalition also works with utilities to develop community microgrid projects that demonstrate that local renewables can provide at least 25% of the total electric energy consumed within the distribution grid while improving grid reliability. The Clean Coalition participates in numerous proceedings before state and federal agencies throughout the United States.

REV is a unique opportunity to support New York's clean energy, resilience, and economic goals by addressing structural biases against DERs. Increasing reliance on and deployment of clean DERs, including local generation, storage, and demand response, will require fundamental shifts in the roles and responsibilities of Distributed System Platform Providers (DSPPs) and the New York Independent System Operator (NYISO). The PSC should allocate to DSPPs responsibilities for planning DER solutions to meet clean energy goals and address both bulk system and local area needs, and managing DERs to meet local reliability targets and reduce reliance on the transmission system. Clarifying the responsibilities of the DSPPs will make it possible to realize the full value of DERs and reduce the operational complexity of managing DERs to meet local and bulk system needs.

First, the Clean Coalition recommends reforming transmission planning so that DERs are treated as primary tools for meeting reliability needs, rather than "alternatives" to transmission and central generation investments. We support the staff report's proposal that DSPPs have responsibility for planning for DERs at the distribution grid level and further recommend that DSPPs be assigned responsibility for proposing integrated DER solutions for meeting transmission system needs as well. REV staff report also calls into question "the assumption that the centralized generation and bulk transmission model is invariably cost effective, due to economies of scale" and proposes that DERs should be used, "not as a last resort but rather as a cost effective, primary tool to manage distribution system flows, shape system load, and enable customers to choose cleaner, more resilient power options."[1] Accordingly, we recommend neutralizing the bias toward capital expenditures by conditioning the rate-based investments in transmission and central generation on a finding by the PSC that such investments will be more cost-effective for ratepayers than investments in clean DERs that could meet the same needs.

Second, the Clean Coalition recommends that the DSPPs be required to submit distribution resources plans for approval by the Public Services Commission. These plans would detail the DSPP's proposed optimal portfolio of DERs to cost-effectively meet state and local goals, such as targeted levels of reliability and demand side management at the substation level, and meet customer desires and needs.

---

[1] Reforming the Energy Vision (REV) staff report at 8.

Third, the Clean Coalition recommends that the PSC clarify the role of the DSPP as the distribution system operator, allocating responsibility to the DSPP for operation of the distribution networks to meet targeted levels of local reliability and resilience. This will enable the implementation of the staff report's vision that "the utility as DSPP will actively coordinate customer activities so that the utility's service area as a whole places more efficient demands on the bulk system, while reducing the need for expensive investments in the distribution system as well."[2] Clarifying the division of responsibilities for maintaining reliability between the distribution operator and the transmission operator will facilitate grid planning for DERs to be a primary means of meeting system needs, reduce the operational complexity of relying on DERs, and unlock opportunities to realize the value of DERs for meeting local needs.

TRANSMISSION GRID PLANNING

Transmission planners across the country view DERs as "alternatives" to transmission and central generation and rarely proactively propose integrated DER solutions to meet needs. Distribution planners, on the other hand, generally fail to account for the value of DER for avoiding investments in transmission and central generation. The REV staff report proposes that DERs should be used "not as a last resort but rather as a cost effective, primary tool to manage distribution system flows, shape system load, and enable customers to choose cleaner, more resilient power options."[3]

An integrated approach to transmission and distribution planning is necessary for moving beyond the current NTA approach toward using DER as a primary means of addressing bulk system operational needs. We support the staff report's proposal that DSPPs have responsibility for planning for DERs at the distribution grid level and further recommend that DSPPs be assigned responsibility for proposing integrated DER solutions for meeting transmission system needs as well. The NYISO would retain responsibility for timely identifying projected bulk system operational needs and working with the DSPPs to model how proposed DER portfolios could meet such needs. However, the DSPPs would have the primary responsibility for developing technically feasible and cost-effective portfolios of integrated DER solutions to meet transmission system-level needs.

---

[2] REV staff report at 9.
[3] REV staff report at 8.

The DSPPs would also have responsibility for ensuring that the approved DER portfolio is acquired and online in time to meet needs.

Further, we recommend neutralizing the bias toward capital expenditures by conditioning the rate-based investments in transmission and central generation on a finding by the PSC that such investments will be more cost-effective for ratepayers than investments in clean DERs that could meet the same needs. As an example of this approach taking shape in another jurisdiction, California's AB 327 takes a major step in this direction by requiring utilities to develop distribution resources plans by July 2015 to guide DERs to optimal locations on the distribution grid and allowing utilities to rate-base only distribution grid investments that yield net benefits for ratepayers.[4] Since the REV staff report explicitly calls into question "the assumption that the centralized generation and bulk transmission model is invariably cost effective, due to economies of scale," we recommend that New York apply this approach to fairly compare transmission and central generation investments with clean DER investments.

In order to compete with transmission and central generation investments on a level playing field and in order to fully comply with state and federal clean energy goals, all values of DERs should be accounted for in cost-effectiveness calculations. DERs provide a number of significant and quantifiable benefits to ratepayers. These benefits include significant locational benefits, including reduction in transmission line losses and opportunities to avoid or defer investments in transmission and central generation.[5] As the Clean Coalition has testified before the California Public Utilities Commission, such locational value especially applies to any portion of the generation that is deemed "deliverable" and does not exceed 100% of the coincident load at the substation, as all such generation avoids the use of transmission system.[6] In addition, compliance with New York's existing greenhouse gas, renewable energy, energy efficiency, and other clean energy goals should be considered quantifiable ratepayer benefits of clean DERs.

---

[4] California Public Utilities Code, Section 769, added by California Assembly Bill 327 (2013).

[5] See the Clean Coalition's Locational Benefits Brief, available at http://www.clean-coalition.org/site/wp-content/uploads/2013/11/Locational-Benefits-Brief-08_tk-6-Nov-2013.pdf.

[6] Opening brief of the Clean Coalition regarding Southern California Edison's application to establish green rate and community renewables programs before the California Public Utilities Commission (May 2014), available at http://www.clean-coalition.org/regulatory-filings/cpuc-opening-brief-on-green-tariff-applications-sb-43/.

Clean DERs have many other benefits that should be accounted for in cost-benefit analyses, including hedge value against rising fuel costs and local economic benefits. For more information, see the Clean Coalition's energy, economic, and environmental benefit analysis performed for the Hunters Point Community Microgrid project in San Francisco.[7]

DISTRIBUTION RESOURCES PLANS

The Clean Coalition recommends that the DSPPs be required to regularly submit distribution resources plans for approval by the Public Services Commission. Each DSPP would propose the optimal portfolio of DERs to cost-effectively meet state and local goals, including clean energy goals and targeted levels of reliability at the substation level, and meet customer desires and needs. These plans would facilitate comparisons of the costs of transmission and central generation investments with DER solutions that would defer or avoid such investments, as required above.

System-wide DER optimization includes optimizing a DER portfolio across a collection of substations in a distribution utility's service territory. This will involve identifying which substations would benefit most from high levels of DERs based on system needs and local goals, such as local clean energy targets or resilience priorities. For example, any distribution grid routinely experiencing or forecast to experience local capacity shortages may be a good candidate for high levels of DERs. Substations with local capacity needs resulting from transmission constraints or congestion will benefit from DERs that can avoid or defer the need for developing costly transmission upgrades. In another example, substations that contain critical infrastructure may be good candidates for islanding campuses within a substation to maintain reliable electrical service to emergency services such as hospitals. However, such islanding capabilities come at far greater costs, which must be weighed against the societal benefit of the resulting increased resilience.

Optimizing a portfolio of DERs within a single substation requires starting from a detailed map of the existing distribution grid, including feeder line topology and transmission capabilities, transformer and switch locations and capacities, and location and time profiles of all existing load and

[7] The Clean Coalition, The Hunters Point Project: A Model for Clean Local Energy, An Energy, Economic and Environmental Benefits Analysis for High Penetrations of Renewable Energy in San Francisco's Bayview Hunters Point Area, available at http://www.clean-coalition.org/site/wp-content/uploads/2012/10/HPP-Benefits-Analysis-Summary-21_gt-26-Mar-2014.pdf.

generation assets within the distribution grid. The optimal locations for a portfolio of DERs can be determined from this map.

Utilities often utilize several different software packages to help with the management of the dynamic and static characteristics of their distribution grid operations. However, these products have historically been designed for the analysis of one-way power flow and typically lack functionality for modeling advanced DER capabilities including, for example, advanced inverter functionality or energy storage operation. Realistically and efficiently creating an optimized DER portfolio requires working with software modeling tool providers to increase visibility of the distribution grid.

The Clean Coalition is currently working on the Hunters Point Project, a Community Microgrid Initiative project in collaboration with Pacific Gas & Electric.[8] This project will serve 25% of total energy consumed at the Hunters Point substation in San Francisco with local renewables, balanced with intelligent grid solutions like advanced inverters, demand response, and energy storage. The Clean Coalition uses sophisticated powerflow modeling and cost-benefit analysis tools to reveal how—and precisely where—DERs can best meet system needs. The Clean Coalition team works with utilities and powerflow and cost optimization modeling tools providers to use existing tools for seeing, and planning enhancements for, the distribution grid. For the Hunters Point project, the Clean Coalition is working with PG&E's powerflow modeling tool provider CYME and its cost analytics tool provider Integral Analytics to show that utilities' favored tools can meet these new challenges once they have the right specifications to move forward. The Clean Coalition is also developing standard specifications for modeling tools providers, so that our lessons learned from this experience can be applied to any other modeling tool.

Existing modeling tools can be used by system planners to locate where a given DER can be located within the distribution grid without requiring costly system upgrades. For example, in areas where the existing distribution feeder lines are robust enough to support incremental power flow without needing network upgrades, additional DG resources may be reliably added to the D-Grid at least cost. The modeling tools can also enable the system planner to examine the operational characteristics of the DER resources that lead to reliable performance. For example, advanced inverter functionality can be used to inject reactive power into the grid during times of high PV generation and low load in order to mitigate power flow and voltage stability issues.

---

[8] For more information, see http://www.clean-coalition.org/our-work/community-microgrids/.

Once least cost, highest value locations for DERs have been determined, the benefits of additional DERs placed on the system must be balanced against their costs.

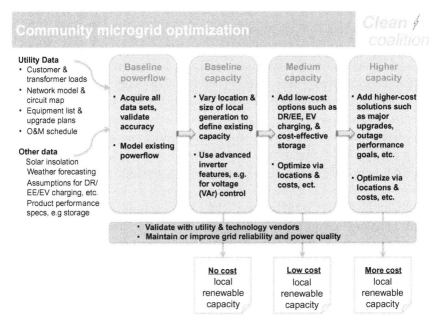

The benefits of higher levels of DER include the following:

- Deferred or avoided transmission and distribution investments.
- Reduced usage and congestion of the transmission system.
- Increased system efficiency, including reduced transmission and distribution line losses and conservation voltage reduction.
- Meeting clean energy goals, such as emissions reductions, and associated regulatory compliance value.
- Increased local reliability—lower frequency and duration of outages.
- Local resilience, including the possibility of islanding a substation or campus.
- Reducing reliance on a few large generators will reduce needs for contingency reserves.
- Increasing independence from the transmission system for energy and energy services.
- Improved power quality.
- Hedging against fossil fuel price volatility.

The benefits of integrated DER solutions are much greater than the sum of the benefits of each individual component. For example, peak PV generation impacts on net load profiles can be mitigated with demand response and daytime electric vehicle charging.[9] Similarly, the value of distributed solar insolation and storage are enhanced by turning on advanced capabilities of the inverters, which can prevent overvoltage due to high levels of distributed solar, prevent blackouts by providing reactive power close to loads, and enable conservation voltage efficiencies.[10]

Long-term optimization of DERs requires long-term load forecasting in order to understand future system needs. Long-term forecasting can best be performed by stochastic simulation that determines design parameters based on a system reliability metric such as a 1 event in 10-year outage rate. This approach requires forecasting likely values for input parameters and imposing correlations between them that reflect physically appropriate behavior. For example, high temperatures are correlated with increased solar insolation and increased PV generation, along with potentially higher loads from air conditioning units. Therefore, high PV generation may be highly correlated with high load in the summer. All such physically justifiable correlations must be quantitatively accounted for in the stochastic modeling process in order to achieve meaningful results.

California is about to embark on the implementation of the AB 327 requirement that each investor owned utility must submit a distribution resources plan by July 2015. The California statute provides that each distribution resources plan must identify "optimal locations" for the deployment of distributed resources, propose or identify mechanisms for the deployment of cost-effective DERs, identify barriers to the deployment of DERs, and propose cost-effective methods to maximize "locational benefits" and minimize incremental costs of DERs.[11]

The importance of identifying optimal locations, maximizing locational benefits, and minimizing incremental costs of DERs is supported by a May 2012 study by Southern California Edison, which found that transmission upgrade costs for their share of the California Governor's goal of 12,000 MW of distributed generation could be reduced by over $2 billion

[9] See the Clean Coalition's presentation to the California Energy Commission, Flattening the Duck (February 2014), available at http://www.clean-coalition.org/resources/february-2014-cec-presentation-flattening-the-duck/.

[10] Craig Lewis, Advanced Inverters—Recovering Costs and Compensating Benefits (October 2013), available at http://www.clean-coalition.org/site/wpcontent/uploads/2013/10/October2013_SolarServer.pdf.

[11] California Public Utilities Code, Section 769, added by California Assembly Bill 327 (2013).

from the trajectory scenario. The lower costs were associated with the "guided case" where 70% of projects would be located in urban areas, and the higher costs were associated with the "unguided case" where 70% of projects would be located in rural areas.[12]

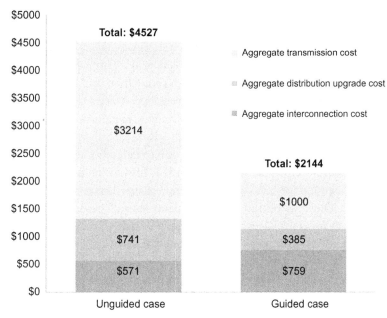

Source: Southern California Edison, The Impact of Localized Energy Resources on Southern California Edison's Transmission and Distribution System (May 2012).

AB 327 also provides that each investor owned utility will propose any spending necessary to accomplish the distribution resources plan as part of the next general rate case, and such spending will be approved if ratepayers would realize net benefits and the associated costs are reasonable. This provision contemplates a major shift from the status quo reactive approach to conducting interconnection studies and requiring DER developers to pay for distribution grid upgrades on a case-by-case basis, to a proactive effort by utilities to identify and rate-base the most cost-effective distribution grid upgrades, and then guide DER development to optimal locations on the grid. This approach can remove the uncertainty around interconnection

[12] Southern California Edison, The Impact of Localized Energy Resources on Southern California Edison's Transmission and Distribution System (May 2012), validated by Navigant Consulting, Distributed Generation Integration Cost Study Analytical Framework prepared for the California Energy Commission (November 2013).

costs and greatly reduce the time required for interconnection studies for DER projects.

Right now (2014), in California, a developer first identifies a potential site for local renewables and then checks a grid map to see if the site may be eligible for fast track interconnection, which is an expedited grid access procedure for the best locations on the grid and is much less time-consuming and expensive than the standard interconnection process. Then, the developer orders a pre-application report, which shows any known interconnection constraints. The developer must also secure site control and then apply for an interconnection study. The developer may then find out that the site was not eligible for fast track interconnection after all and will have to decide whether to go forward with a significantly more expensive project.

The Clean Coalition is currently advising the California policymakers on how to leverage advanced grid modeling tools to help utilities develop interactive distribution resources plans that guide DERs to the best locations on the grid and reduce the time frames and uncertainty involved in grid interconnection. The distribution resources plans should guide developers to optimal locations, instead of forcing them to look through the haystack for these locations, and provide up-to-date information on required grid upgrades and associated costs. This would remove the uncertainty about timelines and costs for developers, greatly reducing their project development costs, translating to lower costs for ratepayers.

DISTRIBUTION SYSTEM OPERATOR

The Clean Coalition recommends that the PSC clarify the role of the DSPPs as distribution system operators (DSOs), allocating responsibility to the DSPPs for operation of the distribution networks to meet targeted levels of local reliability and resilience. In addition to the grid planning role described above, the DSPPs should be responsible for managing the operation of DERs to meet local reliability targets and reduce reliance on the transmission grid, resulting in greater value to customers and lower costs. This will enable the implementation of the staff report's vision that "the utility as DSPP will actively coordinate customer activities so that the utility's service area as a whole places more efficient demands on the bulk system, while reducing the need for expensive investments in the distribution system as well."[13] Clarifying the division of responsibilities for maintaining

[13] REV staff report at 9.

reliability between the DSO and the transmission system operator will reduce the operational complexity of relying on DERs, facilitate grid planning for DERs to be a primary means of meeting system needs, and unlock opportunities to realize the value of DERs for meeting local needs.

The PSC should consider how distribution and transmission operators should interact. Lorenzo Kristov, a director of the California Independent System Operator, asserts that there are two bookend approaches for how the electric grid can reliably evolve.[14] The first approach is a continuation of the top-down, command-and-control structure of the existing grid architecture. Under this approach, the bulk system operator, which currently manages operation of all central station generation and reliability of the flow of bulk power at the transmission system, will take on the added responsibility of managing all DERs placed on the distribution grid. Even with third-party aggregation of DERs, this approach represents a massive shift in the complexity and sheer number of resources that the transmission grid operator will be forced to manage. As DER penetration continues to accelerate, the operational complexity of metering, modeling, and managing real-time operation of DERs will continue to increase. Furthermore, inconsistent objectives between local distribution grid management and transmission grid stability criteria may lead to systemic reliability issues.[15] For example, end-use service reliability may not always be consistent with stability at the transmission level, which tends to optimize system-level stability over distribution-level stability.

The second approach represents a more fundamental change to grid design that would redefine responsibilities between the transmission and distribution system operators and would support prioritization of DER by utilities. Under this new approach, the transmission grid operator continues to manage the electric grid down to the transmission-distribution (T-D) interface and the utility as DSO would now manage all DER resources below the T-D interface. The DSO will manage visibility of DERs and control over their associated reliability services in a manner that will enable it to maintain a more stable and predictable interchange with the TSO at the T-D interface. This will result in less reliance on the TSO to provide energy balancing

---

[14] Lorenzo Kristov, California Independent System Operator, and Paul De Martini, Caltech Resnick Institute, 21st Century Electric Distribution System Operations (May 2014).

[15] CISCO, Paul De Martini, Newport Consulting Group, Ultra Large-Scale Power System Control Architecture: A Strategic Framework for Integrating Advanced Grid Functionality (October 2012).

and other real-time services and more reliability and resilience at the distribution level. The DSO will also be able to aggregate and provide the capabilities of single types of DER and combinations of DER with appropriate performance characteristics to the TSO. In addition, under conditions where sufficiently robust and self-sufficient microgrids have been established—enabled by local generation, energy storage, demand response, and advanced monitoring, communications, and control technologies—elements of the distribution grid may be able to adopt islanding capabilities. The DSO will then have the responsibility and accountability for the reliable real-time operation and balancing of the respective islanded microgrid, along with integration and coordination with other parts of the electric system. The DSO would optimize real-time performance of each distribution network under its control in a similar manner, with each distribution network represented as a single T-D interface to the system operator. The TSO would still be responsible for system-level reliability.

Under high DER penetration, the traditional definition of the T-D interface would become obsolete and would be replaced with a regulatory framework under which the roles and responsibilities of grid operator and DSPP would be developed to reflect the new operational responsibilities. Under the new approach, the distribution network would be operated by the DSPP as a mini ISO below the T-D interface, with appropriate incentives to cost-effectively provide reliable end-use service.

The Clean Coalition recommends that the PSC clarify that DSPPs will be responsible for the operation of the distribution networks below the T-D interface to meet targeted levels of local reliability and resilience. This is necessary for implementing the staff vision that "DSPP will serve as the local balancing authority, forecasting load and dispatching resources in real time to meet customer needs and balance supply with load in real time to maintain reliability."[16] There are several reasons why the DSPPs are the appropriate parties to take responsibility for operation of DERs to meet local reliability and resilience targets. First, the DSPPs should have the most knowledge of their own distribution grids and customer loads and on how to operate DERs within these constraints. Second, the NYISO would not be able to rely heavily on DERs unless it gained much greater visibility into the distribution grids and full control of DER resources. However, NYISO control of DERs is not ideal since NYISO is not primarily interested in

---

[16] REV staff report at 22.

operating DERs in a manner that encourages continued participation in DER programs or providing high levels of end-use customer reliability.

Formalizing the role of DSPPs as DSOs can provide DSPPs with sufficient financial motivation to plan for, procure, and operate integrated DER solutions to meet both local and transmission-level operational needs. The DSO would act as the distribution grid Responsible Interface Party in transactions with the grid operator at the wholesale level. The DSO would be responsible for generating daily load forecasts of the distribution grid downstream of the T–D interface, which would be provided to the NYISO operator for least cost unit commitment purposes. Because the DSO would interface with the wholesale markets at a single T–D interface for each substation under its control, the energy and energy services costs being incurred by each substation will be transparent. Transparent energy and energy service costs for each substation will facilitate development of optimal DER resource placement and management and will provide a strong foundation upon which to develop best practices.

A key benefit of having a transparent (e.g., open and ethical) T–D interface is that it would enable the DSO to realize the value of DERs that are not bid into ISO markets. Although some DERs are bid into TSO markets, requiring all DERs to bid into TSO markets to realize value for contributing to system needs would result in undervaluing DERs since TSO markets are designed to meet bulk system needs, rather than to meet local needs. Further, it would greatly reduce the amount of DERs that are deployed since TSO markets may be overly restrictive about DER performance requirements and have high costs of participation.[17] A DSO can greatly reduce its costs of settlement with the TSO for bulk system services at the T–D interface by relying more on DERs and less on the bulk system for energy and grid services. For example, a DSO could smooth the net load profile and reduce the energy imports of a substation with a combination of distributed solar, storage, demand response, electric vehicle charging, and conservation voltage reductions facilitated with advanced inverters.

DSO operational control of DERs could also reduce the operational complexity of relying on DERs. Rather than giving control of DERs to the NYISO or arranging for the NYISO and the utilities to "share the button" for DERs, the DSOs could control operation of DERs to meet both

---

[17] Olivine Inc., Distributed Energy Resources Integration report to the California Independent System Operator (January 2014), available at http://www.caiso.com/ Documents/OlivineReport_DistributedEnergyResourceChallenges_Barriers.pdf.

local and bulk system needs. For example, the DSO could commit to the TSO that it will trigger certain demand response portfolios in specified circumstances or prove to the TSO that a demand response portfolio will behave in a statistically reliable manner.

The DSO approach would facilitate the transformation of the relationship between transmission and distribution system. The TSO would still be responsible for maintaining bulk system reliability, but the DSOs would be responsible for local reliability and motivated to minimize reliance on TSO energy and grid services.

CONCLUSION

REV is a unique opportunity to support New York's clean energy, resilience, and economic goals by addressing structural biases against DERs. To accomplish fundamental and lasting change, the PSC should realign the roles, responsibilities, and optimization processes for transmission and distribution grid planning to meet state and local goals. The PSC should also clarify the roles of the distribution operators and the transmission operator to make it possible to realize the full value of DERs and reduce the operational complexity of managing DERs for meeting local and bulk system needs.

APPENDIX 7

The World Is Round and Green: It Needs Strong Medicine

Ed Perry

STATUS REPORT (SUMMER 2014)

On June 24, 2014, President Obama laid out his plan to confront climate change, the first president in history to begin the process of breaking our addiction to polluting fossil fuels. His plan will protect not only the future generations but also our country's fish and wildlife resources. As a sportsman and a grandfather, I support his call to action to protect America's outdoor heritage.

Right now, America relies heavily on polluting coal and gas power, with renewable energy sources like solar and wind lagging far behind. Climate scientists have been telling us for years that our continued reliance on fossil fuels would create more extreme weather and cause over 40% of all life on the planet to go extinct in the lifetime of my grandson unless we greatly reduced our emissions of carbon pollution from large industrial sources, like power plants.

Those of us living in the northern part of Pennsylvania are seeing the true cost of satisfying our need for more energy. We are taking the last, best part of Pennsylvania and turning it into a vast industrial forest.

Is anyone out there who cares at all about clean air and water willing to tolerate thousands of natural gas wells in a part of our state that is home to some of the most pristine trout streams and unfragmented forest left in Pennsylvania? It is one thing to drill on private lands; it's quite another matter to turn our beloved state forestlands into a maze of well pads, haul roads, pipelines, and holding ponds. Degraded land, air, and water resources are the price we have to pay if we continue to use coal, oil, and gas to cool and heat our homes and businesses and run our vehicles.

What we really need is a strategy to get off these dirty fossil fuels that ruin our air, land, and water and cut carbon pollution. And the president's latest plan does that. His commonsense plan for meeting our obligation to protect future generations from climate change tackles the largest sources of carbon pollution where it's produced.

At present, coal-fired power plants are allowed to emit as much carbon pollution into the atmosphere as they want. The president plans to not only restrict pollution from new power plants but also reduce pollution from existing power plants.

But he's not stopping there.

To replace these dirty fuels, he proposes to expand responsibly developed renewable energy on public lands. From solar energy on desert lands in the Southwest to offshore wind on the Great Lakes and off the Atlantic Coast, we have limitless potential for safe, clean, American-made energy. To speed the transition to clean green energy, President Obama has directed the Department of the Interior to permit an additional 10 GW of renewable energy on public lands by 2020—enough to power 6 million homes.

Twelve of the hottest years on record have all occurred in the last 15 years. Every decade for the past 40 years has been hotter than the previous decade. We've had 338 consecutive months where the temperature was above the twentieth-century average. Given these facts, isn't it about time we stopped repudiating the overwhelming majority of climate scientists who have been telling us that industrial carbon pollution is causing the planet to heat up and get on with the process of safeguarding our planet for future generations?

We know the steps we need to take to protect wildlife, our communities, and the current and future generations of Americans from climate change. President Obama's plan captures the scale of what is needed to protect our communities and safeguard wildlife and prepare for the climate impacts we're already seeing. His call to action has my full support.

THE GLOBAL PERSPECTIVE

The Intergovernmental Panel on Climate Change released its most stark report (2014) yet on the devastating current and future impacts we all face from climate change, but sadly, many lawmakers in Congress are still actively denying the report's findings on behalf of their big polluter allies and greeting this new report with their eyes and ears covered.

They won't be able to ignore the obvious much longer. The report adds to the overwhelming scientific consensus and mountain of evidence supporting climate change, justifying the great urgency most Americans feel to act before the impacts grow even worse.

According to the IPCC report, climate change will hit the agricultural community hard and threaten food supplies, it will exacerbate poverty and displace millions, and it will be a threat to human health, including from impacts like increased asthma attacks and respiratory disease. Just look at what is happening in California with its extreme drought in 2014 since 2010. The IPCC report also finds heightened global security risks as scarce resources threaten to intensify world conflict.

As the IPCC report shows, climate change and its impacts are not far off in the distant future. They are here now and happening in every part of the world, including right here in Pennsylvania, and will result in even more devastating extreme weather events that affect the safety of our communities and increase our public health risks in the future.

In the United States, President Obama's Climate Action Plan is a major step toward fighting climate change and addressing its impacts. While we can make improvements in our infrastructure and take other adaptation measures, as the climate action plan calls for, the most meaningful step we can take to protect our communities and future generations is to stop climate change at its source. The carbon pollution standards for power plants are a critical step that would address the largest and currently unlimited source of carbon pollution that is fueling climate change.

Sadly, despite the overwhelming evidence backing the need to act, there are still climate deniers in Congress who not only ignore reality but also actively work to block any efforts by the Obama (and others) administration to protect our communities from the impacts of climate change. In fact, in the summer of 2014, the extremists in Congress want to "recall" President Obama for his taking long overdue actions. Big polluters and their allies in Congress remain committed to stopping the EPA from using its authority to set standards for industrial carbon pollution from power plants, which threatens public health and fuels climate change.

The latest IPCC report should be another reminder to our lawmakers that the costs of inaction are too high and the debate on the science is over. It's no longer a conversation about if we act. It's about what actions we will take to slow its impacts. We no longer have any excuse to delay or ignore this issue.

A second report from the IPCC due later in 2014 will outline what governments should do to prepare and combat the effects. In the meantime, the United States can now lead the globe in tackling climate change through commonsense carbon pollution standards. It's time for climate deniers in Congress to realize they are in the minority when it comes to refuting the reality of climate change affecting Pennsylvania and the rest of the globe.

THE LOCAL US STATE PERSPECTIVE: CONSIDER THIS FACTUAL CASE OF THE CLIMATE PROBLEM

Although some are still in denial about global warming, there are no deniers in the natural world. Every species of plant, insect, or wildlife that can move north or to higher elevations is already doing so.

What is most surprising is that a global temperature increase of only 1.5°F in the last 100 years has caused these dramatic changes.

For example, the middle portion of the Susquehanna River, from Sunbury to York Haven, has long been considered one of the premier smallmouth bass fisheries in the Eastern United States. But in 2005, water temperatures in the middle Susky exceeded 91°F, and the bass started dying. Hot water holds less dissolved oxygen, so as the water heats up, the bass become stressed and susceptible to infection called columnaris from a common soil and water bacteria.

The fishery is in such dire straits that the Pennsylvania Fish and Boat Commission has asked the Department of Environmental Protection to declare 100 miles as impaired. Every conceivable type of pollution is being blamed for the die-off. But is it a mere coincidence the fish started dying in the hottest year, in the hottest decade on record, and reoccurs whenever the lethal combination of low flows and high water temperatures coincides?

But global warming also benefits some species. Unfortunately, they are noxious insect pests like ticks and the hemlock woolly adelgid that is attacking our state tree, the hemlock.

Adelgids are small insects closely related to aphids that suck the sap from young branches, causing the needles to drop and branches to dieback. At present, its intolerance of cold weather has prevented its spread north of Massachusetts, but that's likely to change. Studies conducted by the University of Massachusetts predict that the entire northeast will be infested by this century's end. The Fish and Wildlife Service's high-altitude refuge at Canaan Valley does not have an infestation—yet. But biologists report that outside the refuge at lower elevations, "If you drive anywhere in West Virginia other than Canaan, there are hardly any living hemlocks around. It's tragic." Here in Pennsylvania, DCNR is fighting to keep the adelgid from decimating the old growth stands of hemlock in Cook Forest State Park, but right now, their future looks grim.

Our state fish in Pennsylvania, the brook trout, is also in trouble. Development and siltation have already eliminated a third of Pennsylvania's brook trout habitat. Brook trout need clean, cold water to thrive. Above that, they become thermally stressed and vulnerable to pathogens. Hemlocks are an integral part of brook trout habitat. In fact, the two are so closely aligned that brook trout were once called hemlock trout.

Studies have shown that brook trout are three times as likely to be found in streams surrounded by hemlocks because of the shade hemlocks provide during the heat of summer. Scientific literature suggests they can briefly tolerate temperatures above 71°F, but their optimal temperature is 66°F. Scientists forecast temperatures to increase by another 7-11° F by the end of this century unless we take strong action to reduce global warming pollution. If so, it's sayonara to our only native trout and state fish.

Our state bird, the ruffed grouse, is also on its way out of Pennsylvania. Grouse are only abundant in early succession-stage habitat, that is, forests that are 5-15 years old, which are precisely the habitat type deer prefer. Global warming is picking a winner here. The warmer winters and diminished snow pack favor deer survival, and they heavily browse young forests.

Global warming is also altering rainfall patterns. Studies have shown there is less winter precipitation in the form of snow and more in the form of ice and freezing rain, reduced snowpack, and more severe storm events. Due to these changing climatic conditions, grouse fledglings are subjected to torrential downpours and freezing temperatures at a critical phase of their life cycle. As a result, Pennsylvania's grouse population continues to decline. Due to all these changes, Pennsylvania has lost over 28,000 breeding males since 1980.

All these impacts are occurring with just a 1.5°F temperature increase in the last 100 years. Just imagine our world if the temperatures increase to 7-11°F that scientists forecast. It has already started and the data from all sources are conclusive: our planet is in severe jeopardy as well as all our people and other forms of life.

The Obama administration's Environmental Protection Agency is doing all it can to reduce global warming pollution by requiring cars and light trucks to get a fleet-wide average of 54.5 miles per gallon by the year 2025 and developing carbon pollution standards for new power plants.

But the biggest piece is EPA's new rule that was published on June 2, 2014, that requires Pennsylvania to reduce carbon pollution from existing power plants 32% below 2005 levels by the year 2030. As you might expect, the fossil fuel industry and its allies are forecasting gloom and doom, claiming that electric bills will skyrocket and brownouts will become common.

We've all heard that before. This is the same scare tactic they use whenever any sort of pollution control is proposed. It's way past time that pollution controls and protecting public health, and the health of the natural world, should be part of the basic services power plants provide, not something extra they use to gouge us.

SO THE QUESTION IS: WILL WE HAVE THE PRESIDENT'S BACK?

A recent poll by the League of Conservation Voters found that over 62% of Pennsylvanian's want action on global warming, and this includes the coalfields of Pennsylvania. The president has done all he can do through administrative actions because he can't get Congress to act. Now, he needs us to pressure our elected representatives to support EPA's new rule. As one Republican representative told me, "I believe global warming is real, now get my constituents on board and force me to act."

SO THE FINAL QUESTION IS THIS: WHAT ARE YOU PREPARED TO DO?

This past year, America (and the rest of the world) was broiled by the hottest year ever. The southwest has continued in the midst of a decade-long drought that climate scientists say is the new normal; western states continued to set records for wildfires, punctuated by the Waldo Canyon inferno, the most expensive in Colorado history; and farmers in over half the counties in the United States became eligible for disaster assistance because of drought. Scientists tell us we have a chance to curb the impacts of climate change—but will we act fast enough?

What's clear right now is that climate change is the biggest threat to America's wildlife this century. A new report from the National Wildlife Federation documents adverse impacts to wildlife all across our country (nwf.org/climatecrisis).

Here in Pennsylvania, repeat kills of smallmouth bass have decimated the fishery in nearly 100 miles of the middle portion of the Susquehanna River. In 2005, which was then the hottest year on record, my two sons and I were fishing the Susquehanna River on our annual float trip. That weekend, we saw hundreds of fingerling bass floating downstream, killed by a common soil and water bacteria called columnaris. Six miles above Harrisburg, I took a water temperature of 91°F. We didn't know it at the time, but what we were witnessing was the beginning of a nearly annual kill of smallmouth bass. For the past seven years, high water temperatures have lowered oxygen levels in the river, stressing the bass and making them susceptible to bacterial infection.

The agencies are exerting every effort to determine what is killing these fish, but the exact cause remains elusive. There is no question the bass are being affected by multiple stressors, like dissolved inorganic phosphorous. But the fact is that the fish started dying in the hottest year in the hottest decade on record and has continued nearly every year since. Tragically, the bass fishery in nearly 100 miles of the middle portion of the Susquehanna River has collapsed. It appears that climate change may be the final straw for these fish.

The same climate-fueled stressors affecting the bass are killing Minnesota's iconic big game animal, the moose. The state's northwest population has plummeted from over 4000 in 1980 to less than 100 today, while the northeast population has dropped from 8840 animals to about 2700. The decline has been so dramatic and precipitous that Minnesota canceled the fall moose-hunting season indefinitely.

OTHER CLIMATE IMPACTS AND RESULTS: NOT GOOD

Moreover, because winters are milder and growing seasons longer, insect pest populations have exploded. It's not unusual now to find moose with over 70,000 ticks embedded in them. In a desperate attempt to rid themselves of ticks, some moose will literally rub themselves raw, wiping off the hair that protects them from the cold during winter. Some moose in Maine were found to have over 150,000 ticks.

Brook trout are an iconic species in Pennsylvania, and they too are in trouble. According to many fishery scientists, if the temperature goes up as climate scientists forecast, trout will be largely eliminated from the northeast and greatly diminished out west.

Steve Sywensky, owner of Flyfisher's Paradise in State College, reports that in the last decade, the date that aquatic insect hatches occur has changed dramatically, adversely affecting fly-fishing and his business. Fly-fishers enjoy casting to fish rising to feed on insect hatches. However, since the hatches have changed, interest in the sport has declined, causing a distinct drop in business and shortening his "busy" season. He estimates losing $500-1000 per day during a time period when his shop should be full of anglers. Not only does this impact his business, but also it reduces tourism dollars used to purchase food, lodging, etc., in Centre County.

Out in the Western United States, high water temperatures last year forced natural resource agencies to close trophy trout water in Yellowstone

National Park, Montana, and Colorado. These closures are now becoming more like an annual event. A recent study found that the number of thermally stressed days trout experience in the world-class Madison River tripled between 1980 and today.

What is especially alarming about these impacts is that all of this is occurring with just a 1.5°F temperature increase over the last 100 years. If we continue burning fossil fuels at our present rate, climate scientists forecast temperature increases of 7°-11°F in just 80 years. At this rate, our kids and grandkids will never get to enjoy the great outdoor opportunities that our generation has been so blessed to have.

Fortunately, there is reason for optimism. We know what's causing these impacts on wildlife, and we know what needs to be done to chart a better course for the future. We need to say no to dirty energy choices like coal and tar sands while transitioning to clean, more secure energy sources like wind, solar, and geothermal that will help us reduce carbon pollution by 50% by 2030. Moving to renewable energy will not only help protect our natural resources but also give our families better energy choices and put our people back to work, for buildings made in America as the key solution to our energy needs.

All we need is for our elected representatives to put a price on carbon pollution so that we can begin the transition to our clean energy future.

THE GREEN MEDICINE: NOW

For the past 20 years, climate deniers have been claiming that science can't prove anything, that climate science is murky, and that the science is just not settled.

Do these arguments sound familiar? They ought to. These were the same arguments the tobacco companies used so successfully for many years to cast doubt on whether smoking caused lung cancer. In fact, some of the same "scientists" who worked for the R.J. Reynolds Tobacco Company are part of the climate change denier cabal.

But actually, the deniers are right. Science can't prove anything. The best science can do is give us the likelihood that something is true. And based on the scientific evidence, 97% of the world's climate scientists agree that the planet is heating up and our reliance on polluting fossil fuels is the cause. On such a complicated system as our climate, it's hard to do better than that.

For those who think the 3% are right, let's put the question differently. Suppose you were sick and you visited 100 doctors and 97 told you that you had cancer and needed treatment immediately but 3 said you were perfectly healthy. Who would you listen to?

During the tobacco debate, tobacco "scientists" argued we couldn't prove smoking caused lung cancer and the science was not settled. And they were right. After all, we all know someone who smoked well into their 90s. But we all know that the probability of dying from lung cancer is significantly higher if we smoke. No one doubts that any more. And the irony is that the science behind climate change is far more settled than the science behind the connection between cigarettes and lung cancer.

For example, a 2004 study published in Science magazine (a peer-reviewed scientific journal) evaluated every peer-reviewed science journal article published in the previous 10 years that dealt with climate change. Of the 928 articles randomly selected for review (representing 10% of the total number of articles), 75% either explicitly or implicitly accepted the finding that the planet was heating up and our reliance on fossil fuels was the cause. The other 25% of the papers dealt with other issues, taking no position on whether climate change was man-made. None of the papers disagreed with the finding that climate change was happening.

A 2013 follow-up study of over 12,000 peer-reviewed abstracts on the subjects of "global warming" and "global climate change" published between 1991 and 2011 found that of the papers taking a position on the cause of global warming, over 97% agreed that humans are causing it.

So if the science is so settled, why does the public think there is a raging debate among climate scientists? The answer lies in the skill of the denial machine who have convinced the media to give them equal time as if their 3% deserve the same attention as the 97%.

For example, a 2004 study conducted (Boykoff and Boykoff) evaluated all articles published in the previous 14 years on climate change in four major newspapers: the Washington Post, New York Times, Los Angeles Times, and Wall Street Journal. Of a random sample that totaled 18% of all the articles (636 articles), more than half (53%) expressed doubt as to the cause of global warming.

In other words, there is no debate in peer-reviewed scientific journals; the debate is in the popular press, which has confused the public, making them think scientists are uncertain by giving equal time to the scarce few outliers. Which is exactly what the denial industry wants and is a tactic they have used quite successfully for many years. Their strategy is to convince the public that climate science is not settled and not all scientists agree the planet is heating up and that carbon pollution is the cause.

There is good evidence that strategy exists.

A 1998 memo developed by participants from ExxonMobil, Chevron, and the American Petroleum Institute (among others) memorialized an action plan on climate that claimed "victory will be achieved when..." the average citizen and the media understand there are uncertainties behind the science.

Four years later, Republican communication strategist Frank Luntz (who has since accepted the science of climate change) advised Republicans: "Should the public come to believe that the scientific issues are settled, their views about global warming will change accordingly. Therefore, *you need to continue to make the lack of scientific certainty a primary issue in the debate.*"

Both strategies were taken directly from the playbook of the tobacco industry.

In every scientific question, there will be outliers. There are scientists who don't think HIV causes AIDS, and scientists who think vaccinations cause autism in children. These outliers have largely been discredited, but the press continues to give them print as if their views have the same merit as the overwhelming number of scientists who are on the other side.

Given the ability of climate deniers to get equal time in the media, and their strategy of sowing uncertainty, it's no wonder the public gets confused about science.

Creating a Cradle to Cradle World: Executive Summary Cradle to Cradle Products Innovation Institute

Bridgett Luther*, Will Duggan
*Corresponding author: bridgett@c2ccertified.org

Perhaps this is the moment in history when we realize that if today's tragedies and unintended consequences are perpetuated by our everyday acts, then our cultures have a strategy of tragedy. Only by adopting principled strategies of change will human enterprise have a new and enduring strategy of hope.

William McDonough

OVERVIEW

The goal of the Cradle to Cradle Products Innovation Institute is to lead a global transformation of the way we make things. It has been 22 years since our founders William McDonough and Michael Braungart first began their conversations about the flaws and potentials of modern industry; it has been 12 years since the publication of their landmark book, *Cradle to Cradle: Remaking the Way We Make Things*; today, because scores of executives leading the world's top brands have worked with Cradle to Cradle values, principles, and insights, we clearly understand the benefits that companies can receive and how to successfully scale for Cradle to Cradle design to become commonplace.

EXECUTIVE SUMMARY

The Cradle to Cradle Products Innovation Institute is a US educational non-profit organization with its headquarters in San Francisco, California. In May of 2010, William McDonough and Dr. Michael Braungart announced their intention to transfer their proprietary intellectual property for product certification into the public domain, thus giving the Institute the means and mission to exponentially increase the number of product manufacturers and designers who understand and implement C2C certification concepts, so they make safe and healthy products for our world.

Today, the Institute request seeks to dramatically speed worldwide adoption of Cradle to Cradle Certified™ Program. C2C product certification transforms the making of products, so their manufacture is not destructive but is instead beneficial to human and environmental health. C2C certification has five criteria:

- How safe and healthy are a product's ingredients and chemical processes?
- Can the product or its components be easily reclaimed and reused?
- Does the manufacturing process use renewable power?
- Does the manufacturing process produce clean water?
- Is a product made and brought to market in a socially fair way—are workers safe and legal and are they paid a fair "living" wage?

Certification allows companies to answer three dire and urgent needs:

1. The world's economies are stagnated. C2C certification can nurture innovation and economic growth in green chemistry, product design and manufacturing, and resource recovery.
2. Human activity is consuming and wasting natural resources at rate that cannot last. 90% of consumer products end up in waste disposal within 6 months. C2C certification eliminates the concept of waste and reclaims natural and technical resources into perpetually renewing systems of reuse.
3. All life on Earth is being exposed to toxic chemical pollution. C2C certification replaces or eliminates the use of dangerous substances and establishes systems to eliminate exposure during creation, use, and disposal of products.

Our mission with the Institute is to make **possible a faster paced, larger-scale transition to a "Cradle to Cradle World": a peaceful and prosperous place for 9 billion humans and the children of all species for all time**.

THE STATE OF THE C2C WORLD

Today, we believe that Cradle to Cradle values, principles, and insights are widely admired but are not being put in practice at a pace or scale sufficient to meet global economic, social, and environmental challenges. Below, we look at the vision of a "Cradle to Cradle" world, how far that vision has advanced, and the trends and indicators that establish how and why the Institute and our Cradle to Cradle certification program is uniquely poised to greatly accelerate global implementation of C2C practice and product design.

The Big Idea

The beginnings of a world-changing philosophy were born 20 years ago. Two people from very different backgrounds met and discovered that they shared a unique and revolutionary viewpoint: in a world in which the design intention of commerce had a micro focus on profits, these two men proposed a design intention with a macro focus on increasing health, happiness, prosperity, and the bounty of nature.

Ten years later in 2002, these two men—Dr. Michael Braungart, rebel chemical scientist, and William McDonough, independent architect and urban designer—laid out their views in the best-selling book. It proposed a radical new way of looking at the impacts and possibilities of human ingenuity and ambition. The underlying premise of Cradle to Cradle is that today's world, while filled with man-made and natural wonders, is full of problems we have caused.

Our species
- Puts billions of pounds of toxic material into the air, water, and soil every year
- Produces some materials so dangerous that they will require constant vigilance by future generations
- Results in enormous amounts of waste that pollute the air and water;
- Puts valuable materials in holes all over the planet where they can never be retrieved
- Requires thousands of complex laws and regulations—not to keep people and natural systems safe, but rather to keep them from being poisoned too quickly
- Measures productivity by how few people are working;

- Creates prosperity by digging up or cutting down natural resources and then burying or burning them
- Erodes the diversity of species and cultural practices

McDonough and Braungart asked a fundamental question: Are these the results we intended? Have humans, in the quest for economic prosperity, become destructive? The authors also characterized the response from the environmental movement as insufficient. It sought only to

- Release fewer pounds of toxic waste into the air, soil, and water every year
- Measure prosperity by less activity
- Meet the stipulations of thousands of complex regulations to keep people and natural systems from being poisoned too quickly
- Produce fewer materials that are so dangerous that they will require future generations to maintain constant vigilance while living in terror
- Result in smaller amounts of useless water
- Put smaller amounts of valuable materials in holes all over the planet, where they can never be retrieved

Rather than simply poison ourselves and the planet more slowly, McDonough and Braungart proposed a different approach, a model for industrial effectiveness, where companies produce things in ways that actually contribute to nature's bounty. Cradle to Cradle outlined the concepts for "the Next Industrial Revolution":

- Buildings which, like trees, produce more energy than they use and purify their own waste water
- Factories whose water is as clean as drinking water on the way out of the factory
- Products that, when their useful life is over, do not become useless waste but that can be put into the ground, where they will decompose and become food for plants and animals and nutrients for the soil or, alternately, can return to industrial cycles to supply high-quality raw materials for new products
- Billions, even trillions, of dollars' worth of materials that can be used for human and natural purposes
- A world of abundance, not one of limits, pollution, and waste

The impact of the book has been enormous. Time Magazine has called these ideas "a unified philosophy that—in demonstrable and practical ways—is changing the design of the world."

The influential I.D. Magazine called C2C concepts "a landmark in industrial design." Since its publication, reading of the book has been required for millions of students of chemistry, engineering, design, and sustainability.

Governments around the world started to embrace Cradle to Cradle thinking.

- In California, Gov. Arnold Schwarzenegger was struggling with how to regulate toxic products when he met McDonough and began crafting what would eventually become the California Green Chemistry Initiative to guide the world's seventh-largest economy. Among his top policy recommendations was "to move California to a Cradle to Cradle Economy."
- In Europe, a C2C network was established by the European Union's Interregional Cooperation Programme to bring together EU regions to share and capitalize upon regional good practice in implementing C2C principles.
- China's Circular Economy Initiative (2004) and Circular Economy Law (2008) borrow extensively from the Cradle to Cradle design practice.
- In 2006, the documentary television special Waste=Food was broadcast in The Netherlands. The idea that waste could become food for the biosphere and that we could transform current commercial industrial production to become a positive footprint caught the public's imagination.
- Berlin, Germany, celebrated the first Cradle to Cradle Festival with 80 companies and regions that are working to optimize themselves, their products, or their city planning according to the Cradle to Cradle principles. Approximately 6000 people participated at the 80 events.

Cradle to Cradle has also been adopted by businesses ranging from large corporations with a global reach to small, innovative startups, cumulatively representing some trillion dollars in revenue. Hundreds of products have been "certified," receiving a prestigious mark of approval that confirms that both the product and the company have moved beyond "less bad" to "more good" business practice.

And while significant progress has been made since the book's publication and the implementation of the private certification program, Braungart and McDonough's many speeches and working with individual businesses and their products are not enough to create change at the scale or pace that is needed.

Necessary Step for Scaling the Big Idea

In May of 2010—with the direct support of Governor Schwarzenegger—Braungart and McDonough licensed 20 years' worth of proprietary certification methodology to a new US nonprofit organization, the Cradle to Cradle Products Innovation Institute. McDonough and Braungart's thinking: by gifting C2C to the public domain, its ideas and methodologies could spread much farther, much faster to provide the solutions for which the world was desperately searching.

Early investment from The Netherlands-based Stichting Doen Foundation enabled the nonprofit to incorporate, gain status as an educational nonprofit from the United States Internal Revenue Service, transfer the intellectual property of certification and training from a private company to the nonprofit, craft a worldwide protocol, build website and online tools, and raise additional funding.

The Current State of C2C Business Adoption

To date, over 200 companies have had 2000+ products certified as meeting or advancing upon Cradle to Cradle design principles. Today's notable and pioneering companies and products include the following:

Desso

In 2008, the company committed to using 100% "Cradle to Cradle" design by 2020. In 2010, the company launched EcoBase, a carpet tile backing that Desso can completely recycle and use to make new products. Desso's share in the carpet tile market increased from 15% in 2007 to 21% in 2011. The company increased its profitability from less than 1% in 2006 to 9% in 2011.

> We know Cradle to Cradle works. Our earnings (EBIT) have gone up nine-fold between 2007 and 2010. Our customers are very impressed by the products we offer that are Cradle to Cradle® certified. It is fantastic to see how our whole approach is enabling us to be cutting edge both in design and functionality as well as lay the basis for complete sustainability in the use of materials, water and energy.
>
> **Stef Kranendijk, Former CEO of Desso**

Aveda

A leading natural personal care company, Aveda was one of the first to adopt the C2C design practice and certification. Established in 1978, the company is now the largest buyer of organic essential oils in the world. By adopting the C2C design approach, Aveda has gone beyond just making ecological and safe personal care products, pioneering the use of wind energy in the beauty industry, leading by example in the use of recyclable packaging, and supporting organic agriculture. Aveda has received gold C2C designation for seven key products and silver C2C designation for its packaging. Practicing sustainable designs, the company's sales have grown five times in 10 years.

Aveda's adherence to its core environmental values has been nurtured alongside growth in sales and profit…. And these strides have been made without compromising business performance. The company's sales have grown five times in 10 years.

Dominique Conseil, Global Sustainability Manager, Aveda

Steelcase

The global leader in the office furniture industry, Steelcase offers more Cradle to Cradle Certified Products than any other company in any industry globally. Steelcase has assessed over 600 categories of materials spanning across multiple product lines.

We pursue Cradle to Cradle certification of our products because it's a rigorous, holistic approach. It also best mirrors our environmental aspirations and business philosophy.

Angela Nahikian, Director of Global Environmental Sustainability

Shaw Floors

A Berkshire Hathaway company, Shaw was the first manufacturer in the world to offer fully certified Cradle to Cradle flooring products. The ultimate goal of the company is to earn Cradle to Cradle certification for all of their products. More than 80% of Shaw's commercial flooring product sales in 2013 were generated from Cradle to Cradle Certified Products.

Companies today have to consider what kind of impact their decisions will have on both their businesses and the planet—ten, twenty, thirty or forty years from now. And when in doubt, it's wise to err on the side of the planet.

Warren Buffett, Chairman and CEO

Method

Method is one of the first companies to be recognized as a Cradle to Cradle *company*, meaning that they build the Cradle to Cradle design into everything they do, from the ingredients they select for their products to how they manufacture them. Method currently has 100% of their products Cradle to Cradle–certified, among the most of any company in the world, and 80% at the gold–certified level.

Even if concepts are beyond the current relevance of most of the market, it's way better to be ahead of the curve than behind it. For example, Method is a proud Cradle-to-Cradle company and evangelist, even if most mainstream consumers don't yet know what Cradle-to-Cradle means.

Adam Lowry, Cofounder and Chief Greenskeeper

Businesses worldwide face two wholly new sets of challenges as they seek investment and market share:

1. Scarce resources, the rising cost of raw materials, and the impact of more regulation put a premium on a business' ability to innovate.
2. According to research by the World Business Council for Sustainable Development, 75% of a company's asset value is intangible, coming from its ability to innovate, to increase product quality, to eliminate waste, to lower liability and regulatory risk, and to attract and retain the best young talent.

What pioneering early adopters know is that Cradle to Cradle certification helps them address these new challenges and, in the process, raise their profile with investors, regulators, suppliers, and customers.

So where do we go from here? Thousands of people around the world have been inspired by the insights of Cradle to Cradle®, but does it really make a difference? Can we measure it in some kind of scientific way?

So after 4 years of leading the charge to scale Cradle to Cradle certification around the world, the Institute decided to try and find out how the certification standard helps companies make changes in their products and processes and what difference it makes to environmental and social systems and to the bottom line. This study that we conducted represents pilot research designed to contribute the initial measureable evidence base for the environmental, social, and economic benefits of the *Cradle to Cradle Certified Products* Program and to stimulate thought about how the making of things can be transitioned into a positive force for people, planet, and profit. While the study is not intended to provide scientific verification or demonstrate causality, it does provide an initial indication of the very significant economic, environmental, and social potential of the program. More granular research, considerate of a wider sample of companies, is needed to strengthen the pilot findings. The *Impact Study* report series is available to download at www.c2ccertified.org/impact.

From a group of early innovators, the Institute and our partner Trucost selected 10 companies and their certified products to establish a preliminary framework for measuring the business, social, and environmental benefits of achieving Cradle to Cradle certification. Trucost, a leading global environmental data and insight company, has an outstanding track record of helping companies, investors, governments, academics, and thought leaders understand the economic consequences of natural capital dependency. Its global expertise made them an excellent partner in illustrating a more sustainable business model, product, and brand.

These 10 selected companies operating in the US, European, and global markets represented a wide range of product portfolios from carpet tiles to toiletries and had combined revenues of over €6.75bn and employed global workforces exceeding 50,000 people. These 10 offer an early indication of its very significant benefits of the *Impacts of the Cradle to Cradle Certified Products Program*. http://www.c2ccertified.org/impact.

Findings: Impacts of the Cradle to Cradle Certified Products Program

The overall findings of the report revealed that pursuing Cradle to Cradle product optimization and certification helps companies become a frontrunner in the transition to the circular economy, providing them with an astounding competitive edge—an inspiring incentive to pursue Cradle to Cradle product certification.

Business Benefits: Good Design Equals Good Business

An integral aspect of the Cradle to Cradle philosophy is the concept that good design equals good business. The study showed a wide range of

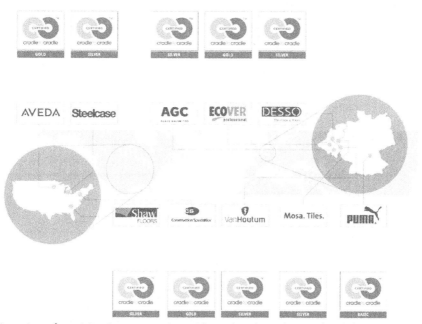

Location of participating companies, with analyzed product and certification levels across product portfolio.

business benefits resulting from the pursuit of product certification and the product optimization inherent to this process. These include reduced costs, improved product value, new revenue streams, and avoided risks.

For example, Shaw Industries, the world's largest carpet manufacturer, received its first Cradle to Cradle Certified Product in 2007 for its Eco-Worx® tile—now, it is the fastest-growing carpet product. Compared to the uncertified version Shaw previously manufactured, energy efficiency and its switch to renewables have cut the environmental cost of making carpet tiles by more than half, along with the amount of water needed to produce a tile. The water and energy savings for total production in 2012 equated to a cost saving of over US$ 4 million.

Environmental Benefits: From Gray to Green

Through the implementation of the certification program, companies are encouraged to work toward the design and production of products that have a positive impact on the environment, making them "more good" rather than "less bad." Ultimately, this means that during the production process, water used is purified instead of polluted, and more renewable, cleaner energy generated than is consumed. The impact study showed that negative environmental impacts were reduced through the use of alternative choices of safer ingredients, leading toward cleaner water and the more renewable energy.

For instance, Puma has developed a biodegradable sneaker called *InCycle*, which holds Cradle to Cradle basic certification. To ensure that product recovery is possible and optimized, PUMA provides collection banks in many stores ran in cooperation with the international recycling company I:CO. The success of these collection banks is yet to be determined, because collection rates are not yet available. However, if all *InCycle Baskets* are composted at end of use, the sneaker has an 87% smaller impact at the end of use compared with conventional sneakers.

Social Benefits: Fair + Healthy = Happy

The Cradle to Cradle Certified Products Program is based on best practice social fairness principles. The impact study found that the majority of participating companies already had high standards of social commitments in place and little additional effort was required to meet the Cradle to Cradle Certified Products Standard. This is likely to be the result of the already strong ethical and social commitment of companies with the desire to certify products.

Ecover, for example, undertakes a wide range of social fairness activities supporting local communities and social projects, including a project at a local home for the mentally disabled—VZW Huize Monnikenheide, Zoersel. Other projects include funding of projects to OKAN in The Netherlands and WaterAid in the United Kingdom and ecological projects that help support local communities dependent on ecosystems. This work is important for the company to move toward a socially positive entity, creating an environment not just capable of minimizing negative impact for its workforce but actually progressing toward bringing value and benefit to the wider community and individuals outside the workspace.

Embracing the Circular Economy

Through the Cradle to Cradle Certified Products Program, companies have a concrete model to produce appealing, safer products made from pure materials designed to be reused. Although a preliminary framework, *Impacts of the Cradle to Cradle Certified Products Program* http://www.c2ccertified. org/impact offers a promising account of impact and value achieved by ten companies who have made steps toward product optimization.

The core questions we ask you to consider are how do we build on their success? What can we bring to the world that would encourage the "early adopters" to take this new way of designed and manufacturing products to the next level and beyond?

We invite you to read the findings here http://www.c2ccertified.org/ impact and consider how the certification program can be a bridge to the circular economy that works to the benefit of all. Let's stop talking and get started. Pioneering companies have already laid the groundwork worldwide. They've built competencies in new product design, new material innovation, and creation of reuse cycles and in increasing their commitment to renewable energy, water stewardship, and social fairness.

Trucost helped gather the information and found some great examples. We all can imagine a Cradle to Cradle world, but these ***innovators*** are showing the way. Change can be exciting. Change can be an opportunity for new growth as more and more companies adopt the Cradle to Cradle principles and start their innovation journey.

An exponentially greater number of companies and consumers need to understand the benefits that come with Cradle to Cradle product certification. This study was a start. In order to have real impact, the number of product manufacturers participating in the system must grow.

There is a huge opportunity for companies to embrace the transition toward the circular economy; and the Cradle to Cradle Certified Products Program guides the path.

APPENDIX 10

The Formidable Fight for Fuel

Ariella S. Lewis[1],*

[1]Ariella S Lews was in United States Air Force, Bioenvironmental Engineering Technician, Senior Airman and now at University of Delaware, Energy and Environmental Policy, Doctoral Candidate.

*Corresponding author: lewisariella@gmail.com

OVERVIEW

As a global community, we have a poisonous dependency on crude oil. This atrocious addiction has been interlaced into societies while dissolving accord between nations. Hostile tensions are fostered between countries as this nonrenewable resource continues to be depleted exponentially. With crude oil being the foundation of countries' industrial and economical development, world leaders have devised strategies to secure future energy demands. The steadfast solution to the oil crisis has influenced dubious foreign policies, while science and green innovation have been snubbed. Conflicts of different magnitudes have been triggered where oil has been the instrumental provocation.

The controversial Iraq War (2003-2011) is an infamous example where oil was a central issue, albeit not the only issue. This deduction derives from Iraq's abatement of protectionist regulations that occurred throughout the war. Prior to the US invasion of Iraq in March 2003, this Middle Eastern nation restricted Western investments in the oil industry. Ten years later, its strict and incapacitating laws had been lifted that boosted the profits of Western oil companies. While Western oil company executives luxuriated in lavish lifestyles, American taxpayers financed their secure investment, and more importantly, American troops expended life and limb for this war-profiteering effort. Environmental deterioration due to our oil dependency is becoming increasingly difficult to ignore. However, are we ignoring the sacrifices made by our brave troops to support our addiction? Throughout this chapter, I will explore our reliance on crude oil, the conflicts that have arisen, the detrimental consequences of warfare, and thus the paramount importance of engineering a future with sustainable energy.

Throughout the history of humanity, armies have been engaging in warfare at the discretion of their powerful governing officials. These heroes camouflage their fear, armor themselves with bravery, and march towards the enemy. They have taken command from and given their lives for thrones, empires, kingdoms, and countries. Despite their conspicuous courage, they are maneuvered as expendable pawns and their noble deaths are considered "a means to an end." In some circumstances, the "end" is not worth the sacrifice of an individual's invaluable life.

The History Channel supports this claim, "Over the years, armies have mobilized and blood has been shed over everything from tragic misunderstandings and perceived slights to petty border disputes and even sporting events" (Andrews, 2013). It referenced a handful of examples to reinforce its argument. These wars included the Pig War (1859), the War of the Stray Dog (1925), and the Pastry War (1838). In the subsequent paragraphs, I will explore other senseless wars stirred by the identical mission of fueling our fossil fuel (oil) dependency and the fearless warriors of the fight.

THE DISGUISED DEMON

The fiend of fossil fuels has lived among us for thousands of years. According to the Northern Mine Research Society, its haunting presence was even felt during Anno Domino times as evidence of coal use can be dated back to AD 43-410. During this time period, it was primarily employed as ballast in the empty grain ships (Hill, 2013). Throughout time, various ingenious applications for fossil fuels were devised by civilizations. By the fourteenth and sixteenth centuries, coal was used for forges, smithies, limeburners, and breweries and became the major heating source for buildings and homes (EIR, 2014). Different forms of the ghoulish fossil fuel continued to take shape and propelled many countries towards industrialization. Although coal was the first form of fossil fuel that impacted societies, our modern lifestyle is indebted to oil. "Oil brought about the most profound social transformations in history...."

Today, the most common products derived from oil are found in the energy sector: gasoline, heating oil, natural gas, aviation fuels, and diesel fuel. "Oil is also the key ingredient in tens of thousands of consumer goods, including ink, plastic, dishwashing liquids, crayons, eyeglasses, deodorants, tires, ammonia, and heart valves" (EIR, 2014). The revenue influx triggered by industrialization reinforced our dependency on oil; thus, its spirit continues to haunt our society and economy.

THE CONCEALED PRICE WE PAY

A natural resource is typically derived from the Earth and is considered non-renewable if it cannot be promptly replenished since its creation requires millions of years. The rarity of nonrenewable resources has made them a precious commodity, evidenced by two unlikely counterparts, grimy oil and the exquisite diamond. The law of supply and demand is an important economic tool used to determine pricing strategies. By considering the diamond industry, this economic theory can be explained effortlessly and, consequently, be applied to the oil industry. Diamond, a mesmerizing and rare gem, comes with a high price tag to correlate with its high demand. The demand to possess this rare metal (nonrenewable resource) has prompted not only a preposterous price tag but also the term "blood diamond," dubbed to reflect the questionable ethics of diamond trade.

This phenomenon is also evident with fossil fuels, an overpriced nonrenewable resource that sparked an immoral industry. Fossil fuels are derived from decomposed prehistoric plants and animals that inhabited the Earth hundreds of millions of years ago. The limited and dwindling supply, coupled with an outrageous demand, has resulted in a high price, both literally and figuratively. Feeding our addiction to fuel has cost us the cleanness of our planet, the existence of species, the health of our people, the peace among countries, and the lives of our military heroes.

THE SQUABBLE FOR NONRENEWABLE RESOURCES

A snapshot of the past century would affirm that foreign policy negotiations between nations have been brutish. When discord persisted, the brutish behavior quickly became belligerent. International discussions, in regard to oil, have similarly sparked warfare and thus drawn bucketfuls of blood. In the 1930s, heightened tensions between two landlocked South American nations, Bolivia and Paraguay, resulted in the Chaco War (1932-1935). Bolivia provoked this war in attempt to dominate geographically questionable land between the two countries. International oil companies suspected the territory was saturated with oil reserves and pressured the Bolivian government to gain control. After 3 years of war, Bolivia's intended land expansion was successfully halted by Paraguayan troops. Ironically, the oil companies' suspicions proved to be false as the oil reserves contained minute amounts of oil. 100,000 soldiers sacrificed their lives in this futile war (Lindsay, n.d.).

Adolf Hitler, an infamous dictator in world history, was committed to securing Germany as a world power. Conquering Russia was an essential ingredient to feed his desire for dominion over Europe. In 1942, he devised a military strategy to invade Russia at an oil-rich city, Stalingrad. Consequently, he underestimated the city's military presence. Despite the annihilation of his troops, he refused to surrender as the oil fields located at Stalingrad tantalized him and were vital to his mission (Gero, 1998). The 6-month Battle of Stalingrad left Hitler embarrassingly defeated. His naive tactics killed his armies by the multitudes; 1.5 million German, Hungarian, Romanian, and Italian soldiers died in the month long battle (Proyect, n.d.). Saddam Hussein, notoriously described as a tyrant, served as the president of Iraq from 1979 to 2003 (Saddam Hussein Abd al-Majid al-Tikriti, 2014). During his presidency, he invaded two neighboring nations, Iran and Kuwait. In both instances, he encroached in regions reputable for their extensive amounts of oil fields. However, his attempts to achieve control over these oil-endowed lands were unavailing. From the onset of Iraq-Iran War (1980-1988) to 1984, 250,000 Iraqi troops were killed or wounded (Military, n.d.). The Iraq-Kuwait War (1990-1991) left 25,000 soldiers dead (Iraq invades Kuwait, n.d.).

Dissent and tensions breathed life into Sudanese warfare similar to the South American, European, and Middle Eastern cases analyzed previously; however, it contrasts slightly. The two warring factions were not from opposing nations; rather, it was one nation with deviating ideals. In July 2011, South Sudan gained independence from Sudan after decades of unrest and civil war (South Sudan, n.d.). One year after agreements, the oil fields located on the border of both countries continued to trigger war (Walid, 2012). In 2014, the South Sudan government lost control of its major oil hub when it was seized by rebel forces (South Sudan rebels 'seize' oil hub, 2014). An estimated 1.5 million people have been killed in the cross fire since the beginning of the 1983 civil war (Sudan country profile, 2014).

THE REVOCATION OF HINDERING REGULATION

In 1972, Saddam Hussein nationalized Iraqi oil production, giving all rights to the Iraq Petroleum Company Limited (Iraq: Law Nationalizing the Petroleum Company, 1972, p. 846). This severed Western oil producers' entitlements to the country's appealing oil reserves. However, the 2003 American invasion of Iraq amended this Hussein-implemented regulation. On March 22, 2003, President George W. Bush addressed the American public to announce the gravity of Operation Iraqi Freedom. He outlined the mission in three simple points, "to disarm Iraq of weapons of mass destruction, to end Saddam Hussein's support for terrorism, and to free the Iraqi people" (Office of the Press Secretary, 2003). While the announced mission was abridged, a central detail was omitted from his speech.

According to General John Abizaid, the 2007 head of the Iraq War's US Central Command and Military Operations, the war was predominantly concerned with oil. The widely respected, Alan Greenspan, made a similar proclamation in his memoir (Juhasz, 2013). Regardless of the US motive to invade Iraq, the outcome was favorable for Exxon, Chevron, BP, Shell, Halliburton, and other war contractors and embezzlers. Halliburton, along with foreign oil companies, was thirsty for Iraqi oil, but the nightmarish 1972 restrictions left them quenchless.

Upon the removal of Iraq's leader, the United States and the United Kingdom established the Coalition Provisional Authority (CPA), an organization whose purpose was to ensure the seamless restoration of Iraq by implementing a Unite States-biased strategy. The CPA implemented a plethora of regulations, orders, memorandums, and public notices that revamped economic, investment, banking, finance, and oil distribution laws. When the government of Iraq (GOI) was established in 2006, the government's constitution contained obscure direction on CPA regulation (Toone, 2013). While the newly instituted investment law allowed foreigners to invest in Iraqi markets, an exception existed; oil distribution companies noted "Oil constitutes a strategic and lucrative market for foreign investors."

Foreign direct investment "in the oil sector is unique in comparison to other sectors of the Iraqi economy insofar as there are a variety of additional legal requirements that foreign investors must meet in order to participate in oil and gas extraction and its attendant industries" (Toone, 2013). Despite the vague regulation, Iraqi oil was no longer nationalized by the war's conclusion. At the termination of the war, studies were conducted to determine if the United States was triumphant in the war against Iraq. Contradicting deductions were the result as analysts were incapable of conceiving a transparent conclusion due to the war's opaque mission.

THE FEARLESS FIGHTERS

Despite the controversy and the political friction generated by the Iraqi invasion, the American military served with honor. They combined the unique strengths of the Air Force, Army, Marines, Navy, and Coast Guard and fought together with skill and resilience. They are a brotherhood that uncritically obey the orders of the US president and others appointed over them. They obey these orders, without question but with trust, evident in the ambiguous military operations in Iraq that lasted 8 years, 9 months, and 12 days. Over 1.5 million brave troops deployed to the area and committed their lives to the mission, the opaque mission. 4474 troops made the ultimate commitment. 32,226 service members' lives were drastically impacted due to severe injury (Iraq by the numbers, 2011). These American heroes lost blood. They lost time spent with loved ones. They lost limb, eyesight, and mental stability. They lost their lives.

THE ACCRUAL OF BLOOD MONEY

The winners of the Iraq War were the companies that were awarded contracts to support the Iraqi military operations. The Business Insider approximated that $75 billion was expected to go to these American-contracted companies, the largest of all Halliburton (Kelley and Ingersoll, 2013). "Halliburton is one of the world's largest providers of products and services to the energy industry. With more than 80,000 employees, representing 140 nationalities in approximately 80 countries, the company serves the upstream oil and gas industry throughout the lifecycle of the reservoir— from locating hydrocarbons and managing geological data, to drilling and formation evaluation, well construction and completion, and optimizing production through the life of the field" (Corporate profile, n.d.). Exxon, Chevron, BP, and Shell were other winners post-Operation Iraqi Freedom. Ridding Iraq of Saddam Hussein and his extreme protectionist restrictions provided oil companies with the world of opportunity. The hurdle was removed and the path was paved for discussions and contracts between oil companies and Iraqi oil fields.

THE FIGHT FOR A FINER FUTURE

Assuming that General John Abizaid and Alan Greenspan were sincere when they revealed the hidden intentions for invading Iraq, the aforementioned "winners" and our government officials selfishly sacrificed our selfless military. A military that serves our nation pridefully boasts to be the best in the world yet humbly honors the heroes that have sacrificed before them. They are the one percent of Americans that voluntarily uttered an oath for the love of this nation. These valiant individuals and their families deserve more than to be pawns in devious oil schemes.

Operation Iraqi Freedom was intended to secure our energy future. However, investing the 1.7 trillion war dollars spent by the US Treasury Department through Fiscal Year 2013 (Iraq by the numbers, 2011) in renewable energy could have secured our energy future and above all prevented 4474 honorable deaths. Investing in a future with renewable energy is paramount, as this type of resource would tread lightly on our environment and curtail senseless warfare. The sustained supply of renewable resources and its ability to be replenished would solve the striking shortcoming of our oil dependency.

CONCLUSION

Lessons of the past have had little influence on foreign policy, as oil continues to be an underlying cause of international political turbulence. Millions have lost their lives, and millions of dollars have been lost to this fight for fuel. Reassessing our energy supply would yield a rewarding future. Grasping onto outdated technologies and ideologies inhibits innovation and consequently the green industrial revolution.

REFERENCES

Andrews, E., 2013. 6 Wars fought for ridiculous reasons. History.com (December 31). Retrieved July 19, 2014, from http://www.history.com/news/history-lists/6-wars-fought-for-ridiculous-reasons.

Corporate profile, n.d. Halliburton. Retrieved July 20, 2014, from http://www.halliburton.com/en-US/about-us/corporate-profile/default.page?node-id=hgeyxt5p.

EIR Global, www.eirglobal.eu/el_sustainable_development.html, 2014.

Gero, 1998. Why Stalingrad? Stalingrad. Retrieved July 20, 2014, from http://www.stalingrad.net/introduction/introduction.

Hill, A., 2013. Coal—a chronology for Britain. British Mining.

Iraq by the numbers, 2011. Democratic Policy & Communications Center. Retrieved July 20, 2014, from http://www.dpc.senate.gov/docs/fs-112-1-36.pdf.

Iraq invades Kuwait, n.d. History.com. Retrieved July 21, 2014, from http://www.history.com/this-day-in-history/iraq-invades-kuwait.

1972. Iraq: law nationalizing the petroleum company. International Legal Documents 11(4), 846. Retrieved July 20, 2014, from the JSTOR database.

Juhasz, A., 2013. Why the war in Iraq was fought for big oil. CNN (April 15). Retrieved July 21, 2014, from http://www.cnn.com/2013/03/19/opinion/iraq-war-oil-juhasz/.

Kelley, M., and Ingersoll, G., 2013. By the numbers: the staggering cost of the Iraq war. Business Insider (March 20). Retrieved July 20, 2014, from http://www.businessinsider.com/iraq-war-facts-numbers-stats-total-2013-3.

Lindsay, R., n.d. The Chaco War. Retrieved July 19, 2014, from http://www1.american. edu/ted/ice/chaco.ht.

Military, n.d. Iran-Iraq War (1980–1988). Retrieved July 19, 2014, from http://www. globalsecurity.org/military/world/war/iran-iraq.htm.

Office of the Press Secretary, 2003. President discusses beginning of Operation Iraqi Free-dom. The White House (March 22). Retrieved July 20, 2014, from http:// georgewbush-whitehouse.archives.gov/news/releases/2003/.

Proyect, L., n.d. The battle of Stalingrad in film and history. University of Columbia. Retrieved July 20, 2014, from http://www.columbia.edu/~lnp3/mydocs/culture/ BattleofStalingrad.

Saddam Hussein Abd al-Majid al-Tikriti, 2014. The Biography.com website. Retrieved Jul 20, 2014, from http://www.biography.com/people/saddam-hussein-9347918.

South Sudan rebels 'seize' oil hub, 2014. BBC News (April 15). Retrieved July 20, 2014, from http://www.bbc.com/news/world-africa-27037336.

South Sudan, n.d. United Nations High Commissioner for Refugees. Retrieved July 21, 2014, from http://www.unhcr.org/pages/4e43cb466.html.

Sudan country profile, 2014. BBC News (July 5). Retrieved July 20, 2014, from http:// www.bbc.com/news/world-africa-14094995.

Toone, J., 2013. Foreign direct investment in post-war Iraq: an investor's introductory guide to the legal framework. BYU Int. L. Manage. Rev. 9 (141), 140–173, Retrieved July 19, 2014, from the LexisNexis database.

Walid, K., 2012. In Sudan, a race to occupy oil fields. International Edition. Retrieved July 20, 2014, from the LexisNexis database.

Smart Green Energy Communities: A Definition and an Analysis of Their Economic Sustainability in Europe

OVERVIEW

The next energy revolution is a "green and smart" one in which the energy systems evolve towards a new paradigm based in the "energy community" concept, which refers to consumers who decide to make common choices related to the their energy needs satisfaction, through the implementation of technological solutions to be "smart" in generating and managing energy (Energy & Strategy Group, 2014).

In this sense, energy communities represent one of the new key actors— like prosumers, that is, energy users who self-produce a significant share of their energy needs, and electric vehicles—of the increasingly important smart grid paradigm. The result is smart green energy communities (SGECs) that are beginning to become more and more prevalent all over the European Union.

The issue of energy efficiency and saving has gained increasing interest at international level, for both policy makers and end users. The former— especially in Europe, where the awareness towards environmental issues is historically high—have defined obligations and incentives to the adoption of energy efficiency initiatives. The latter have increasingly become aware of the benefits of proper management of the energy variable, through which they can reduce one of the most significant expenditure within their budget and consequently increase their competitiveness, which is particularly relevant in the industrial sector (Energy & Strategy Group, 2013).

However, energy users historically have invested efforts and money in intelligent energy management on an individual basis. The novelty underlying the SGEC approach lies in the transition of the energy users from an "individual" approach to a "collegial" one.

The categories of energy users (consumers), which are potentially interested in developing a smart green energy community, range from residential users (e.g., condominiums), to industrial firms (e.g., industrial districts), to service firms (e.g., shopping malls or university campuses). Irrespective of the type of aggregation (which can be homogeneous, if the energy users within the community belong to the same category, or heterogeneous), the energy users can achieve several benefits thanks to the realization of an SGEC, ranging from a reduction in energy expenditure—compared to the cost of traditional supply sources (i.e., the purchasing of electricity from the grid and the on-site thermal energy production through traditional technologies)—to an improvement of the power quality (in terms of continuity of service and voltage quality; Energy & Strategy Group, 2014).

The different categories of energy users give different importance to such benefits. For example, residential users are usually more interested in reducing their energy expenditures, which are typically one of the most relevant expenses within a family budget, while industrial firms—with particular reference to specific industries like mechanical or food and beverage sectors—need a huge power quality in order to guarantee the right functioning of their production assets. Figure A1 shows the relative importance of such benefits for the different categories of energy users. This influences the most appropriate configuration of the SGEC in terms of technological architecture.

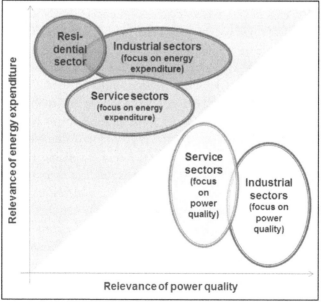

Relative importance of SGEC benefits for different energy users. *Adapted from Energy & Strategy Group (2014).*

IMPLEMENTATION OF SMART GREEN ENERGY COMMUNITIES

In order to implement a smart green energy community, a set of technologies must be adopted, which can be classified into three clusters: (i) technologies for energy production, storage, and usage; (ii) technologies for power flow management and control; and (iii) technologies for energy distribution and communication infrastructures (Energy and Strategy Group, 2014).

The first cluster refers to technologies that allow to (i.a) produce *in situ* a relevant portion (up to 100%) of the energy needed by the energy users within the smart green energy community—from both renewable energy sources (such as solar, wind, and bioenergy) and traditional sources (such as natural gas cogeneration or trigeneration); (i.b) to store energy through electrochemical energy storage systems (the so-called battery energy storage systems (BESS)) or other types of energy storage (e.g., electrical, mechanical, or thermal); and (i.c) to consume this energy in a smart and efficient way, through the adoption of systems and devices like building automation systems, smart appliances, and efficient lighting systems.

The second cluster refers to technologies (both hardware and software) that allow to remotely control the energy production/distribution/storage/consumption assets within the smart green energy community and to manage the energy flows within the same. In particular, the software for energy flow management and control allows to plan the "optimal" operation of the energy production/storage/consumption assets within an SGEC (from a technical and/or economic point of view), thanks to consumption forecasts of the energy users and the production forecasts of the generation plants sourced by nonprogrammable renewable energy sources within an SGEC, while in real time, it allows to optimize the operation of the SGEC based on the real operating conditions of the assets, eventually interacting with the electricity system (in particular with the distribution network operator). The hardware for energy flow management and control allows to assign the right operational mode to the energy production/storage/consumption assets within an SGEC, based on the choices made by the software and the real operating conditions of the assets.

The third one refers to technologies that allow to distribute the energy and information flows among the production/storage/consumption assets and the power flow management and control systems within an SGEC. In particular, the former refer to electricity grid and district heating, while the latter refer to wired or wireless communication networks (e.g., optic fiber, power-line communication, and digital subscriber lines) and related protocols (e.g., Ethernet, ZigBee, and Wi-Fi).

The largest part of these technologies is already mature and being implemented, even though individually. A small number of technologies are currently not mature, among which are energy storage systems (at least with reference to the most promising technologies, e.g., lithium-ion batteries or supercapacitors), for both their impact on the overall investment in an SGEC and the essential functional role they hold in this kind of application.

Economics and finance of SGECs are a critical factor in their growth throughout the EU. While it is still in its early paradigm stages of development, the shared and coordinated approach through which they are adopted in the smart green energy community is spreading quite rapidly in the EU. To enable the transition from an "individual" approach to energy management to a "collegial" one, a simultaneous adoption of a set of enabling technologies is required, to produce energy and consume it in an efficient way, as rather widespread today. What is necessary is a widespread adoption of technologies for managing and distributing the energy and information flows between the energy users and other assets within the SGEC.

Given these additional investments—which would not be necessary or would be carried out on a much lower scale in the "individual" approach—thanks to the "collegial" approach, several benefits can be achieved, related to the increased investment size (economies of scale) and to the synergies that can be exploited through the combination of a set of energy users, such as the ability to leverage energy locally that would otherwise be lost or injected into the grid at a lower economic value.

The diffusion of such energy communities that are smart and green depends upon several factors, the most relevant one related to the energy system regulatory framework—in terms of the possibility to develop such aggregation of customers and the eventual incentive mechanisms available to promote them—and to the economic sustainability of the investments, which is influenced by the abovementioned regulatory framework and the enabling technology performance (from both a technical point of view and an economic point of view).

THE CURRENT STATE OF SGECs

The number of SGECs currently under development or operational in Europe is still rather small, although increasingly rapidly as mentioned above. Consider, for example, the case of Italy—which can be considered as a frontrunner in smart grid project development, especially (but not only) thanks to an incentive program set in 2010 by the Italian Regulatory Authority for Electricity, Gas, and Water for the promotion of pilot projects in this field (AEEG, 2010a). The limited diffusion, in absolute terms, of SGECs stems primarily from the fact that the current regulatory framework lacks an overarching definition of SGEC (Energy and Strategy Group, 2014). In particular, there are only two typologies of configurations (i.e., models of SGEC) that are already defined and regulated. They are the so-called internal user networks (belonging to the category of Closed Distribution Systems, defined by the European Directive 2009/72/EC and currently defined and regulated in other European countries) and the Historical Cooperative. Although they are acknowledged in the current regulatory framework and they are rather widespread in Italy (with 73 and 77 cases, respectively), it is not possible to develop new ones (AEEG, 2010b; CONFCOOPERATIVE-FEDERCONSUMO, 2014).

Looking at the most recent initiatives under development in Italy—overall, there are eight cases, at different stages of development—they mainly refer to experimental projects, primarily focused on the industrial and service sectors.

A first relevant example of SGECs in the service sector refers to the "SCUOLA" (Smart Campus as Urban Open Labs) project, started in June 2013 by the Politecnico di Milano in collaboration with the University of Brescia, A2a, and 12 other companies (Delfanti, 2014). The aim is to develop an energy community within a university campus (the Leonardo Campus of the Politecnico di Milano) in order to make it more sustainable and to test innovative technological solutions designed within university laboratories. The total investment—around 10 million euros, partially funded by the Lombardy region (4 million euros) as part of the support to research and development projects for smart cities and energy communities sectors—includes the adoption of a wide set of SGEC-enabling technologies: from energy production plants, from both renewable energy sources and traditional ones (e.g., photovoltaic plants, solar thermal plants, and heat pump), to energy storage systems (in particular, a lithium-ion battery), to electric vehicles and charging infrastructure, in addition to the necessary hardware and software to manage the energy and information flows.

A second relevant example of SGECs in the service sector refers to the "Smart Polygeneration Microgrid" project, started in November 2012 and officially launched in February 2014 by the University of Genoa, in collaboration with Siemens, within the Savona Campus of the same university (Zanellini, 2014). It also aims to create an energy community within a university campus with the aim to experiment the adoption of innovative smart technologies that can potentially be implemented in urban areas and to reduce the energy consumptions and the environmental impact of the campus. The total investment—around 9 million euros, partially funded by the Italian Department of Education (2.4 million euros) as part of the special interventions for the energy sector development—includes the adoption of a wide set of energy community-enabling technologies (as shown in Figure A2): from energy production plants, from both renewable energy sources and traditional ones (e.g., photovoltaic plants, solar thermal plants, micro-gas turbines, and heat pump), to energy storage systems (in particular, a sodium nickel chloride battery), to electric vehicles and charging infrastructure. Finally, hardware and software technological solutions for energy and information flows management are developed and provided by Siemens.

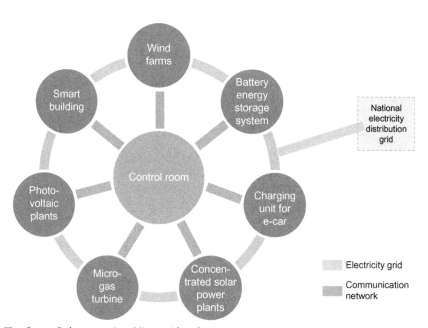

The Smart Polygeneration Microgrid architecture.

A third relevant example of SGECs—which can be classified as "heterogeneous" because it involves residential, industrial, and service applications—refers to the "Leaf Community" project, started in 2008 by the Loccioni Group in collaboration with around 50 partners, that aims to achieve the first integrated energy community in Italy in order to experiment the technical feasibility of this type of initiative (the project is expected to be completed by 2015; Romiti, 2013). The total investment—of several million euros, without any public financial support—includes the adoption of a wide set of SGECs, especially for energy production from renewable energy sources (e.g., photovoltaic plants, mini-hydro power plants, and solar thermal plants) and storage (both battery energy storage systems and chemical energy storage system). The hardware and software to manage the energy and information flows are currently under development. Interestingly, within the Leaf Community, the so-called Leaf House is located, a six-apartment residential building in which several technological solutions for energy production and management are adopted, such as photovoltaic plant, solar thermal plant, geothermal heat pump, thermal and acoustic insulation, building automation systems for the monitoring of household appliances energy consumption and photovoltaic plant energy production, sensors for the living comfort control, rainwater storage system, and energy storage systems (hydrogen fuel cells and lithium-ion batteries, for the storing of electricity in excess produced by the photovoltaic plant, in order to achieve a level of independence from the grid equal to around 20 hours per day). Compared with a traditional house, the energy requirements of the Leaf House for cooling are 30% lower (20 kWh/m^2 compared with 30 kWh/m^2 in a traditional house) and, considering only heating, 70% lower (27 kWh/m^2 compared with 100 kWh/m^2).

The common denominator of the initiatives currently in place in Italy—which involve investments of several million euros—is represented by the close cooperation between research institutions (universities and public research organizations), firms (technology providers and system integrators and utilities), and public administrations (central and local authorities), which often help fund a relevant portion of whole investment. They have been promoted to assess the technical feasibility of multitechnology integration rather than to earn an economic advantage out of them.

Several initiatives are also underway at the European level, among which are the projects in Spain and Germany, countries that—like Italy—have seen comparable dynamics within their energy systems, primarily due to the massive diffusion of the nonprogrammable renewable energy sources, especially photovoltaic and wind power.

SPECIFIC NATIONS IN THE EU

Spain

Two examples in Spain refer to the "i-Sare" project and the "Design, development and implementation of microgrids in Navarre" project. The first one was started during 2012 and launched at the end of 2013 by the GAIA-Cluster TEIC and the IK4 Research Alliance, with the collaboration of several other firms and research institutions located in the Gipuzkoa area, led by the Jema Group (Corzo et al., 2013). The project, developed within an industrial area (the Enertic building located in the industrial zone Polígono 27), aims to foster the integration of various renewable energy sources, through the development and testing of innovative technologies, in order to save energy, reduce costs, and increase the power system reliability. The total investment—around 6.5 million euros, partially funded by the Provincial Council of Gipuzkoa and Ministry of Science and Innovation (4 million euros)—includes the adoption of a wide set of energy community-enabling technologies: from energy production plants (e.g., photovoltaic plants, wind plants, combined heat and power group, diesel generator, and fuel cell), to energy storage systems (batteries, flywheels, and supercapacitors), to electric vehicles and charging infrastructure.

The second project in Spain ("Design, development and implementation of microgrids in Navarre" project) is an example of SGECs in the industrial sector (Aguado Alonso, 2012). Developed by the Renewable Energy Grid Integration Department of CENER (a technology center specialized in applied research and development and the promotion of renewable energies) in collaboration with the Public University of Navarre, the facility is located on the premises of the Wind Turbine Test Laboratory of CENER, in the industrial estate of Rocaforte de Sangüesa (Navarra). The overall project aims to manage the power generated within the community to maximize the self-consumption from renewable energy sources, to increase the power quality for the energy users, and to experiment innovative technologies for energy production and storage and control strategies. The initiative, partially funded by the Government of Navarre and the European Union through its FEDER funds, includes the adoption of several energy community-enabling technologies. Regarding the technology for energy production, it includes photovoltaic plants, wind plant, diesel engine, and gas micro-turbine, in addition to electrical energy storage systems (e.g., lead-acid battery and vanadium flow battery), electric vehicles, and charging infrastructure. Finally, hardware and software to manage the energy and information flows are adopted.

Germany

Considering Germany, among the many SGEC initiatives currently underway, it is worth mentioning the recent "Enka" project as an example of residential SGECs (BOSCH, 2014). Started in mid-2014 by Süwag Erneuerbare Energien GmbH with the collaboration of several other firms like Deutsche Reihenhaus AG (the construction company) and Bosch (the energy storage systems provider) and the scientific support of the Frankfurt University of Applied Sciences, the project—developed within a housing complex, which comprises 180 townhouses in Kelsterbach (near Frankfurt)—aims to make the complex self-sufficient in its energy needs and to reduce the energy bill. The total investment—partially funded by the Hessian Ministry of Economic Affairs, Energy, Transport and Regional Development with funds from the European Regional Development Fund—includes the adoption of a wide set of SGEC-enabling technologies for energy production: photovoltaic plants, combined heat and power group, and peak load boilers. Both electricity (lithium-ion batteries) and thermal storage systems are installed in order to ensure the matching between energy production and consumption. In the near future, the construction of a business park within the area is also expected.

Although overlooked at this stage, the economic feasibility is a prerequisite for the large-scale deployment of SGECs. In order to understand whether and to what extent the SGEC model is applicable, five different archetypes of the community are identified and evaluated, which refers to the different categories of energy users potentially interested in the creation of the SGEC and characterized by different sets of enabling technologies (Energy & Strategy Group, 2014). In particular, our simulations assume that the investments for the implementation of the SGEC are in charge of the energy users within the same *pro-quota* according to each energy need and that none of the enabling technologies are already in place before the setup of the smart green energy community.

Simulations conducted on the SGEC models—supposing the realization in Italy—show very remarkable internal rate of return (IRR) of the investments. The highest returns are in the industrial and service SGEC models, which show an IRR between 8% and 40%, although the payback time (PBT) of the investments are higher than the acceptability threshold usually taken into account by such investors for comparable initiatives (e.g., for investments in energy efficiency). On the other hand, the residential model turns out to be the least sustainable from an economic point of view, with an IRR of around 4% and a PBT of around 15 years.

ECONOMICS: HOW THE SCEG IS FUNDED

Economic sustainability is strongly dependent on the assumptions regarding the regulatory framework (as showed in Tables A1 and A2 with particular reference to the payment of the so-called general charges and network charges, which account for around 40% of the whole electricity bill, slightly variable according to the different categories of energy users). Indeed, the allocation of general charges and network charges on the whole amount of electricity that is consumed by the energy users within the SGEC ("2nd scenario" in Tables A1 and A2)—and on the amount of electricity taken from the grid ("1st scenario" in Tables A1 and A2), that is, excluding the amount of electricity self-produced within the SGEC—has a significant negative impact on the economics of the investments. In particular, this aspect increases the payback time of around 30-50% for the different models (as shown in Table A1) and decreases the internal rate of return of around 40-70% (as shown in Table A2). This effect is particularly emphasized in those cases in which the implementation of the SGEC allows to significantly increase the level of independence from the electricity grid.

SGEC economic evaluation measured through the payback time

SGEC model	PBT 1st scenario (year)	PBT 2nd scenario (year)	IRR variation 1st vs. 2nd scenario (%)
Residential sector	15	19	+27
Service sector: focus on energy expenditure	8	10	+25
Industrial sector: focus on power quality	4	6	+50
Service sector: focus on power quality	6	8	+33
"Urban" sector[a]	8	12	+50

[a]It refers to a heterogeneous aggregation of energy users (residential users and service firms).
Source: Adapted from Energy & Strategy Group (2014).

SGEC economic evaluation measured through the internal rate of return

SGEC model	IRR 1st scenario (%)	IRR 2nd scenario (%)	IRR variation 1st vs. 2nd scenario (%)
Residential sector	4	1	−66
Service sector: focus on energy expenditure	16	10	−37
Industrial sector: focus on power quality	38	21	−45
Service sector: focus on power quality	22	14	−37
"Urban" sector	15	8	−48

Source: Adapted from Energy & Strategy Group (2014).

In addition to the economic perspective, the SGECs may also be assessed from an "energetic" point of view, that is, in terms of the reduction in energy needs for the users within the SGEC (compared to the situation before the implementation of the SGEC) and in terms of the reduction of electricity exchanges with the grid (as the sum of energy withdrawals from the grid and energy injections to the grid). Simulations on the same SGEC models show that, on the one hand, the implementation of the SGEC represents a potential driving force for the energy efficiency promotion. The expected energy savings are on average equal to, or greater than, 10% in different SGEC models. On the other hand, the implementation of the SGECs can reduce the "impact" of the energy users on the electricity grid by an amount equal to or greater than 50% compared to the situation prior to the implementation of the SGEC. The only exception is represented by the residential model, where as a result of the "electrification" of heat consumption—thanks to the adoption of the heat pump and the adoption of photovoltaic plants—the total electricity exchanges with the grid increase, despite the adoption of an energy storage system (Energy & Strategy Group, 2014).

POLICY MAKERS: FROM ELECTED OFFICIALS TO REGULATORS

The spread of the SGECs—to which a huge "theoretical" potential is associated, around 500,000 units just taking into account Italy (Energy & Strategy Group, 2014)—depends on a variety of factors. The first one is for sure, as mentioned above, related to the regulatory framework. It is of paramount importance that policy makers define a supportive regulatory framework, also taking into account the significant benefits that can be achieved at a "systemic" level from the diffusion of SGECs. Such benefits primarily relate to the possibility that SGECs contribute to the electricity system safety and to an increase of the electricity system capability to accommodate growing amount of nonprogrammable renewable energy source plants, which result in a reduction of several cost items that are currently afforded by the electricity system, for example, the costs for the dispatching activity incurred by the network operators for the supply of balancing services or investments for transmission and distribution network infrastructures development. Second, there are other types of benefits that can be achieved, particularly relevant at the European level for their potential impact on the European competitiveness. They refer to the reduction of energy dependence from abroad—considering that the current European energy dependence is more than 50% of its overall energy need (EUROSTAT, 2013)—the reduction of polluting emissions (thanks the possibility to further increase the spread of renewable energy sources and of technologies for energy efficiency), and the development of national value chains around the technologies involved in any SGEC.

Based on of the abovementioned benefits, it may also be reasonable that the policy maker will define specific supporting mechanisms to foster the SGECs diffusion, especially for those characterized by the most significant "systemic" impacts, which are less sustainable from an economic standpoint.

In addition to the policy maker's role, it is equally important that energy users overcome the barriers that hinder the development of these SGECs (Energy & Strategy Group, 2014). The first type of barrier is related to a lack of awareness, which currently prevents them from a rational evaluation of the benefits of a shared approach to energy management. The second one refers to the ability to raise the financial resources that are necessary to establish any SGEC, both internally or involving third parties, such as financial institutions. Finally, the third one is related to the ability of energy users to make common decisions and their stability over time. The three barriers have different relative importance depending on the categories of energy users considered; however, the most critical one across the categories of end users is related to the raising of financial resources. In this direction, other stakeholders may play an important role, especially financial institutions and energy service companies.

An emerging model for the realization of SGECs that is capturing growing interest refers indeed to the so-called Microgrid-As-A-Service. According to this model, similar to the *modus operandi* of the energy service company in the energy efficiency sector, a third party (i.e., a subject outside the community) takes care of the development of the SGEC—including the raising of the necessary financial resources—and the subsequent management of the same. Such third party can be, for example, energy service companies or other players involved in the energy industry like the so-called aggregator, another emerging player within this sector, which is currently defined as a demand service provider that combines multiple short-duration consumer loads for sale or auction in organized energy markets (European Parliament and the Council, 2012).

From the point of view of banks—which have provided substantial amounts of funding for investments in renewable energy and, to a lesser extent, in energy efficiency (e.g., about 25 billion euros between 2007 and 2012 in Italy)—the SGEC represents emerging business opportunities, also because of an expected contraction of their investment in other "green" opportunities such as renewable energy plants. However, they perceive it only as an opportunity for the medium term and long term. This is due, on one hand, to the incompleteness and instability of the regulatory framework and, on the other hand, to a series of critical issues that characterize this business, some of which are similar to those already found for the energy efficiency financing: in particular, the guarantee of stability over time of the "target" activity (which, unlike the "traditional" energy efficiency measures, in the case of SGECs, concerns a cluster of energy users instead of a single entity) and the technical evaluation of the investment in an SGEC (which, unlike the "traditional" energy efficiency measures, concerns the joint adoption of several technologies).

Finally, the technological issue does not appear to be the most critical one for the SGEC diffusion, since the majority of the enabling technologies are already mature, as discussed above. Nevertheless, technology providers—with particular reference to the technologies that are not yet mature, such as energy storage systems and some other technologies for small-scale energy production like mini-hydro, mini-wind, and mini-/micro-CHP—are required to reduce investment and operational costs of such technologies and to improve technical performance in order to ultimately improve the economics of investing in an SGEC.

CONCLUSIONS

The smart green energy communities (SGECs), defined as consumers who decide to make common choices related to their energy needs satisfaction, represent one of the new key actors of the increasingly important smart grid paradigm. The novelty underlying the SGEC approach lies in the transition of the energy users—for example, residential users, industrial firms, or service firms—from an "individual" approach regarding the energy management to a "collegial" one, which enables to achieve potential benefits like a reduction in energy expenditure and an improvement of the power quality, through the implementation of technological solutions to be "smart" in generating and managing energy.

The diffusion of SGECs depends upon several factors, primarily related to the energy system regulatory framework and the economic sustainability of the investments. Regarding the first one, it should be amended in order to give the possibility to develop such aggregation of customers, also considering the significant benefits that can be achieved at a "systemic" level from the diffusion of SGEC, like the reduction of energy dependence from abroad. Regarding the second one, which is influenced by the technical and economic performance of the enabling technological solutions—the largest part of them is already mature and being implemented, even though individually—and by the regulatory framework as well, simulations conducted (supposing the realization in Italy) show very remarkable internal rate of return of the investments, the highest ones observed in the industrial and service SGEC models, with a strong impact related to the assumptions regarding the regulatory framework (with particular reference to the payment of the so-called general charges and network charges).

The number of SGECs currently under development or operational in Europe is still rather small, mainly related to experimental projects aiming to assess the technical feasibility of multitechnology integration. In order to enable a widespread diffusion of SGEC, in addition to the policy makers' and technology providers' roles (with particular reference to technologies that are not yet mature, like energy storage systems), it is also important that energy users overcome the barriers that hinder the development of these SGEC: (i) a lack of awareness regarding SGEC potential benefits, (ii) the ability to raise the financial resources to develop an SGEC, and (iii) the ability to make common decisions and to be stable over time. To raise the necessary financial resources—which appears to be the most critical issue—third parties can be involved, for example, banks, or an innovative implementation model can be taken into account, like the so-called Microgrid-As-A-Service.

REFERENCES

AEEG. 2010a. Delibera ARG/elt 39/10. Home Office, Milan.
AEEG. 2010b. Delibera ARG/elt 52/10. Home Office, Milan.

Aguado Alonso, M. 2012. An example for microgrid in national renewable industrial environment: CENER's microgrid [PowerPoint Presentation]. Microgrid—A building block for smart grids, 27 January, Brussels.

BOSCH. 2014. Independently-generated power: optimum energy supply. Bosch supplies energy storage system for pioneering housing complex. [Press Release] (accessed 27.06.14). Available from: http://www.bosch-presse.de/presseforum/details.htm?locale=en&txtID=6641.

CONFCOOPERATIVE-FEDERCONSUMO. 2014. FEDERCONSUMO in cifre [Leaflet]. CONFCOOPERATIVE, Trento.

Corzo, L.G., Cerro, I., Sansinenea, E., Santamaria, G., Zubizarreta, J., Arrizubieta, L., 2013. i-Sare. The future grid. In: International Conference on Renewable Energies and Power Quality (ICREPQ'13), 20-22th March 2013, Bilbao.

Delfanti, M. 2014. Cosa sta facendo il Politecnico: il caso del campus Leonardo [PowerPoint Presentation]. Stati Generali dell'Efficienza Energetica: al via la consultazione. Il contributo dell'efficienza energetica al Sistema Paese, 6th May, Politecnico di Milano.

Energy & Strategy Group, 2013. Energy Efficiency Report, third ed. Politecnico di Milano—Dipartimento di Ingegneria Gestionale—Collana Quaderni AIP, Milan.

Energy & Strategy Group, 2014. Smart Grid Report, third ed. Politecnico di Milano—Dipartimento di Ingegneria Gestionale—Collana Quaderni AIP, Milan.

European Parliament and the Council Directive 2012/27/EC of 25 October 2012 on energy efficiency, amending Directives 2009/125/EC and 2010/30/EU and repealing Directives 2004/8/EC and 2006/32/EC.

EUROSTAT, 2013. Energy, Transport and Environment Indicators, 2012 ed. European Union, Luxembourg.

Romiti, G. 2013. Adattamenti urbani oltre il retrofit edilizio. Soluzioni Tecnologiche [PowerPoint Presentation]. The Innovation Cloud, 9 May, Fiera Milano.

Zanellini, F., 2014. Un esempio di microgrid: la Smart Polygeneration Grid di Savona. Qualenergia, University of Genoa, Italy. Available from: http://www.qualenergia.it/sites/default/files/articolo-doc/smart-polygeneration-grid_savona_zanellini_0.pdf (accessed 04.07.14).

Potential of Offshore Wind in the Republic of Mauritius

Mohammad Khalil Elahee[1,*], Ackshay Panray Jungbadoor[1]

[1]Mohammad Khalil Elahee and Ackshay Panray Jungbadoor, Faculty of Engineering, University of Mauritius, Reduit, Mauritius.
*Corresponding author: elahee@uom.ac.mu

INTRODUCTION

Offshore wind is a growing industry in the field of renewable energy and is subject to a fueling interest around the globe for energy production. The United Kingdom and Denmark are leaders in terms of installed capacity with 1.34 and 0.85 GW, respectively, in 2010 (GBI Research, 2011). In all, the offshore installed wind capacity is currently near 3.9 GW and a forecast of 9.8 GW installed worldwide by the end of 2014 is expected.

The islands in the Republic of Mauritius are found to be largely exposed to high wind speeds found in the Indian Ocean. During summer, there is an average wind power density of about 450 W/m^2 for Mauritius, but in winter, offshore wind power density reaches up to 800 W/m^2 (Jet Propulsion Lab (JPL), 2008). Embracing the worldwide green policy in energy production, the untapped energy source of offshore wind could reduce the reliance on imported fossil fuels and their volatile prices while securing affordable energy for local consumers (Ministry of Renewable Energy and Public Utilities (MoREPU), 2009).

Energy Demand in the Republic of Mauritius

The CEB and IPP cater to the electricity demand of Mauritius. In 2011, 1096 GWh, representing 44.9% of the total electricity generated, came from thermal and hydro power plants of the CEB (2013a). The main sources of electricity in Rodrigues, from the CEB, come from thermal power plants and wind farms. The total wind farm capacity at Trefles and Mont Grenade, in Rodrigues, sums up to 1.3 MW and the whole thermal capacity is 11.4 MW. In 2011, about 3 GWh of electricity in Rodrigues was obtained from the wind farms and the remaining 30 GWh was from fossil fuels (CEB, 2013b).

Agalega and St Brandon islands have been excluded from this study as an offshore wind farm because these islands would not be economically feasible due to the low energy demand. Agaléga has a net population of 285 according to the 2011 Housing Census (Economic and Social Indicators (ESI), 2011) and feasibility studies by the MRC have shown that the best sources of renewable energy for the islands are solar and biofuel (Tylamma, n.d.). St. Brandon is mainly used as a fishing base and does not have any permanent residents despite having about 63 fishermen present on the island (Central Statistics Office (CSO), 2001).

So, a potential offshore wind farm in the Republic of Mauritius would be considered only for Mauritius and Rodrigues.

IMPACTS OF OFFSHORE WIND FARMS IN THE REPUBLIC OF MAURITIUS

In spite of supplying considerable energy in a sustainable way, offshore wind energy has numerous environmental and social impacts. The main impacts of a potential offshore wind farm in the Republic of Mauritius are hence assessed.

Visual Impact

With the growing land-based wind farm projects around the globe, visual issues are often the main reason of public objection (DTI, 2005). Offshore wind farms situated far from the coastline do away with this problem as compared with onshore wind farms. However, apart from the distance offshore, the wind farms are also visually influenced by the colors of the wind turbine. The distance from the coastline determines the level of perception of offshore wind farms as depicted in Table A1.

In order to further reduce the visual impact of an offshore wind farm, the most appropriate color need to be selected for the wind turbine. Thus, the

Table A1 Visual Impact of a Wind Farm (University of Newcastle (UON), 2002)

Distance (km)	Perception
Up to 2	Likely to be a prominent feature
2–5	Relatively prominent
5–15	Only prominent in clear visibility
15–30	Only seen in clear visibility

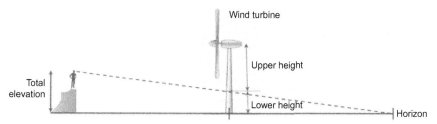

Figure A1 Lower and upper wind turbine height.

apparent height of the wind turbine is to be determined when viewed from the coast, as well as the height before the predominant background of the wind farm to find out the color required (Figure A1).

Apparent Height

When the distance between the wind turbine and the viewer increases, the wind turbine appears to be smaller as the visual angle is reduced too. This implies that from the coast, the wind turbine would have an apparent height that is much smaller than its actual height. In order to determine the visual effect of the wind turbine, this apparent height has to be determined.

Assuming that the lower section of the offshore wind turbine would merge with the marine environment, the apparent height of the upper part is calculated according to the equation (Anon, 2013):

$$h = \frac{a}{d} \tag{A1}$$

where h is the apparent height ratio, a the actual size of the object (m), and d the distance of the object (m).

Noise Impact

As offshore wind farms are sited far from human population, the noise generated has nearly no effect on the people. Nevertheless, marine life is more susceptible to the noise generated by wind farms during construction, operation, or decommissioning. The maximum noise emitted during construction phases reaches about 196 dB. This noise level is temporary and would last only during pile drilling. For the operational stage, there would be a steady noise emission in the range of 90 dB at the hub height (EWEA, 2009).

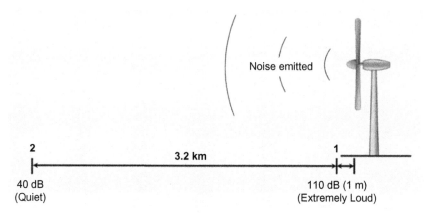

Figure A2 Distance to dissipate 110 dB emitted during operational phase.

Distance to Dissipate Noise Emitted During Operation

Assuming a maximum sound level of 110 dB emitted by the wind turbine at point 1 (Figure A2), the distance for the sound level to reach 40 dB (quiet) at point 2 is found to be 3.2 km.

Impact on the Wildlife

Fish

The islands in the Republic of Mauritius form part of the fishing banks in the Indian Ocean, which stands for 70% of the total fish production for direct consumption in Mauritius (BMT Cordah Limited, 2010). With the high abundance of fish in the fishing banks, fish aggregating devices (FADs) are placed off the coast to attract pelagic fishes such as tuna, dolphin fish, wahoo, and marlin (Anon, 1998). The foundations of the wind turbines consist of several components extending from the seabed to the sea level and would hence act as FADs and artificial reefs (Wilhelmsson et al., 2012).

Birds

Offshore wind farms pose as a huge threat to birds during their flight, leading to mortality. The main causes that contribute to the death rate are

- direct collisions with the blades of the turbines during flight and
- disruption of feeding grounds and migration routes due to offshore wind farms (Wilson, 2007).

Mauritius forms part of the West Asian-East African Flyway, which is an intermediate resting and feeding country for migratory birds that annually move from breeding to nonbreeding grounds (The Partnership for the East

Asian-Australasian Flyway (EAAFP), 2010). The resting area for the migratory birds in Mauritius is Rivulet Terre Rouge Estuary Bird Sanctuary, located northwest of the island. So, the appropriate mitigation procedure should be considered during siting to avoid disrupting the migratory route.

Environmental Impact

Offshore wind energy provides a safe, environment-friendly, and competitive alternative for conventional sources of energy. The exploitation of this renewable energy source could meet many environmental and energy policy goals like the reduction of greenhouse gases. According to a study from EWEA (2011), it is predicted that by 2020, offshore wind energy would meet over 4% of the European Union (EU) total energy demand with 148 TWh annually. In the process, up to 102 million tons of CO_2 emissions would be avoided (Figure A3).

Cost Breakdown

Even with greater wind resource availability offshore that would eventually payback the investment costs, the costs are, however, higher than onshore wind farms. The approximate cost of energy (in MRs/MW), from recent offshore wind farms, is illustrated in Figure A4. The development of offshore wind farms depends on several factors in the construction and operation and maintenance phases that contribute to the high overall capital cost of the projects.

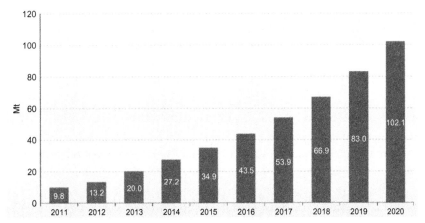

Figure A3 CO_2 emissions avoided in offshore wind farms installed in 2011-2020 (EWEA, 2011).

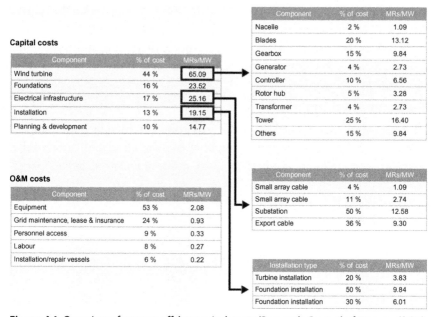

Figure A4 Overview of current offshore wind costs (Research Council of Norway (RCN), 2010).

INTERFERENCE AND EXTREME WEATHER CONDITIONS

Cyclonic Risks to Offshore Wind Farms

Offshore wind energy is garnering interest around the world and has been widely exploited in Europe. However, as compared with Europe, certain countries experience extreme weather conditions that would likely reduce the life span of an offshore wind farm. This is particularly the case for countries lying in the Indian Ocean, which is a cyclonic zone. Hence, wind farms sited in the open sea around Mauritius or Rodrigues would be highly vulnerable to tropical cyclones. Due to cyclone intensities, the wind turbines selected play a vital role. They depend on the wind classes defined by the International Electrotechnical Commission (IEC) standard and are considered according to the wind parameters at a site. So, a wind turbine would be based on the average annual wind speed and the speed of extreme gusts for a period of 50 years, mainly due to cyclones (Table A2) (Vestas, n.d.b).

Offshore wind turbines are designed for extreme wind conditions, up to 70 m/s (252 km/h). In the southwest Indian Ocean, very intense tropical cyclones usually have wind speed over 59 m/s (212.4 km/h) and the IEC I wind turbines can cope with such situations.

Table A2 IEC Wind Turbine Classes Versus Tropical Cyclone Intensities (Vestas, 2011; Météo-France, 2012)
1-min Sustained Winds

m/s	km/h	Type	Turbine Class
<17	<61.2	Tropical depression	IEC I, II, III
17-32	61.2-115.2	Moderate to severe tropical storm	IEC I, II, III
33-42	118.8-151.2	Tropical cyclone	IEC I, II, III
43-49	154.8-176.4	Tropical to intense tropical cyclone	IEC I, II
50-58	180-208.8	Intense tropical cyclone	IEC I, II
59-70	212.4-252	Very intense tropical cyclone	IEC I
>70	>252	Very intense tropical cyclone	–

Tropical Cyclones in the Republic of Mauritius

Offshore wind farms in cyclonic region have higher probability of encountering damages due to extreme wind speed despite the use of mitigation techniques. Passive and active techniques may help to reduce harmful consequences but cyclones would nevertheless be a huge threat. In order to eliminate or at least reduce the risk associated with cyclones, offshore wind farms should be sited in regions that are less likely to be hit by tropical cyclones. So, an assessment of cyclonic risk in the Republic of Mauritius to determine the best region for a potential offshore wind farm is carried out.

Wind Gusts

In Mauritius, the maximum cyclonic gust recorded was 280 km/h, in 1975, and for Rodrigues, no cyclones had attained 254 km/h since tropical cyclone Fabienne, in 1972. Also, no cyclonic gusts exceeding 221 km/h (61.4 m/s) have been recorded after 1979 for both Mauritius and Rodrigues.

Frequency and Direction of Incoming Tropical Cyclones

The regions experiencing the most tropical cyclones approaching Mauritius and Rodrigues over the last 52 years are determined as well to find the most appropriate location for the wind farm. The northwest and southeast regions for Mauritius and Rodrigues are continuously battered by tropical cyclones over the years. The only major tropical cyclone that approached Mauritius, in the south, dated back to 1989. The cyclone had wind gusts of 150 km/h (41.6 m/s), which is within the design parameters of most wind turbines, nowadays. So, an offshore wind farm is deemed to be most suitable in the south of Mauritius and southwest of Rodrigues to reduce the cyclonic risk associated with their locations (Figure A5).

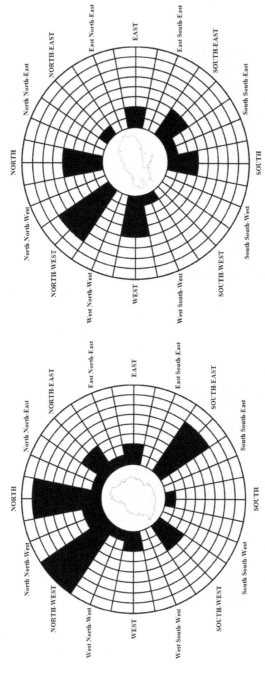

Figure A5 Direction and frequency of major tropical cyclones in Mauritius and Rodrigues.

Interference with the Aviation Industry

Siting an offshore wind farm near an airport may interfere with the aviation industry. A prominent example is the Bigara onshore wind park in Mauritius, which had to be resited twice due to the potential interference with aircraft communication and landing systems (Hansard, 2012). This has lead to a waste of resources as the capacity of the wind farm is now 29.4 MW at Plaine Sophie, while it was initially supposed to be about 40 MW at Bigara.

Hence, for the airport at Grand Port (Mauritius) and Plaine Corail (Rodrigues), the three problems associated with a potential offshore wind farm are obstruction, radar interference, and electromagnetic interference (Table A3).

According to the Plaisance Airport (Building Restrictions) Act 1964 and considering the major problems associated with the aviation industry in the Republic of Mauritius, an offshore wind farm (160-m wind turbine) should be located (Figure A6).

1. beyond 17.2 km, from the runway, along the interference line, and
2. beyond a radius of 11 km from the airport control tower.

Table A3 Noninterference Zones of Problems Associated with Aviation Industry

Problem	Noninterference Zone
Obstruction	The potential location of the offshore wind farm should be beyond 6.3 km from the runway, which is the minimum takeoff horizontal distance of an aircraft
Radar interference	Radar interference is not an issue in the Republic of Mauritius as no radar is used
Electromagnetic interference	
• Near-field effects	The maximum near-field distance of commonly used radio waves and microwaves is found to be approximately 720 m from a telecommunication tower, and hence, a distance of 1 km is assumed to avoid near-field effects
• Shadows	At a transmission distance of about 50 km, a 160-m high wind turbine would cast a shadow of 1224 m, but as the cruise altitude is about 10 km, vertical shadow effect would not cause any problem. Similarly, horizontal shadow zones would also have no impact due to the cruise altitude
• Reflection/scattering	So as to avoid signal scattering and reflection, a wind turbine (160 m high) should be sited 14.1 km from the airport to provide a second Fresnel zone radius of 167 m

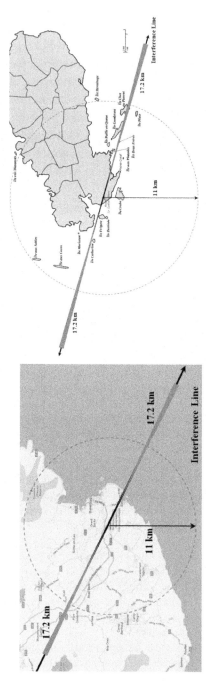

Figure A6 Minimum distance from the airport for siting a wind farm in Mauritius and Rodrigues (Google Maps, 2012; Kingroyos, 2011).

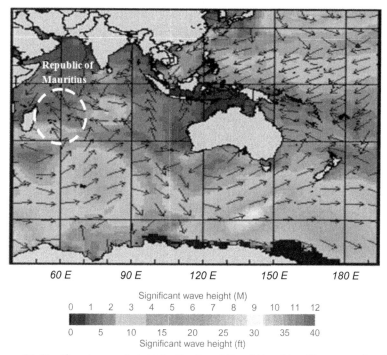

Figure A7 Significant wave height in the Republic of Mauritius (Oceanweather Inc., 2012).

Wave Loads

Wave heights increase in areas of strong winds that blow in a particular direction, such as the trade winds in the south of the Indian Ocean where the Mascarenes Islands are located (Anthoni, 2000). According to Figure A7, the significant wave height for Mauritius and Rodrigues is about 2-2.5 m.

Extreme Waves
Cyclonic Waves

The Mascarenes Islands is prone to tropical cyclones originating in the tropics and subtropics throughout the year. Tropical cyclones approaching Mauritius and Rodrigues generate swell waves, and the wave heights produced are even greater due to a phenomenon called raz-de-marée. Based on the previous raz-de-marées affecting Mauritius and Rodrigues, the probability of occurrence of swell waves bearing a height 4 m and beyond 6 m is 0.5 and 0.3, respectively. In general, all the raz-de-marées are created near Crozet Island, and therefore, the southwestern coasts of Mauritius and

Rodrigues are the most vulnerable. A maximum significant wave height of 7.0 m is therefore assumed to affect Mauritius and Rodrigues due to the coupling effect of subpolar lows and cyclones.

Tsunami

Tsunamis consist of waves formed due to the deformation of the seafloor by earthquakes. The 2004 Indian Ocean earthquake that took place off the western coast of Sumatra had a moment magnitude scale of 9.1 and released about 1.1×10^{17} J of energy. As a result, the waves traveled several kilometers across the planet and crashed across the shore of various countries, with wave heights up to 10 m (U.S. Geological Survey (USGS), 2004). Mauritius and Rodrigues were spared from the Indian Ocean tsunami as the waves generated had significantly decreased in height and speed before striking the coast. The maximum wave height of the Indian Ocean tsunami was about 0.4-0.5 m around Mauritius and Rodrigues, and this wave height is much lower compared with the 20 m waves that struck South Asia. Moreover, until now, there has been no record of any considerable tsunami affecting Mauritius and Rodrigues (Mauritius Meteorological Services (MMS), 2010).

Effect of Wave Loads

Breaking waves have the highest amplitude and dissipate the greatest loads. So, only the impact of breaking wave on a wind turbine in the Republic of Mauritius is calculated under extreme wave conditions. For a maximum significant wave height of 7 m, the breaking wave produced is 420.6 kN and its equivalent wind load on the foundation, 17 m above the sea level, would occur for an offshore wind speed of 56 m/s. As stated in Section "Cyclonic Risks to Offshore Wind Farms", there are different offshore wind turbine classes that can withstand such wind speed. Similarly, for a monopile foundation of 7 m diameter, it would be able to withstand breaking wave up to 10.9 m high.

SITING

Wind Resource in the Republic of Mauritius

In the Republic of Mauritius, the most comprehensive data for wind speed come from the Mauritius Meteorological Services (MMS). The weather prediction maps and data available can be used to determine the wind resource potential over the geographic areas of Mauritius and Rodrigues. Normally, offshore wind speeds are recorded by anemometers from offshore meteorologic stations, weather buoys, and ocean vessels (Stephens, 2000) but, when it comes to the open sea, there is a lack of consistent wind data from the MMS. The only wind data available are from wind stations located around the islands, and hence, wind measurements near the coastal regions are considered to estimate the offshore wind speed.

According to the wind rose plotted for Mauritius, the wind speed during the last 5 years is found to be much higher in the southern part of the island. Thus, only the wind speeds in the south, southeast, and southwest of Mauritius are considered to draw the maximum power output for an offshore wind farm. However, the seabed drops to about 86 m, a few meters off the coastline, in the south and southwest of the island at Souillac and Le Morne, respectively. In the southeast, at Plaisance, the sea depth is in the range of 30-40 m, beyond 5 km from the coast. The sea depth at Pointe Canon is also less than 40 m (Albion Fisheries Research Centre (AFRC), 1996).

As a result, only the scenario at Plaisance (scenario 1) and Pointe Canon (scenario 2) would be assessed in this study since the present technology for offshore wind foundations can only cope with water depth up to 45 m.

Offshore Wind Speed in the Republic of Mauritius

The wind speeds obtained from on-land wind stations would differ from the actual offshore wind speeds that are less turbulent (Vestas, n.d.a). Based on various measurements over land and water wind speeds, Stephens (2000) suggested that offshore wind can be correlated with onshore wind by the equation

$$U_{SEA} = 1.62(m/s) + 1.17U_{LAND} \qquad (A2)$$

where U_{SEA} is the offshore wind speed (m/s) and U_{LAND} is the onshore wind speed (m/s).

This equation, however, excludes factors such as roughness, topography, and fetch (Stephens, 2000). The wind speeds obtained from the MMS (Mauritius Meteorological Services (MMS), 2012) are therefore correlated to determine the offshore wind speed for Plaisance and Pointe Canon.

As portrayed in Figure A8, the wind speed from wind stations along the coast is correlated with the offshore wind speed, at the same elevation, and as a result, the offshore wind speed at the required hub height can be determined for each site.

Mean Offshore Wind Speed

The mean offshore wind speed at each site is modeled using a Weibull distribution, assuming a 90-m hub height (Table A4).

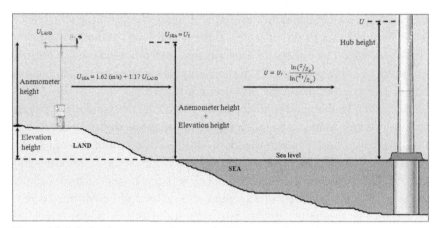

Figure A8 Relation between onshore and offshore wind speeds.

Table A4 Annual Long-Term Monthly Average Offshore Wind Speed

	Plaisance (Scenario 1)	Pointe Canon (Scenario 2)
Elevation (m)	50	58
Anemometer height (m)	10	10
Total elevation (m)	60	68
U_{SEA} at total elevation (m/s)	6.42	8.42
U at hub height (m/s)	6.63	8.72

Wind Turbine Selection

The offshore wind turbines considered are from the top manufacturers around the world. The selection criteria are based on the cost of the wind turbines, capacity factor, and cost of power generated. The best offshore wind turbine for both scenarios 1 and 2 is found to be Vestas V112–3.0 MW offshore wind turbine owing to its highest capacity factor of 0.60 and lowest cost of power output of 109.3 Rs/W.

Site Assessment

The various drawbacks and benefits associated with the two scenarios are assessed to determine the best location for an offshore wind farm in the Republic of Mauritius. The factors considered are as follows:
- Location impacts
- Organization
- Cost and power output

Comparative Analysis

A comparative analysis is presented in Table A5 to evaluate the two scenarios and to determine the potential location for an offshore wind farm in the Republic of Mauritius. The economic analysis (Section "Economic Analysis") is taken into account as well.

The two scenarios vary considerably but scenario 2 has higher economic and energy output. The economic payback period of scenario 2 is below 50% than that of scenario 1, with a net revenue of over Rs 18.5 billion. According to the wind speeds experienced at the site in the 2010 and 2011, scenario 2 is

Table A5 Comparative Analysis for Scenarios 1 and 2

	Plaisance (Scenario 1)	Pointe Canon (Scenario 2)
Site analysis		
Offshore wind speed	6.63 m/s	8.72 m/s
Distance from the coast	3.81 km	6.93 km
Wind turbine	Vestas V112–3.0 MW	Vestas V112–3.0 MW
Capacity factor	0.27	0.60
Impacts		
Turbulence	Low	Low (offshore)
Aviation	Outside interference zone	Outside interference zone
Navigation	None	Safety distance of 1.85 km kept
Cyclone frequency	Low (east southeast)	Low (north northeast)
Visual impact	Prominent in clear visibility only	Lower than ships at harbor
Population	Low	Low
Birds	None	None
Fishing industry	Outside protected area	Outside protected area
Organization		
Number of wind turbines	28	13
Capacity	78 MW	39 MW
Economic analysis		
Total cost	Rs 15.6 billion	Rs 7.8 billion
Cost of electricity	Rs 4.30/kWh	Rs 1.99/kWh
Economic payback	20 years	9 years
Net revenue	Rs 399 million	Rs 18.5 billion
Energy analysis		
Electricity output	181.6 GWh/year	196.5 GWh/year
Capacity factor	26.6%	57.5%

assumed to have a high capacity factor of 57.5% as it is the case for Burradale and Horns Rev 2 wind farms, with 57.9% and 46.7% capacity factors, respectively (International Energy Agency (IEA) Wind, 2012). Hence, scenario 2 is deemed to be the best location for the offshore wind farm.

ECONOMIC ANALYSIS

In order to determine whether the proposed scenarios are economically feasible, the payback period and revenue of each scenario are evaluated. Assuming a life span of 20 years and a similar annual power generated by the wind turbines throughout the lifetime of the offshore wind farm due to yearly maintenance, the economic payback period is evaluated.

Electricity Cost

Due the fact that 83.8% of the energy consumed in the Republic of Mauritius came from nonrenewable sources in 2011 (Ministry of Finance and Economic Development (MOFED), 2011), the electricity tariff would eventually increase in the upcoming years with the depletion and extensive demand of these resources around the world. Hence, the cost of electricity for the next 20 years is predicted based on the sales of electricity tariff over the last decade to predict future electricity tariffs (Table A6).

Payback Period

The payback period is determined by considering the total energy generated annually and by predicting the revenue per kilowatt-hour obtained from the sales electricity for the upcoming years. Initially, the remaining cost of the offshore wind farm would be equal to the total cost of the system, but as the years go by, the value would decrease until it becomes zero. This period is known as the payback period. Once the total cost of the system has been recovered, the corresponding revenue can be determined. By the end of the 20th year, the net revenue of the system would be generated.

The payback period is 20 years with a net revenue of Rs 399 Million for scenario 1 and 9 years with net revenue of Rs 18,524 Million for scenario 2 (Figure A9).

Table A6 Energy Generated and the Corresponding Tariff for the Different Scenarios

Scenario	Total Energy Generated (GWh)		Total Cost (MRs)	Cost of Energy Generated (Rs/kWh)
Scenario 1	181.6	3631.8	15.6 billion	4.30
Scenario 2	196.5	3929.4	7.8 billion	1.99

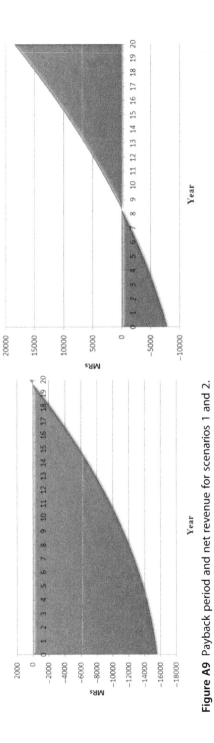

Figure A9 Payback period and net revenue for scenarios 1 and 2.

Out of the two scenarios, scenario 2 is more feasible economically, due to the smaller payback period and higher revenue. However, the economic payback period can be further brought down by proposing additional mechanisms such as a polynomial variation of electricity prices and with the carbon finance mechanism. The outcome is a new economic payback period of 7 years, with a net revenue of Rs 41.0 billion.

ENERGY ANALYSIS

Life-Cycle Assessment

Scenario 2 was deemed to be the best alternative in terms of economy and energy output. However, the environmental issues of the proposed 39 MW offshore wind farm at Pointe Canon need to be assessed. Thus, an LCA, having objectives of evaluating the potential environmental impacts associated with the entire lifetime of the system and determining the energy payback period, is carried out. The whole life cycle of the wind farm is considered during the study, that is, from the raw material extraction stage to the end-of-life stage. The maintenance processes are also taken into account as they form part of the life cycle too (Figure A10).

Life Cycle Impact Assessment

According to ISO 14044, the life cycle impact assessment (LCIA) is aimed at evaluating the implication of potential environmental impacts associated to a system over its lifetime. The LCIA is carried out by Sustainable Minds that is composed of an eco-concept modeling and life-cycle assessment (LCA) software.

The functional unit selected for the offshore wind farm is 1 kWh of electricity generated as this would ease comparison between other LCA results

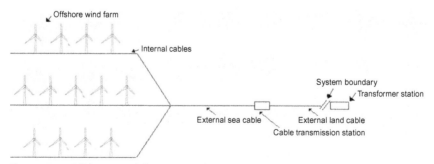

Figure A10 Offshore wind farm model for the LCA.

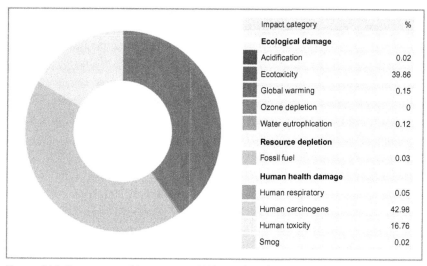

Impact category	%
Ecological damage	
Acidification	0.02
Ecotoxicity	39.86
Global warming	0.15
Ozone depletion	0
Water eutrophication	0.12
Resource depletion	
Fossil fuel	0.03
Human health damage	
Human respiratory	0.05
Human carcinogens	42.98
Human toxicity	16.76
Smog	0.02

Figure A11 Percentage of impacts per kWh produced (Sustainable Minds, 2013).

from Sustainable Minds. With an expected 3.93 TWh of electricity generated over the lifetime of the wind farm, the impacts per functional unit are therefore 0.06 mPts/kWh (Figure A11).

According to the LCIA performed through Sustainable Minds database, human carcinogen is found to produce the highest environmental impact, followed by ecotoxicity and human toxicity, due to incineration and land filling of some components during the end-of-life phase of the system. Based on the functional unit, the impacts of the offshore wind farm are much lower as compared with other sources of energy such as natural gas power (0.14 mPts), photovoltaic electricity (0.47 mPts), and nuclear electricity (0.22 mPts). This implies an even smaller effect on the ecology and the human health.

Carbon Credit

In 2005, the Clean Development Mechanism (CDM) was set up, under the Kyoto Protocol, allowing industrialized countries to reduce their emissions of greenhouse gases by funding sustainable projects in developing countries. Such projects earn a certified emission reduction unit that is generated for each equivalent ton of carbon dioxide reduced (Ministry of Environment and National Development Unit (MOENDU), n.d.a).

With an emission factor of 0.95 $kgCO_2eq/kWh$ for Mauritius (Ministry of Environment and National Development Unit (MOENDU), n.d.b) and a carbon footprint of 6.20 g/kWh (Sustainable Minds, 2013) for the proposed 39 MW offshore wind farm at Pointe Canon, the CO_2 equivalent is calculated. The offshore wind farm is expected to thwart the emission of 3,708,559 tons of carbon dioxide, and according to the CDM of the Kyoto Protocol, it would be eligible for carbon credits. For CDM projects in Africa (Ventures Africa, 2012), carbon credits are worth about €4/ton. Assuming the project is able to claim carbon credits at Rs 152.92/ton (Currency UK, 2013), it would be able to benefit from the carbon finance scheme and liable to Rs 567.11 Million.

Energy Payback

In order to determine the total energy consumed, GaBi 6 software is used to find the energy associated with the different phases of the offshore wind farm. The results are listed in Table A7.

According to the LCA of the 39 MW offshore wind farm, the energy payback period is found to be 4.5 months.

RECOMMENDATIONS

In 2011, the total energy consumed in Rodrigues was 33 GWh. However, the proposed wind farm at Pointe Canon was based on the optimum off-shore energy that could be harnessed and the result was about 196 GWh yearly energy output. In order to cope with this surplus energy output, the following scenarios are proposed:

Table A7 Life-Cycle Inventory Data of Energy Consumed to Produce 1 kWh (GaBi, 2013)

Resource Consumption	Wind Farm (MJ)	Transport (MJ)	End of Life (MJ)	Total (MJ)
Nonrenewable energy				
Crude oil	6.88×10^{-3}	2.57×10^{-2}	-2.54×10^{-5}	3.26×10^{-2}
Hard coal	2.26×10^{-2}	–	-1.14×10^{-3}	2.15×10^{-2}
Lignite	6.28×10^{-3}	–	6.79×10^{-5}	6.34×10^{-3}
Natural gas	7.03×10^{-3}	–	7.79×10^{-5}	7.11×10^{-3}
Renewable energy				
Renewable fuels	1.52×10^{-5}	–	-1.60×10^{-8}	1.52×10^{-5}
Wood	5.63×10^{-5}	–	3.60×10^{-8}	5.63×10^{-5}
Total energy consumed (MJ)				6.77×10^{-2}

$$\text{Energy payback period} = \frac{\text{Energy consumed for 20 years}}{\text{Energy produced/year}} \qquad (A3)$$

Scenario A: Downscaling of the Wind Farm Capacity

The annual energy generated by the offshore wind farm exceeds the yearly electricity demand of Rodrigues due its capacity. So, the same location is considered with a reduced capacity to provide an energy output of 30 GWh per year.

Scenario B: Exportation of Excess Energy

With the emerging technology of submarine power cables, the additional energy generated from the wind farm could be exported to Mauritius, where about 41% of the 2730 GWh annual energy generated comes from the CEB.

Scenario C: Meeting the Future Energy Demand

In the upcoming years, the energy demand would eventually increase and the actual wind farm could be downsized by a smaller degree to meet the future energy demand of the island, hence going toward to a new Rodrigues island.

Scenario D: New location for Offshore Wind Farm

The trade winds blow in the south eastern direction over both Mauritius and Rodrigues. In Rodrigues, the wind data considered were based solely at Pointe Canon due to the lack of wind stations in other parts of the islands. The region of Trou d'Argent therefore would normally have a lower turbulence and higher wind speed as it is located in the east of Rodrigues. Hence, to improve the energy and economic outcome, the scenario at Trou d'Argent could be considered as well. Due to the low population level and minimal interference, the offshore wind farm could be sited closer to the island, hence leading to lower cable cost.

From the different situations considered, scenario A that involves downsizing the wind farm to 6 MW to meet Rodrigues' yearly energy demand is recommended (Table A8).

CONCLUSION

In light of the above exposé, the offshore wind farm proposed at Pointe Canon is deemed to be the scenario in the Republic of Mauritius, based on current technology and practices. The extreme weather conditions and the impact of interference would also be minimal provided the safety distances are respected and proper standards used. Implementing the

Table A8 Scenario A—Downscaled Offshore Wind Farm
Benefits and Drawbacks

Capacity	6 MW
Total cost	Rs 1.3 billion
Annual energy output	30.2 GWh
Economic payback period	9 years
Net revenue	Rs 2.7 billion
Economic payback period (with carbon finance and polynomial variation of electricity tariff)	7 years
Net revenue (with carbon finance and polynomial variation of electricity tariff)	Rs 6.2 billion

offshore wind farm would generate a net revenue of Rs18.5 billion over its lifetime of 20 years while avoiding a total amount of 3.7 million tons of CO_2 equivalent. These figures could be further amplified by considering additional mechanisms such as carbon credit finance or by predicting the variation of electricity prices during the upcoming years. The outcome is a payback period of 7 years with a net revenue of Rs 41 million.

In Rodrigues, the total electricity required for the island is furnished by the Central Electricity Board. The annual energy output of the proposed wind farm is 196.5 GWh, which is beyond the 33 GWh yearly energy demand of the island, in 2011. Even though the cost of electricity is expected to rise in the future, the wind farm could easily meet the whole energy demand of the island while providing energy security for the future generations in the midst of the price volatility of oil and fossil fuels. Owing to the surplus energy generated, the exportation of electricity could be considered in the future as the submarine power transmission is still a growing technology. Currently, the project capacity could be simply downscaled to provide energy for the island only.

CONCLUSION: FURTHER WORK

The project was solely based on the potential of an offshore wind in the Republic of Mauritius. However, islands in the Indian Ocean are subjected to the high wave potential as well. Mauritius and Rodrigues could therefore take advantage of the combined wind and wave energy. A prominent example is the hybrid design proposed by Wavestar (Marquis et al., 2012), where a 5-MW wind turbine was implemented on a 2.4-MW wave device structured to attain a total capacity of 7.4 MW. Hence, a further assessment comprising both wind potential and wave potential in the Republic of Mauritius would help meet the increasing energy demand in a clean and sustainable way.

REFERENCES

Albion Fisheries Research Centre (AFRC), 1996a. Carte des Fonds Marin de la Zee Mauricienne—Ile Maurice. Mauritius Stationery Manufacturers Ltd., Mauritius [Accessed 06 November 2012].

Anon, 1998. Marine resources. Available at: http://iels.intnet.mu/marine_mau.htm, Online, [Accessed 15 December 2012].

Anon, 2013. Perspective (visual). Maths 1202 to 1800 AD [online]. Available through: Scribd database http://www.scribd.com/ [Accessed 21 January 2013].

Anthoni, J.F., 2000. Oceanography: waves. Available at: www.seafriends.org.nz/oceano/waves.htm, Online, [Accessed 26 December 2012].

BMT Cordah Limited, 2010. Part two, addendum to Chapter 18—Marine impact assessment. Supplemental environmental impact assessment for a 2 x 55 MW coal fired power plant in mauritius. Available at:http://www.gov.mu/portal/goc/menv/files/ctp_add_coal/PART_3_Supplemental_EIA_Marine_Addendum_MASTER%5B1%5D.pdf, Online, [Accessed 25 August 2012].

CEB, 2013a. Chapter 5: Power Generation Plan 2013-2022. Integrated Electricity Plan 2013–2022. Available at: http://ceb.intnet.mu/CorporateInfo/IEP2013/Chapter5_Power%20Generation%20Plan.pdf, Online, [Accessed 16 March 2013].

CEB, 2013b. Chapter 8: Electricity Demand-Supply in Rodrigues. Integrated Electricity Plan 2013–2022. Available at: http://ceb.intnet.mu/CorporateInfo/IEP2013/Chapter8_Demand-Supply%20in%20Rodrigues.pdf, Online, [Accessed 16 March 2013].

Central Statistics Office (CSO), 2001. Introduction. Available at: http://www.gov.mu/portal/sites/ncb/cso/report/hpcen00/disa/intro.htm, Online, [Accessed 18 December 2012].

Currency UK, 2013. Mauritius rupee currency exchange rate. Available through: Currency UK Database, http://www.currency.me.uk/rates/mur-mauritius-rupee, Online.

DTI, 2005. Guidance on the assessment of the impact of offshore wind farms: seascape and visual impact report. Available at: http://www.catpaisatge.net/fitxers/guies/eolics/file22852.pdf, Online, [Accessed 13 August 2012].

Economic and Social Indicators (ESI), 2011. 2011 Housing Census—Main Results. Available at: http://www.gov.mu/portal/goc/cso/ei915/esi2011.pdf, Online, [Accessed 18 December 2012].

EWEA, 2009. Wind energy—the facts. Earthscan, UK and USA. Available through: Scribd database, http://www.scribd.com/, Online, [Accessed 12 August 2012].

EWEA, 2011a. Wind in our sails. Available at: http://www.ewea.org/fileadmin/ewea_documents/documents/publications/reports/23420_Offshore_report_web.pdf, Online, [Accessed 18 March 2013].

GaBi, 2013. GaBi 6: Software-System and Databases for Life Cycle Engineering [software]. Version: 6.0.1.0. PE INTERNATIONAL: Leinfelden-Echterdingen, Germ.

GBI Research, 2011. Global offshore wind market could be 80 GW by 2020. Available at: http://www.windpowerengineering.com/construction/offshore-renewable-energy-to-2020-a-report/, Online, [Accessed 20 July 2012].

Google Maps, 2012. Indian Ocean [Maps]. Available through: Google Maps database, https://maps.google.mu/, [Accessed 20 September 2012].

Hansard, 2012. Parliamentary Debates. No. 17 of 2012. Available at: http://www.gov.mu/portal/site/AssemblySite, Online, [Accessed 06 August 2012].

International Energy Agency (IEA) Wind, 2012. 2011 Annual Report. Available at: http://www1.eere.energy.gov/wind/pdfs/iea_wind_2011_annual_report.pdf, Online, [Accessed 01 April 2013].

Jet Propulsion Lab (JPL), 2008. Ocean wind power maps reveal possible wind energy sources. Available at: http://www.jpl.nasa.gov/news/news.cfm?release=2008-128, Online, [Accessed 26 August 2012].

Kingroyos, 2011. Rodrigues locations named. Available at: http://upload.wikimedia.org/wikipedia/commons/thumb/b/bc/Rodrigues_locations_named.svg/2000px-Rodrigues_locations_named.svg.png, Image, [Accessed 16 October 2012].

Marquis, L., Kramer, M.M., Kringelum, J., Chozas, J.F., Helstrup, N.E., 2012. Introduction of Wavestar Wave Energy Converters at the Danish offshore wind power plant Horns Rev 2. Available at: http://nodc.intnet.mu/Excel/Rapport%20Raz%20de%20Maree%202012%20May%202007.pdf, Online, [Accessed 21 December 2012].

Mauritius Meteorological Services (MMS), 2010. Tsunami Warning System and Other General Info. Available at: http://metservice.intnet.mu/?cat=30, Online, [Accessed 21 December 2012].

Mauritius Meteorological Services (MMS), 2012. Meteorological data request. Available from: ackshay.jungbadoor@umail.uom.ac.mu, Email, [Accessed 10 November 2012].

Météo-France, 2012. Tableau de définition des vents des systèmes tropicaux. Available at: http://www.meteo.fr/temps/domtom/La_Reunion/TGPR/PagesFixes/GUIDE/GuideAlerteCyclonique.html#tableaudanger, Online, [Accessed 30 September 2012].

Ministry of Environment & National Development Unit (MOENDU), n.d., What is CDM [online]. Available at: http://www.gov.mu/portal/sites/cdmmauritius/cdm.htm [Accessed 15 February 2013].

Ministry of Environment & National Development Unit (MOENDU), n.d., UNDP CDM Project [online]. Available at: http://www.gov.mu/portal/sites/cdmmauritius/undp.htm [Accessed 23 February 2013].

Ministry of Finance & Economic Development (MOFED), 2011. Energy and Water Statistics—2010. Available at: http://www.gov.mu/portal/goc/cso/ei900/energy.pdf, Online, [Accessed 05 December 2012].

Ministry of Renewable Energy & Public Utilities (MoREPU), 2009. Draft long-term energy strategy 2009–2025. Available at: http://www.gov.mu/portal/goc/mpu/file/finalLTES.pdf, Online, [Accessed 05 August 2012].

Oceanweather Inc., 2012. Significant wave height with wave direction. Available at: http://www.thaiwater.net/Tracking/Now/wave.php, Image online, [Accessed 21 December 2012].

Plaisance Airport (Building Restrictions) Act 1964. (10), Mauritius. [Accessed 14 November 2012].

Research Council of Norway (RCN), 2010. Full cost of offshore wind power. Offshore wind assessment for Norway. Available at:http://www.nve.no/Global/Energi/Havvind/Vedlegg/Annet/Offshore%20Wind%20Asessment%20For%20Norway%20-%20Final%20Report%20-%20190510%20with%20dc.pdf, Online, [Accessed 08 September 2012].

Stephens, A., 2000. Long-term variability in offshore wind speeds. Thesis (Master), University of East Anglia.

Sustainable Minds, 2013a. LCIA of Offshore wind farm at Pointe Canon. Available through: Sustainable Minds websitehttps://app.sustainableminds.com/homepage, Online, [Accessed 23 February 2013].

The Partnership for the East Asian-Australasian Flyway (EAAFP), 2010. Flyways. Available at: http://www.eaaflyway.net/flyways.php, Online, [Accessed 28 August 2012].

Tylamma, S., n.d. Prospects of Renewable Energy on Agalega Islands [online]. Available at: www.mrc.org.mu/Projects/WREDay111.pdf [Accessed 18 December 2012].

U.S. Geological Survey (USGS), 2004. USGS Energy and Broadband Solution off W Coast of Northern Sumatra. Available at: http://neic.usgs.gov/neis/eq_depot/2004/eq_041226/neic_slav_e.html, Online, [Accessed 24 December 2012].

University of Newcastle (UON), 2002. Visual assessment of windfarms best practice. Available at: http://www.snh.org.uk/pdfs/publications/commissioned_reports/f01aa303a.pdf, Online, [Accessed 17 December 2012].

Ventures Africa, 2012. African businesses can earn carbon credits through new regulations. Available at:http://www.ventures-africa.com/2012/07/african-businesses-can-earn-carbon-credits-through-new-regulations/, Online, [Accessed 12 March 2013].

Vestas, 2011. Vestas in the Caribbean. Available at: http://www.carilec.com/members2/uploads/RE2011_Presentations/OrchidRoomSessions/3_JCNavarrete_VestasintheCaribbean.pdf, Online, [Accessed 28 September 2012].

Vestas, n.d.b. Wind turbine classes [online]. Available at: http://www.vestas.com/en/wind-power-plants/wind-project-planning/siting/wind-classes.aspx#/vestas-univers [Accessed 28 September 2012].

Vestas, n.d. Onshore or offshore wind power plants [online]. Available at: http://www.vestas.com/en/wind-power-plants/wind-project-planning/on-or-offshore.aspx#/vestas-univers [Accessed 27 September 2012].

Wilhelmsson, D., Malm, T., Öhmana, M.C., 2012. The influence of offshore windpower on demersal fish. Available at: http://icesjms.oxfordjournals.org/content/63/5/775.full, Online, [Accessed 15 December 2012].

Wilson, J.C., 2007. Offshore wind farms: their impacts, and potential habitat gains as artificial reefs, in particular for fish. Thesis (Degree), University of Hull.

A Case of Community Involvement in Wind Turbine Planning

Jonas Krogh Jensen[1], Lisa Blenstrup Nielsen[1], Tor Zipkin[1],*, Maria Eftychia Vestarchi[1], Zaklin Dasyra[1]
[1]Aalborg University, Sustainable Energy Planning and Management, Department of Development and Planning, Skibbrogade 5, Aalborg, Denmark.
*Corresponding author: tzo1029@gmail.com

INTRODUCTION

The advent of wind energy has gained much momentum since its beginning as an inefficient and expensive technology. Worldwide, price decreases, higher efficiency, and both public support and governmental support have stimulated somewhat exponential growth of the technology, with it expected to play a major role in the world's electricity makeup into the future. Denmark is arguably at the forefront when it comes to wind development, being the first country as a whole to truly capitalize its wind resource. By the end of 2013, 4792 MW of wind turbines was installed and contributed to more than 50% of Danish electricity consumption in 1 month (Energinet.dk, 2014).

Yet, there are still issues concerning wind turbines and their development. Economic issues such as high upfront costs per MW compared to fossil fuel technologies, technical issues such as difficulty of grid integration due to the intermittent nature of the technology, and social issues such as NIMBYism[1], among other things, all work against wind development (Energinet.dk and The Danish Energy Agency, 2012). Whereas economic issues and technical issues can most likely be cured through advancements in technology, social issues do not hold the same luxury. One approach when dealing with societal issue is going straight to the source, that is, directly to the people.

Involving people directly when planning turbines in their respective area and in the ownership of said turbines has proved to be successful in

---

[1] NIMBYism: Not In My Backyard theory (see Chapter 3).

decreasing the amount of opposition to turbine development in the past. In countries such as the United Kingdom, Denmark, and the United States, the concept of community ownership, where the surrounding community is directly involved in the ownership of turbines, has shown to even be a way of generating support for wind turbine development (Warren and Mcfadyen, 2008). In Denmark, there was such involvement from local communities that helped wind power grow to be as successful as it is today (Karl et al., 2010).

History of Development and Ownership

Denmark is a special case when it comes to wind turbine development. Not only was the country one of the first to capitalize on the technology, but also it went about development in a manner unique to that of the common commercial development seen around the world today. The concept of community ownership of wind turbines has been employed in Denmark since the beginning of turbine development, taking off in the 1970s after the famous oil crisis throughout the world. While early turbine development was mainly due to Danish farmers, as time passed, more and more people began investing in wind turbines, coinciding with turbines getting bigger and more efficient (Miguel et al., 2009). As turbines got better, they also became more expensive, resulting in the need for more investors as well as a larger variety of investors within a project, from small groups of people to municipalities. This trend continued and resulted in the formation of cooperatives made of multiple investors, which helped drive the wind turbine growth throughout the 1980s and 1990s. Public interest in the technology, as well as governmental support such as subsidies and feed-in tariffs (FITs), resulted in over 175,000 households owning 80% of all wind turbines in Denmark on either an individual basis or in the form of cooperatives (Miguel et al., 2009).

The early 2000s brought political change in Denmark and wind development slowed immensely due mostly to the policy changes (Miguel et al., 2009). However, renewable energy development returned back to the political spheres of politicians in 2008, resulting in the adoption of new goals, such as 50% of Denmark's electricity consumption supplied by wind by 2020 (Danish Energy Agency, 2013). Such ambitious goals call for the development of large offshore and near shore parks and the replacement of old small onshore turbines with larger more efficient turbines. This has caused a shift in the ownership of wind turbines. High investment costs, the structure of the Nord Pool Spot market where wind electricity is sold, and certain policies are all disincentives for cooperatives and local people to invest in turbines within their local vicinities. Commercial developers tend to now be the main investors, a complete opposite from how it was 20 years ago, yet by law, they are required to offer a 20% stake in the turbine to the local community, showing there is still some degree of community ownership of wind turbines in Denmark (Energinet.dk, 2013a,b,c).

Definition of Community Ownership

Community ownership is not a strictly limited definition and can be seen through multiple approaches to turbine ownership, with Danish turbine development highlighting this. In the beginning, farmers would individually own turbines. In this sense, this can be considered community ownership for it was the local resident who owned the turbine. The next popular form of community ownership that came about was the formation of a cooperative that would own a turbine. Cooperatives can have an array of investors from very few to very many; however like individual owners, the revenues are distributed correspondingly to the distribution of the members' investments.

Yet there are other forms of community ownership that are not as simple as an individual making an investment to simply generate a return, as is done in a cooperative or with individual investors. Many of the community-owned projects of present day in Denmark are set up so revenues from the investment are used directly to help the community where the turbines are located, such as with the creation of foundations to own wind turbines or through municipal ownership. Such forms of community ownership will be discussed in more detail later.

This project focuses on compiling the most common forms of community ownership, for the concept of community benefit through turbine development has helped in the successful implementation of renewable energy throughout the past, and it is hoped to do the same in the future.

Advantages of Community Ownership

There are multiple reasons that when developing wind turbines, it is useful to include some level of community ownership. Not only in Denmark but also in other places such as the United Kingdom, when asked about wind turbine development, people are more receptive when community ownership is involved. A UK study conducted on a Scottish island reported that 45% of residents would be more supportive of a wind farm if it was community-owned, with no one expressing a negative attitude towards the project. On another island, 65% of people said their support would go down if a wind farm was to be developed by a commercial company (Warren and Mcfadyen, 2008). Denmark's wind energy goals call for large amounts of repowering, replacing old small turbines with new larger turbines. Whereas this would mean there would be fewer turbines due to the increased capacity, the turbines that would be installed would be larger, hence more visible. It is very important to minimize opposition for these new turbines, one of the best ways being community ownership.

It is also important to increase acceptance and generate support for wind turbines, especially in Denmark, because it can be economically advantageous on both a microlevel and a macrolevel, that is, on a local community level and on a broader societal one. On a community level, revenues from turbines can be recycled within the community. This can help the development of areas that are in need of economic growth, such as in western Denmark where the wind resources are good yet economies are depressed. One can say that people are more receptive towards wind turbines not only because they own shares of them but also because they or their community is fairly compensated. Depending on the type of community ownership, they are again more receptive towards a project and in many cases their acceptance turns into support. The recycling of revenues within a community from turbine development is believed to be a major draw to the people who are directly affected by such a project, which is why it can be such a helpful tool when attempting to develop a wind project.

It is not only on a microlevel that involving the community in turbine development can prove to be economically advantageous but also on a societal level. Increased acceptance of onshore wind turbines would decrease the amount of expensive offshore turbines that would be built. Hvelplund et al. (2013) predicted that if "local and municipal ownership makes it possible to replace 550 MW offshore capacities with 800 MW onshore capacities, society and electricity consumers will annually save around 140 million euro." Including locals in the ownership of both large wind projects and small wind projects is key to gaining the acceptance needed to realize goals of the future. Successful projects involving community ownership can also serve as examples for other projects.

Problem Description

There are natural visual and economic thresholds individuals hold that determine whether they support a project, are indifferent towards it, or oppose it.

Whereas people might not mind a 2 MW turbine 500 m away, they might oppose a 3.5-MW turbine 500 m away or a 2-MW turbine 300 m away. Opinions on visibility nuisance caused by wind turbines vary between individuals, and it is important to understand these thresholds when planning a project. The same goes for the economic benefits of a project. Naturally people will be more receptive towards a project if they are compensated more, and if they don't feel they are compensated enough, their support will wane. This also goes for how money will potentially be spent by the community, where one might think it is alright to use revenues for a library expansion, yet another citizen would only give support for a project if it was used to create a football field.

This is the importance of community ownership for it changes peoples' willingness to accept wind turbine development in a positive direction. However, it is unknown to what degree it changes people's acceptance and is most likely subjective depending on the characteristics of the involved community. That is why it is important to look at specific areas to better understand how community ownership of wind turbines affects project development in that said area, in hopes to extend such knowledge to similar communities in the future.

Whereas community ownership may seem like an excellent way to increase support for wind turbine development, there are still instances of failed development or indifference to projects regardless of the inclusion of community ownership. It can be hard to determine why this happens on a countrywide level, yet by examining the approaches taken in regard to development, that is, how projects are introduced and how locals are involved in the planning process, one can better understand why such a project failed despite involving the community with ownership. With this information, future project developers can be better prepared when approaching the community they wish to involve with development. Again, it is important to look at individual cases similar to a development project one wants to pursue, to both ensure continuity and maximize success.

Research Question

Wind turbines not only are a way to loosen the tight grip fossil fuels have on electricity production but also can be lucrative investments, yet due to their intrusive visual impact, there may be opposition to turbine development from those who may encounter them. However, the involvement of local communities when planning turbines can negate these oppositions, which is why this should be examined in more detail, leading to the research question:

How can local individuals be involved in the planning and ownership of wind turbines in small communities to best decrease opposition towards development?

In order to answer this question, the case of Vinderød will be investigated to answer the following subquestions:

Subquestions

1. How can the community surrounding Vinderød be best involved in the ownership structure of the potential turbines?
2. What approach is best to take when introducing the project to the community, and how can this process be continued in the town of Vinderød?

In this context, Vinderød is considered as a small community. It is defined as a town where there is a high degree of interaction between the community members, which the low population contributes to.

Project Description

The Halsnæs Municipality is located, as shown in Figure A1, in the northwest part of the Danish island Sjælland, in which a small town Vinderød has the opportunity to install a couple wind turbines. Local investors, who want to install the turbines on their property, would like to see the development take place. The potential area of turbine development is relatively small. Summerhouses are located just north of the site, and a small village is located south. The turbines, if developed, would also be quite close to existing farm houses.

It is chosen to look at Vinderød case for this community is thought to share similar characteristics with many other areas that may in the future plan for wind turbines. Mainly, it is a small community with local investors wanting to install wind turbines, and there is the potential for opposition towards such development.

The goal of this project is to help plan and introduce the turbine development to the local community, and those potentially affected, so the least amount of opposition from them will arise. This includes finding the least intrusive siting of the turbines and examining what would be the best way to approach the community and how the community can be involved

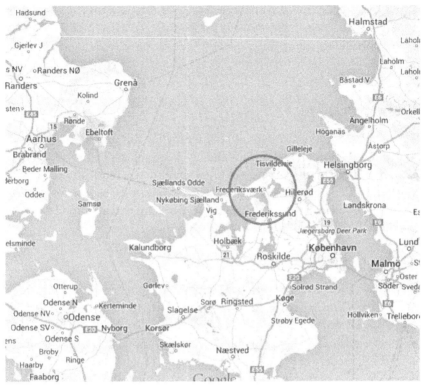

Figure A1 Sjælland and the area of Frederiksværk (Halsnæs Municipality) (Google, 2014).

in the ownership, the second two things receiving the main attention. By better understanding this community in particular, it is hoped that other communities with similar characteristics can use this report as a helpful guide when they may be planning turbine development.

The nature of the project deals with a certain degree of social research, in that it plans to interact with those in the community of focus in order to achieve its desired goals. It should be noted that interactions with the community during the project, such as interviews with stakeholders, could affect the projects outcome before the project is completed; yet this is not believed to cause different outcomes had such interactions not happened at all. The completed project is meant to have real effects of the success of turbine development.

While this project examines how community opposition towards turbine development can be decreased, it does not look specifically at ways to increase support towards such projects. While it is assumed generating support also decreases opposition, as can be the case when offering ownership opportunities, the project mainly focuses on decreasing opposition (Figure A2).

METHODOLOGY

This chapter describes the methods that are used in this project in an attempt to answer the research question and the subquestions from Chapter 1. These methods include data collection through interviews, literature study, case study, and stakeholder analysis, as well as a wind site assessment of the area and financial calculations of the project. It also examined how valid and reliable the collected information and the used methods are.

Data Collection

This project uses primary and secondary data to build and answer the problem. Primary data are collected as qualitative data through interviews with actors, considered having relevant connection or knowledge in the challenges of community-owned renewable energy projects, in this case wind turbines. Secondary data are mainly collected through literature study from sources found on the Internet and Aalborg University Library, but also quantitative data through the software of WindPRO, in the form of mapping and wind statistics data, and through the software of Microsoft Excel, in the form of financial data (Andersen, 2009). The method of case study as well as a stakeholder analysis is also used as a way of data collection and will be analyzed later in this section.

Literature Study

Literature study helps gain secondary data, and a large part of the project is based on it. It is important for the research process in any project and creates an understanding of the topics that are looked at through the whole process. Even though all data gained from literature are secondary data, that is, not produced for this project, they are still considered useful and important information for the basis of this project. This literature study is based on scientific articles, reports, books, websites, and information from lectures. All sources, especially those from the Internet, are assessed in relation to their credibility and to the authors' objectivity. The literature study is mainly used

Sub questions   Chapters and Outcome   Applied Methodes

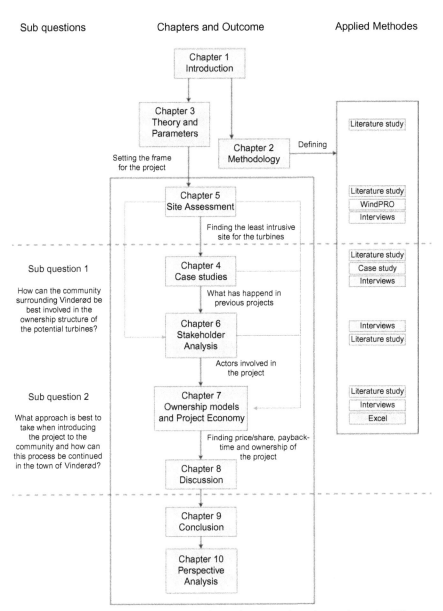

Figure A2 Structure of the project. The relation between the subquestions, the different chapters, and the applied methods. It also shows the purpose and outcome of each chapter in the whole context of this study.

in the first chapters of the project, but knowledge gained from these chapters is used to create the later chapters; that is, the literature study corresponds to the basic foundation of the project.

Interviews[2]

Interviews are conversations between individuals with a subject of mutual interest. The interviews are used in this project in order to understand this project's problem from the interviewee's point of view, that is, to gain knowledge on the situation from a point of view that the group members of this project don't necessarily have.

Questions are asked by the interviewer to be able to achieve new knowledge, facts, opinions, etc., from the interviewee. Interviews are made to collect empirical data from actors, with probable different views on the subject. Opinions and views on planning and placing wind turbines in a certain area can be very different, and the different views are important for the project to be able to answer the research question, that is, the problem, with a more objective point of view (Kvale, 1996).

There are four different types of interviews: structured, semistructured, unstructured, and focus group. The interviews conducted in this project are considered semistructured and qualitative, since the interviewer can speak beyond the questions, elaborate on them, and clarify them in order to achieve an in-depth understanding (May, 2011).

Before every interview, a flexible guide was prepared. The guide contains rough topics, and some specific questions that should be covered. It depends on the interviewer's judgment and the interview interaction whether or not it should be followed (Kvale, 1996). Moreover, permission by two of interviewees was granted in order to record the interviews. Also, only one person conducted the interview, in order not to confuse the interviewees and the rest of the group members can take notes. Afterward, the interviews were transcribed and elaborated throughout the project.

An attempt was made to contact people that are experienced on their field and have dealt with wind turbine planning before. As a result, the first contact person for this project was Lea Vangstrup. She is the founder of Wind People, which is an international humanitarian organization aiming to reduce CO_2 emissions by promoting renewable energy and more specifically wind projects owned by a community or municipality to ensure that the project's profit will be used to benefit vulnerable communities

---

[2] An overview of all interviews is found in Appendix A.

(Windpeople, 2014a, 2014b). Two interviews were conducted with Lea. The first one was more of an informal meeting, where she gave us insight into different ownership models and she introduced the specific case of Vinderød. The second one was a semistructured interview about the involvement of Wind People in the case of Ærø, which will be analyzed further in Chapter 4.1.

Afterward, an interview with Søren Hermansen, who is the director of Samsø Energy Academy, was held via Skype, but it was not recorded. As a result, a summary from notes (see Apendix A.A2) taken during the interview was created and sent back to Søren in order to be reviewed. As Ærø, the case of Samsø will be presented in Chapter 4.2.

Finally, a visit to Vinderød occurred in order to interview Ole Østergaard, who is the initiator of this project and a potential investor, as he owns part of the land where the turbines would be placed. This last interview was made to get a better understanding of a stakeholder's point of view.

Case Study

Case study is a form of qualitative research (Briggs et al., 2012). It is defined as an analytic inquiry of a particular subject, an event, or a collection of records (Bogdan and Biklen, 1982). There are different methods to perform a case study like investigating the written material that exists concerning the case, interviewing or making a conversation with the key actors, and visiting a relevant site (Wellington and Szczerbinski, 2007).

The results in a case study can be used as an educational example for similar cases as they are close to reality. Often, they indicate matters that need to be examined further in the future as well. According to Flyvbjerg (2006), it is a common misunderstanding that general content-independent knowledge has a higher value than concrete and practical knowledge like case studies. This is however a simplified statement; case studies can be used as contribution to scientific development and building foundation for new theories. Case studies are therefore considered a valuable and useful tool for the further studies in the project.

In this project, it is decided to do two case studies, one for the island of Samsø and one for the island of Ærø, as these are two cases of involving the community in the ownership structure of wind turbine schemes. The information gained from the literature study, the interviews, and the visit to one of the islands is considered as useful resources for gaining knowledge about how to plan the wind turbines in the area of Vinderød. Knowing how the ownership models were done in the Samsø and Ærø cases is helpful when creating ownership models for the turbines in Vinderød. The findings of these case studies are presented in Chapter 4.

Stakeholder Analysis

The purpose of the stakeholder analysis is to best understand the local context surrounding wind development and to guide best practices and suggestions to decrease the amount of opposition towards wind development in the target area.

The method used for the stakeholder analysis follows closely the *World Health Organization's Guidelines for Conducting a Stakeholder Analysis* as well as drawing inspiration from other sources (Schmeer, 1999; Worldbank, n.d.). The steps are as follows:

Step 1: Identify the purpose of the stakeholder analysis. Here is where it is explained why the stakeholder analysis is being done and what is the goal of the analysis, that is, to best understand the context of the project location, to best suggest ownership models and future approaches towards the area.

Step 2: Identifying key stakeholders. Here is where key stakeholders are introduced and discussed, which is done by using the prior knowledge from interviews. This section also assigns certain levels of interest, knowledge, power, etc., to the stakeholders, which are discussed in more detail later.

Step 3: Stakeholder table. Here, a stakeholder table is made, loosely following the guidelines of the WHO's respective table. The stakeholder table is meant to organize the project actors and their characteristics in order to best make conclusions on such actors and how they affect the projects outcome.

Step 4: Analysis of the table/stakeholders. Here, the most important actors based on the stakeholder table and how they can influence the project are discussed. The purpose of this section is to understand how certain actors could potentially decrease the success of turbine development, in order to provide suggestions as to how these actors can be approached in accordance with the future turbine development process.

According to these steps, a stakeholder analysis is undertaken for the area of interest this project is focusing on (see Chapter 6). The information gathered beforehand from the literature study and the case studies, plus an interview conducted with the potential wind developer in the area, provides the knowledge needed to examine and then come to conclusions on the actors involved in this project.

Wind Site Assessment

Modeling wind turbines or wind farms require knowledge in different planning areas, for example, knowledge on not only the terrain and meteorologic conditions in the chosen area but also the design of the chosen turbines and how this can affect nearby residents. A wind site assessment is done to investigate if putting up turbines in the area is a feasible solution. The model can show if a certain number of turbines or size is better than another, according to price, production, and environmental issues. A tool to handle with these issues is the software WindPRO, which was introduced to the writers during a university lecture and was chosen by them because they were familiar with it.

WindPRO is developed by EMD International A/S, which is mainly used to simulate wind farms. It is a tool that calculates the energy production of the wind turbines taking into consideration the wind resources in a specific site. It is also used to optimize the specific location of the wind turbines in a spot and calculate both the capacity and the costs for their connection to the grid. In addition, the noise impact and flickering and shadow effects can be calculated and documented (EMD, 2014). This software consists of different block modules that all have the purpose of making a model of a number of wind turbines in a chosen area. In some cases, the DTU-developed software WAsP is needed for the modeling, when orography and roughness on the maps are taken into consideration.

The software is mainly used in this project to model the noise and shadow/flicker from a few turbines in the chosen area near Vinderød. This is done by loading needed maps, elevation data, wind data, and chosen turbines into WindPRO. The model is thereafter used as a helping hand for creating the ownership models and financial calculations, that is, payback time.

Financial Calculations

Financial calculations are made for the turbine investments with the purpose of understanding the economy of the different ownership models. These calculations are made with the software Microsoft Excel and concern the investment costs of the turbines, the income that is derived from the sold electricity, and the payback time of the investment for the different owners.

In order to realize the distribution of the income over the years, a net present value (NPV) calculation is made with a chosen interest rate based on the economic situation of the investors and the currently applied interest rate in Denmark. NPV calculations highlight the difference between the value of the money today and the value of the money in the future.

The reason why Microsoft Excel is chosen is because it is a software where the user can define the calculations made and the steps taken are easy to follow. Moreover, it is a tool that everybody in the study group is familiar with, which is important considering the time limitations of the project. More details about the formulas and the exact calculations that are made are given in Chapter 7.1.

Validity and Reliability of Data

In social research, validity and reliability are two important concepts that are used in order to establish the trustworthiness and accuracy of the gathered information as well as the results from a study. Validity is an indicator of the legitimacy and the quality of a research, while reliability is an indicator of the consistency of the results (Yin, 2009).

The quality of an interview is very hard to be determined and can be quite decisive. The validity and the reliability of the answers should be examined since interviewers may be biased or not fully aware of the question asked. As a result, the acquired data can be misleading. So it is essential to be critical towards every single answer and information obtained by an interview (Bryman, 2012).

Because of time constraints, only a small number of interviews were done. Some useful knowledge may be missing, for example, some local people's view, yet the project relies on the knowledge from literature study, case studies, and an interview with a project initiator instead. The lack of interviews also affects the stakeholder analysis, since only one stakeholder was interviewed, limiting the scope of the stakeholder analysis. If it was possible, interviews with more stakeholders, for example, summerhouse owners and community members, would be made.

The validity and the reliability of WindPRO are well established since it is recognized worldwide and used by manufacturers, consultancy and engineering companies, and planning authorities to conduct feasibility studies and environmental impact assessments (EMD, 2014). However, it should be considered that different sources are used for the wind, elevation, roughness, etc., of the area, so the validity of the results that are generated from WindPRO depends also on the validity of the input data.

THEORY AND PARAMETERS

This chapter describes the various theoretical terms and concepts used in this project as a basis for the understanding of the problem raised in the research question. The chapter will focus on the different types of community, the need of community ownership, and the legal framework in relation to a wind project like Vinderød.

The Need of Community Ownership and the Term Social Acceptance

One of the reasons why it is decided to focus on community ownership stems from the nature of renewable energies, specifically wind energy. RE technologies are distributed not only over rural areas but also often near residential areas resulting in local resistance due to visibility and sound nuisances.

Moreover, in order to achieve the renewable energy targets that a government sets, *social acceptance* needs to be examined. There is no specific definition for this term, but according to Wüstenhagen et al. (2007), there are three dimensions of it: sociopolitical acceptance, market acceptance, and community acceptance. In addition, there are some basic characteristics of renewable energy technologies in comparison with the conventional ones that influence social acceptance.

The decisions about where the wind farms should be placed are quite important since the available spots are limited (Vangstrup, 2014a, 2014b). As the energy density of the renewable energy sources is lower than the non-renewable energy generators, total visual impacts are increased. As it is mentioned above, the energy production can often occur closer to communities, so people feel more effected by the technology.

Sociopolitical acceptance refers to the acceptance by the public, the key stakeholders, and the policy makers (Wüstenhagen et al., 2007). An EU study found that 96% of Danes consider it important their government set targets for renewable energy (Special Eurobarometer, 2014). This should be examined very carefully by policy makers because *moving from global to local* can create resistance towards implementation of a renewable project (Bell et al., 2005).

Market acceptance refers to acceptance by consumers and investors. Within sustainable communities, consumers can be investors, as is seen in community ownership of wind turbines (Wüstenhagen et al., 2007).

The acceptance of specific renewable energy schemes by local actors, namely, the citizens of a region and their authorities, describes *community acceptance*. Closely connected to this dimension of social acceptance is the NIMBY theory (Wüstenhagen et al., 2007). According to this theory, some people support wind energy projects, but when they are to be placed in their "backyards" or really closely to their residence, they are against them. In addition, some people claim that a wind farm can negatively afflict tourism and property prices (Wolsink, 2007a).

There are a number of reasons that can influence the community acceptance of a wind farm. These reasons, which can alter based on the current circumstances of a planning project, depend on the national regulations and laws, the local perception of financial impacts, trust, and fairness among the participants during the planning process and the completion of the project. Moreover, these factors are considered very subjective and influenced by the culture of each case (Wolsink, 2009). Thus, the principal element towards people's negative attitude is the visual effects of a wind turbine in the surroundings (Wolsink, 2007b).

Surveys Concerning Attitudes and Acceptance

Many surveys have shown that people's attitudes towards wind turbines follow a U-shaped progression through time. In the beginning, the responses are positive as long as the wind farm is not planned nearby their houses. Then, during the planning process, attitudes turn to doubt and mistrust due to the lack of information, which are then followed by positive attitudes when the project is realized to not be so intrusive as far as visual and noise impacts are concerned (Wolsink, 2007a).

As it was said, the main cause of opposition or acceptance towards wind energy projects is the visual impacts of the wind turbines within a landscape (Warren et al., 2005). A lot of studies have been made concerning this factor and the way it influences community acceptance. An interesting survey was conducted by Ladenburg and Dahlgaard (2012) between 1086 people living in different regions in Denmark. The sample was a typical one of the Danish population concerning age and gender, but the levels of household income and education were higher. The main topic of the survey was people's opinion about the number of onshore wind turbines that they daily see. The purpose was to find a threshold number of turbines that made their attitudes negative towards them. The result of the survey was that as the number of wind turbines seen increased, it negatively affects people's opinion towards them, the threshold number being six or more turbines. Interestingly, after 6 turbines, people were equally negative of how many turbines they saw, regardless if it was 10, 20, etc.

These results can be useful for the initiators of wind projects because they propose that if the distribution of the wind turbines within an area is made in a way that people see less than six turbines a day, then their acceptance for the plans will not wane. However, further research is definitely needed, because the survey did not include any data about the height, the distance, or the exact number of the turbines that a respondent can see from his or hers house (Ladenburg and Dahlgaard, 2012).

Another analysis of 18 case studies in England, Wales, and Denmark deduced some useful outcomes concerning acceptance after interviewing important actors like planners, landowners, and people for and against the projects. It was found out that if the community participates highly in the planning phase of a wind project, this is quite beneficial. Although this participation does not certainly lead to an equal level of acceptance, it is expected that people who are familiarized with the project early feel more comfortable and accept it. Moreover, it was noticed that it is more probable for a project to get a planning permission if there is a high degree of community acceptance than the opposite (Loring, 2007).

Other case studies from European countries indicated that if people are economically involved in the project, then it is more likely to be accepted (Davidmusall and Kuik, 2011). In addition, opposition to be decreased is possible, if shares of ownership of the wind turbine scheme are offered to the community (Warren and Mcfadyen, 2008). On the other hand, a commercial ownership model is proved to achieve less degree of acceptance than a community one (Davidmusall and Kuik, 2011).

Types of Community

To better understand the concept of community ownership, and how it can be successful in generating support towards a project, it is important to examine what it means to be a community and what makes a community. One can then see the interactions between groups of people and how this affects project development and outcome.

When considering turbine development and the community associated with the project, one must take a step back and look at the area of interest from a distance. Communities can be large and can be small, both of which have influence over turbine development. For example, the European Union is a community of European nations with the goals of peace, equality, and affluence and sets rules and regulations that all member states must abide by. At the same time, the group of farmers who live in proximity of one another is also a community. Entities that share something in common, again be it large or small, can make up a community.

When discussing community ownership, many degrees of community are involved, some more active than others, as well as holding different roles, and having different opinions when it comes to a project. The different communities that affect a small wind turbine development project extend beyond the immediate area of development. These degrees of community and how they affect a project can be the following:

Danish State

It sets a series of regulations that affect where turbines can be placed as well as policies to help with the development of turbines including FITs and subsidies, both affecting the financial feasibility of a project. Tax regulation set by the state also affects the financial feasibility of a project as well as how revenues can be distributed among a community.

Municipality

The municipality has its own rules about turbine development, for example, height restrictions, and placement regulations. Before turbines can be developed, they need to be approved by the municipality.

Large Communities within a Municipality

This can include villages, associations, and other large groups of people within a municipality who share something in common. Such examples include towns, villages, a large collection of summer homes, students, and the elderly, depending on the size of the population compared to the total municipal population. A large group is subjective and varies throughout municipalities.

Small Communities Within a Municipality

These small communities are made up of a small number of individuals that is more quantifiable than the larger local communities. They can be local investors, a small group of farmers, the inhabitants of a particular road, etc. Like large communities, to be considered a small community is subjective form area to area.

It is important to know the different levels of community, because interactions between them affect the outcome of a project. This project will focus on the lowest levels of community, that is, the municipality and below, and the interaction between them, for these interactions are relevant and constantly affecting each other.

To be more specific, within a municipality, there are multiple communities that have the potential to affect the outcome of turbine development. It is within the municipality that opposition that could stop turbine development arises, the goal of this project being to reduce such opposition. It is also within the municipality that support can be generated for a project if a specific community is affected in a positive way, for example, if the elderly or students benefited somehow from such development (Vangstrup, 2014a, 2014b).

Ownership Models in Denmark

When looking at ownership structures that best limit the potential opposition towards a wind development project, it is important to look at who the potential actors of these ownership models may be. For the context of this project, the foundation for ownership can be divided between two categories, *commercial ownership* and *community ownership* (Vangstrup, 2014a, 2014b).

Commercial or private ownership refers to making an investment, that is, investing in a wind turbine, where profitability is the main reason. This is more often seen not only in the form of large multinational companies, such as DONG Energy and Vattenfall, but also in the form of individuals, such as a group of farmers.

Community ownership is harder to define, as there are many degrees and forms it can take. In this project, community ownership refers to an entity owning a turbine, be it locals, a municipality, or a foundation, where profits are reinvested within the community. Advantages to this form of ownership are the financial benefits received by the community and the decrease of opposition towards the project (Vangstrup, 2014a, 2014b).

These two forms of ownership are not mutually exclusive, in that wind development projects can take aspects of both and utilize a *mixed ownership model*, where private and community interests are both represented. In these mixed ownership models, individuals still have the potential for financial benefit, as does the community. That being said, this inclusion of community in turbine development is beneficial to the for-profit developer, because it has potential to decrease the amount of local opposition towards a project, making project success more likely.

The possibility of mixing ownership is common and has been seen throughout Denmark. The Middelgrunden wind farm and Samsø are both examples of successful projects with multiple forms of ownership comprising the overall ownership scheme (Energiakademiet, 2014; Vikkelsø et al., 2003). It is important to realize that ownership structures are contextual in that they vary from location to location, and when developing a wind turbine, one has to understand the local people and their opinions in order for project success. Having a flexible and open-minded approach when developing turbines is therefore important, for people's opinions can change quickly and can be swayed towards acceptance if presented with what they perceive as the best forms of ownership.

Looking at the actors that can be a part of the ownership of turbines helps one better understand mixed ownership structures and how the community can be involved with turbine development. These actors might be the following:

Individual Investors

Individual investors fall under the category of private owners and can comprise only one individual or a group of people. Common examples of individual investors are farmers placing wind turbines on their land. These individuals are the sole investors in the turbines and can do whatever they please with the turbine revenues. The motivations can vary as to why they want to develop turbines; some may want to produce clean electricity, while others may be interested in profits (Sperling, 2007).

Individual investors are unique in that they have a special relationship with the community they live in. Many individual investors come from the area in which they want to construct the wind turbines, as can be seen in the history of Danish wind development (Miguel et al., 2009). This strengthens the relationship they have with those that may be potentially affected by such turbines. Coming from the said community, it can potentially be easier to talk with community members. This can be advantageous to project development, especially if aspects of community benefit from turbine development are included in the individual investor's project goals.

Private Companies

Private companies have the main goal of generating returns when investing in wind turbines. They have little relationship with the community their turbines may affect, and what relationship they may have can potentially be viewed as impersonal. Whereas their goals are strictly economic, they may find incentive to work with a community in order to generate support for a development project. As a company, considering mixed ownership models and involving the community more in the planning process may bring higher project success and should therefore not be overlooked. An example of a private company that could be interested in investing in a project like Vinderød could be Vattenfall and DONG Energy.

Cooperatives

Cooperatives are historically the main forms of community ownership in Denmark and can be defined as *an autonomous association of persons united voluntarily to meet their common economic, social, and cultural needs and aspirations through a jointly-owned and democratically-controlled enterprise* (International Co-operative Alliance, 2014). Cooperatives are set up so local communities can benefit from the development of wind turbines where they are placed. Members of a cooperative own small amounts of a wind turbine through the purchasing of shares, resulting in a turbine being technically owned by a group of many individuals. The rules of a cooperative are set by the bylaws created when the cooperative is formed.

An example of a wind cooperative is the Middelgrunden cooperative in Copenhagen. The ownership of ten wind turbines was split into 40,500 shares and in total had 8552 owners, 38% of owners coming from the city (Vikkelsø et al., 2003). As one can see, 62% of the owners of the cooperative were not from Copenhagen, raising testament that cooperatives are not necessarily purely local endeavors. This is interesting for one cannot assume such distant ownership would be as well received in smaller communities, which is counterintuitive towards the goal of this project. With distant ownership comes less community benefit where the turbines are located. Therefore, in this project, it is necessary to ensure that cooperatives' shares will be bought by the local community.

Municipalities

It is possible for a municipality to own wind turbines, or shares of wind turbines, as has been seen in the case of Middelgrunden and Samsø. All profits generated from such wind turbines are kept within the community, providing the opportunity to stimulate the local economy (Energiakademiet, 2014; Vikkelsø et al., 2003).

Municipalities have the right to take part in municipal companies that produce, transport, or sell. These companies are those responsible for the implementation and ownership of the wind projects, when and if a municipality wants to be involved in turbine ownership. Profits from such turbine ownership are untaxed if used for energy-related projects, yet they need to be approved by the Danish Energy Regulatory Authority. For all other projects not approved by the Danish Energy Regulatory Authority, profits are taxed (Brejnholt, 2013). While having a municipality own a wind turbine keeps money within the community, some might be disinterested in this due to the regulations regarding how the money can be utilized (Vangstrup, 2014a, 2014b).

Municipalities are also involved in the granting of permits for turbine development. Including municipalities in the ownership of turbines may allow for easier access to development permits, for such municipalities might want to make investments in wind turbines.

Foundations

The creation of foundations can be a way to secure that the profits from a wind project will stay within the community. A foundation is by definition *"a non-governmental entity that is established as a non-profit corporation or charitable trust, with a principal purpose of making grants to unrelated organizations, institutions, or individuals for scientific, educational, cultural, religious, or other charitable purposes"* (Grant Space, 2014).

A fund is by law considered as a legal person and cannot be owned by any person or company, meaning that profits can never go back to the founders (Erhvervsstyrelsen, n.d.). The profits can only be used in ways the bylaws of the foundation dictate they can be.

There are various types of foundations, all with different purposes described in the bylaws of the given fund, humanitarian foundations being one of them. In order to establish a fund, the board of the foundation must first of all register the fund at the Danish Business Authority. In addition, it needs to create the bylaws that state the purpose of the fund and what the potential profits may be used for. After insuring all these, the fund obtains its rights and responsibilities (Retsinformation, 2010).

The creation of a foundation that owns a wind turbine is an innovative way to keep profits from turbine development within a community, the foundation set up in Hvide Sande being an example of this. The fund was established by Holmsland Klit Turistforening, who paid the stating cost of the fund, which consists of various bylaws that dictate how the fund will function and how revenues will be utilized. Foundations are interesting in that "the founders do not hold the ownership and thus have no special rights regarding the revenues of the industrial foundation." The purpose of the fund, that is, how revenues will be used, is agreed upon by the cofounders. In Hvide Sande case, revenues were allocated to the development of the local harbor. Whereas turbine development was at one time stalled due to local complaints, the creation of the foundation increased local acceptance and ended with project success (Brejnholt, 2013).

Renewable Energy Laws

When planning a wind project like Vinderød, one needs to be aware of the laws that have influence on the planning process, on the economic situation, and in general laws about RE. This section describes some of the key laws in renewable energy. Detailed laws about the planning process, for example, how far away from the road a turbine needs to be placed, will be explained in Chapter 5. Tax laws for different types of investors will be described in Chapter 7.1.

The Four Re Schemes

This section will contain a general description of the four RE schemes that exist in Denmark as a consequence of the consolidated law on the promotion of renewable energy (Bekendtgørelse af lov om fremme af vedvarende energi) (Energinet.dk, 2013a, 2013b, 2013c; Retsinformation, 2013). Only laws of what is estimated as being relevant for this project are described.

The loss-of-value scheme (Værditabsordning) concerns § 6-12 and secures that neighbors to planned wind projects can get compensation if their estate loses value due to the installation of the turbines. This scheme is not elaborated since it is not the main focus of this project.

The buyer's right scheme (Køberetsordning) concerns § 13-17 and gives local citizens the opportunity to buy shares of the wind turbine project.

As the main investor of a new wind project, one needs to be aware of the buyer's right scheme. This obliges the main investor/investors of onshore wind projects with turbines taller than 25 m to offer 20% of the projects' shares to the local citizens within an area of 4.5 km from the wind turbine/turbines.

The citizens that can apply for the shares of these 20% are split up into three groups. Only citizens with a CPR register address in the municipality of the wind site can buy shares:

- Group 1: Citizens living within 4.5 km of where the turbines are placed
- Group 2: Citizens living more than 4.5 km away from the installation site but still live within the border of the municipality
- Group 3: Citizens living more than 4.5 km away from the installation site but in a municipality that has a coastline within 16 km of the site (only concerns offshore turbines)

The first group has the first priority to purchase of up to 50 shares, whereas the two last groups have the second priority in terms of purchasing shares. Note that the first group can buy more than 50 shares. The number of shares depends on the demand for shares, but the citizens in the first group are secured at least 50 shares each compared with the citizens outside the 4.5-km border. That is to say, if a citizen within the 4.5-km border has bid on more than 50 shares, the 51st share and upward will be divided evenly between the bids from all three groups (Figure A3).

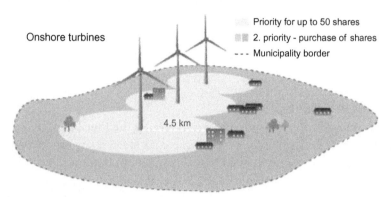

Figure A3 Priority of the different groups of shareholders in terms of ownership (Energinet.dk, 2013a, 2013b, 2013c).

Shares under these 20% cannot be bought by citizens living in second homes, for example, summerhouses even though they live within the 4.5-km border, because they are not CPR-registered at that address. Every Danish citizen has a CPR number, which is their Social Security number. Neither companies nor municipalities can own shares of these 20% of the whole wind project.

The remaining 80% of the shares can be owned by anyone, that being a private investor living a 100-km away, a municipality, private companies, etc. Rules about who can own these remaining shares can be made as in the case of Ærø, where a shareholder had to own an address on the island in order to become a shareholder (Schmidt, 2014). The main initiator, in this case of Vinderød being Ole Østergaard, can decide what he wants to do with them, for example, offer more percentages to the local community or other interested investors.

It is important to state that if these 20% of the shares offered to the local community are not purchased by the community due to the lack of interest, the wind project can still be done. The shares are only an offer, and if the community does not want to invest, the shares will go to whoever wants to invest.

The green scheme (Grøn ordning) concerns § 18-20 and gives the municipalities, where the turbines are placed, the opportunity to apply for grants for different types of construction works and cultural and informative activities in local association, in order to increase the acceptance of RE energy sources in the community.

The value of grants is calculated from 0.4 øre per kWh of 22.000 full load hours per turbine onshore. This corresponds to 88.000 DKK/MW installed capacity.

The warranty fund (Garantifonden) concerns § 21 and secures that the initiative group gets a guaranteed loan for the financing of prestudies (feasibility studies) in connection with new wind turbine projects.

The initiative group can apply for up to 500.000 DKK. This fund can be used in prestudies, for example, environmental impact assessments (EIA), survey of the installation site, and technical and economical evaluations of the project, among other things. The initiated group needs to be consisted of at least ten individuals, where the majority has a CPR-registered address in the municipality where the wind project is being realized at a maximum of 4.5 km away from where the turbines are being installed. If the project for some reason is not realized, this fund does not have to be paid back, unless the project in whole or in parts is being transferred to others.

Summary

The purpose of this chapter was to understand why and how community ownership is helpful, the types of community and how they can potentially affect each other, and how community can be involved in ownership of wind turbines and then go over some of the parameters of Danish law that affect such forms of community ownership of wind turbines. Understanding the theory and applications of community ownership provides knowledge that is necessary to better understand how community ownership can be used in a specific case, that is, Vinderød, and then be extended to a broader situation, that is, small communities. The information in this chapter can be used to create new theories and suggestions as to how community ownership can be applied, which is what this project strives to do.

CASE STUDIES

In this chapter, two successful cases regarding community ownership are presented, highlighting not only the division of ownership but also the difficulties concerning the public involvement in the planning phase and the procedure to reassure the acceptance of local people. The specific cases of Ærø and Samsø are chosen because these islands are examples with the successful application of community ownership.

Ærø

Ærø is a Danish island located on the Baltic Sea. It covers an area of approximately 88 km^2 and has a population of 6684 inhabitants. The exact location of the island can be seen in Figure A4.

This island is an example of a proactive community interested in renewable energy. During the mid-1980s, the Ærø's Wind Energy Cooperative was funded, which financed and owned Denmark's largest wind park at the time (11 × 55 KW) with a total of 128 shareholders, providing construction jobs to the locals (ÆRØ Energy and Environment Office, n.d.). Afterward, during the late 1990s, another community organization was formed, the RE-Organization Ærø, again highlighting the island commitment to sustainability and community, members including the island's two mayors, the local chairmen for electricity and heat supply, and other community leaders. It was a group widely supported by the inhabitants on the island (Wind Energy Local Financing, n.d.).

In 1997, Ærø's Wind Energy Cooperative and Ærø Renewable Energy Organization decided that the old turbines should be replaced by newer, bigger, and fewer turbines, in an attempt to cover the island's electricity

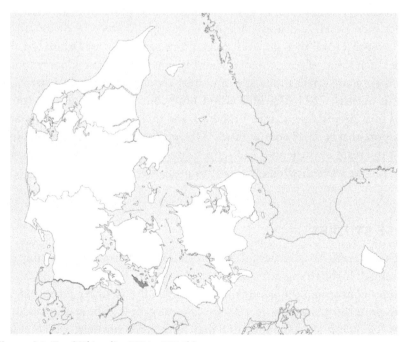

Figure A4 Ærø (Wikipedia, 2014a, 2014b).

consumption. The initial plan was to install nine turbines of 1.5 MW, which was afterward revised to two parks each of three 2 MW turbines. Each turbine was expected to cover the annual consumption of 1700 households, around 6788 MWh/year (Wind Energy Local Financing, n.d.).

A meeting between the two municipalities on the island, the Danish Energy Agency, the Forest and Nature Agency, and the former county of Fyn was held in order to discuss and agree upon the details of the project and more particularly whether or not there was space for both parks. Based on wind conditions and low population density, the siting of the turbines was chosen (Wind Energy Local Financing, n.d.).

From 7 August to 13 September 2002, people either living on Ærø or owning a summerhouse on the island were able to buy shares of the turbines. Each person could only buy up to 20 shares; afterward, the buyer could buy shares that would be on sale from a pool of leftover reserves (Wind Energy Local Financing, n.d.). As a result, two ownership organizations were created in order for one of them to benefit from the tax regulations in Denmark[3]. The first consisted of 6 people, owning one turbine, and the second consisted of 550 members, owning the other 2 (Vangstrup, 2014a, 2014b). A nonprofit foundation named "Fonden Ærøs Vedvarende Energipulje" was also established to purchase any leftover shares and use the profits to promote renewable energy and energy conservations measures on the island.

In regard to the second small wind park, the cooperative board of administration consisting of five local peoples initiated it. The only difference was that only people living on the island could be shareholders, which wasn't the case earlier. Consequently, although the model of ownership was exactly the same in both wind parks, the second one had only 200 members and 50 of them were repeated from the first project (Vangstrup, 2014a, 2014b).

In both cases, most people bought 20-50 shares as they were tax-free. Exceptions existed, where people owned 300-400 shares; however, they pay higher taxes on their share revenues. These shares can be sold but only to other permanent inhabitants on the island. Banks provided people with loans to buy shares, with the shares themselves serving as collateral. For those who took out loans to buy shares, the revenues from their shares went to the banks to pay off their loans, which in this case took 8 years. This was important because many houses had a low valuation and therefore couldn't be used as collateral when taking out a loan (Schmidt, 2014).

---

[3] If a company consists of 10 members, then a company tax structure is applied, which offers more benefits than a company with more than 10 members (Vangstrup, 2014a, 2014b).

Both parks are considered very efficient, and their energy production reaches the level of productivity that offshore wind parks do. The payback time of the first park was around 8 years. The six turbines cover 130% of the island's electricity, exporting the excess production to the mainland (Schmidt, 2014).

It can be seen that many inhabitants of the island are a part of the turbine ownership; at the moment, all wind turbines are owned 100% by locals through the island's wind cooperatives, with more than 650 people on the island owning shares, approximately 10% of the population (Schmidt, 2014; Vangstrup, 2014a, 2014b). Such a degree of involvement however did not come without opposition. There was a lack of effort to involve the public in the planning process, where they were not involved until they decided to involve themselves. This on top of no community meetings regarding the turbines in the beginning of the planning phases and only one period of two meetings regarding public ownership, opposition towards the planned development was able to form. A group consisting of 46 people, some living close to the planned turbines and others new residents who didn't want to spoil the island's history, were able to collect 1000 signatures, 1/6 of the island's population, opposing the planned turbines, misinformation and other scare tactics being used to generate such opposition. Yet interestingly, in the end after public discussion, opposition groups were no longer against turbine development; they just wanted turbines to be below 100m, some of such opposition even decided to purchase shares.

Samsø

Samsø is a Danish island with an area of 114km^2 and approximately 4000 inhabitants (Samsø Erhvervs–og Turistcenter, 2014). It has a large number of Danish and foreign tourists, and it has managed to become—in 10 years' time—100% self-sufficient in electricity produced by wind turbines and 70% self-sufficient in heat consumption produced by local renewable energy sources (Energiakademiet, 2014). It was an ambitious project that won a competition between other Danish islands organized by the Danish Energy Agency in 1997, and it turned Samsø into an international demonstration project. The purpose was not to test RE technologies, but to prove that when there are local involvement and emphasis on the development of the local community, an energy system based on renewable energy sources can be self-sufficient up to a very big percentage (The Danish Energy Authority, 2003). The location of the island is seen in Figure A5.

Figure A5 Samsø (Wikipedia, 2014a, 2014b).

The result of this was a project consisting of 11 onshore wind turbines with a total capacity of 11 MW, built in 1999-2000, and ten offshore wind turbines with a capacity of 2.3 MW each (construction starting in 2002). Along with these installations, the residents of the island were encouraged to install solar collectors, biomass heating, and heat pumps with the aim of taking advantage of a national subsidy that was given until 2002. Moreover, they could ask for a free energy assessment of their house that included recommendations for energy efficiency and utilization of renewable energy (Energiakademiet, 2014).

From the beginning of this project, local authorities, utility companies, local organizations, and the public were involved. The Samsø Energy Company coordinated the preparation and implementation of the onshore and offshore wind turbines. Numerous information meetings were held to inform the public and to generate support. The public was invited to these meetings from the very beginning of the planning process, where various community benefits from such development were explained and discussed. Those meetings were held around once a month with 20-40 people taking part for each (Energiakademiet, 2014; Søren Hermansen, 2014).

In the beginning, people rejected the project due to their lack of knowledge and because they were afraid of the costs. They were also worried about the size of the turbines. Financial figures provided by the banks and visualizations of the wind turbines in potential sites were presented during the information meetings, helping people better understand the project and increasing support towards it. Most of them were convinced that the project would be beneficial for local development, in part because each individual was explained the benefits they would receive (Søren Hermansen, 2014).

Everybody was invited to those meetings—local residents, banks, the local nature conservancy association, etc. The involvement of the banks was useful for the needed economic analyses of the project, as was the involvement of the nature conservancy association, in order to ensure decreased resistance from their side, done simply through including a representative in the planning process (Energiakademiet, 2014; Søren Hermansen, 2014).

The ten offshore turbines comprised five being owned by the municipality of Samsø, which gave a grant for the creation of the Samsø Energy Academy[4], and uses the revenues from the turbines for other energy-related projects. Three are owned by independent private groups, and the other two are owned by two cooperatives. The one is local and it consists of 450 people owning shares, and the other one is national and it consists of 1100 people owning shares. As for the 11 onshore wind turbines, nine of them are owned independently by local farmers and the other two by two local cooperatives with 230 people as shareholders (Søren Hermansen, 2014).

These two cooperatives owning the onshore turbines consist of 5400 shares; there was a period of 6 months that these shares were offered to the public. One had the possibility to buy packages of 1, 10, or 30 shares in order to join the cooperative. The role of the banks was very helpful in this phase because they agreed to give loans to the people with the guarantee that in case they couldn't pay them back, then the shares would belong to the bank (Søren Hermansen, 2014).

The ownership structure seen on Samsø has been said to be key to the project's success (Søren Hermansen, 2014) (Figure A6).

---

[4] The Samsø Energy Academy is an exhibition center of all the renewable energy projects that have been implemented on the island. It was created in 2006. Scientists, students, and tourists with a particular interest in renewable energy and energy efficiency can visit it and attend a lecture or a guided tour. It also serves as a conference center for energy-related topics (Energiakademiet, 2014).

Ownership

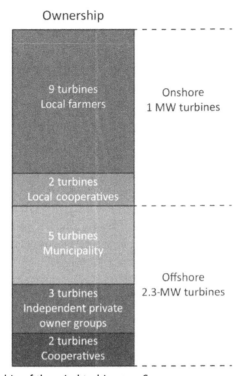

Figure A6 Ownership of the wind turbines on Samsø.

Summary

In this chapter, two case studies of the islands of Ærø and Samsø are presented. It was meant to compile information on the techniques and difficulties when transitioning to a sustainable energy system, focusing on the development of the existing wind turbines and the role of the public concerning the planning process and the ownership of the turbines.

In particular, it has been seen that the public should be involved from the beginning of the planning process in order to minimize opposition towards such projects. Everybody should be invited to the information meetings, and the focus of the meetings should be on local development and community benefit. In addition, the sharing of information between residents about the benefits of turbine development is important for the social acceptance of such projects.

It is also interesting that the approach towards resistance is noticed in these two cases. Some of the opposed ones changed their minds when they became part of the information meetings, like the nature conservation association on Samsø, while others were convinced to buy shares although they were completely opposite and even signed against it in the beginning of the project, as it is seen on case.

Moreover, it is useful to see how the planning process may change until the implementation of a project. This is what happened on Ærø, where after the initial resistance against the turbines, they were planned to be less than 100 m tall so that the local residents did not feel disturbed by them.

SITE ASSESSMENT

The purpose of this chapter is to examine if there is a feasible potential for wind turbines in the Vinderød area. This is done by investigating the land area for the turbines and its surroundings and by modeling the production from wind turbines in this area through WindPRO.

Area Introduction

The potential area is situated in the northwestern part of Region Hovedstaden in Halsnæs Municipality, between the town Vinderød and the summerhouse area Karsemose Syd. The municipality has around 30,000 inhabitants, with Frederiksværk being the biggest city with around 12,000 inhabitants. Frederiksværk is an old industrial city because of a canal between the lake Arresø and Roskilde Fjord. This made it possible for using water mills in the area, which started an industry especially for cannons and other war materials. Since 1940, a large iron work has been situated in Frederiksværk, still with the canal in use, for cooling waters. Because of the large iron works, Frederiksværk experienced a large growth and went from 2200 inhabitants in 1940 to 18,000 in 1990 but has since been going down (Industrimuseet, n.d.).

The small town Vinderød, which is the town nearest the wind site, has 1000 inhabitants and is a so-called satellite town to Frederiksværk (Danmarks Statistik, 2014).

The chosen area can be seen in Figure A7 and is the land area east of the farms marked with black dots. The nearest affected houses are all farms or houses in open land and summerhouses. The town in the southern part is Vinderød.

Halsnæs Municipality has 8400 summerhouses, which is an important resource for the municipality. The summerhouse area close to the potential

Figure A7 Vinderød area (Miljøministeriet, n.d.).

wind site is called Karsemose and has around 300 summerhouses, where a small number have a status as normal residential houses. Certain summerhouse areas in the municipality require special materials, for example, wood fences (Halsnæs Kommune, 2014a, 2014b). Even though this is a summerhouse area, some of the municipality's goals and rules still apply, for example the city council wanting to

promote renewable energy including in the form of wind turbines.
Halsnæs Kommune (2014a, 2014b)—translated from Danish

The owners of the summerhouses are mostly not from this local area. It is chosen to place turbines as far from these summerhouses as possible and still be able to make it feasible relative to wind flow. Placing the turbines as far from the summerhouses as possible leads to a placement closer to some of the farms (Østergaard, 2014).

The lake Arresø, which is situated just east of the area, is the biggest lake in Denmark and is protected by, for example, Natura 2000 and certain other nature protection agreements. A large area in northern Sjælland has a proposal for becoming a national park called National Park Kongernes Nordsjælland. The chosen wind site is a part of this area, possible for becoming a national park. This will however not change the rules for placing wind turbines:

The creation of a national park does not introduce new restrictions on agriculture and forestry—and not for other professions.
Danmark Nationalparker (n.d.)—translated from Danish

The possible part owners of the project are farmers or house owners situated on the roads Helsingevej and Søfrydvej, which can be seen in Figure A7 as the road with farms marked with black dots and the small area in the south-ern part with two farms and three houses. The project will possibly contain main investors from four farms, where two or three are the owners of the land where the turbines are going to be placed.

Site Overview

The area, where the turbines will be placed, is a hill that leads down towards the lake Arresø, as can be seen in Figure A8. It contains mainly fields for cattle and horses and small wetland areas close to the lake. The top of the hill is obviously the best position for the turbines according to the wind pro-file, but options are limited because of distance to the road, as described in the following section.

WindPRO Model Analysis

In this section, the WindPRO simulation of the planned wind farm is pre-sented. More details about the way it is designed, the regulations that are based on, and the noise and flickering impacts are included.

WindPRO has the ability to directly download online maps, wind, etc., data by a tool in the program. Maps for the Frederiksværk area are from online data at the Danish Kort og Matrikelstyrelsen, in different formats and sizes. Mainly, a topographical map in 1:25000 is used, but orthophotos and other maps are also used for validation correction of the data.

Along with the topographical map, orography and roughness for the chosen area are also needed. Both data are downloaded through WindPRO as well but come from different sources. The elevation map is a surface map that describes elevation and height of all buildings and vegetations. Two

Figure A8 Wind site from the top of the hill (Jensen, 2014).

maps have been used, one downloaded from Danmarks Højdemodel (DMH) and one from the digital terrain model, based on the Shuttle Radar Topography Mission (SRTM). Both have been used, but the SRTM map is considered the most detailed map. Roughness data are obtained from www. dataforwind.com, which is a service developed for wind farm management (Poglio and Ranchin, 2014). These data are imported into WindPRO via an online tool in the program, like the other maps.

The most important part for choosing placement and calculating actual production from the turbines is the wind speed. The data for wind speed are collected for a 20-year period (1994-2014), to be able to make a more reliable analysis. This is done in WindPRO through a METEO object. The METEO object in WindPRO refers to the wind data: both mean wind speed and wind direction. Wind data of the area around northern Zealand are downloaded through WindPRO from EMD's server. When talking about wind speeds, it is also necessary to have data for the surface in the chosen area. This is especially important in areas with hills, where the wind speed can change drastically. These calculations, along with the roughness of the surface, are very complex and are calculated with the WAsP software from DTU (White et al., 1996).

Wake interaction is where wind speed is decreased and turbulence is increased, which results in reduced kinetic energy. Turbines operating too close to each other will operate in each other's "wakes," which results in lower energy outputs (White et al., 1996). Because of the limited amount of turbines in this project (2-3), further measurements of the wake effect are not done.

Visual and audible influence

For this project, turbines will be placed in an area close to farms, permanent residencies, and summerhouses. Visual considerations like shadow and flicker effect from them are considered very important for their implementation. If wind turbines have to be accepted in the community, the visual issues, especially flicker, have to be looked into and follow certain regulations. It should be noticed that these effects vary depending on whether the surrounding landscape is an open field, a coastal area, a forest, the top of the hill, or a flat land and also on the size and number of the turbines. Shadow flicker is as problem where the blades from the turbines cast shadows on, for example, residential areas. Flicker effect can be calculated in a way where WindPRO shows the amount of hours where flicker occurs in a chosen area.

Along with visual influence, noise from the turbines is also considered a serious issue. It occurs from the friction between the wing blades. In addition, it is a function of many parameters, such as wind direction, the landscape, the surroundings, the distance from houses, and the background noise (Pedersen, 2003). The noise levels are calculated using WindPRO and presented on a map of the area. These calculations have also an effect on the final placement of the turbines.

The exact location of the wind turbines had to comply with many different regulations concerning their distance from roads and houses, as well as the noise and flickering effects on people living in the area. The distance regulations point out that a wind turbine must not be placed closer than "4 × the total height of the turbine" to the nearest neighbor. Also, the distance to all roads or railways needs to be at least "1 × the total height of the turbine," which is measured from the ground to the tip of the wing (Danmarks Vindmølleforening, 2013).

The space among them should be taken into consideration. Across the prevailing wind direction, at least "3 × rotor diameter" distance is recommended, while along the wind direction, a minimum of "5 × the rotor diameter" is preferable, as the front turbines will create a bigger wind shadow. Studies showed that spacing between the turbines of "3-4 × the rotor diameter" seems more harmonious, if they are placed in a row. At a spacing larger than 5 × the rotor diameter, the turbines aren't connected as one unit, because the distance between them seems to be too large (Danish Nature Agency, n.d.).

In this project, different types of wind turbines were tested before choosing the final one. In the beginning, 3 turbines from Siemens SWT-2.3-108 with a capacity of 2.3 MW, hub height of 80 m, and rotor diameter of 108 m were placed in an area that is located up to 320 m from the houses and 80 m from the roads. The space between the turbines was 300 m. This attempt failed because of the shadow effects, since the nearest house had to face 1 h and 25 min of shadow every day, which doesn't comply with the recommendation from the Ministry of Environment and Energy, that a neighbor's exposure to shadow casting must at maximum be 10 h of "real value" each year (Danish Energy Agency, 2013).

Due to the space limitation and the negative noise and shadow effects, the number of wind turbines was reconsidered, and a scenario with only two turbines was made. Also, a different smaller size was chosen for this specific case. The simulation is now carried out with 2 turbines from Vestas V80-2 MW with hub height of 78 m and rotor diameter of 80 m. Their

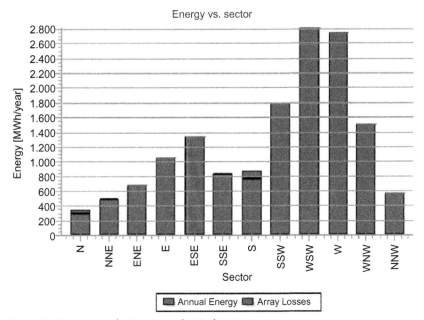

Figure A9 Energy production in each wind sector.

location was chosen carefully in order to apply to all the above-mentioned regulation.

Two Vestas V80-2 MW turbines will in this certain area have a yearly energy production of 14,878.8 MWh, running for 3719 h at full load equivalent combined.

As shown in Figure A9, most wind comes from the west. The red zones in the graphs are the array losses from the wake effect. Since the turbines are placed in a north/south direction of each other, energy losses are most significant in the southern and northern wind directions. Because the main wind direction in this area is from the west, these are not considered as large losses.

All nearby buildings (farms, summerhouses, etc.) are mapped as noise-sensitive areas (NSAa) in WindPRO, for making an estimate over the noise levels from the turbines. A NSA is a term in WindPRO to mark a specific area, for the program to analyze noise levels from turbines in that particular spot.

As shown in Figure A10, the areas are marked in red. According to WindPRO, none of the areas will have low-frequency noise above 18.2 dB from the turbines at 8 m/s.

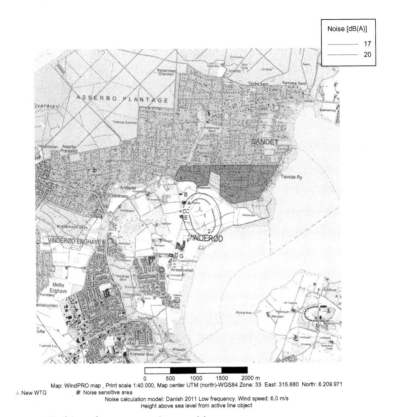

Figure A10 Turbine placement, noise-sensitive areas.

The red circle in Figure A10 represents a noise level of 20 dB, while the blue circle represents a level of 17 dB, both at wind speed of 8 m/s. As can be seen, the blue circle slightly touches the summerhouse area. This can result in issues and complaints from the owners. It is however possible to move the turbines to a slightly more southern position. This will however bring area F on the figure closer to the noise levels, but this area is the farm and houses the primary investors own and is therefore not considered to be an issue. Areas A, D and E are inside the blue 17 dB zone. These house owners have also shown interest in owning shares of a wind turbine and are not considered an issue in this case.

Like the NSA's in WindPRO, zones of visual influence (ZVI) are mapped as well. These zones are considered the same as the noise-sensitive areas and are made to be able to map the shadow and flicker effect from the turbines in these areas. Figure A11 shows the shadow and flicker effect from

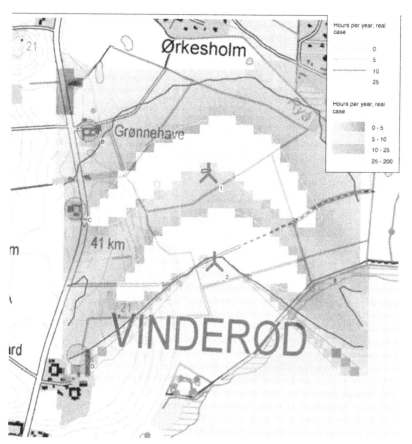

Figure A11 Shadow and flicker effect from the turbines.

the turbines, with the yellow area being the most affected and the blue the least. The reason why the shadows have the "boomerang" shape is because of the sun movement and elevation on the sky. The sun is lower in the morning and evening where it is in east or west, which results in longer shadows casted from the turbines in these hours.

Areas B and C on the map are the ones with most shadow flicker, with a worst-case scenario of 56 h of shadow flicker per year in area C. The expected scenario though is only 14 h, and if or when flicker occurs, the turbine is shut down in these hours. There is again a small interference with the summerhouses, but again, a solution for this is to move the turbines slightly south. This will also remove some of the flicker from area C but increase it on area D. Area D is on the ZVI map, the area owned by the main investors. As with the noise, the biggest issue with shadow and flicker effect is the summerhouse area. As mentioned before, a solution is to turn the turbine off in these hours where flicker occur. If the turbines cast too much shadow, a more southern position is possible.

STAKEHOLDER ANALYSIS

Stakeholder analysis takes ideas stemming from the case studies as well as an interview with a local actor with high interest in the project. It will be used to help suggest an ownership structure and approach specific to the area of interest.

The purpose of conducting a stakeholder analysis is to best understand the context of the local area where wind turbines have the potential to be placed, in the form of assigning importance to each stakeholder. By understanding the local actors who would be affected by such development, it is possible to plan for the future, that is, how to approach certain actors and how to qualm whatever opposition they may have towards such project. The goal of the stakeholder analysis is to help provide information that can be used to dictate what ownership structures can be used for the wind turbines, in the specific location in Vinderød. Information from the stakeholder analysis can also be used to formulate a "best approach" that can be used during the planning process towards the local community, in order to again minimize any opposition that could potentially arise. The stakeholders that will be the focus of the project are those who interact within small communities, that is, within the municipality, including the municipality itself (Schmeer, 1999). It is chosen to mainly focus on local stakeholders, for they are those who have direct effect on project success, be it through their desire for project realization or through their ability to disrupt project success through opposition.

Stakeholder Characteristics

In order to best assign importance to the individual stakeholders, it is necessary to examine a series of characteristics relating to importance. These characteristics, when compiled for an individual stakeholder, make it possible to make relative conclusions in regard to their importance to the project. These characteristics are as follows (Schmeer, 1999):

Interest in project: This explains why a stakeholder is interested in the proposed project, for example, for financial reasons and aesthetic reasons.

Internal/external: Internal means the stakeholder is involved with the planning and promotion of the project. External is given to those stakeholders who are not a part of this process and currently have no knowledge of the project.

Stakeholder position: This is the degree of support a stakeholder has towards a project, which is determined through interviews and previous case studies. This degree of support is theoretical for all actors except the private investors, yet it is justified through outside knowledge. There are five levels to this category: opponent, moderate opponent, neutral, moderate supporter, and supporter.

Stakeholder knowledge: This describes how much knowledge the stakeholder has on the subject of wind turbine development, which in turn affects their understanding of the project. It can be broken into four categories: high knowledge, moderate knowledge, low knowledge, and no knowledge.

Resource: This category is split into two parts: quantity, that is, how many individuals make up an actor, and ability to mobilize, that is, how well can the said actors organize themselves together to achieve their desired goals. They are measured from low, to moderate, to high. Whereas resource can be related to financial resource, technological resource, etc., it is chosen to focus on human quantity here.

Power: This category is also split into two parts: influence, that is, how much social influence the stakeholder has within the community and how much influence the stakeholder has on the project outcome, and legal power, that is, how much legal power the stakeholder has when determining the outcome of the project. Legal power is a representation of the degree a stakeholder can affect the project legally. They are measured from low, to moderate, to high (Wilson, n.d.).

Key Stakeholders

Ole Østergaard/Primary Investors

Ole Østergaard is a local farmer, who has property that he would like to have wind turbines placed on or near it. Next to him are 4 other farms, of which two or three have owners who also have property that could see the potential development of wind turbines. Their motivation for the project stems from financial gain, as well as the desire to produce their own electricity and decrease the amount of fossil fuel usage. Naturally, they would like to receive the highest possible financial gain from the project; therefore, they do not want to involve bank loans to finance the project if not needed. They are open to involving the community in the ownership of the wind turbines to help finance the project. They are very motivated to see the project transpire; however, they need help with the project planning (Østergaard, 2014).

Their roles in the project include being the landowners and the main investors. This also makes them the only actors aware to the turbine development at this stage; naturally, they are very supportive of the project. They have limited knowledge of the best practices in the development planning process and best practices when approaching the community. There consist of a small number of local investors who would spearhead the project, and they are very capable of joining and working together. Due to their status as locals within the community, they do have limited influence over others. As landowners, they also have direct control over project completion. However, they have no legal power (Østergaard, 2014).

Vinderød Community

Vinderød is a small town within the Halsnæs Municipality; it has a population of approximately 1000 (Danmarks Statistik, 2014). Due to the potential visual impact some community members might encounter from turbine development, certain Vinderød community members may be very interested in the project. Because Vinderød can be included in the ownership of turbines, this may also incite interest towards the project within the community. Interest in the project can be either positive or negative, depending on person to person. At present, they are external to the promotion and planning of the project and therefore have no knowledge at all of the project. They are also expected to have limited knowledge of turbine development and best planning practices. There are many members of the Vinderød community, and it is assumed they will have dialog about turbine construction when they are informed, which could potentially lead to small opposition groups. Because this is a small community and there is potential for opposition groups, Vinderød does hold some influence over the project. However, as individuals of Vinderød, their legal power is limited.

Municipality

The Halsnæs Municipality is where both the potential turbine location and the town of Vinderød are located. As mentioned in Chapter 5.1, the municipality is interested in promoting renewable energy sources, for example, wind turbines. The opportunity to be included in the ownership of turbines can also generate interest towards the project, within the municipality and within the group of municipal leaders. Like the town of Vinderød, at present, they are external to the planning process and therefore have no knowledge of the potential development. The municipality also collects taxes through turbine revenues, providing some form of positive incentive as well. Due to the stated interest in renewable projects, the municipality can be assumed to have some moderate support towards turbine development. As a municipality, they are knowledgeable about turbine development and specific rules associated with it. There are a large number of individuals within the municipality, yet due to their larger population and expansive proximity to the turbines, it is likely they have low capabilities in collaborating together. The municipality's opinions on turbine development may be irrelevant to most, yet they are a very powerful actor in that they approve wind turbine development, and therefore, it is imperative to have their support for the project.

Environmental Interest Group, for Example, Danmarks Naturfredningsforening

As has been seen in the projects mentioned before, environmental conservation groups are very interested in wind turbine development for the threat it can potentially pose to the local environment (Søren Hermansen, 2014). They are interested in the project for reasons other than financial; therefore, it is unlikely including them in the ownership of turbines would increase their support towards them. At present, these groups have no knowledge of the potential turbine development, and it can be assumed that when development proposals are introduced, environmental conservation groups will be some of the first to oppose such projects, again based on what has been seen before, such as in Samsø. It can therefore be assumed that environmental groups will somewhat oppose turbine development. They are aware of the local environment and if there are any threatened species that could inhibit turbine development and would use this in their opposition towards turbine development. While there are not many people involved with environmental interest groups, they are very capable of beginning opposition campaigns against potential wind turbines. They have moderate influence on others within the community, for they can be community members and have the best of intentions. They have limited legal power, yet it can be assumed they would use whatever legal power they have.

Summerhouse Owners

Located near the potential turbine placement site are summerhouses. These houses are inhabited during certain periods throughout the year. It can be assumed that people go to these summerhouses partly for the aesthetic value of the surrounding area. A select number of summerhouses are within the bounds of noise and visual effects caused by the turbines. Summerhouse owners will potentially be very interested in the project, as has been seen in previous projects where they have formed opposition groups towards wind turbine development (Søren Hermansen, 2014). There is no knowledge of the potential development for this category of actors. Their position towards the project will most likely be opposition, because turbine development has the potential to visually disturb them and there may be unwanted sound effects. They most likely have little knowledge about turbine development. There are a moderate number of summerhouses, yet due to their nature, it can be assumed that they would have incentive to collaborate against turbine development in the form of opposition. As has been seen in previous cases. they can hold moderate influence within a community (Søren Hermansen, 2014). However, it can be assumed they have little legal power.

Nonprofit Consultant

Consultants are important actors in that they help the local investors realize their wants, such as turbine development. Their interests in the project can stem from numerous reasons, such as academic and business. Their position is supportive as they are working for their client. Their knowledge of turbine development is high. They work in small numbers but are very capable at collaboration. Their influence is high for they are professionals. Yet they hold little legal power.

Stakeholder Table[5]

A	B	C	D	E	F		G	
Stakeholder	Interest in project	Internal/ external	Stakeholder position	Stakeholder knowledge	Resource		Power	
					Quantity	Ability to mobilize	Influence	Legal power
Ole Østergaard /Primary investors	Financial	Internal	Supporter	Low knowledge	Low	High	High	Low
Vinderød	Aesthetic	External	Neutral	No knowledge	Large	Low	Low	Low
Halsnæs Municipality	Image	External	Moderate supporter	Moderate knowledge	Large	Low	Low	High
Environmental groups	Aesthetic, Environmental preservation	External	Moderate Opponent	High knowledge	Low	High	Moderate	Moderate
Summer house owners	Aesthetic	External	Opponent	Low knowledge	Moderate	High	Moderate	Low
Non-profit groups	-	Internal	Supporter	High knowledge	-	-	High	Low

[5] Stakeholder table is meant to organize the information in the previous section.

Stakeholder Table Analysis

The "importance" of stakeholders is defined as their ability to affect the implementation of the policy (project).

Schmeer (1999)

Using the information that has been collected, it can be determined which stakeholders are most imperative to project success and which have a low potential to affect the outcome of the project. When compiling the characteristics of a certain stakeholder, for example, Ole Østergaard/primary investors being supporters with financial incentive, who have a high capacity to work together and also have high influence over the project, their subjective importance can be compared with other stakeholders'. Doing this for each stakeholder shows the relative degrees of importance they may hold, as discussed below. Figure A12 shows the actors involved and the relation to each other in terms of importance.

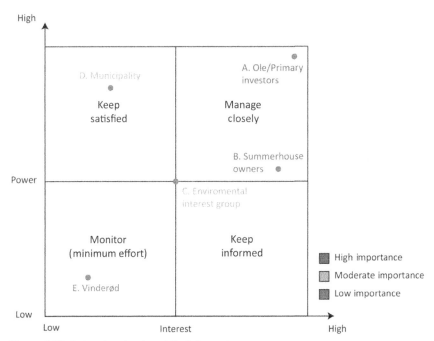

Figure A12 Actors involved and their importance.

High Importance
Ole Østergaard/Primary Investors

Because the primary investors are the owners of the land for the proposed turbine development, they are in direct control of the project. At any point, they can back out or change their mind about the location of the turbines, thereby eliminating any possibility of the project whatsoever. These primary investors also have direct control over the ownership structure of the turbines, which has been seen to affect project outcome immensely. While it can be assumed that they will not change their mind about constructing turbines on their property, it is important that they decide the ownership structure of the turbines in a way that would maximize support and decrease opposition towards them.

It also falls on the primary investors to introduce the project to the community and plan the project according to those a part of the local community, a very important aspect of project success. The primary investors must know how to include the community in the planning process and how to potentially deal with whatever opposition that may arise throughout the planning process. If the introduction and planning of the turbines are done in a way that does not involve the community, there is opportunity for opposition groups to grow larger than they would have had, if they were included in the planning process and properly informed about the benefits of turbine development, a technique that was employed on the island of Samsø, discussed in Chapter 4.2.

As can be seen, the direct control of the primary investors makes them the most important stakeholder to the project. It is important they are as informed as possible, in this case by the outside consultants, so they can make the right decisions that will maximize the probability of project success.

Summerhouse Owners

Summerhouse owners are less important than the primary investors, yet they do pose the largest amount of potential opposition towards turbine development, which is why they are considered a very important stakeholder in regard to the project. Made up of approximately 300 summerhouses that have a large incentive to oppose turbine development for aesthetic reasons, this group of the community can be assumed to be the most vocal opposition group towards the turbines, as has been seen in previous cases such as in Samsø. This could take the form of antiturbine campaigns that can spread throughout the larger community and seriously reduce the probability of project success. Summerhouse owners can also be assumed to not always be as motivated by financial gains in their support for the project as compared with those who live in the area year round; therefore, offering shares may or may not have any effect on the acceptance towards turbine development.

It is important to include the summerhouse owners in the planning process directly from the beginning so at no point do they feel left out and at no point are misinformed about the project. Constant dialog is also necessary so all actors can voice their concerns and again receive the correct knowledge in regard to turbine development.

Moderate Importance
Environmental Interest Group, for Example, Danmarks Naturfredningsforening

As has been seen in previous cases, such as the case on Samsø, nature conservancy organizations are very interested in turbine development for the potential negative effects it can have on their surrounding nature. They are very knowledgeable about the environment and how it can be disrupted through turbine development; this combined with their passion for nature makes them potentially strong opponents to turbine development if they see it as environmentally disruptive. While they may not be composed of many members, they are highly capable of working together and can be very vocal about their concerns. This could slow or even halt project completion if not taken into account during the planning phases of the turbines. While they have little legal power, when and if they have an opportunity to use it, say for the protection of a protected species, it can be assumed that they will use it. As is true with the summerhouse owners, financial gain is assumed to have little effect on these actors, nulling the potential for offering shares as a means to increase acceptance towards the turbine development. Nature conservancy groups, despite being small, can have influence within a community and can affect project success; therefore, it is important that they are included in all planning processes.

Municipality

While there are many members of the municipality that can be assumed to be indifferent towards turbine development located far away from their residences, the municipal government is important. The municipality's importance to project success lies in the legal power it holds over the permitting of turbine development. This direct control over project realization makes them a moderately important actor. There are potential financial gains the municipality and its inhabitants can receive through turbine development, yet at the same time, the municipality deals with complaints about turbine development. It can be said that the politics of turbine development affect what decisions are made in regard to whether or not a project is realized within the municipality. Depending upon whether or not members within the municipal government feel they can gain political support or lose political support determines the decision they make. As far as municipal ownership is concerned, it is assumed the municipality will not form any company to involve itself in turbine ownership. This is due to the relatively small nature of the individual project.

What the municipal residents say about turbine development dictates how the municipality determines if it permits turbine development. This means that such municipal residents who may oppose turbine development are those that should be included in the planning, in order to best maximize potential for project success. It is also necessary to explain to the municipality the potential benefits it can receive through turbine development, as far as local development, to better increase the chances of including them within the project and getting approval for such development.

Low Importance
Vinderød

The town of Vinderød is seen as having low importance to the project success for numerous reasons. Vinderød is a small community not much larger than the summerhouse community. There are minimal visual effects the town may experience due to turbine development, and these visual effects can be mitigated with the offering of shares to such individuals experiencing them. As a neutral actor, the offering of shares is assumed to generate some support towards the project, yet it is unlikely that the community will be campaigning for the turbines. As a small community with little incentive to be opposed or supporters towards turbine development, the town of Vinderød can be assumed to be the least important actor. This being said, it is still important that they are included in the planning phases of the turbine development and they are fully informed about the project.

Miscellaneous
Consultant

Consultants are very important for project success, yet they are not included in the analysis because of the role they play as developers. It can be assumed that it is in their power to help complete the project they are working on, yet they have limited invested interest in the project beyond whatever financial compensation they may receive from the client. For this, they will not be examined any further.

Summary

The reason for the stakeholder analysis is to better understand the local context where potential turbines may be developed. It is meant to inform how to best include the community in turbine ownership, that is, who are the most important stakeholders to include. It is also meant to help with the successful planning of such turbines. Through stakeholder analysis, it can be known which actors may provide the most opposition towards a project and who are the most important to keep satisfied. This information will be used to help answer how it is best to approach the community about turbine development and how is it best to include them within the planning process of said turbines.

OWNERSHIP MODELS AND PROJECT ECONOMICS

The purpose of this chapter is to present the economy of the project, that is, the investment cost and the income of the different owners, and to outline the different ownership structures including both the primary investors and local community. These ownership models are a recommendation to the primary investors after having followed the methods that are described in this chapter.

Economics

In this section, the assumptions and the calculations that were made in Microsoft Excel in order to estimate the payback time of the wind turbines for each category of investors are described. This is achieved by calculating first of all the investment cost of the wind park, then the income produced by the electricity that is sold to the grid, and finally the taxes that are applied to this income depending on the investor.

Costs

Starting with the calculation of the costs, this refers to the investment costs of the wind turbines. According to Energinet.dk and The Danish Energy Agency (2012), the nominal investment cost for onshore wind turbines in Denmark is considered to be 10.45 MDKK/MW. This means a total investment cost of 41.8 MDKK for the two Vestas V80 turbines of 2 MW each.

Table A1 Cost structure for A Typical Medium Wind Turbine (Lemming et al., 2008)

	Share of Total Cost %	Typical Share of Other Costs %
Turbine, ex works[a]	74-82	–
Foundation	1-6	20-25
Electric installation	1-9	10-15
Grid connection	2-9	35-45
Consultancy	1-3	5-10
Land	1-3	5-10
Financial costs	1-5	5-10
Road construction	1-5	5-10

[a]Ex works means that no balance of plant, i.e., site work, foundation, or grid connection costs are included. Ex works costs include the turbine as provided by the manufacturer, including the turbine itself, blades, tower, and transport to the site (Energinet.dk and The Danish Energy Agency, 2012).

These costs are divided into different shares, where the actual turbine cost only corresponds to 74-82% of the costs, as it is shown in Table A1.

The limitation for this table is that it is for turbines up to 1.5 MW. Since the Vestas V80 turbine is 2 MW, these shares could be less accurate. However, in order to calculate the total investment costs, assumptions are made about the exact percentage of each share and numbers are chosen according to the following table.

As it is seen in Table A1, the grid connection covers a share of 2-9%, and the price for the land covers a share of 1-3%. The grid connection in this case is covered by Energinet.dk, and the turbines are installed on already owned land areas in Vinderød. Therefore, the grid connection and the land price are subtracted from the total investment costs, that is, the total investment is calculated as 39.29 MDKK.

Income

In order to calculate the income, a NPV calculation is made using the formula below:

$$\mathrm{Net}\,p_{resent_{\mathrm{value}}} = \sum_{i=0}^{n} NP_t \cdot (1 + \mathrm{Discount}_{\mathrm{Rate}})^{-t}$$

where NP_t is the net payment at time t, NP_0 the invested amount, and n the duration of the investment (Lund and Østergaard, 2010).

The discount rate is chosen to be 2% for the primary investors and 4% for the rest of them. This decision is based upon the knowledge that the primary investors can fund the investment with their own capital, so there is a low risk for them. The number of the 4% is based on the average discount rate that is applied currently in Denmark (ycharts.com, 2014). The duration of the investment is considered to be equal to the lifetime of the wind turbines, that is, $n = 20$ years from 2016 to 2035 (Energinet.dk and The Danish Energy Agency, 2012).

The production is assumed to be stable during the years, and it is taken as a result from the WindPRO calculations as 14878.8 MWh/year. The spot market price is also assumed to be stable during the lifetime of the wind turbines. The price that is used in the calculations is the average spot market price for 2013 in east Denmark found from Energinet.dk, and it is 35.123 øre/kWh = 351.23 DKK/MWh (Danmarks Vindmølleforening, 2014).

Additionally, the FIT price for the electricity produced from onshore wind turbines, from 01.01.2014, is 250 DKK/MWh, as long as it does not exceed 580 DKK/MWh when combined with the spot price. Because in this case the sum of the two prices, spot market price plus FIT, is always higher than 580 DKK/MWh, the price of 580 DKK/MWh is applied for the years that the FIT price is applicable (Danish Energy Agency, 2012).

The FIT is granted for the first 6600 full load hours plus the production of 5.6 MWh per square meter rotor area. The rotor area is 5.027 m^2 for each 2-MW Vestas V80, which means

$$2\text{MW} \times 6600\text{h} + 5027\text{m}^2 \times 5.6\frac{\text{MWH}}{\text{m}^2} = 41351.2\text{MWh}$$

So in this case, the FIT price is applied for 41251.2 MWh produced by each turbine, which means for 2.78 years. The total income that is generated by selling the produced electricity is calculated to be 113.98 MDKK (without the taxes) and 93.76 MDKK (without taxes and discounted) (Retsinformation, 2013).

Taxation

The next step in the calculations is to apply taxes in order to estimate the profit generated from the wind turbines. An assumption is made that one share is equal to 1000 kWh of production, which sums up to a total of 14,879 shares in total for the two turbines. Different rules of taxation apply to different types of investors. In this case, there are three types of investors: individual investors[6], cooperatives, and foundations. The revenues of a

[6] As mentioned earlier, individual investors are defined as private people wanting to invest in the wind project. In this case, it could be the farmers owning the land where the turbines will be placed or private people from one of the three groups that are explained in Chapter 3.

private investor are taxable. The investor has two options when it comes to taxation according to the law: the schematic scheme and the business scheme (dkvind.dk, 2014).

The schematic scheme is primarily for private investors that have invested a small amount of money in the turbines. The first 7000 DKK of the revenue is tax-free. If the investor has revenue of more than 7000 DKK, it is taxable by 60% (Skat.dk, 2014a, 2014b).

The business scheme is considered as a commercial investment. This scheme applies to either private investors wanting to form a company to invest in the turbines or an already formed company. The business tax by 2014 is 24.5% of all revenues (tax.dk, 2014).

For the 20% mandatory offering to the locals, it is assumed that they will not own more than a total amount of 12 shares per investor, which is equal to an income lower than 7000 DKK. In that way, these investors get some tax incentives because they will not have to pay any taxes for revenue less than 7000 DKK.

The investment made by the primary investors, in this case the farmers, follows the business scheme, as it is assumed that the farmers will create a business for the investment of the turbines. This means that the revenues of the farmers are taxable by 24.5%.

In addition, it is assumed that cooperatives will be founded by anyone interested in joining and that the members, as in the case of the 20 % offered local ownership, will not own more than 12 shares each, in order for the revenues to be tax-free.

The taxation of the foundations depends on the type of it. In this case, it is assumed that a humanitarian foundation will be founded like in the case of Hvide Sande (Windpeople, 2014a, 2014b). Humanitarian foundations are not taxable; therefore, no tax will be applied to the revenues of the foundations (Table A2) (Skat.dk, 2014a, 2014b).

Dynamic Payback Time

Also, the payback time is calculated for each investor. The calculation of it is made by using the following formula:

$$\sum_{i=0}^{n} NP_t \cdot (1 + Discount_{Rate})^{-t} = Investment$$

where t the number of the years that needs to be found and it represents the years needed in order for the investment to be paid off by the produced income (Østergaard, 2013).

Table A2 Summary of the Assumption Made in Order to Calculate the Payback Time

Category	Assumption	Results
Total investment cost (DKK)	The discount rate is assumed to be 2% for the main investor and 4% for the three other investors	39,290,496
Total number of shares	1000 kWh = 1 share	14,879 shares
Cost/share (DKK/share)		2640.70
Tax		Tax rate
Main investor	All income is taxed	24.5%
Local people	To remain tax-free, this investor buys a maximum of 12 shares	0%
Cooperatives	To remain tax-free, this investor buys a maximum of 12 shares	0%
Foundation	Humanitarian foundation = tax-free	0%

Ownership Models

This section provides suggestions about what the best forms of ownership may be for the primary investors of the Vinderød project. Because it is unknown how much capital the primary investors are able to invest, it is necessary to provide multiple ownership suggestions, each with different levels of private investment and community involvement. The degree of primary investor ownership and community ownership is based on the information that has been compiled throughout the project.

Three ownership structures are presented and can be seen in Figure A13. This allows the primary investors to see their potential degrees of ownership of the project and the corresponding involvement of the community. Each ownership structure, with different degrees of community ownership, will be commented on as to how sufficient it may be at reducing opposition towards the potential wind development.

80% Primary Scheme

Assumptions

This model is an example of the minimum amount of ownership that is obligated by law to be offered to the community; therefore, only 20% of the turbine investment is being offered through shares to the surrounding community. This ownership structure assumes that the primary investors, that is, Ole Østergaard and his partners, would be able to cover the full investment cost; however, they decide to finance it and that local people do take

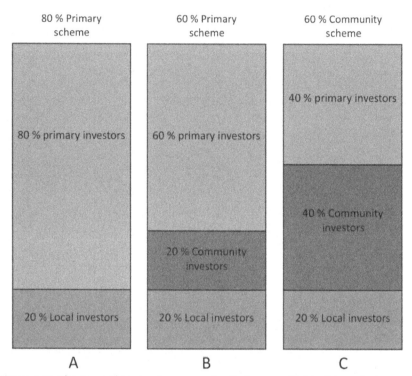

Figure A13 The ownership structure scenarios that are applied in this project.

advantage of the full 20% turbine investment that is offered to them. Ole Østergaard, in his interview, stated that he couldn't finance the project alone and that he does not want to take out loans in order to fund his investment, yet the level and method of financing the investment for the other primary investors are unknown; therefore, the full ownership of 80% of the turbine is a valid assumption. The 80% investment is broken up by the private investors, that is, Ole Østergaard and his partners, and is purely up to them, based on how much they individually want to invest.

Investment/Returns

It is further assumed that those buying shares offered through the mandatory public offering do not purchase enough shares to receive over 7000 DKK in revenues per year in order for the revenues to be tax-free, as mentioned above. This would equate to an investment of 31,870 DKK, that is, 12 shares, which is thought to be the highest amount of money that a community member will invest. This assumption for the mandatory public offering

Table A3 Results from Financial Calculation for the 80% Primary Scheme
80% Primary Scheme

	Local People	Private Investors
Investment (MDKK)	7.86	31.4
Taxes	0%	24,5%
Income after tax (MDKK)	16.6	58.2
Payback time	6 years and 5 months	8 years and 10 months

will be made for each scenario. The full cost for 20% of the turbine investment is 7,858,099 DKK, and the total return after 20 years is 16,598,237 DKK. The payback time for the local people for this investment is 6 years and 5 months.

To take full advantage of turbine revenues, it is suggested that the primary investors set up a company to own the turbine. Turbine revenues would then be taxed 24.5%, as it was mentioned earlier. This is to be suggested for all scenarios. After creating a company, the total cost of the turbine investment for the primary investors would be 31,432,396 DKK, with a return of 58,255,702 DKK over 20 years after taxation. The payback time would be 8 years and 10 months. In Table A3, the results from the calculations are presented. The creation of a company has costs associated with it that are not elaborated on due to their subjective nature.

Discussion

This ownership scheme is beneficial to the primary investors, because they invest the most and would therefore receive the highest financial gains, which is something Ole Østergaard as an investor has stated he would like to see, and it can be assumed that the other investors would react in the same way. As investors with high financial gain, this provides the primary investors, the most important stakeholders, maximum incentive to see project completion. With the 20% mandatory offering to the surrounding community, this also still provides for local people to be involved in the ownership of the turbines.

Regarding the 20% mandatory offering to the surrounding community, as explained earlier in Chapter 3.4, one has to be specially registered, that is, have a CPR number, associated with that area. Therefore, any chance of reducing summerhouse opposition by including them in the ownership of turbines is diminished. A 20% offering of ownership may also not be enough in the eyes of community members. This can potentially lead to a community, which due to their lack of incentive, may form opposition groups.

60% Primary Scheme
Assumptions
Scenario B assumes that the turbines will be 60% owned by the primary investors, while the other 40% will be owned/offered to the community. Half of this 40%, that is, 20% of the investment, will, by obligation, be offered to the community based on the regulations stated in Chapter 3.4, whereas the other 20% will be offered to the community through other ownership methods. It is suggested that 10% of the investment will be paid through the creation of a wind cooperative, while the other 10% will be paid through the creation of a foundation. This is suggested for simplicity purposes as a reference point that can be used by those interested. This scenario assumes the primary investors cannot or do not want to invest in more than 60% of the project cost, that the 20% community obligation will be completely taken advantage of, and that there are still those who would want to invest in the potential wind turbines.

Investment/Returns
As 60% of the owners of the wind turbines, the primary investors would invest 23,574,297 DKK with a return of 43,691,776 DKK over 20 years after taxation. The 20% mandatory community offering would remain the same. The cooperative that would be created would consist of 1488 shares with a cost of 2640.7 DKK per share. Each share would generate 5577.8 DKK over the 20-year turbine life span. Similar to the mandatory public offering, those joining the cooperative are assumed to not buy an excess number of shares that would generate a yearly return of 7000 DKK, making it tax-free based on Danish regulation.

As mentioned earlier, the foundation as a humanitarian one will not be taxed. The foundation's turbine investment equals 3,929,049 DKK, and the return over 20 years equals 8,299,118 DKK. The payback time is 6 years and 5 months. The foundation can make this investment through private investment, that is, charity, or through bank loans. Because it is unknown how this will occur, and because of their subjective nature, bank loans are not included in the financial analysis. In Table A4, the investment, the income, and the payback time for the investors in this case can be seen.

Table A4 Results from Financial Calculation for the 60% Primary Scheme
60% Primary Scheme

	Local People	Private Investors	Cooperative	Foundation
Investment (MDKK)	7.86	23.6	3.93	3.93
Taxes (%)	0	24.5	0	0
Income after tax (MDKK)	16.6	43.7	8.29	8.29
Payback time	6 years and 5 months	8 years and 10 months	6 years and 5 months	6 years and 5 months

Discussion

Unlike the first scenario, this scenario extends ownership to the community beyond what is obligated by law, and it allows the community to be more involved in the ownership structure. The formation of a wind cooperative allows for the summerhouse owners and other persons who might not have been able to take part in the 20% offering to now be a part of turbine ownership. The potential for more members of the community to be involved in turbine ownership can reduce opposition. More potential for ownership can equate to increased opportunity for the community members to benefit financially from the project.

The creation of a foundation, where revenues can be used for local development purposes, also has great potential of reducing opposition towards turbine development, depending on how the turbine revenues would be spent. At the same time, the primary investors are still majority owners, technically owning 1 and 1/5 of the turbines. Because it is unknown at this time how much capital the primary investors can invest, this option provides them with a standard they can follow when determining the ownership structure of the potential turbines while still acting as majority owners.

However, this model of ownership has some disadvantages, since it is not capable of pleasing a lot of members of the community. It may not offer enough ownership shares of the turbines to the community, as they would like to. The same case may be for the primary investors.

60 % Community Scheme

Assumptions

Scenario C assumes that 40% of the turbine investment is owned by the primary investors, whereas 60% is offered to the community. 1/3 of this 60%

will be offered by obligation to the public following the same rules seen in the other scenarios. The other two-thirds of this 60%, that is, 40% of the total investment, can be broken down in a variety of ways, yet it is chosen for ease of economic calculation to split it evenly between cooperative ownership and foundation ownership. This scenario assumes that 40% ownership is enough for the primary investors and that the 20% obligation is fully taken advantage of. It also assumes that there will still be demand for turbine shares.

Investment/Returns

As 40% of the owners, the primary investors would invest 15,716,198 DKK to generate a return over 20 years of 29,127,851 DKK (after taxation) and a payback time of 8 years and 10 months. Cooperative shares would cost 2640.7 DKK each and would return 16,598,237 DKK in total with a payback period of 6 years and 5 months. There would be a total of 2976 shares offered to the public through this cooperative. It is assumed that everybody is buying no more than 12 shares, which sums up to 248 investors. This corresponds to approximately ¼ of Vinderød. It needs to be kept in mind that the whole Halsnæs Municipality can buy these shares. The foundation would invest 7,858,099 DKK to receive 16,598,237 DKK over 20 years. The public offering is again the same as in the first scenario, as it is the formation of a company for the primary investors, and the other assumptions regarding turbine shares and taxation. In the Table A5, financial results for each investor are presented.

Discussion

With the most amount of community ownership, this scenario can be assumed to have at least the same amount of community acceptance as

Table A5 Results from Financial Calculation for the 60% community scheme
60% Community Scheme

	Local People	Private Investors	Cooperative	Foundation
Investment (MDKK)	7.86	15.72	7.86	7.86
Taxes (%)	0	24.5	0	0
Income after tax (MDKK)	16.6	29.1	16.6	16.6
Payback time	6 years and 5 months	8 years and 10 months	6 years and 5 months	6 years and 5 months

scenario B, if not more. More people can be involved in the ownership structure, and with the community as the majority owner, they can be certain they are not being undermined financially.

The issue with this scenario is that the level of financial involvement for the primary investors might not be sufficient in their eyes. This may result in a lack of support for the project from the primary investors, which can result in the stoppage of any further turbine planning on their property. It is also unknown how much the community would like to be involved in turbine ownership, potentially making this scenario less considerable.

Sensitivity Analysis

To conduct a sensitivity analysis, different discount rates, that is, 0-5%, and an increase and a decrease of ±5%, ±10%, ±20%, and ±30% in the spot market price are applied, while the other factors remain unchanged. It is assumed that the spot market will increase/decrease once by 2020 and 2 years after the FIT will expire, and then it is considered stable for the rest of the years. The discount rate of 0% is examined considering the fact that the exact owners of the wind turbines are unknown and so is the risk of the investment.

As it can be seen in Figure A14, in the case of the second scenario when the discount rate is increasing, the income of each investor is decreasing following a different inclination for each one of them. The bigger decrease happens in the case of private investors where the income drops by 32% from

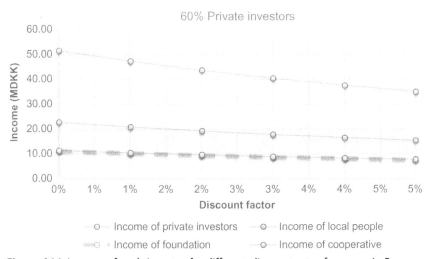

Figure A14 Income of each investor for different discount rates for scenario B.

51.63 MDKK with a discount factor of 0% to 35 MDKK with a discount factor of 5%. However, this is expected since they will own a bigger share of the turbines. As far as it concerns the payback time, it follows the same trend as the income, where the 8 years and 2 months becomes 10 years and 2 months for the private investors and the 5 years and 8 months becomes 6 years and 7 months for all the other investors. The same conclusions are made for the other two scenarios, where the income follows similar curves for all the investors (Figure A15).

When the spot price decreases by 30%, the payback period increases by 2 years. When the spot price increases by 30%, the payback time of the primary investors is decreased by almost 1 year. Similar trends occur with the other fluctuations in spot price, but they are just less pronounced. Because it is impossible to predict the spot-price change, there is a high degree of uncertainty.

Summary

The goal of this chapter is to provide theoretical ownership examples for the specific case of Vinderød, breaking down attitudes held by a community that could be extended to other communities of similar characteristics. By looking at the specific case of Vinderød, and how community ownership can be

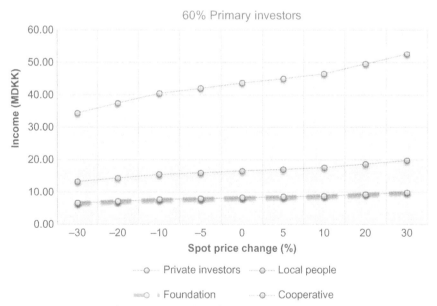

Figure A15 Income of each investor for different spot prices for scenario B.

applied to different degrees, other areas of similar characteristics can use these models for their individual situation.

In particular, by creating three ownership models that could potentially be utilized in the case of Vinderød, and then discussing the pros and cons of each one individually, as well as attaching financial characteristics to them, it can be better understood how the final suggestion should be formed and what ownership structure should be used in the specific case the project is dealing with.

By understanding the degrees in which each ownership model would affect community acceptance, and then keeping in mind the interests of the certain actors in control of the project, an all-encompassing suggestion can be formulated (see Chapter 8). Discussing how actors feel in each case allows for a better understanding of the degree that the community owner-ship may affect their view towards the project.

DISCUSSION

This chapter answers the subquestions. Based on the pros and cons of the model scenarios in Chapter 7, an ownership model is suggested. A best approach towards the community is suggested also in various stages.

Proposed Ownership Model

In order to answer the research question, it is necessary to examine a certain area holding such characteristics. Answering questions about such an area provides information that can be extended to other communities with sim-ilar makeup. In the case of Vinderød, based on what has been previously seen through case studies and stakeholder analysis and after examining different ownership structures, a suggestion can be made as to how to answer the first subquestion, "How can the community surrounding Vinderød be best involved in the ownership structure of potential turbines to decrease oppo-sition?," in order to decrease opposition towards such turbines, for this specific area.

Starting with the primary investors and Ole Østergaard, based on stake-holder analysis, it is important that they are majority investors. As has been seen in the case studies, private investors, in this case the primary investors, can be majority owners of wind turbine while project success is maintained. As 80% of the owners, the community has the potential to feel left out, whereas at 40%, the primary investors are no longer majority owners. Therefore, it is suggested in the case of Vinderød that the primary investors start out owning 60% of the turbines, with the other 40% being offered to the community or owned through a foundation. Because it is unknown how much community interest there will in fact be towards turbine ownership, primary investors still have the potential to own an even larger percent of the turbines. Based on stakeholder analysis, this should both please the primary investors and satisfy those of the local community who may want to be a part of the turbine ownership structure. There is no difference in each investor's payback time, for either the primary investor or the community member for all three scenarios.

After the mandatory public offering stated by Danish law, there is 20% left of the project that can be offered to the public. Because it is unknown how much demand there will be for shares, it is necessary to be flexible with this second community offering. In the case of high demand, where community members want to buy many shares, cooperative ownership is recommended for the remainder 20%. Whatever shares are not sold should be bought by a humanitarian foundation, with the purpose of not only using turbine revenues for community benefit but also selling and buying shares from community members for continued satisfaction. This foundation can be created through bank loans or through private investment. In the case of low demand, where community members are uninterested in the project, it is recommended to create a foundation owning 10% of the turbine, with the same buy/sell purpose as before. The remaining investment can then be made by the primary investors. If there is no demand to buy shares from the community, including the mandatory offering, a larger foundation is recommended to be created owning 20% of the project, so community improvement and benefit from such a foundation will be more evident throughout the community. The remainder of the investment would then be made by the primary investors.

When planning the ownership structure of the potential turbines in Vinderød, it is important to consider the unknown variability regarding such. Because of this, it is necessary to be flexible. The dynamic ownership structure above is subjective based on the community, where its flexibility best decreases whatever opposition could be decreased through financial inclusion in the project, regardless of the degree of community involvement.

Approach Towards the Community

In the same way the first subquestion helps to answer the research question, answering the second subquestion, "What approach is best to take when introducing the project to the community and how can this process be continued in the town of Vinderød?" can be used to better understand small communities, their reactions to turbine development, and how to plan accordingly. It is important to understand the community and how they will react to turbine development, for their local peoples can limit turbine development with their opposition. Therefore, developing an approach to an individual community that results in the least amount of opposition when carrying out turbine development is a key to project success. Such an approach and accompanying suggestions can be extended to other small communities with similar characteristics.

This theoretical best approach, despite not being tested, is based upon what has been witnessed in other instances of turbine development such as on Samsø and Ærø, an on stakeholder analysis. One can therefore have confidence that what this approach consists of is relevant and can indeed be extended to other similar areas.

The suggested best approach is divided into stages that cover the most important steps when approaching a community about turbine development. This is done in order to provide flexibility, for this is needed whenever dealing with human resources. These stages can take the form of public information meetings, with the number of meetings depending on the progress made in previous meetings.

Stage of "Knowledge Gathering"

The first step, before any meetings occur, should consist of gathering all information about the local community where the turbines will be placed and compiling information about the planned turbine development characteristics. Information about the community, such as in the form of a stakeholder analysis, is imperative because it allows the potential developer to understand the different members and groups within a community and plan accordingly around such people and groups. Understanding the local community is meant not only to understand who the potential opponents may be but also to see whether there are potential actors who support the project and can contribute to project success.

It is also just as important to have all information about the potential turbines, such as where they will be placed, how big they will be, all visual and sound effects they may have, the costs and benefits for those who may be interested in buying shares, etc. As could be seen in the Samsø case, it was important to be able to answer the peoples' questions, especially individual questions that only pertained to that person in particular, so that everyone would be as informed as possible (see Chapter 4). Having information that might be of use to inform the local community is important in order to minimize any spread of opposition due to miss or lack of information about turbine development. It should be noted that in the case of Vinderød, there is a limited area where potential turbines could be located; however, if this were not the case, it would be important to get information on the various potential sites and then introduce them to the community. As was done in Samsø, this helps make the community feel a part of the project and not like things are being forced upon them.

The stakeholder and financial analysis, as well as the site assessment of this report, are examples of what should be compiled before introducing the project to the community. They prepare the potential developers for whatever questions or oppositions that may arise in the initial phases after introducing the project to the community, and they help in planning the next stages of a best approach.

Stage of "Introducing the Project to the Community"

The next stage for successful turbine development is a successful introduction of the project to the surrounding community. When introducing plans for turbine development to the community, it is important to focus on a number of things. It is important to include as many community members in such introduction meetings as possible, especially those who may be opposed to turbine development. It is also important to stress the degree of community involvement the ownership structure of the potential turbines would like to achieve.

Providing information about the potential turbine development to as many community members as possible is important so the community is well informed with correct information about the turbines. It should be noted that this is an ongoing process that can consist of multiple meetings. All members of the community should be aware of such meetings, especially those who may be opposed to turbine development, in the case of Vinderød, the summerhouse owners and nature conservancy groups. Making potential opposition groups feel a part of the planning process is a way to decrease opposition from such groups, as it is seen on the Samsø case with the nature conservancy group (see Chapter 4.2).

It is also important to stress the importance of community involvement and the benefits the project can provide. By explaining the formation of a cooperative, the costs and benefits of purchasing shares, the creation of a foundation, and any other possible ways of community benefit, the community can see that the project is not one of "selfish" financial gain on the behalf of the primary investors. It is necessary to be thorough and transparent when explaining how the community can benefit from turbine development so the potential spread of misinformation and opposition throughout the community will not occur. Sufficient information gathering, as is described in the first step, is needed so that whatever questions those may have about the project can be answered.

"Preconstruction" Stage

By this stage, the community should be sufficiently informed about the project, as that is the main goal of the previous stage. By this stage, any opposition and any support for the project should have developed. If the second stage is done in a thorough and flexible manor, as stated above, it is assumed whatever opposition may have arose towards the project will be minimized. As this is a "best approach" and meant to reduce opposition, it is impossible to guarantee zero opposition towards the project, something potential developers should keep in mind. The purpose of this stage is to convey finalization to the community, in the form of an EIA and deadline for share purchasing. Once this is completed, a timeline for project construction should be announced to the public.

Both holding public meetings regarding an EIA and announcing a deadline for involvement in the turbine cooperative are things that must happen before project completion. These two aspects can be regarded by the community as quantifiable and give the project more a sense of reality. Making these two aspects of the planning process clear ensures transparency, so that no last-minute opposition may arise. Regarding the EIA, it is a formalized account of whatever environmental impacts the turbine development may have, something that should have already been discussed within the community in the previous information sharing stage. Finally, explaining and announcing the construction process can be seen as the conclusion to the project.

Final remarks: Suggestions

In the case of Vinderød, certain suggestions can be made, and certain occurrences can be planned around, so that project success is realized, a matter of prime importance being those owning summerhouses close to the potential turbine location. As it is seen in the stakeholder analysis, such actors can be considered unmotivated to see project completion and can form opposition towards the project (see Chapter 6). As it is also seen in the case studies, groups opposed to turbine development can gain a lot of support, such as in Ærø, by taking advantage of the lack of proper information about turbine development (see Chapter 4). While it might be impossible to diminish any opposition that may arise from summerhouse owners, it is possible to spread the correct information about the potential turbine development to the surrounding community, as it was said before, helping prevent the spread of opposition. Also, while it is unknown how summerhouse owners will react to the project, yet assumed they will be opposed, meetings and information sessions can be planned around them, for example, holding meetings in the summertime, when they would most likely be present, again in order to make the planning of such turbines as transparent as possible.

As it is mentioned in the suggested ownership model, flexibility is of high importance. Flexible community meetings, flexible ways of involving the community in the ownership of the turbines, and flexibility with the proposed structure of the turbines, that is, height and number, are all important to ensure that the least amount of opposition will arise towards the project.

CONCLUSION

The growing importance of renewable energy in the global energy mix means the increase in the development of wind turbines, a technology some might regard as visually disruptive and noisy. To ensure satisfied electricity consumers and that development ensues into the future, those living in close proximity to the turbines are required to hold no major opposition towards such. Many small communities may be faced in the future with potential wind development; therefore, it is important to understand the methods for decreasing such opposition within these communities.

The concept of community ownership, or inclusion of community members in the ownership of turbines, and the importance of the approach towards the community when introducing turbine development are given most focus in this project for they are methods that can be utilized throughout the world. In order to better understand how such methods could be applied, it is best to look at an individual case, for it would provide the most accurate depiction of a real planning process. This project decided to look into the small town of Vinderød, where the opportunity to develop wind turbines presented itself. Hypothetically, planning such turbine construction by answering the subquestions A and B provides information that not only can be used by those in Vinderød but also serves as an outline to the information and steps that are needed to successfully plan turbines in other similar areas.

The research question of this project is as follows:

How can local individuals be involved in the planning and ownership of wind turbines in small communities to best decrease opposition towards development?

And it can be answered by dividing it into two parts, one concerning the involvement of the community in the planning process and the other one concerning their involvement in the ownership.

When introducing turbine development to small communities, one thing that is very important is to include the local community members from the beginning of the planning process. By including such community members in the beginning through public information meetings, they feel a part of the planning process and this will decrease any chances of false information spreading about the turbines that may result in opposition. This gives rise to the second most important part regarding introducing turbine development, making sure the local community is fully informed about turbine development and the goals of the ownership structure. By ensuring the community is informed about the importance and advantages of involving them in the ownership of turbines, as well as the characteristics and whatever potential noise and sound effects the turbines may have, whatever opposition that may arise can be addressed and the community will feel the project is not for the gain of only a small number of individuals. By including the local community in a flexible and transparent planning process, the least amount of opposition will arise and whatever opposition does arise can be dealt with.

When involving local communities in the ownership of wind turbines, it is best to include them in more than just the mandatory 20% offering dictated by the Danish state. This again prevents community members from thinking the wind turbines will only be for the profit of a small number of individuals and also provides financial incentive for development, hence decreasing opposition. It is important to again be flexible with how to involve the community in turbine ownership, be it through cooperative or foundations. The opportunity for choice is a draw for locals that again provides transparency and reduces opposition.

PERSPECTIVE ANALYSIS

As mentioned in Chapter 3.4, the 20% mandatory public offering shares for a wind turbine cannot be bought by citizens whom are not CPR-registered in the municipality where the turbine is installed. This leads to some controversy when, for example, summerhouse owners have interest in the ownership of a turbine. A person does not necessarily have to live all year round in the particular area with wind turbines to be disturbed by them or have interest in being part owners of them. This is the case in Vinderød where around 300 summerhouses are situated just north of the potentially installed turbines in this project. These owners cannot necessarily be part owners, since they most likely have their primary address outside of Halsnæs Municipality, but certainly can have interest, both positive and negative, in the potential turbines. The rule of not being able to be a part of the turbines if not CPR-registered in the municipality can also lead to other controversies. An example of this is shown in Figure A16.

In this case, two turbines are situated in municipality A, and zero in municipality B. Here, it is possible for house 1 to buy shares, because it is situated in the same municipality as the turbines. However, house 2 cannot buy share because of the municipality rule, even though house 2 is situated closer to the wind turbines than house 1. It is obvious that house 2 is affected more from noise, sight, vibration, etc., from the turbines than house 1 and therefore is more likely to gain opposition, especially because the owner is not allowed to be a part owner.

It is therefore suggested that the CPR registration rule is removed, and distance rule is changed, letting all affected persons to gain an opportunity

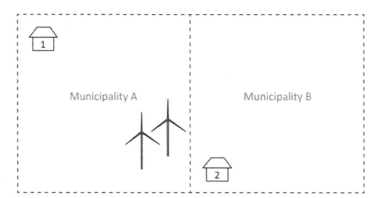

Figure A16 CPR registration in two different municipalities.

from the advantages of owning parts of a renewable energy source. To complete the social research this project has touched upon, it would be necessary to see how the community of Vinderød reacts to the suggestions given in this report. It could then be understood how valid the suggestions of this project are, and new suggestions could be made based on what was seen.

REFERENCES

ÆRØ Energy and Environment Office, n.d., Ærø—a renewable energy island. Available: http://www.aeroe-emk.dk/eng/index.htm [03.05.14].

Andersen, I., 2009. Den Skinbarlige Virkelighed. fourth ed.

BELL, D., GRAY, T., HAGGETT, C., 2005. The 'Social Gap' in wind farm siting decisions: explanations and policy responses. Environ. Politics 14 (4), 460–477.

BOGDAN, R.C., BIKLEN, S.K., 1982. Qualitative Research for Education: An Introduction to Theories and Methods. Allyn and Bacon, Boston, Mass.

Brejnholt, B.K., 2013. Wind energy as a lever for local development in peripheral regions. Nordic Folkecenter for Renewable Energy.

BRIGGS, A.R.J., COLEMAN, M., MORRISON, M., 2012. Research Methods in Educational Leadership & Management, third ed. SAGE Publications Ltd., London, England.

BRYMAN, A., 2012. Social Research Methods, fourth ed. Oxford University Press, New York.

Danish Energy Agency, 2012. Subsidies for Wind Power. Available: http://www.ens.dk/en/supply/renewable-energy/wind-power/facts-about-wind-power/subsidies-wind-power [19.05.2014]. Last update.

Danish Energy Agency, 2013. Energy Policy Report 2013. Ministry of Climate, Energy and Building, Danish Parliament, Denmark.

Danish Nature Agency, n.d. Danish Ministry of the environment. Available: http://www.sns.dk/udgivelser/2007/978-87-7279-751-9/html/kap08.htm # 8.1) [08.05.2014].

Danmark Nationalparker, n.d. Nationalparker—for erhvervsdrivende. Available: http://danmarksnationalparker.dk/om-nationalparker/beboere-og-erhverv/for-erhvervsdrivende/2014.

Danmarks Statistik, 2014. BEF44: Folketal 1. januar efter byområde. Available: http://www.statistikbanken.dk/BEF442014. Last update.

Danmarks Vindmølleforening, 2013. Afstandskrav. Available: http://www.dkvind.dk/html/planlagning/krav_afstand.html2014. Last update.

Danmarks Vindmølleforening, 2014. Faktisk Afregning for vindmøller på markedsvilkår. Available: http://www.dkvind.dk/html/nogletal/afregning_faktisk.html [23.05.14] last update.

DAVIDMUSALL, F., KUIK, O., 2011. Local Acceptance of Renewable Energy—A Case Study from SOUTHEAST GERMANY. Institute for Environmental Studies, Amsterdam.

Den Store Danske, 2014. Ærø. Available: http://www.denstoredanske.dk/Danmarks_geografi_og_historie/Danmarks_geografi/Danske_småøer/Ærø?highlight=ærø2014. Last update.

Dkvind.dk, 2014. Vindmøller og skat. Available: http://www.dkvind.dk/html/okonomi/ejer_skat_moms.html [26.05.14] last update.

EMD, 2014. WindPRO. Available: http://www.emd.dk/ [10.04.14] last update.

Energiakademiet, 2014. RE-Island [Homepage of Energiakademiet], [Online]. Available: http://energiakademiet.dk/en/ [14.04.14] last update.

Energinet.dk, 2013a. De fire VE-ordninger. Available: http://energinet.dk/DA/El/Vindmoeller/De-fire-VE-ordninger/Sider/De-fire-VE-ordninger.aspx2014. Last update.

Energinet.dk, 2013b. Hvem kan købe andele. Available: http://energinet.dk/DA/El/Vindmoeller/De-fire-VE-ordninger/Koeberetsordningen/Borger/Sider/Hvem.aspx2014. Last update.

Energinet.dk, 2013c. Køberetsordningen. Denmark: energinet.dk.

Energinet.dk, 2014. 2013 was a record-setting year for Danish Wind power. Available: http://energinet.dk/EN/El/Nyheder/Sider/2013-var-et-rekordaar-for-danskvindkraft.aspx [08.04.14] last update.

Energinet.dk, The Danish Energy Agency, 2012. Technology Data for Energy Plants. Energinet.dk.

Samsø Erhvervs-OG Turistcenter, 2014. About Samsø [Homepage of Samsø Erhvervs- og Turistcenter], [Online]. Available: http://www.visitsamsoe.dk/en/om-samso/ [14.04.14] last update.

Erhvervsstyrelsen, n.d. Fonde. Available: http://erhvervsstyrelsen.dk/fonde2014.

Flyvbjerg, B., 2006. Five misunderstandings about case studies. Qual. Inq. 12 (2), 219.

Google, 2014. Google Maps. Available: https://www.google.dk/maps/@55.9960892,11.9604692,9z2014. Last update.

Grant Space, 2014. What is a foundation. Available: http://grantspace.org/Tools/Knowledge-Base/Funding-Resources/Foundations/what-is-a-foundation2014. Last update.

Halsnæs Kommune, 2014a. Kommuneplan 2013. Available: http://www.kommuneplan.halsnaes.dk/dk/forside.htm2014. Last update.

Halsnæs Kommune, 2014b. Rammeområde 2.S3. Available: http://kommuneplan.halsnaes.dk/dk/rammer/enkeltomraader/vedtaget/2s3.htm2014. Last update.

HVELPLUND, F., MÖLLER, B., SPERLING, K., 2013. Local ownership, smart energy systems and better wind power economy. Energy Strategy Rev. 1, 164–170.

Industrimuseet, n.d. Frederiksværks historie.

International Co-operative Alliance, 2014. Co-operative identity, values & principles [Homepage of ICA], [Online]. Available: http://ica.coop/en/whats-co-op/co-operative-identity-values-principles [30.03.14] last update.

Jensen, J., 2014. Field photo.

KARL, S., FREDE, H., BRIAN VAD, M., 2010. Evaluation of Wind Power Planning in Denmark—Towards an Integrated Perspective. Aalborg University, Aalborg.

KVALE, S., 1996. The interview situation. In: Interviews: An introduction to Qualitative Research Interviewing. Sage Publications, London, p. 124.

LADENBURG, J., DAHLGAARD, J., 2012. Attitudes, threshold levels and cumulative effects of the daily wind-turbine encounters. Appl. Energy 98, 40–46.

LEMMING, J., MORTHORST, P.E., CLAUSEN, N., JENSEN, P.H., 2008. Contribution to the Chapter on Wind Power Energy Technology Perspectives 2008. Risø National Laboratory for Sustainable Energy, Technical University of Denmark, Roskilde, Denmark.

LORING, J.M., 2007. Wind energy planning in England, Wales and Denmark: factors influencing project success. Energy Policy 35, 2648–2660.

Lund, H., Østergaard, P.A., 2010. Fundamental Investment Theory. Aalborg University, Aalborg, Denmark.

May, T., 2011. Social Research: Issues, Methods and Process, fourth ed. Open University Press, England.

MIGUEL, M., STEPHEN, L., FREDE, H., 2009. Stability, Participation and Transparency in Renewable Energy Policy: Lessons from Denmark and the United States. World future council and Department of development and planning and Renewble Energy World. Com, England, Denmark and United States.

Miljøministeriet, n.d. Vindmølleplacering. Available: http://miljoegis.mim.dk/cbkort? profile=miljoegis-vindmoeller [19.05.14].

Østergaard, P.A., 2013. Energy and Environmental Tools and Project Evaluation Lecture 1: Fundamental Investment Theory. Aalborg University, Aalborg, Denmark.

Østergaard, O., 2014. Interview with Ole Østergaard. Vinderød.

PEDERSEN, E., 2003. Noise Annoyance from Wind Turbines—A Review. Swedish environmental protection agency, Stockholm.

Poglio, T., Ranchin, T., 2014. DataForWind: Services for Professionals in Wind Energy. Available: http://www.dataforwind.com/2014. Last update.

Retsinformation, 2010. Bekendtgørelse af lov om erhvervsdrivende fonde. Available: https://www.retsinformation.dk/Forms/r0710.aspx?id=1317322014. Last update.

Retsinformation, 2013. Bekendtgørelse af lov om fremme af vedvarende energi. Available: https://www.retsinformation.dk/Forms/r0710.aspx?id=159159 [19.05.14] last update.

Schmeer, K., 1999. Guidelines for Conducting a Stakeholder Analysis. http://www.who. int/management/partnerships/overall/GuidelinesConductingStakeholderAnalysis.pdf. edn.

Schmidt, R., 2014. Ærø guided tour. Ærø.

Skat.dk, 2014a. Beskatning af fonde og visse foreninger. Available: https://www.skat.dk/ SKAT.aspx?oId=573512014. Last update.

Skat.dk, 2014b. Den skematiske metode for skat af vedvarende energianlæg. Available: http:// www.skat.dk/SKAT.aspx?oId=1973870&vId=0&lang=DA [May 26] last update.

Søren Hermansen, 2014. Interview with the Director of the Samsø Energy Academy, Søren Hermansen. Aalborg University.

Special Eurobarometer, 2014. Climate Change. European Comission, Brussels, Belgium.

SPERLING, K., 2007. Large-Scale Wind Power in Denmark. New Perpectives on Ownership. Aalborg University, Aalborg, Denmark.

Tax.dk, 2014. Skatteberegning 2014. Available: http://www.tax.dk/jv/cc/C_C_5_2_10. htm [21.05.14] last update.

The Danish Energy Authority, 2003. Renewable Energy Danish solutions. The Danish Energy Authority, Copenhagen, Denmark.

Vangstrup, L., 2014a. Conversation about ownership of wind turbines.

Vangstrup, L., 2014b. Interview concerning Ærø. Aalborg University, Aalborg, Denmark.

VIKKELSØ, A., LARSEN, J.H.M., SØRENSEN, H.C., 2003. The Middelgrunden Offshore Wind Farm. A Popular Initiative. Copenhagen Environment and Energy Office (CEEO), Copenhagen.

WARREN, C., MCFADYEN, M., 2008. Does Community Ownership Affect Public Attitudes to Wind Energy? A Case Study from South-West Scotland. School of Geography & Geosciences, University of St Andrews, Scotland, United Kingdom.

WARREN, C., LUMSDENA, C., O'DOWDA, S., BIRNIEB, R., 2005. 'Green on Green': Public Perceptions of Wind Power in Scotland and Ireland. Journal of Environmental Planning and Management, Scotland.

WELFI-Wind Energy Local Financing, n.d. Ærø. Welfi.

WELLINGTON, J., SZCZERBINSKI, M., 2007. Research Methods for the Social Sciences. Continuum International Publishing Group, London.

White, G., Garrad, A., Tindal, A., 1996. Integrated design methodology for wind farms. pp. 62.

Wikipedia, 2014a. Ærø. Available: http://en.wikipedia.org/wiki/%C3%86r%C3%B8 [03.05.14] last update.

Wikipedia, 2014b. Samsø. Available: http://en.wikipedia.org/wiki/Sams%C3%B8 [06.05.14] last update.

Wilson, J., n.d. Four types of stakeholder power. Available: http://www.ehow.com/info_12105818_four-types-stakeholder-power.html2014.

Windpeople, 2014a. Consulting and design of municipally owned energy [Homepage of Windpeople.org], [Online]. Available: http://windpeople.org/ [02.052014] last update.

Windpeople, 2014b. Hvide Sande. Available: http://windpeople.org/projekter/hvide-sande/2014. Last update.

WOLSINK, M., 2007a. Wind power implementation: the nature of public attitudes: equity and fairness instead of 'backyard motives. Renew. Sust. Energ. Rev. 11 (6), 1188–1207.

WOLSINK, M., 2007b. Planning of renewables schemes: deliberative and fair decision-making on landscape issues instead of reproachful accusations of non-cooperation. Energy Policy 35 (5), 2692–2704.

Wolsink, M., 2009. Planning: problem 'carrier' or problem 'source'? Planning Theory & Practise 10 (4), 521–547.

Worldbank, n.d. What is stakeholder analysis? http://www1.worldbank.org/publicsector/anticorrupt/PoliticalEconomy/PDFVersion.pdf.edn.

WÜSTENHAGEN, R., WOLSINK, M., BÜRER, M.J., 2007. Social acceptance of renewable energy innovation: an introduction to the concept. Energy Policy 35, 2683–2691.

Ycharts.com, 2014. Denmark Long Term Interest Rate. Available: denmark_long_term_interest_rates [May 26] last update.

YIN, R., 2009. Case study research: design and methods, fourth ed. Sage, Los Angeles.

eHealth for Sustainable Health Care in Serbia

Jane Paunkovic[1,*], Nebojsa Paunkovic[2]

[1]Faculty of Management Zajecar, Megatrend University.
[2]Poliklinika Paunkovic, Zajecar, Serbia.
*Corresponding author. jane@fmz.edu.rs

INTRODUCTION

In a sustainable world, people entail a healthy environment, a strong economy, and a responsible society with healthy population, not just efficient health care services. The Human Development Report (Oxford University Press, 2001) commissioned by the United Nations Development Programme at the beginning of the twenty-first century argued that information and communication technologies (ICTs) offer an unprecedented opportunity for achieving development, particularly poverty reduction, health care, and education: "People all over the world have high hopes that these new technologies will lead to healthier lives, greater social freedoms, increased knowledge and more productive livelihoods."

Technological solutions in health care have already raised life expectancies even in poor countries without much health infrastructure. They contribute to the improvement of quality in health care and can potentially save money, increase economic productivity, and save lives. Technology in medicine can enable faster and more accurate diagnostic procedures, improve efficiency of health providers, and improve the quality of life for patients.

The terminology in health-related information technology has been constantly changing over the years. The World Health Organization (WHO) is addressing the use of ICTs for health, also known as eHealth. Examples of eHealth may include treating patients, conducting research, educating the health workforce, tracking diseases, and monitoring public health (http://www.who.int).

The European Commission "eHealth Action Plan 2012-2020: Innovative healthcare for the 21st century" defined eHealth as the use of not only ICTs in health products, services, and processes combined with organizational change in the health care systems but also new skills in order to improve the health of citizens, the efficiency and productivity in health care delivery, and the economic and social value of health. eHealth covers the interaction between patients and health care service providers, institution-to-institution transmission of data, or peer-to-peer communication between patients and health care professionals. The "eHealth Action Plan 2012-2020: Innovative healthcare for the 21st century" outlines that "eHealth can benefit citizens, patients, health and care professionals but also health organizations and public authorities. eHealth—when applied effectively—delivers more personalized 'citizen-centric' healthcare, which is more targeted, effective and efficient and helps reduce errors, as well as the length of hospitalization. It facilitates socio-economic inclusion and equality, quality of life and patient empowerment through greater transparency, access to services and information and the use of social media for health" (http://ec.europa.eu/).

ICTs in Europe have been found to be of great assistance to health care professionals in the process of diagnosis, treatment, monitoring, medication prescription, referral, information retrieval and communication, documentation, and transactions (Economic Impact of Interoperable Electronic Health Records and ePrescription in Europe 01-2008/02-2009).

Clinical applications of ICTs are possible in all areas of patient care (Johnson et al., 2001; Harno, 1999) and include real-time and/or store-and-forward technologies (Houston et al., 1999; Loane et al., 2000) ranging from telephone and fax machines, to e-mails, to chat rooms, to discussion boards, to audio- and videoconferencing, to mobile technologies in recent years (Free et al., 2013).

Administrative applications include recording (Walsh, 2004; Schoenberg and Safran, 2000) and sharing of billing summaries and electronic connections to pharmacies. Remote medical instruments include various types of imaging technologies (Weinstock et al., 2002), pressure sensors, haptic feedback devices, and robotics. Educational applications focus on continuing medical education for professionals and patients including telementoring.

THE SERBIAN PERSPECTIVE ON HEALTH CARE AND THE NEED FOR ICTS

The need for ICTs in the Serbian health care system was recognized by the Government of Serbia at the beginning of this century. The health care status in Serbia was, at that time, comprehensively analyzed in the World Bank's Project Appraisal Document:

> *Despite all the difficult factors during the 1990s (economic crisis, war, sanctions, bombing) in the former Republic of Yugoslavia (FRY) (excluding Kosovo), all vital indicators improved during that time period according to data based on house-hold surveys conducted by UNICEF in 2000. Under five mortality rate decreased by 29.5 percent while infant mortality rate decreased by 31.5 percent to 11.23 deaths per l000 live births in 2000. Today, life expectancy at birth is estimated to be 69.8 years for males and 74.5 years for females. Access of the population to improved drinking water sources and sanitary means of excreta disposal is almost universal and vaccine preventable diseases are under control. When looking at causes of death, the picture is clearly one of a developed and transitional country with high levels of heart disease, strokes, and cancer. Smoking is estimated to cause 30% of the mortality in Serbia. Poor nutrition is another major risk factor.*
>
> *Some minor declines in health status have been reported recently, however, and although not well documented, are of concern given the other conditions in the health sector and experiences in other countries in the region where health status has deteriorated significantly. A high annual incidence of tuberculosis (39 per 100,000 population) indicates a need to continue to be vigilant about infectious diseases, particularly given the living situation of the most vulnerable population such as IDPs and refugees and the affordability of drugs. The Government's view that there has been a deterioration in health status (Government of Serbia (GOS), Interim Poverty Reduction Strategy, June 2002) has not been documented by reliable data, which is in itself an issue.*
>
> **(World Bank, 2003)**

In that same document, the Serbian health care system was characterized as with high inefficiencies that jeopardized macroeconomic stability.

In 2002, the Government of Serbia declared a health care system reform as one of the priorities and prepared a health care statement that indicated directions to be pursued in reforming the health sector. That was the basis for the World Bank support through the Serbia Health Project (SHP). The objectives of the SHP (World Bank, 2003) were defined as

> *to build capacity to develop a sustainable, performance oriented health care sys-tem where providers are rewarded for quality and efficiency and where health insurance coverage ensures access to affordable and effective care.*

The World Bank's health country assistance strategy was endorsed by both the World Bank management and the Government of Serbia. It identified the following priorities for World Bank assistance, which are reflected in the activities planned for the SHP:

- Health information systems development, which is required to improve the accuracy and timelines of health data available for policy making and will support efficient operations of the HIF (Health Insurance Fund)
- Health financing, where the objective will be to help the HIF regain fiscal sustainability through a combination of measures on the revenue and expenditure side and to ensure that out-of-pocket spending does not become a financial burden for the poor
- Health service restructuring, where the objective will be to improve the quality and efficiency of service delivery
- Human resources, where the objective will be to introduce adjustment programs that will help Serbia achieve the optimal labor force in terms of distribution, skills, and affordability

A crucial objective was that the health service remains accessible for the poorest segments of the population and those that are adversely affected by transition. The monitoring indicators for the project encompass the distribution of resources, access, and out-of-pocket payments by the poor. Coordination between reforms in the health sector and other related sectors is recognized also to be an important strategy to improve health and access to health services for the poor and to mitigate the impacts of restructuring of health services.

After the approval of the SHP, the loan became effective in October 2003. The SHP and the Serbia Health Project Additional Financing (SHPAF) focused on comprehensive reform process in Serbia and continued until 2012.

Most of the processes activated by the SHP and SHPAF became the standard in the delivery of health care in Serbia: *Health Care Development Plan for 2010-2015* adopted by the Government of the Republic of Serbia, *Poverty Reduction Strategy paper for Serbia*, *Strategy for the Development of the Informatics Society in the Republic of Serbia to 2020*, and *e-Health 2015 IT Strategy*.

ICT strategy for health, prepared by the SHP team, was one of first IT sector strategies in Serbia. Standards and accompanying regulation established for the software used in health institutions were unprecedented in a wider region, and it is now used to set an example. The implementation of the hospital information system is done with minimal cost and in combination with EU funding (World Bank, 2012).

In accordance with the Government of Serbia's strategy for improving the efficiency of health care delivery while maintaining quality, ICTs have the potential to improve health care through the use of ICT in the following ways:

- To improve access to high-quality specialty care, especially in rural and poor communities
- To improve service and quality of health care
- To improve productivity and efficiency in the health sector
- To use the opportunities of IT to distribute information to the general public and health care professionals and to increase the level of knowledge
- To improve working conditions and personal planning for health care professionals

At the same time, the European Commission dedicated many sections to health care and funded many health information projects through "The European Fourth Framework Programme for Research and Technological Development" and "The Global Healthcare Applications Project" (European Commission, DG XIII, Telecommunications, Information Market, and Exploitation of Research, 1994; European Foundation for the Improvement of Living and Working Conditions, 1994). Numerous programs had already demonstrated the feasibility and utility of health information systems (Nielsen and Jorgensen, 1996; Johnson et al., 2001). Health information programs have been very successful in the United States (Kokesh et al., 2004) and throughout the world especially in developing countries (Edworthy, 2001) and their clinical and economic utility has been demonstrated in neighboring Croatia and in Kosovo (Kovai et al., 2000; Latifi et al., 2004).

ELECTRONIC THYROID NETWORK TO IMPROVE PATIENT REFERRAL SYSTEM: "THYRO-NET SERBIA"

At the beginning of 2005, a group of distinguished telemedicine experts, led by Professor Robert Doktor from the College of Business Administration, University of Hawaii; Victoria Garshnek from Telehealth Research Institute, University of Hawaii; Robert Whitton, Technical Innovations, Inc., Ka'a'awa, Hawaii; and Lawrence Burgess from John A. Burns School of Medicine, University of Hawaii, in cooperation with the authors of this chapter, has developed a project proposal: "An Electronic Thyroid Consultation Network to Optimize Patient Care" (Paunković et al., 2008). That project proposal was submitted for approval to the SHP representatives on 31 May 2005.

Proposed Project Objective

The overall project objective of the proposed eHealth project in Serbia was to develop a cost-effective, highly efficient thyroid consultation and referral system from primary care to secondary and tertiary institutions. It was supposed to be a readily accessible Web-based store-and-forward system developed for use in Serbian institutions and in compliance with the goal of optimizing the relationship between primary, secondary, and tertiary levels of care. The aspects of security and privacy for the individual patient would also be implemented and assured.

Rationale

Patient referral system to specialist clinics is one of the major problems for health care system in Serbia. Some 50-80% of all patients referred to specialist consultations are deemed unnecessary or ill-timed by the specialty care providers and detract from time and effort available to treat the remaining appropriate referrals. All those consultations present a superfluous burden to the specialist services and enormously increase costs in diagnostics. Consultations are also expensive and time-consuming for both the health care service and patients in terms of travel costs, lost days of work, etc. That is especially evident for chronic, often lifetime, diseases such as diseases of the thyroid gland.

Diagnostic and therapy procedures for thyroid diseases are costly and present considerable burden to the health care system. According to the Third National Health and Nutrition Examination Survey (NHANES III), the thyroid diseases are as follows:

- Common: Approximately 5-20% of the American population have some thyroid abnormality depending on the indicator chosen; many may not be aware of their condition (Dunn, 2002).
- Prevalent: More than diabetes (Hollowell et al., 2002).
- Disruptive: Impair physical and mental performance, produce morbidity, and pose special risks for pregnancy and the developing fetus and neonate (Dunn and Delange, 2001; Glinoer and Delange, 2000).
- Expensive: Thyroid hormone T4 is among the most commonly prescribed medications in the United States; testing of thyroid function is a routine laboratory procedure costing millions of dollars annually; the effects of iodine deficiency on the thyroid alone cost one country (Germany) an estimated annual $1 billion (Pfannenstiel, 1998).

- Treatable: Highly satisfactory therapies exist for all the common problems—hyperthyroidism, hypothyroidism, nodules, cancer, and iodine deficiency.
- Preventable in some cases: The consequences of iodine deficiency are readily avoided by optimal iodine nutrition (Delange, 2000); appropriate diagnosis and treatment can control the effects of hypothyroidism on human development; avoidance of excess iodine can prevent many of its complications, including goiter, hypothyroidism, hyperthyroidism, and autoimmune disease.

Thyroid disease prevalence in Serbia has not been systematically investigated. Some calculated extrapolations of various prevalence or incidence rates against the populations of a particular country or region estimated around 215,000 patients with autoimmune thyroid disease (some 50% of all thyroid disease patients) in Serbia (based on a population of 10,825,900, at that time). Relevant data from the Serbian region Timok available from the thyroid register established in 1970 (Paunkovic et al., 1998) provide information about the annual incidence of thyroid disease with an estimated prevalence of around 8% (Paunkovic and Paunkovic, 2004).

Thyroid disease can be especially suited for electronic data collection and transfer, due to the large number of diagnostic tests required and the requirement of these values for diagnostic determination. Large and potentially complex imaging workups are usually not necessary, nor are lengthy case history presentations, both of which require additional time and technical complexity in the already overburdened primary care setting. Referral necessity can be determined most often through the completion of a 10- or 15-question consultation form, which provides the consultant with enough information to decide on referral to tertiary care or continued management in the primary care setting.

Pilot Project "Zlatibor"

An important first step in this project was the development of a pilot program implementing Web-based consultation in a patient referral system. The pilot program proposed an implementation of an electronic consultation and referral system between the Thyroid Gland and Metabolism Institute (now "The Special Hospital"; http://www.cigota.rs/en/strana/special-hospital), Zlatibor, and the Endocrinology Clinic, University of Belgrade, Serbia. The Thyroid Gland and Metabolism Institute in Zlatibor, at that time, consisted of hospital facilities (400 beds) and outpatient units (ambulatory part). It was a combination of a local (regional) primary health institution (for outpatients) and specialized hospital (second and tertiary levels). Thyroid disease patients from the surrounding area were referred to the Thyroid Gland and Metabolism Institute in Zlatibor.

The Thyroid Gland and Metabolism Institute in Zlatibor exemplified an excellent model for the rationalization of a patient referral system, since almost 100% of the patients that enter the institute are being referred to specialist consultations. Almost all procedures routinely used in thyroid disease diagnostics were available at the Thyroid Gland and Metabolism Institute in Zlatibor (clinical biochemistry, radio-immune tests, ultrasound, nuclear medicine, etc.). Almost all therapeutic procedures were also available (medications, radioactive iodine, etc.). Unfortunately, physician and staff education levels and experience at the institute were at the primary care level. Specialty care consultations are provided by the referent (Endocrinology Clinic, University of Belgrade, Serbia) institutions from Belgrade (300 km distance). These specialists travel to Zlatibor to review practically all inpatients and outpatients during their weekend visits. They usually have more than 100 consultations per visit (averaging two minutes per patient). Thus, the quality of this type of consultation is less than ideal. The Thyroid Gland and Metabolism Institute in Zlatibor serves a population of around 2,000,000 people (northern Montenegro and eastern Bosnia included) and is estimated to have over 15,000 thyroid patients. The number of annual specialist consultations was around 20,000 and the number of diagnostic tests was around 60,000. In order to improve the situation at the Thyroid Gland and Metabolism Institute in Zlatibor in terms of quality and efficiency of health care, this project proposed the implementation of a Web-based store-and-forward telemedicine, eHealth system, that would connect the Thyroid Gland and Metabolism Institute in Zlatibor with a tertiary Endocrinology Clinic, University of Belgrade, Serbia, in the initial phase and subsequently with other existing specialized thyroid institutions in Serbia.

We have proposed a pilot project for the installation and configuration of an electronic Web-based thyroid consultation and patient management system, based at the Ministry of Health in Belgrade (see Figure A1).

In the case of a referral, treatment provided by the specialists would be documented in the system for archival purposes and reviewed by the patient's primary care physician.

After installation of the hardware and software, specialists in thyroid disease would create the Web-based protocols for referral of patients to tertiary care centers for thyroid treatment. A group of physicians at the primary care

Electronic thyroid consultation (year one)

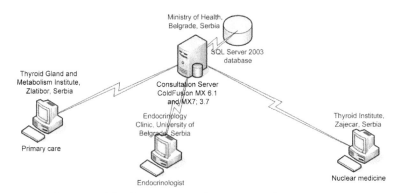

Figure A1 Schematic of the first-year consultation network.

center at the Thyroid Gland and Metabolism Institute in Zlatibor would receive training on using the system and filling out forms for patients whom they feel will require consultation and possibly referral to the specialists. Cases that are submitted via the system would be reviewed by a clinical case manager, to ensure they are completed properly, and then triaged and referred to available thyroid specialists. The specialists will review the findings and interact electronically and securely with the referring provider and other specialists as necessary.

Upon successful completion of the first year of the program, four additional regional hospitals would be included in the consultation and referral network (see Figure A2).

Trainers will work with primary care staff to instruct them on the use of the system. Additional specialists will be included to respond to the increased consultation load. By the third year, all thyroid institutions will be included, as will all major primary care centers in Serbia (see Figure A3).

ORGANIZATIONAL CHANGE/MONITORING SYSTEM UTILIZATION

All new technical implementations impact also the social system of which they become apart, and this is particularly true of health informatics systems (Whitten et al., 2002; Aas, 2002). The technological innovation and the

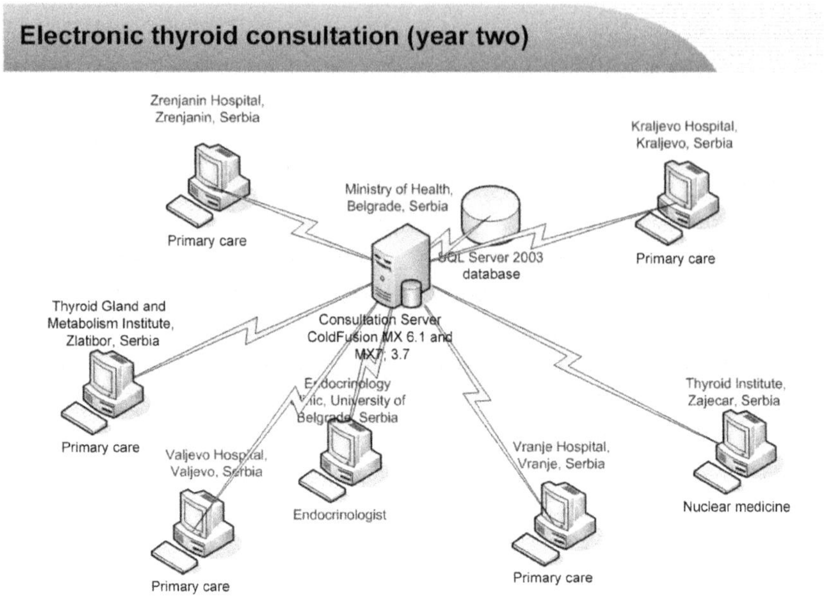

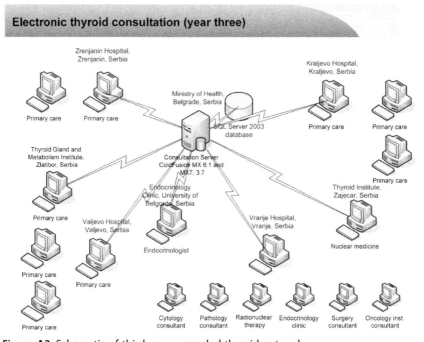

Figure A2 Schematic of second-year expanded consultation network.

Figure A3 Schematic of third-year expanded thyroid network.

desire to diffuse it into the medical profession must strategically engage stakeholders within its project scope and be sensitive to cultural beliefs and the local values system. Effective change requires that people not only believe that change is necessary but also understand how change will come about and what the consequences will be. Change management is about people, not about changing technology or processes. For change to work, it needs participation from all sides. It should involve all stakeholders, but the message, the training, and the involvement should be tailored to the needs of each individual group. For change to last, it has to be reinforced (Dawson, 2003). And the more control people feel they have over the change, the less stressed they become. According to Bashshur (2002),

> when technological innovations are not accepted or implemented properly, generally the failure may be traced to a poor fit between the nature of the innovation and the vested interests, resources, and expectations of its major gatekeepers.

Therefore, the design of a health informatics system needs to include both a priori organizational change effort to overcome latent resistance within the social system and an early-on and continuing evaluation component to assure accurate feedback of utilization rates on a continuing basis. Of fundamental importance is a clearly articulated and hierarchically supported purpose that facilitates the change process. Continuous feedback loops maximize stakeholder input, enhance the recognition for the need to change, and indicate the level of commitment to the process.

eHealth programs hold not only the potential of increasing patient access and enhancing the quality and timelines of patient care but also the potential of altering the flow of clinical information and the loci of clinical decision-making that may hold the roots of social resistance to the utilization of eHealth. The degree of this resistance varies from case to case but is always present and more often than not results in low utilization rates of health IT solutions.

Therefore, this project included both an organizational change component during the initial implementation phase and a continuous monitoring of system utilization. Additional organizational change interventions were to be scheduled as dictated by downward trends in utilization rates. The organizational change program should involve group sessions with both primary and secondary caregivers and their staffs at the onset of the pilot project and thereafter as necessary. The monitoring effort is continuous during the entire length of the project and includes on-site observation during at a 3-day period during the 0- to 3-month, 11- to 12-month, 23- to 24-month, and 35- to 36-month timeline periods. Qualitative methodologies are to be used as primary data collection techniques, with special reliance on structured interviews and field observations. Local professionals and students are engaged to assist in the implementation of the methodologies to facilitate cross-cultural communications. Analysis of system utilization data will allow quantitative data cross validation of the qualitative results.

SUSTAINABLE UTILIZATION OF EHEALTH IS INFLUENCED BY ORGANIZATIONAL AND CULTURAL FACTORS

From the first quarter of 2005 in a process of strengthening monitoring activity by the World Bank and the Ministry of Health, the SHP was restructured. Although the development objectives, the general architecture of the project, and the total budget remained unchanged, the scope and the number of activities were reduced. The Ministry of Health was allowed to focus its efforts on a more manageable number of activities and to enable the project to be completed in a more effective manner (World Bank, 2003).

Due to the restructuring of the SHP and the establishment of the new priorities by the Ministry of Health and the HIF in October 2005, our eHealth project proposal for the optimization of patient referral system was not approved for implementation.

Nevertheless, after more than 7 years, we have continued our research on organizational and cultural factors in Serbian health care system. We were following the information technology implementation in Serbian health care and conducted research in several health centers.

Our aim was to investigate if implementation problems in clinical eHealth programs are organizational in their origin and if the nature of the successful design of a health care organization depends upon the values of the society, or national culture, as we hypothesized.

From the previous work of Professors Doktor and Bangert and their team, we have learned that when introducing a new technology that changes the core processes of an organization, such as an eHealth initiative, it is important that the structural design and culture of the organization are aligned with the predominant national culture in which the organization is embedded. When a harmonious alignment is achieved, speedy and effective organizational learning can occur. This, in turn, promotes effective utilization of the new technology (Doktor et al., 2005).

With reference to the implementation of eHealth programs, Hu suggested that cultural and professional organization variables may be more explanatory of ICT use than perceived usefulness or perceived ease of use (Hu et al., 1999).

Many authors suggest that a major reason why organizational change efforts usually fail to materialize as planned is the frequent neglect of aspects of organizational culture (Balogun and Johnson, 2004). Culture is often seen as either the key issue to be changed or something that is crucial to be taken seriously in order to make change possible (Alvesson and Stefan, 2007; Balogun, 2006). As Carl-Henric Svanberg, CEO of Ericsson, stated: "culture always defeats strategy."

We have also learned that cross-cultural analyses are important to show that what may work in one culture may not be appropriate in another. The culture as defined by Hofstede is perceived as "the collective programming of the mind which distinguishes the members of one human group from another" and its building blocks include "systems of values." Values, in this case, are "broad preferences for one state of affairs over others," and they are mostly unconscious. In that context, culture is conceptualized and measured through different value dimensions that can be defined as latent (non-observable) collections of several interrelated values with which to compare groups (Hofstede, 1980).

Many different cultural dimensions have been identified over the years (Triandis, 1995, 2004; Schwartz, 2012; Inglehart, 1997; Trompenaars and Hampden-Turner, 1998), but the most replicated and with high practical value according to the opinion of numerous researchers are cultural dimensions defined by Hofstede (1980).

Based on attitude survey of 117,000 employees within the subsidiaries of IBM in 40 countries and 3 regions, Hofstede proposed four basic cultural dimensions, largely independent of each other:

- *Individualism versus collectivism*, measured by individualism index (IDV) ranging from 0 (low individualism, high collectivism) to 100 (high individualism)
- *Power distance index*, measured by the power distance index (PDI) ranging from 0 (small PD) to 100 (large PD)
- *Uncertainty avoidance*, measured by the uncertainty avoidance index (UAI) ranging from 8 (lowest UA country) to 112 (highest UA country)
- *Masculinity versus femininity*, measured by the masculinity index (MAS) ranging from 0 (low masculinity) to 100 (high masculinity)

Additionally, research by Michael Bond performed in 1991 led him to adding the fifth dimension called *long- versus short-term orientation* (Hofstede and Bond, 1988).

In 2010, the sixth dimension has been added called *indulgence versus restraint*, based on research performed by Michael Minkov on World Values Survey (WVS) data analysis for 93 countries (http://www.geerthofstede.nl/) (Hofstede, 2011).

Hofstede's original research included Yugoslavia as the only eastern European country. After the disintegration of the former Yugoslavia, Hofstede adapted the original data by dividing it into data on the national cultures of Slovenia, Croatia, and Serbia (Hofstede, 2001). According to Hofstede, the Serbian national culture is characterized by high PDI, 86; high UAI, 92; collectivism-low IDV, 25; and high to medium femininity-low to medium MAS, 43.

Bangert and Doktor (2005) found the work of Geert Hofstede insightful when considering the organizational designs for successful eHealth implementation. They argued that only through a harmonious match of organizational structure and culture effective and efficient organizational learning can emerge. And it is only through organizational learning that new technologies can be effectively utilized.

Research Objective

Our research was based on Hofstede's work on cultural dimensions, and our hypothesis was that for Serbia with high PDI (86), high UAI (92), and collectivism-low IDV (25), successful organizational design of eHealth projects has to be strongly supported by leadership, but with strong collectivistic aspect. If eHealth technology is to be successfully and sustainably implemented in societies with high UAI, such as in Serbia, then more mechanistic organizational design (high in complexity, formality, and centralization) should be adequate in organizations in health care.

In an attempt to understand these problems, we have investigated a number of organizational characteristics and made an effort to associate them with particular cultural dimensions according to Hofstede's work. Our intent was to explore the optimal organizational design for eHealth projects correlated with the predominant (Serbian) national culture.

Study Design

Our investigation was conducted in health centers in Serbia (Vranje in 2009, Pirot in 2010, and Zajecar in 2011) in a process of implementation of eHealth programs. These organizations were implementing software "HELIANT"—health information system developed at the Faculty of

Electro-Techniques, University of Belgrade, requested by the Republic Health Insurance Fund of Serbia. HELIANT is characterized by a multilayer architecture with centralized database. During its development, open-source technologies were exclusively used. It is a Web application, developed in Java EE programming language. The applicative server JBoss is also used. Business logistic layer has been implemented through EJB 3.0 technology use. Any of the operational systems can be installed on the PCs. The software known as HELIANT was primarily created for the secondary health care but has been successfully adapted for primary care use (Stojkovic et al., 2009).

Our research was performed by using questionnaires and unstructured interviews to assess participants' views on optimal organizational design in health care organizations in reference to the implementation of eHealth programs. The investigation included 97 employees in health centers in Pirot, Vranje, and Zajecar (72 female and 25 male) with different educational backgrounds (50 with high school education and 47 with higher education) and with work experience of over 5 years as a rule.

The control group consisted of employees from various professions from local institutions including education, business, and administration (58 female and 33 male), with mostly higher education (64) and with work experience over 5 years. Investigation was conducted in the same towns in Serbia and in the same period of time (2009-2011).

Participants in the survey were asked to grade (1, not important; 5, very important) particular statements about some characteristics of their organization. These organizational characteristics were our own modification, made after the initial pilot investigations (Stojkovic et al., 2009; Paunković et al., 2010), of the organizational characteristics found in the literature to correlate with culture and structure of organizations (Doktor et al., 2005).

The results of the investigation of organizational characteristics conducted using questionnaires are presented in Table 1.

After considering the results from the questionnaires, unstructured interviews, and on-site research observations, we could conclude that the participants in our study have delineated the following as the most important: "Support from superiors," "Clear instructions from superiors," "Good communication with superiors," and "Good working relations with colleagues."

"Support for continuing education," "Career advancement through individual performance," "Decision-making in own line of work," and "Independence in choosing own work style" were found less important.

Table 1 Average Marks for Investigated Organizational Characteristics

	Organizational characteristics	Average mark Investigated group $n = 97$	Average mark Control group $n = 91$
1	Support from superiors	4.62	4.4
2	Involvement of superiors	4.03	3.5
3	Clear instructions from superiors	4.67	4.4
4	Independence in choosing own work style	4.23	4.3
5	Decision-making in own line of work	4.33	4.4
6	Good working relations with colleagues	4.60	4.8
7	Good communication with superiors	4.74	4.4
8	Acknowledgment of individual performance through salary	4.12	4.1
9	Career advancement through individual performance	4.31	4.3
10	Support for continuing education	4.4	4.4

The least important for the participants were "Acknowledgment of individual performance through salary" and "Actual involvement of superiors."

These findings correlated with our hypothesis that the organization of the eHealth project has to be strongly supported by leadership (in a national culture with high PDI (86)), with dominant collectivistic conduct (Good working relations with colleagues and Good communication with superiors), and minor individualistic performance (Acknowledgment of individual performance through salary and Career advancement through individual performance). Nevertheless, some of the characteristics that could be associated with individualism (Independence in choosing own work style) were sometimes ranked high. Emphasis on the need for clear instructions from their superiors is also in correlation with the high UAI (92) and deserves special attention in the implementation of any new and unfamiliar practice like the introduction of information technology and should be of special concern in future studies.

The authors have previously reported some earlier versions of these findings on the implementation of information technology projects in health care (Paunković et al., 2010, 2011). In both studies, the participants have delineated interdependence and teamwork along with acknowledgment of individual performance and highlighted clear instructions from superiors, acknowledgment of individual performance and independence in choosing their way of working. The least important for the participants was the involvement from the leadership although they have recognized the need for their support, which per se may be indicative of the culture and climate in the organizations and deserves further investigation.

DISCUSSION

A successful eHealth program is the product of careful planning, appropriate management support, dedicated health care professionals and support staff, and a commitment to appropriate funding. It also requires the adequate combination of multiple technologies such as medical devices, network computing, videoconferencing, software, and telecommunications. A large number of ICT programs are actually more predominately change programs, or service improvement programs, with the implementation of technology being only the part of the solution (Legris and Collerette, 2006). Furthermore, concentrating on the technological aspects of these programs can lead to less effective results. Lucas and Spitler (1991) had demonstrated that in real field settings, organizational variables, such as cultural norms and the nature of the job, were far more important in predicting the use of technology than the potential user's perception of likely usefulness or ease of use. According to Bashshur (2002),

> when technological innovations are not accepted or implemented properly, generally the failure may be traced to a poor fit between the nature of the innovation and the vested interests, resources, and expectations of its major gatekeepers.

Buntin et al. (2011), in the recent review of the literature on the benefits of health information technology, had shown predominantly positive results. Her investigation was methodologically following systematic review by Chaudray et al. (2006). Buntin had pointed out: "In fact, the stronger finding may be that the 'human element' is critical to health IT implementation."

She also emphasized the need for studies that document the challenging aspects of implementing health information technology more specifically and how these challenges might be addressed.

In an interpretive review on organizational issues surrounding technology implementations in health care settings, Cresswell and Sheikh (2013) emphasized that these issues are crucially important, but have as yet not received adequate research attention that may in part be not only due to the subjective nature of factors on individuals and organizations but also due to a lack of coordinated efforts towards more theoretically informed work.

CONCLUSION

In our investigation of organizational aspects of eHealth program implementation in Serbia, we have found Hofstede's cultural dimensions very useful for both understanding the problems in organizations and hypothesizing the solutions. We are completely aware of the limitations of our work in terms of the numbers of institutions and members of organizations to make some distinctive conclusions, but since it has been performed in a country where cross-cultural approach to organizational problems has not been studied frequently (except for the pioneering work of Jovanovic and Langovic-Milicevic, 2006 and Bogićević Milikić (2009)), we hope that it may be the motivation for others to understand and utilize Hofstede's cultural dimensions framework in their own domain. In this context, we agree with the argument of Mats and Sveningsson (2007) that getting a rich and detailed picture, sensitive of local context and the meanings of the people involved, is necessary in order to understand the phenomenon and to learn something that can encourage more reflective and realistic change work.

For the conclusions about the possible outcome of "Electronic Thyroid Network to Improve Patient Referral System: 'Thyro-Net Serbia,'" the authors are relying on previous experience of Professor Doktor's team that the earliest inclusion of administrative and clinical leadership is imperative to ensure a smooth transition from donor-funded effort to institutionally supported program. Lasting and meaningful organizational change ultimately comes from the individuals who are the building blocks of the organization. Prior to the first year of implementation, sponsorship or sense of participatory ownership must be cultivated, strengthened, and shared with each facility to ensure the success of the pilot program and a successful transition from pilot deployment to the broader integrated delivery of care. Budget estimates must be codeveloped and discussed within the first year. To allow for the allocation of funds, planning for inclusion of the program's continued costs should begin prior to the second year.

The costs of maintaining the thyroid network will be focused around the case managers and administrative positions, as these individuals are crucial for ensuring timely response to consultation and referral requests and ensuring all aspects of the systems are utilized as trained. Technical continuing support costs are generally quite low, with possible hardware upgrades to the server and server software within 5 years and annual maintenance fee to receive upgrades and service support being the only forecasted requirements. The true key to sustainability would depend upon the successful implementation of the thyroid network, where primary care clinicians receive timely and accurate responses to their requests and specialists are able to better screen and prepare for the most appropriate referrals. This success will occur with the inclusion of the Ministry of Health leadership, the hospital leadership, and the consulting clinicians, in all aspects of the program, to ensure the focus is retained on the improvement of the health of the Serbian people.

An important document titled "US: National Telemedicine Initiatives: Essential to Healthcare Reform," published in 2009 (Bashshur et al., 2009) "reflects the strongly held views and perspective of a diverse group of healthcare academicians, researchers, providers, and industry representatives from across the country [the United States] who share a belief in the necessity of healthcare reform and the centrality of telemedicine—or information technology-enhanced healthcare—in that reform." It outlined that a broader focus on telemedicine (also frequently referred to as telehealth or eHealth) is a more prudent and effective approach:

> While not a panacea, telemedicine offers significant opportunities to address the issues of inequities in access to care, cost containment, and quality enhancement. Telemedicine not only provides the potential to address structural issues of the health system, but it also promotes transparency and evaluation to drive further improvement.

It also stated that telemedicine connections between primary care providers and specialists would lend greater economic benefits and social prestige to primary care and patients would be less likely to get "lost" in the complexities of fragmented and unconnected medical providers and health systems.

In the more recent paper, Bashshur et al. (2013) emphasized that sustaining the promise of telemedicine rests on the three pillars of care: improved access, enhanced quality, and cost containment. It relies on understanding telemedicine's contribution to the full spectrum of care and the manner in which it enables an appropriate balance. Thus, the considerable promise of telemedicine in addressing the issues of quality, efficiency, cost, and access to care should be placed at the forefront of national effort to reform health care.

Presently, in Serbia, the Second Serbian Health Project approved by the World Bank in February 2014 (Johnston, 2014) has recognized the need for addressing crosscutting issues, such as strengthening the referral system between primary and hospital care.

In that context of renewed interest in telemedicine eHealth programs in the European Union and worldwide and the commitment from the World

Bank to improve patient referral system in Serbia, it may be a new opportunity for "Thyroid Consultation Network to Optimize Patient Care in Serbia: Thyro-Net Serbia."

The authors are hoping that this publication may also contribute to the realization of this, in our opinion, extremely important project, since this referral method is applicable in other medical disciplines and could be readily adopted for all referrals from primary to secondary and tertiary institutions in general and could contribute to the reform and sustainability of the Serbian health care system.

ACKNOWLEDGMENTS

The authors wish to thank the following people for their contribution to this project: Professor Robert Doktor, Robert Whitton, and Victoria Garshnek for their professional guidance and support in the process of creation and writing of the project proposal for electronic consultation network; Nenad Crncevic, Ivica Stojkovic, Dusica Stojanovic, and Zoran Rosko for collecting the data in Serbian health care organizations; and my colleagues and students for their technical support. Special thanks should be given to Dr David Bangert and Dr Linda Harris Bangert for their useful discussions and constructive recommendations for this work.

REFERENCES

Aas, I.H., 2002. Telemedicine and changes in the distribution of tasks between levels of care. J. Telemed. Telecare 8, 152.

Alvesson, M., Stefan, S., 2007. Changing Organizational Culture. Cultural Change Work in Progress. Taylor & Francis e-Library.

Balogun, J., 2006. Managing change: steering a course between intended strategies and unanticipated outcomes. Long Range Plann. 39, 29–49.

Balogun, J., Johnson, G., 2004. Organizational restructuring and middle manager sensemaking. Acad. Manag. J. 47 (4), 523–549.

Bangert, D., Doktor, R. (Eds.), 2005. Human and Organizational Dynamics in e-Health. Oxford/Seattle, Radcliffe.

Bashshur, R.L., 2002. Telemedicine and health care. Telemed. J. e-Health 8 (1), 5–12.

Bashshur, R.L., et al., 2009. National telemedicine initiatives: essential to healthcare reform. Telemed. J. e-Health 15 (6), 600–610.

Bashshur, R.L., Shannon, G., Krupinski, E.A., Grigsby, J., 2013. Sustaining and realizing the promise of telemedicine. Telemed. J. e-Health 19 (5), 339–345.

Bogićević Milikić, B., 2009. The influence of culture on human resource management processes and practices: the propositions for Serbia. Econ. Ann.. LIV (181), April–June.

Buntin, M.B., Burke, M.F., Hoaglin, M.C., Blumenthal, D., 2011. The benefits of health information technology: review of the recent literature shows predominantly positive results. Health Aff. 30 (3), 464–471.

Chaudray, B., Wang, J., Wu, S., et al., 2006. Systematic review: impact of health information technology on quality, efficiency, and costs of medical care. Ann. Intern. Med. 144, 742–752.

Cresswell, K., Sheikh, A., 2013. Organizational issues in the implementation and adoption of health information technology innovations: an interpretative review. Int. J. Med. Inform. 82 (5), e73–e86.

Dawson, P., 2003. Understanding Organizational Change: The Contemporary Experience of People at Work. Sage, London.

Delange, F., 2000. Iodine deficiency. In: Utiger, R.D., Braverman, L.E. (Eds.), The Thyroid. A Fundamental and Clinical Text. Lippincott, Philadelphia, pp. 295–316, 108–112.

Doktor, R., Bangert, D., Valdez, M., 2005. Organizational learning and culture in the managerial implementation of clinical e-Health systems: an international perspective. In: Proceedings of the 38th Hawaii International Conference on System Sciences— (HICSS'05)—Track 6.

Dunn, J.T., 2002. Guarding our nation's thyroid health. J. Clin. Endocrinol. Metab. 87 (2), 486–488.

Dunn, J.T., Delange, F., 2001. Damaged reproduction: the most important consequence of iodine deficiency. J. Clin. Endocrinol. Metab. 86, 2360–2363.

Edworthy, S.M., 2001. Telemedicine in developing countries. BMJ 323, 524–525.

European Commission, 1994. DG XIII—Telecommunications, information market and exploitation research. Telematics Applications Programme (1994–1998), 10 November.

European Foundation for the Improvement of Living and Working Conditions, 1994. Telehealth and Telemedicine: Executive Summary of a European Foundation Research Project. Loughlinstown House, Shankill, Co., Dublin, Ireland, 18 pp.

Free, C., Phillips, G., Galli, L., Watson, L., Felix, L., et al., 2013. The effectiveness of mobile-health technology-based health behaviour change or disease management interventions for health care consumers: a systematic review. PLoS Med. 10 (1), e1001362.

Glinoer, D., Delange, F., 2000. The potential repercussions of maternal, fetal and neonatal hypothyroxinemia on the progeny. Thyroid 10, 871–887.

Harno, K.S., 1999. Telemedicine in managing demand for secondary-care services. J. Telemed. Telecare 5, 189–192.

Hofstede, G., 1980. Motivation, leadership and organization: do American theories apply abroad? Organ. Dyn. 9 (1), 42–63.

Hofstede, G., 2001. Culture's Consequence. Sage Publications, Thousand Oaks, CA.

Hofstede, G., 2011. Dimensionalizing cultures: the Hofstede model in context. Online Read. Psychol. Cult.. 2 (1). http://dx.doi.org/10.9707/2307-0919.1014.

Hofstede, G., Bond, M.H., 1988. The Confucius connection: from cultural roots to economic growth. Organ. Dyn. 16, 4–21.

Hollowell, J.G., Staehling, N.W., Flanders, W.D., Hannon, W.H., Gunter, E.W., Spencer, C.A., Braverman, L.E., 2002. Serum TSH, T4, and thyroid antibodies in the United States population (1988 to 1994): National Health and Nutrition Examination Survey (NHANES III). J. Clin. Endocrinol. Metab. 87, 489–499.

Houston, M.S., Myers, J.D., Levens, S.P., McEvoy, M.T., Smith, S.A., Khandheria, B.K., Shen, W.K., Torchia, M.E., Berry, D.J., 1999. Clinical consultations using store-and forward telemedicine. Mayo Clin. Proc. 74 (8), 764–769.

Hu, P., Chau, P., Sheng, O., Tam, K., 1999. Examining the technology acceptance model using physician acceptance of telemedicine technology. J. Manag. Inf. Syst. 16 (2), 91–112.

Inglehart, R., 1997. Modernization and Postmodernization: Cultural, Economic, and Political Change in 43 Societies. Princeton University Press, Princeton, NJ.

Johnson, P., Andrews, D.C., Wells, S., de-Lusignan, S., Robinson, J., Vandenburg, M., 2001. The use of a new continuous wireless cardiorespiratory telemonitoring system by elderly patients at home. J. Telemed. Telecare 7 (Suppl. 1), 76–77.

Johnston, T.A., 2014. Integrated Safeguards Data Sheet (Appraisal Stage)—Second Serbia Health Project—P129539. World Bank, Washington, DC.

Jovanovic, M., 2004. Interkulturni menadzment. Megatrend, Beograd.

Jovanovic, M., Langovic-Milicevic, A., 2006. Interkulturni izazovi globalizacije. Megatrend University, Belgrade.

Kokesh, J., Ferguson, A.S., Patricoski, C., 2004. Telehealth in Alaska: delivery of health care services from a specialist's perspective. Int. J. Circumpolar Health 63 (4), 387–400.

Kovai, L., Lonari, S., Paladino, J., Kern, J., 2000. The Croatian telemedicine. Stud. Health Technol. Inform. 77, 1146–1150.

Latifi, R., Muja, S., Bekteshi, F., Reinicke, M., 2004. Use of information technology to improve quality of th healthcare: Kosova's telemedicine project and international virtual e-hospital as an example. Stud. Health Technol. Inform. 104, 159–167.

Legris, P., Collerette, P., 2006. A roadmap for IT implementation: integrating stakeholders and change. Project Manag. J. 37 (5), 64–75.

Loane, M.A., Bloomer, S.E., Corbett, R., Eedy, D.J., Hicks, N., Lotery, H.E., Mathews, C., Paisley, J., Steele, K., Wootton, R., 2000. A randomized controlled trial to assess the clinical effectiveness of both realtime and store-and-forward teledermatology compared with conventional care. J. Telemed. Telecare 6 (Suppl. 1), S1–S3.

Lucas, H.C., Spitler, V.K., 1991. Technology use and performance: a field study of broker workstations. Decis. Sci. 30 (2), 291–311.

Nielsen, T.M., Jorgensen, H.D., 1996. Factors to consider when establishing a europe-wide network for exchange of health information. In: Medical Informatics Europe '96—Human Facets in Information Technology. IOS Press, Amsterdam, pp. 8–12.

Paunkovic, N., Paunkovic, J., 2004. Thyroidology—Topics and Chronology. Megatrend, Belgrade.

Paunkovic, N., Paunkovic, J., Pavlovic, O., 1998. The significant increase in incidence of Graves' disease in eastern Serbia during the civil war in the former Yugoslavia (1992 to 1995). Thyroid 8, 37–41.

Paunković, J., Doktor, R., Paunković, N., Whitton, R., Garshnek, V., Crnčević, N., 2008. Information technology for improvement of patient referral system from primary care to secondary and tertiary care in Serbia: 'Thyronet'-electronic thyroid consultation network. Facta Univ., Ser.: Econ. Organ. 5 (1), 71–81.

Paunković, J., Jovanović, R., Stojković, Z., Stojković, I., 2010. Sustainable implementation of information and communication technology in health care. Case study of organizational and cultural factors. Sibiu Alma Mater Univ. J. Ser. A. Econ. Sci. 3 (3), 1–8.

Paunkovic, J., Stojkovic, I., Stojkovic, Z., Zikic, S., 2011. Awareness of organizational culture is important for sustainable implementation of e-health. In: International Scientific Conference Management of Technology—Step to Sustainable Production June 2010, Rovinj, Croatia., ISBN 978-953-7738-09-9.

Pfannenstiel, P., 1998. The cost of continuing iodine deficiency in Germany and the potential cost benefit of iodine prophylaxis. IDD Newsletter 14, 11–12.

Schoenberg, R., Safran, C., 2000. Internet based repository of medical records that retains patient confidentiality. BMJ 321, 1199–1203.

Schwartz, S.H., 2012. An overview of the Schwartz theory of basic values. Online Read. Psychol. Cult.. 2 (1)http://dx.doi.org/10.9707/2307-0919.1116.

Stojkovic, Z., Stojkovic, I., Paunkovic, J., 2009. Improvement of health care efficiency using the new information technologies in primary health care. In: Management of Technology—Step to Sustainable Production, Sibenik, Croatia, June 10-12.

Triandis, H.C., 1995. Individualism and Collectivism. Westview, Boulder, CO.

Triandis, H.C., 2004. The many dimensions of culture. Acad. Manag. Exec. 18, 88–93.

Trompenaars, F., Hampden-Turner, C., 1998. Riding the Waves of Culture: Understanding Cultural Diversity in Global Business, second ed. McGraw-Hill, New York.

Walsh, S.H., 2004. The clinician's perspective on electronic health records and how they can affect patient care. BMJ 328, 1184–1187.

Weinstock, M.A., Nguyen, F.Q., Risica, P.M., 2002. Patient and referring provider satisfaction with teledermatology. J. Am. Acad. Dermatol. 47 (1), 68–72.

Whitten, P.S., Mair, F., Haycox, A., May, C., Williams, T., Hellmich, S., 2002. Systematic review of cost effectiveness studies of telemedicine interventions. BMJ 324, 1434–1437.

World Bank, 2003. Serbia and Montenegro—Serbia Health Project. World Bank, Washington, DC.

World Bank, 2012. Serbia—Additional Financing for the Health Project. World Bank, Washington, DC.

WEBSITES

http://documents.worldbank.org/curated/en/2003/04/2291228/serbia-montenegro-serbia-health-project.

http://documents.worldbank.org/curated/en/2012/09/16828470/serbia-additional-financing-health-project.

http://documents.worldbank.org/curated/en/2014/01/18809589/integrated-safeguards-data-sheet-appraisal-stage-second-serbia-health-project-p129539.

http://dx.doi.org/10.9707/2307-0919.1014. Online Read. Psychol. Cult. 2 (1).

http://dx.doi.org/10.9707/2307-0919.1116. Online Read. Psychol. Cult. 2 (1).

http://ec.europa.eu/digital-agenda/en/news/ehealth-action-plan-2012-2020-innovative-healthcare-21st-century.

http://ec.europa.eu/information_society/activities/health/docs/publications/201002ehrimpact_study-final.pdf.

http://hdr.undp.org/sites/default/files/reports/262/hdr_2001_en.pdf.

http://web.worldbank.org/external/projects/main?pagePK=104231&piPK=73230&theSitePK=40941&menuPK=228424&Projectid=P077675.

http://www.cigota.rs/en/strana/special-hospital.

http://www.geerthofstede.nl/dimensions-of-national-cultures.

http://www.who.int/topics/ehealth/en/.

http://wwwds.worldbank.org/external/default/WDSContentServer/WDSP/IB/2012/10/15/000386194_20121015011358/Rendered/PDF/ICR20640P0776700disclosed0100110120.pdf.

http://wwwds.worldbank.org/servlet/WDSContentServer/WDSP/IB/2002/11/15/000094946_02111404.

APPENDIX 15

UC Davis West Village Energy Initiative Annual Report 2012-2013

UC DAVIS WEST VILLAGE
ENERGY INITIATIVE ANNUAL REPORT
2012 - 2013

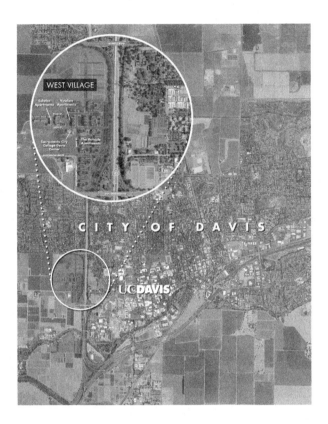

FOREWORD

The most important advances in new technologies come through the process of building, debugging, and continually learning and improving technologies in use.

Andrew Hargadon, UC Davis Graduate School of Management; Charles J. Soderquist Chair in Entrepreneurship and Professor of Technology Management

In his remarks at the UC Davis West Village ribbon cutting ceremony in October 2011, Andrew Hargadon likened the West Village Energy Initiative to the invention process of Thomas Edison. For example, before Edison broke ground on the first central power plant ever built in the United States, he had produced 14 patents involving electric light. Once he began the construction, he produced 368 more.

The UC Davis West Village Energy Initiative follows on Edison's tradition. From its inception as an "environmentally responsible campus housing project" to its current status as the nation's largest planned zero net energy (ZNE) community and home to the university's Energy and Transportation Center, UC Davis West Village has steadily contributed to the practical knowledge of how to plan, construct, operate, and improve upon a large-scale, sustainable, mixed-use neighborhood.

This first annual report provides an overview of the energy initiative and results to date. We describe lessons learned and next steps in the ongoing process. UC Davis and its developer partner, West Village Community Partnership, LLC (WVCP), are committed to continuing progress towards our mutual energy goals. We are extremely proud of how far we have come, even as we recognize that there is still much work to be done to achieve our ultimate goal of ZNE.

From the earliest planning stages, UC Davis West Village aspired to extraordinary goals. As the project has progressed, it has moved from being extraordinary to being transformational—as a campus neighborhood and as an experience for all involved in its creation and evolution. West Village started as a public-private partnership to develop much-needed housing for UC Davis students, faculty, and staff and is now home to nearly 2000 students in 663 apartments in a district anchored by recreational amenities and a community college. Along the way, UC Davis and WVCP embraced the aspirational goal of making West Village the largest planned ZNE community in the United States.

Through catalytic grants from the US Department of Energy, California Energy Commission, and the California Public Utilities Commission, UC Davis West Village is poised to be a road map around the technological, financial, and regulatory barriers that projects face in striving to be ZNE. It already is becoming a living laboratory for energy efficiency and renewable energy research providing not only valuable data but also a test bed for new technologies and business models related to ZNE.

About UC Davis West Village

ABOUT UC DAVIS WEST VILLAGE

A New Campus Community

UC Davis West Village (West Village) is a new campus neighborhood designed to be the home for approximately 3000 students and 500 staff and faculty families. Located on the UC Davis campus, the overarching goals for the community are the following:

Quality of place—to create a great community and desirable place to live that will help UC Davis recruit the best and brightest students, faculty, and staff; to let them live within walking or cycling distance of the campus; and to participate fully in campus life

Affordability—to enable faculty and staff to purchase new homes locally at below market prices and to expand the choices for students to live near campus

Environmental responsiveness—to develop the site and buildings according to sound environmental principles so as to reduce reliance on cars, limit energy consumption, enable renewable energy production, and contribute to a healthy environment

The community was developed through a public-private partnership between UC Davis and West Village Community Partnership, LLC, a joint venture of Carmel Partners from San Francisco and Urban Villages from Denver.

The West Village Energy Initiative

THE WEST VILLAGE ENERGY INITIATIVE

Through the collaborative design process with WVCP, UC Davis expanded upon its core principle of making UC Davis West Village environmentally responsive and launched the WVEI. Working together, the UC Davis and

WVCP team first looked for ways to make West Village as energy-efficient as possible. In 2007, UC Davis commissioned a study with its own UC Davis Energy Efficiency Center and local consulting firm, the Davis Energy Group, to help identify deep energy efficiency measures that could be included in the design of the student housing and single-family residences to be built as part of West Village. The results of this study demonstrated that by adopting deep energy efficiency measures, WVCP could reduce consumption in West Village by nearly 50 percent compared to the California energy efficiency building code.

With this result, WVCP and UC Davis realized that a much larger goal was within reach—the goal of making West Village a ZNE community. In 2008, WVCP engaged Chevron Energy Solutions to evaluate the financial feasibility of achieving a ZNE goal defined as "zero net electricity from the grid measured on an annual basis."

In 2009, WVCP and UC Davis decided to strive for this goal. Because West Village had to be accessible for UC Davis faculty, students, and staff, the ZNE goal had to be balanced against the goal of affordability. In response to these competing principles, WVEI was created and the following principles were adopted by the team:

- West Village would strive to use ZNE from the grid measured on an annual basis.
- ZNE needed to be achieved at no higher cost to the developer.
- ZNE needed to be achieved at no higher cost to the consumer.
- West Village would adopt deep energy efficiency measures to reduce energy demand.
- ZNE would be achieved through multiple renewable resources developed on-site at a community scale.
- West Village would be used as a living laboratory for further energy-related topics.

Energy Design And Performance

ENERGY DESIGN AND PERFORMANCE

West Village's first-phase components have achieved a remarkable 87 percent of the initial ZNE goal, years ahead of the full completion of the community. Along the way, a number of challenges and issues emerged and, with them, lessons learned.

The following report includes

1. major milestones in the construction of West Village and its energy systems,
2. progress towards ZNE,
3. lessons learned about implementing ZNE,
4. UC Davis West Village living lab,
5. outreach and awards,
6. next steps towards achieving ZNE.

Other aspects of the WVEI reported here include UC Davis existing and ongoing energy-related research and teaching activities related to West Village.

REPORTING PERIOD

The reporting period for this initial annual report is March 2012 through February 2013. This period was chosen because it is the first 12-month period for which both electricity demand (consumption) and generation (production) data were available for a significant portion of the community. Prior to this period, some apartments were occupied, but solar panels that are part of the renewable electricity generation capacity had not been installed and commissioned or were installed but the apartments were not yet occupied. Thus, this is the first meaningful period where progress towards the ZNE goal could be documented. Other aspects of the WVEI initiative that occurred through August 2013 also are reported.

Future annual reports will describe results by prior leasing year (September through August). The second WVEI annual report scheduled for winter 2015 will report on results from September 2013 to August 2014.

Major Milestones in the Construction of UC Davis West Village

MAJOR MILESTONES IN THE CONSTRUCTION OF UC DAVIS WEST VILLAGE

The Phase I Ramble Apartments (192 units with 654 student beds) were occupied in September 2011. The mixed-use buildings around the Village Square, including the Viridian (123 units with 192 beds) on the second through fourth floors, were occupied in September 2011. The Viridian apartments are a higher-end product consisting of one or two bedroom units and are occupied by faculty and staff as well as students. The first floor of these mixed-use buildings includes approximately 42,500 square feet of office/retail space. This space was unoccupied until January 2013 when UC Davis established its Energy and Transportation uHub at West Village and moved the associated centers and institutes into approximately two-thirds of the ground floor office space. However, the solar panels were operating prior to occupancy.

The Phase II Ramble Apartments (192 units with 630 student beds) opened in September 2012. The final phase of student housing built by WVCP, the Solstice Apartments (156 units with 504 beds), opened in September 2013. Construction of single-family homes is expected to start in 2014 with a spring opening of the first models. The first building of the Sacramento City College West Village Center opened in January 2012. The community college currently is not a participant in the ZNE goal.

A total of 2.1 megawatts of photovoltaic solar panels (PV) was installed to serve the Phase I Ramble Apartments and Viridian apartments, Viridian commercial spaces, and the leasing recreation center. An additional 1.1 megawatts was installed to serve Phase II Ramble Apartments. PV systems were sized to serve common area and parking lot lighting. PV systems were installed on building rooftops and on carport shade structures. To comply with net metering laws applicable at the time, each Ramble Apartment has its own utility meter and associated PV system.

Installed Photovoltaic Capacity: Phase 1

Facility	Installed PV (rated kW STCDC)
Phase 1 Ramble Apartments	1072
Viridian	872
Leasing recreation center	154

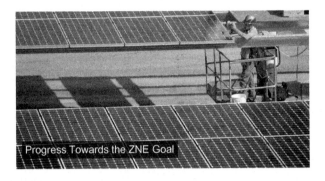

Progress Towards the ZNE Goal

PROGRESS TOWARDS THE ZNE GOAL

For the purposes of the WVEI, ZNE is defined as "ZNE from the grid measured on an annual basis." More specifically, this means the community would produce enough energy on-site to offset its annual consumption. It is connected to the regional electrical grid and during peak hours will be feeding electricity into the grid, while at night, it will be drawing electricity from the grid.

In spring 2013, WVCP engaged the Davis Energy Group to perform a comprehensive evaluation of the West Village energy consumption and production for March 2012 through February 2013 reporting period. This early "snapshot" of the community's energy performance against its ZNE goals was to inform UC Davis and WVCP of progress to date and to help with future implementation of the WVEI. The study was also a way to share with other developers, policy makers, and institutions how they could learn from the West Village experience.

Energy Production and Consumption: March 2012 Through February 2013[1]

Facility	Production (MWh)			Consumption (MWh)			
	Modeled (MP)	Actual (AP)	Percent (AP/MP) (%)	Modeled (MC)	Actual (AC)	Percent (AC/MC) (%)	Percent (AP/AC) (%)
Phase 1 Ramble Apartments							
Apartments	1024	1110	108	1127	1377	122	81
Common areas[2]	471	451	96	390	602	155	75
Total	1495	1561	104	1516	1979	131	79
Viridian							
Apartments	519	530	102	530	515	97	103
Common areas	321	314	98	141	432	306	73
Total	839	844	101	672	947	141	89
Viridian commercial areas							
Total	415	358	86	377	84	22	424
Leasing and recreation center							
Building	225	218	97	225	292	130	75
Pool/outdoor	0	0	n/a	0[3]	109	n/a	0
Total	225	218	97	225	402	178	54
Total	2974	2981	100	2790	3412	122	87

[1] Includes buildings/meters with full year of occupancy and production. Los Rios Community College District is not part of the ZNE goal and so not included in this analysis.
[2] Common areas include central heat pump water heaters, common area lighting, and elevators, and common area meters also have parking lot and path lighting loads attached.
[3] Pool equipment and outdoor lighting were not included in the original modeling.
Source: Evaluation of UC Davis West Village Phase I Energy Use and PV Production. Davis Energy Group, Inc. September 19, 2013

OBSTACLES IDENTIFIED AND LESSONS LEARNED

While achieving an exceptional 87 percent of our initial ZNE goals, years ahead of fully completing the community, a number of challenges and issues emerged during that time period. Each brought with it important lessons. Key reasons that electrical consumption exceeded production during the first reporting period are summarized below.

COMMISSIONING

Commissioning newly constructed systems to work as designed is critical to achieving optimum performance. Commissioning issues surrounding the central heat pump water heaters (HPWHs) contributed to West Village falling short of its ZNE goal for this first measurement period.

The heat pump water heaters installed for the student housing initially failed to perform according to specification, causing the heating control system to automatically shift to less energy-efficient, backup resistance-type electrical heaters to meet the water heating demand of the residents. During commissioning, this problem was identified and has been resolved by the WVCP team. Nonetheless, reliance on the backup water heating systems added substantial unanticipated power demand to the project.

The initial commissioning challenges have been addressed, and the project has improved its commissioning practices as more solar PV is installed and additional student housing is built.

Lessons Learned

- With new technologies and strategies, it is important to test and commission building systems to ensure that they are operating as designed.
- Provide alarms on the heat pump water heaters that can notify operations staff when a system goes down. WVCP has worked with the manufacturer to install alarms on all heat pump water heater systems.
- Implement ongoing commissioning of the more complex building systems, such as the central heat pump water heaters and the mechanical systems at the leasing recreation center.

MODELING AND DESIGN

Reviewing consumption data for the first measurement period, the Davis Energy Group discovered that some of the earlier modeling assumptions were not consistent with actual consumption data. The modeling assumptions for student apartments were based on several published sources for multifamily projects[1]. Evident from the consumption data collected for the first measurement period is a distinct difference among students compared to residents of other multifamily projects within the more general population in terms of miscellaneous electrical loads (plug loads). In a typical multifamily setting, there may be only one or two computers for the household, one gaming system and other multiuser appliances. In contrast, a four-bedroom student housing apartment turns out to resemble four separate households, each with its own computer, smart phone, gaming system, television, and other separate appliances. WVCP, working with researchers at UC Davis, is developing educational programs to encourage students to conserve energy. Additional solar PV alternatives to offset overconsumption are being evaluated.

Original modeling assumptions also assumed plug load consumption reductions based on the implementation of plug load controls through a "one-switch" device and energy consumption displays in each apartment. The proposed "one-switch" device would enable the occupant to turn off noncritical plug loads when not needed, while the energy consumption display would provide occupant feedback on real-time energy use, allowing

---

[1] End-use profiles for different end uses were based on two sources: Department of Energy's Building America House Simulation Protocols (NREL, 2008) and the California Residential Appliance Saturation Study (KEMA, 2004).

them to better reduce apartment energy use. Due to challenges with finding cost-effective products to serve these needs, neither of these devices has been implemented in the project at this time.

In the initial design modeling for the community, energy demand for the recreational swimming pools and common area lighting were not included. The modeling focused on the buildings that make up the student apartments. These demands are, however, included in this reporting of progress towards the ZNE goal. Strategies to reduce overall consumption and other ways to offset overconsumption due to incorrect modeling assumptions are being developed.

Lessons Learned
- Apartment plug load energy use assumptions should be higher to account for the higher number of electronic devices found and the increased use of these devices in student housing.

OCCUPANT BEHAVIOR VARIABILITY

Individual apartment consumption for high-energy-use apartments was up to three times higher than low-energy-use apartments. This large difference suggests significant behavioral variability between occupants. High energy use is primarily occupant-related and most likely due to occupant-supplied plug loads. Since occupants do not directly pay for their utilities and do not have access to records of how much energy they consume, there is little awareness of their consumption habits or how their behavior affects energy use. As mentioned above, WVCP is developing an educational program to encourage students to conserve energy.

Lessons Learned
- Develop and implement community engagement strategies to encourage energy conservation. Educate the community to the occupant's role in a ZNE building and better translate the ZNE vision. Strategies could include incentives for apartments that deviate from the targets, contests, and awareness campaigns.
- Identify "high-use" apartments early in the school year and develop a strategy to discourage excessive (larger than estimated) energy consumption. Provide means for occupant feedback and control of consumption. Evaluate currently available products on the market to determine if there are cost-effective solutions to providing occupant feedback and control.

UC DAVIS WEST VILLAGE LIVING LAB

UC Davis West Village is now the home for the first UC Davis-based "university hub" or "uHub"—a prototype for future "innovation hubs" aimed at better fostering collaboration among related research units, enhancing interaction with the private sector, and accelerating the transfer of university inventions from the lab to the marketplace. The Energy and Transportation uHub at West Village is now the physical home for these UC Davis research centers:

- Center for Water-Energy Efficiency
- China Center for Energy and Transportation
- Energy Efficiency Center
- Institute of Transportation Studies
- Plug-in Hybrid & Electric Vehicle Research Center
- Policy Institute for Energy, Environment and the Economy
- Program for International Energy Technologies
- Sustainable Transportation Energy Pathways program
- Energy Institute
- Urban Land Use and Transportation Center
- Western Cooling Efficiency Center

The co-location of these research centers at West Village supports the uHub concept. It also creates the opportunity to use West Village as a "living lab" to evaluate and develop energy and transportation technologies and solutions at the building and community scales. In addition to monitoring activities relating to ZNE for West Village, several other research investigations are under way, including the following:

Battery-buffered electric vehicle charging station: The project combines on-site solar energy, a high-voltage lithium-ion battery pack, and electric vehicle charging stations to store PV energy and charge vehicles, day or night, without adding loads to the grid. The system's primary source of energy is a PV tower located in the Village Square but is also grid-connected. This allows the system to be operated in a number of ways including mitigating electric vehicle charging loads, reducing demand during peak hours and load shifting services for vehicle charging and building loads.

Multifamily hybrid solar demonstration: To demonstrate advanced ZNE technologies, a 24-panel PV-plus-thermal hybrid solar system has been installed at Solstice student housing. The system provides PV electricity to one apartment and supplies thermal energy to the building's central hot water system. The PV-plus-thermal system was integrated with the central hot water system to interact with the high-efficiency air-to-water heat pump to investigate how to optimize both water heating systems for multifamily applications.

Single-family hybrid solar and demand-side management retrofits: Since single-family homes are not yet available at West Village, these technologies are being investigated at Aggie Village, an existing university faculty and staff community located in downtown Davis. The technologies in the test home include a PV-plus-thermal hybrid solar system, a lithium-ion second life battery, and a home energy management system. The energy management system monitors and controls plug loads, appliances, battery storage charging and discharging, and electric vehicle charging. Future plans include installing a ground source heat pump system for heating and cooling of the home. The goal of the project is to achieve ZNE in a retrofit environment and transfer learned lessons to future single-family homes in West Village and elsewhere.

AEC behavior study: Architectural Energy Corporation, in conjunction with PG&E and Sustainable Design + Behavior, is working closely with the West Village management team to create a monitoring and outreach program for the community. This program includes monitoring approximately 140 apartment units to understand end-use loads and provide specific monthly intervention messages to those units, based on their observed consumption. The program has initiated a community-wide outreach program that provides messaging and information on energy consumption. This includes regular outreach activities, such as contests, that promote education and energy conservation within the community.

Honda Smart Home demonstration project: This unique, high-tech sustainable home will demonstrate an approach to meeting California's goal of requiring all new residential construction to be ZNE by 2020. Technologies featured in the home include a solar power system, a smart-grid Honda energy management system, direct solar PV-to-vehicle charging, and high-efficiency HVAC (heating, ventilation, and air-conditioning) and lighting systems designed by UC Davis.

OUTREACH AND AWARDS

Tours

Since its opening in 2011, West Village has attracted tremendous interest from around the region and the world. UC Davis faculty and staff and WVCP employees host several tours of West Village each week with visitors ranging from researchers to environmental scientists, energy regulators, and elected officials. Foreign visitors have come from as far away as the Netherlands, Saudi Arabia, Abu Dhabi, and Lithuania. Since 2011, there have been approximately 250 on-site tours of West Village with thousands of participants.

Media Coverage

West Village and its ZNE goals have been featured in the New York Times, CNN International, The Wall Street Journal, National Public Radio, Sunset magazine, Forbes, San Francisco Chronicle, Los Angeles Times, and multiple mainstream and trade publications.

Awards

- 2013 PCBC Gold Nugget Awards for Best Community Site Plan and Judges Special Award of Excellence
- 2012 Sacramento Business Journal Green Leadership Award for a Game Changer Project
- 2012 Breathe California Clean Air Award for Innovative Strategy
- 2012 Sacramento Valley American Planning Association California Chapter Local Vision Award
- 2012 Green Dot Award for the West Village Square

Next Steps

Next Steps

Several major milestones will occur during the next reporting period, which ends in August 2014.

- The Solstice Apartments opened in September 2013 and bring on both new energy demand and new renewable energy supply.
- Ground floor commercial/office areas around the Village Square will be nearly fully occupied. These will bring on new demand without adding new supply. The demand will include UC Davis energy-related laboratories that were not anticipated when the original modeling for the ZNE goal was adopted.
- An evaluation of opportunities to install additional solar PV to offset overconsumption at existing facilities will be undertaken.
- WVCP will begin implementation of its energy use and efficiency educational program for West Village residents.

- Renewable Energy Anaerobic Digester at the UC Davis landfill will open off-site. At this biodigester, organic waste will be used to generate biogas that will be used to produce approximately 4 million kWh of electricity per year. This project was a spin-off of the original West Village ZNE planning, and the renewable energy produced may be credited towards West Village consumption to achieve the ZNE goal.

ACKNOWLEDGMENTS

The WVEI has truly been and will continue to be a coordinated effort of the University of California and its development partners who are committed to making West Village more sustainable and extending results of the project to the broader community. This project would not have been possible without the commitment and efforts of the following:

- West Village Community Partnership, LLC (Carmel Partners and Urban Villages)
- Nolan Zail Consulting
- US Department of Energy
- California Energy Commission
- California Public Utilities Commission
- Pacific Gas and Electric Company
- Chevron Energy Solutions
- Davis Energy Group
- Energy + Environmental Economics
- Itron
- UC Davis Energy Efficiency Center
- UC Davis California Lighting Technology Center
- UC Davis Institute of Transportation Studies
- UC Davis Energy Institute
- UC Davis Western Cooling Efficiency Center

WESTVILLAGE
UCDAVIS

1580 Jade Street
Davis, CA 95616
www.ucdaviswestvillage.com

UCDAVIS
ENVIRONMENTAL STEWARDSHIP
AND SUSTAINABILITY

436 Mrak Hall
University of California, Davis
One Shields Avenue
Davis, CA 95616
www. sustainability.ucdavis.edu

APPENDIX 17

Achieving Fossil-Free Homes through Residential PACE Financing

Elisabeth N. Radow[1,2]

[1]Elisabeth N. Radow, Esq., is the managing attorney of Radow Law PLLC. Radow serves as chairwoman of the Committee on Energy, Agriculture and the Environment of the New York State League of Women Voters. She is also a member of the Town of Mamaroneck, NY, Sustainability Collaborative. www.radowlaw.com
[2]Copyright © 2014 by Elisabeth N. Radow All rights reserved.

The race is on. Climate change, which brings droughts, floods, and sea-level rise, among other extreme effects, has to be stopped in its tracks. To prevent more intense interference with the climate system, greenhouse gas (GHG) emissions from worldwide energy systems have to be phased out by the second half of the twenty-first century. Initiatives that transform the built environment to function independently of fossil fuel will play a central role in achieving this goal.[1] All American homeowners can begin now to reduce their household-

---

[1] Architecture 2030, *Roadmap to Zero Emissions* (June 4, 2014) (amended version) (http://architecture2030.org/multimedia/publications). See also *Sources of Greenhouse Gas Emissions*, U.S. ENVTL. PROT. AGENCY, http://www.epa.gov/climatechange/ghgemissions/sources.html (last visited August 15, 2014). According to the EPA-based reports, the primary sources of greenhouse gas in the United States are the following: (1) electricity production (34% of 2010 greenhouse gas emissions)—electricity production generates the largest share of greenhouse gas emissions. Over 70% of our electricity comes from burning fossil fuels, mostly coal and natural gas. (2) Transportation (27% of 2010 greenhouse gas emissions)—greenhouse gas emissions from transportation primarily come from burning fossil fuel for our cars, trucks, ships, trains, and planes. (3) About 90% of the fuel used for transportation is petroleum-based, which includes gasoline and diesel. (4) Industry (21% of 2010 greenhouse gas emissions)—greenhouse gas emissions from industry primarily come from burning fossil fuels for energy and greenhouse gas emissions from certain chemical reactions necessary to produce goods from raw materials. (5) Commercial and residential (11% of 2010 greenhouse gas emissions)—greenhouse gas emissions from businesses and homes arise primarily from fossil fuels burned for heat, the use of certain products that contain greenhouse gases, and handling of waste. (6) Agriculture (7% of 2010 greenhouse gas emissions)—greenhouse gas emissions from agriculture come from livestock such as cows, agricultural soils, and rice production.

generated GHG emissions powered by fossil fuel sources such as coal, oil, and natural gas with affordable shifts in activities of daily living. One strategy involves incorporating into the house energy efficiency measures; another strategy involves installing renewal energy retrofits. The technology from the building sector exists as do money-saving energy conservation standards for household appliances; yet, accessing financing that easily facilitates these essential shifts requires federal residential mortgage financing policy to become consistent with the Obama administration's public recognition of climate change and America's need to reduce pollution. This chapter will address how to move beyond current federal residential mortgage financing policy and practice that favors America's fossil fuel production over, and potentially at the expense of, financing policy and practice, which could facilitate a swifter shift to energy efficiency and renewable energy alternatives for homeowners.

America's 129,950,000 households[2] emit 5% of the nation's GHGs by cooking, heating, and cooling our homes with fossil fuel; households also contribute a portion of GHG emissions from waste sent to waste treatment facilities and landfills. In addition, households contribute 32% of the nation's GHG emissions attributable to electrical consumption.[3]

---

[2] *Table 989 Housing Units by Unit in Structure and State: 2009, Statistical Abstract of the United States:* U.S. CENSUS BUREAU, 2012, https://www.census.gov/compendia/statab/2012/tables/12s0989.pdf (last visited August 15, 2014).

[3] *Sources of Greenhouse Gas Emissions-Commercial and Residential,* U.S. ENVTL. PROT. AGENCY, http://www.epa.gov/climatechange/ghgemissions/sources/commercialresidential.htm (last visited August 15, 2014). Combustion of natural gas and petroleum products for heating and cooking needs emits carbon dioxide (CO_2), methane (CH_4), and nitrous oxide (N_2O). Emissions from natural gas consumption represent about 79% of the direct fossil fuel CO_2 emissions from the residential and commercial sectors. Coal consumption is a minor component of energy use in both of these sectors. Organic waste sent to landfills emits CH_4. Wastewater treatment plants emit CH_4 and N_2O. Fluorinated gases (mainly hydrofluorocarbons (HFCs)) used in air conditioning and refrigeration systems can be released during servicing or from leaking equipment. *Sources of GHG emissions* from the electricity sector involve the generation, transmission, and distribution of electricity. Carbon dioxide (CO_2) makes up the vast majority of greenhouse gas emissions from the sector, but smaller amounts of methane (CH_4) and nitrous oxide (N_2O) are also emitted. These gases are released during the combustion of fossil fuels, such as coal, oil, and natural gas, to produce electricity.

The United States' population of 322,983,889[4] people composes approximately 4.5% of the world's of more than 7 billion people; yet, as of 2008, the United States ranked second, behind China, by contributing 19% to the Earth's GHG emissions.[5] As climate change results in more extreme and expensive droughts, storms, and floods, water and energy consumption, historically a matter of exclusive private choice, now presents multiple mounting community and societal costs when left unchecked. In 2012 alone, climate or weather disasters involving droughts, hurricanes, wildfires, and other severe weather cost the American economy more than $100 billion.[6] A home represents a family's most valuable financial asset and source of personal comfort. In the interest of stabilizing the climate by balancing private and societal interests regarding water and energy use, federal policy can broaden existing financing options for homeowners who seek to streamline household expenses and consume less water and energy.

FINANCING OPTIONS

The range of current financing options that homeowners can use when upgrading their home's energy efficiency or installing renewable energy retrofits consists of the following:

For energy efficiency projects of limited scope and cost, streamlining household expenses can free up the funds necessary to pay for the improvement and is interest-free and without the risk of foreclosure that looms with a delinquent secured loan.

For moderate to expansive water and energy efficiency upgrades and renewable energy retrofits, homeowners can consider secured financing, such as a conventional mortgage or a loan that uses the equity in the home (that portion of the home that is not already subject to a mortgage). Secured loans are extended to a borrower with acceptable credit and a house with an appraised value in excess of the loan amount, typically at least 10 percent. A mortgage serves as a security instrument that is recorded in the public records, constitutes a lien against the home, and is intended to facilitate full

---

[4] *Current World Population (Ranked)* http://www.geohive.com/earth/population_now.aspx (statistics as of August 19, 2014).

[5] *Global Greenhouse Gas Emissions Data*, U.S. ENVT'L PROT. AGENCY, http://www.epa. gov/climatechange/ghgemissions/global.html (last visited August 14, 2014).

[6] President Barak Obama, *Climate Change and President Obama's Action Plan; see generally, the effects of climate change.* http://www.whitehouse.gov/climate-change (last visited August 20, 2014).

repayment of the loan from the proceeds of a foreclosure sale if the home-owner defaults and cannot otherwise repay the debt.

Financing a mortgage or secured home equity loan entails a credit check of the borrower, a property appraisal, review and approval by the lender of the homeowner's financial condition based upon two or more years of tax returns, and in-force homeowner's insurance covering the lender. Closing costs for a conventional mortgage or secured home equity loan typically involve paying for a title search, lender's title insurance policy, a bank fee equal to one or more percentage points of the loan amount, attorneys' fees, and recording costs, all of which can collectively amount to several thousand dollars. In addition, the loan approval process can take a period of months. This can be an expensive and cumbersome process for a homeowner seeking to make these types of home improvements; however, unlike an unsecured loan, the interest is tax-deductible.

Another alternative is a home equity line of credit, which is similar to a home equity loan but uses a revolving line of credit based upon the homeowner-borrower's equity in the home that the homeowner can bor-row against on an as-needed basis. It is secured by a mortgage and typically has a variable rate of interest, which can become expensive; however, the interest is tax-deductible.

Homeowners with good credit but no home equity can obtain an unse-cured loan that does not involve the time or closing costs associated with a secured mortgage or home equity loan. The interest rate, which will be higher than a secured loan and lower than a credit card, typically corresponds to the borrower's credit rating; the better the borrower's credit score, the lower the interest rate. The term of an unsecured home equity loan is shorter than a secured loan, usually no longer than 8 years.

Credit cards present a borrowing alternative for modest household improvements or for homeowners who do not qualify for the other financ-ing options. While credit card debt does not involve the risk of foreclosure associated with a mortgage loan, the interest rate varies, can become exor-bitant over time, and is not tax-deductible.

To avoid ongoing debt, homeowners who purchase with a credit card may select, for example, an appliance such as an air conditioner or heating unit, which costs less than a more energy-efficient model but will incur the long-term expense of higher utility bills and lower energy savings. This short-term approach undermines the ultimate point of energy efficiency upgrades and renewable energy retrofits, namely, to save money over the long term, conserve on water and energy use, and emit as little GHG as possible.

Individual home energy efficiency and renewable energy providers offer customized financing options to their customers. For example, homeowners can purchase energy efficiency upgrades such as insulation and energy star appliances with no money down and a low interest, short-term loan, and lease solar panels through a 20-year agreement with monthly payments equal to a fixed utility bill payment. These combined financing options help to fill the funding gap, but more is needed if American homeowners are going to do their part to phase out household-generated GHG emissions over the next few decades.

To make possible for homeowners a swift and effective shift away from fossil fuel consumption to energy conservation and renewable energy sources to power activities of daily living, accessible and affordable financing is needed that can be paid for over the useful life of the improvement. The secondary mortgage market plays an indirect, but critical, role in how quickly Americans can respond to this call to action. Current federal secondary mortgage market policy favors fossil fuel production over access to funding for energy efficiency upgrades and renewable energy retrofits. Emerging financing options are enabling a limited number of homeowners to phase out their GHG emissions cost effectively. Yet, significantly more progress can be achieved, provided that current federal mortgage policy concerns regarding this financing can be addressed and resolved.

RESIDENTIAL PACE FINANCING

Property-assessed clean energy (PACE) financing, which was inspired by century-old bond financing used to construct municipal infrastructure such as sewers, represents an emerging option with promise. In the current form, municipal bonds enable property owners to access low-interest funds to install energy efficiency upgrades such as insulation or appliances or home energy retrofits such as energy-efficient windows and solar panels, with no out of pocket payment, by treating the financing as a property assessment and paying for it gradually on the property tax bill. While the debt is not characterized as mortgage debt, the interest is tax-deductible. Additional benefits of PACE financing include a low interest rate, immediate positive cash flow, greater long-term property value attributable to enhanced energy efficiency, and preservation of the homeowner's borrowing capacity.[7]

---

[7] PACE*Now* 2013 Annual Report, pages 5-9.

More than thirty states and the District of Columbia have passed legislation to implement PACE programs, including residential PACE financing.[8] PACE financing enables the assessment to "run" with the property on a sale so that the original homeowner does not have to pay off the loan when delivering the property deed to the new owner. This setup creates an incentive to homeowners seeking to optimize their home's long-term energy and cost efficiency without the obligation to remain in the home until the improvement is fully paid for. It also attracts investment capital since PACE financing is considered a tax lien levied against the property, which takes precedence over other liens; therefore, any unpaid PACE assessment would be paid for before other liens, including a residential mortgage.

RESIDENTIAL PACE FINANCING AND THE FEDERAL HOUSING FINANCE AGENCY

The lien priority that makes residential PACE financing attractive to capital investors lies at the crux of the Federal Housing Finance Agency's (FHFA) objection to this financing option. The FHFA was established in the wake of the recent financial crisis by the Housing and Economic Recovery Act of 2008 (HERA) and "is responsible for the effective supervision, regulation, and housing mission oversight of the Federal National Mortgage Association (Fannie Mae), the Federal Home Loan Mortgage Corporation (Freddie Mac), and the Federal Home Loan Bank System, which includes 12 Banks and the Office of Finance."[9] These institutions collectively provide more than $5.5 trillion in funding for the US mortgage markets and financial institutions.[10] The FHFA has an understandable responsibility to protect and promote the likelihood that the loans in the federally insured secondary mortgage market will be repaid, especially since taxpayers bailed out the mortgage market in the aftermath of the 2008 mortgage crisis. Yet, whether lien priority afforded by residential PACE financing presents an actual risk to the secondary mortgage market deserves attention, particularly in light of the threats that unbridled water and energy consumption poses to the

---

[8] PACE*Now*, supra note 6.

[9] FHFA, FED. Hous. Fin. Agency, *The 2014 Strategic Plan For The Conservatorships Of Fannie Mae And Freddie Mac*, May 13, 2014, at page 3. http://www.fhfa.gov/AboutUs/Reports/ReportDocuments/2014StrategicPlan05132014Final.pdf (last visited August 16, 2014).

[10] About FHFA, FED. Hous. Fin. Agency, http://www.fhfa.gov/AboutUs (last visited August 20, 2014).

uninsurable mounting costs and potentially irreversible effects associated with climate change.[11]

In a speech on May 13, 2014,[12] the FHFA director Melvin Watt announced the release of a new strategic plan for the conservatorship of Fannie Mae and Freddie Mac. The strategic plan embodies three goals: Strategic Goal 1: *MAINTAIN, in a safe and sound manner, foreclosure prevention activities and credit availability for new and refinanced mortgages to foster liquid, efficient, competitive, and resilient national housing finance markets*; Strategic Goal 2: *REDUCE taxpayer risk through increasing the role of private capital in the mortgage market*; and Strategic Goal 3: *BUILD a new single-family securitization infrastructure for use by the enterprises and adaptable for use by other participants in the secondary market in the future*. At the near conclusion of the speech, Director Watt summed up, "[s]ince any stumbles along the way could have ripple effects in the $10 trillion housing finance market, there's a lot at stake in getting this right. As a result, our decision has been to "de-risk" this project." As indicated in Strategic Goals 2 and 3, the FHFA strategic plan reflects the Obama administration's intention to significantly scale back the role of the government-sponsored secondary mortgage market and replace it with private investment capital.[13] Thus, decisions made by the FHFA with respect to residential PACE financing will involve not undermining sources of future mortgage capital while also preserving the performance of existing mortgage loans.

The role of the FHFA as conservator operates in the context of a complex, evolving finance market, federal government, and nation with competing interests. Nowhere are the competing interests more apparent than at the intersection of the housing finance system, America's current national

---

[11] On March 24, 2014, H.R. 4285 was introduced in the US Congress and referred to the US House Committee on Financial Services. This bill would facilitate local government financing of energy efficiency, water conservation, and renewable energy generation improvements on private property that would be paid for on the property owner's tax bill. If adopted, this act would direct FHFA to permit residential PACE financing, provided that the individual PACE program, the homeowner, and the improvements meet the requirements of the act. See https://www.govtrack.us/congress/bills/113/hr4285.

[12] FHFA, FED. Hous. Fin. Agency, *Prepared Remarks of Melvin L. Watt at the Brookings Institution Forum on the Future of Fannie Mae and Freddie Mac: Managing The Present: The 2014 Strategic Plan For The Conservatorships Of Fannie Mae And Freddie Mac*, May 13, 2014, http://www.fhfa.gov/Media/PublicAffairs/Pages/Watt-Brookings-Keynote-5132014.aspx.

[13] Jackie Calmes, *Obama Backs Limits to U.S. Role in Mortgages*, N.Y. TIMES, Aug. 7, 2013, at A3.

struggle with its energy policy and a destabilized climate, which has no patience for politics. In July 2010, the FHFA issued a statement asserting that residential PACE financing would pose substantial risks to lenders and the loans "do not have the traditional community benefits associated with taxing initiatives." The FHFA also expressed skepticism about the performance of the retrofits and the resulting energy savings.[14] In August 2010, Fannie Mae and Freddie Mac issued letters to lenders stating that they would no longer purchase mortgages securing homes with residential PACE financing if the homeowner obligation constitutes a first lien priority,[15] that is, a lien senior to a federally insured mortgage.

More recently, following litigation brought by the state of California, which challenged the FHFA but was dismissed by the Ninth Circuit for lack of jurisdiction, the state established a $10 million loss reserve fund to compensate Fannie Mae and Freddie Mac for losses in any foreclosure resulting from the payment of any PACE assessment paid while in possession of the property and in any forced sale for unpaid taxes or special assessments from PACE assessments paid before the outstanding mortgage;[16] however, this measure failed to change the FHFA's position. Alfred Pollard, general counsel to the FHFA, wrote to California officials: "The Reserve Fund does not sufficiently address the risks to the Enterprises [Fannie Mae and Freddie Mac] that we have previously described, and FHFA will continue our policy of not authorizing the Enterprises to purchase or refinance mortgages that are encumbered by PACE loans in a first lien position."[17]

The FHFA has an ambitious mission to fulfill, as embodied in the three main goals of its strategic plan. Yet, the question arises as to whether lien priority has to be an all-or-nothing proposition. The approximately $5.5

---

[14] Kat Friedrich, *Residential Pace Energy Programs Pursue Innovative Approaches,* CLEAN ENERGY FINANCE CENTER, Sept. 6, 2013, http://www.renewableenergyworld.com/rea/news/article/2013/09/residential-pace-energy-programs-pursue-innovative-approaches.

[15] Joe Kaatz, Scott J. Anders, *Residential and Commercial Property Assessed Clean Energy (PACE) Financing in California,* UNIVERSITY OF SAN DIEGO ENERGY POLICY INITIATIVES CENTER, March 2013, page 54, http://energycenter.org/sites/default/files/docs/nav/policy/research-and-reports/PACE%20in%20California.pdf.

[16] Letter from Edmund G. Brown, Jr. Governor of California to Edward DeMarco, Acting Director, Federal Housing Finance Agency, Sept. 23, 2013.

[17] Stephen Lacey, *Why Residential PACE Is Growing in Spite of Opposition to Federal Housing Lenders,* PACENOW FINANCING ENERGY EFFICIENCY July 24, 2014, http://pacenow.org/why-residential-pace-is-growing-in-spite-of-opposition-from-federal-housing-lenders/.

trillion portion of the secondary mortgage market under consideration by the FHFA dwarfs the residential PACE loan market with the latter's per residence loan comprising a limited percentage of the home's full market value. Accordingly, lien priority associated with residential PACE financing should matter only to the extent it presents an actual risk to the mortgage lender and secondary mortgage market investors. In the event the homeowner seeking the residential PACE financing holds an equity interest in the residence that exceeds, by a reasonable percentage, the outstanding principal balance of the mortgage loan *plus* the amount of the PACE assessment, this should eliminate the risks to the lender and the secondary market. The same should hold true of a purchaser who maintains an equity interest in the same residential property at least equal to the original residential PACE borrower. For example, the proposed federal legislation, H.R. 4285, introduced on March 24, 2014, would require "the property owner to have equity in the property of not less than 10 percent of the estimated value of the property calculated without consideration of the amount of the PACE assessment or the value of the PACE improvements" and a total PACE assessment that cannot exceed 15 percent of the estimated value of the property. [18] Whether 10 percent net equity reflects the appropriate owner interest to protect the secondary mortgage market remains a matter for the FHFA and all interested parties to discuss and resolve; but conceptually, net equity presents an approach through which all parties should be able to find common ground.

The FHFA's assertion that residential PACE loans "do not have traditional community benefits associated with taxing initiatives" has merit, but arguably only with respect to the use of the word "traditional." Municipal sewer pipes, paid for by a method similar to PACE financing, represent a tangible asset with a useful life. A loan program that has the effect of stabilizing the climate over the useful life of the installed home improvement likewise produces community benefits, albeit, perhaps, not traditional. PACE financing is marketed to assist homeowners to reduce their costs; but recent developments in California underscore how residential PACE financing represents a tool of necessity: to facilitate water conservation in a state with a crippling drought associated with climate change. Facilitating water conservation by a critical mass of California homeowners will benefit communities throughout the state, perhaps in a nontraditional way, but in a material way nevertheless.

---

[18] *See* H.R. 4285, *supra* note 10 at Sections 5(g) (2) and (3).

Despite the FHFA's warnings, residential PACE financing is taking place through different programs, albeit with the limitations imposed by the FHFA. For example, the California-based Home Energy Renovation Opportunity (HERO) Program partners with local governments to provide low-interest-rate financing to help homeowners reduce water and energy consumption. Homeowners are notified of the limitations imposed by the FHFA and agree to pay off the PACE loan if they move or refinance their home mortgage. In response to the FHFA's concerns that residential property-assessed loans can increase risks to mortgage lenders, J.P. McNeill, CEO of Renovate America, which administers the PACE loans, explained, "The purpose of the HERO program is to help homeowners select and finance products which save them money by lowering their utility bills. Having lower utility bills has two benefits to a mortgage lender. First, a homeowner has more money to pay for their mortgage. Second, a home is worth more money. We are able to gather data and measure the impact HERO has in terms of local jobs created, energy and water savings, lower emissions and the impact on mortgages. So far, HERO has had a positive impact on all."[19]

Residential PACE loans are typically limited to single-family, two-family, and three-family residences and permanently attached mobile homes. Residences with 4 or more units qualify for commercial PACE financing, which is not subject to oversight by the FHFA. According to this breakdown, the current pool of potential households, which could seek residential PACE financing, equals approximately 103,960,000 or 80% of American households;[20] yet, only the households with mortgages held by Fannie Mae and Freddie Mac have relevance to a risk assessment of these PACE loans. As the secondary mortgage market shifts more toward involving private capital, the limitations on residential PACE financing may likewise disappear. Time will tell. In addition, future residential construction is likely to incorporate energy efficiency and renewable energy design and materials into the home's structure, thereby streamlining the need for residential PACE financing. What remains is the need for current, ready access to cost-effective residential funding that transforms as large a segment as possible of the country's housing stock into water- and energy-saving homes that function independently of fossil fuel.

---

[19] Morgan Lee, *Clean Energy loans to benefit homeowners: Renovate America offering property-assessed loans for clean-energy upgrades,* UTSANDIEGO, August 3, 2014, http://www.utsandiego.com/news/2014/Aug/03/conscious-improvement/?#article-copy.

[20] See U.S. Census, *supra* at note 2.

IMPACTS OF UNCONVENTIONAL GAS DRILLING ON THE SECONDARY MORTGAGE MARKET

The FHFA's policy with respect to residential PACE financing should not be viewed in isolation. The fossil fuel industry has expanded its unconventional drilling—that combination of high-volume hydraulic fracturing and horizontal drilling operations (aka "fracking") for oil and natural (methane) gas extraction—onto residential property across a majority of the lower 48 states presenting mounting, unmonitored threats to the secondary mortgage market, threats that add "risk to lenders and secondary markets and altered valuation of mortgage backed securities because of uncertainty surrounding potential foreclosures, diminution in value at sale, increased risk of delinquency[21]. . .and [impacts to] loan-to-value ratios," in short, the very concerns expressed by the FHFA with respect to residential PACE loans.

Standard mortgages used in the secondary mortgage market prohibit the transfer of an interest in the real property (which includes entering into a gas lease) without lender consent. In addition, standard mortgages prohibit the presence of hazardous materials and hazardous activity consistent with the practices characterized by unconventional gas drilling operations. For, example, mortgages prohibit the presence, use, storage, or release of any hazardous substances on, under, or about the mortgaged property, including gasoline, other flammable or toxic petroleum products, volatile solvents, and radioactive materials, among other things. The heavy industrial drilling and fracturing process involves use of explosives at various stages, millions of gallons of chemically treated water for fracking at the well site, and on-site use and temporary storage of toxic materials, including radioactive fracking waste. Standard mortgages also prohibit borrowers from committing waste, damage, or destruction or causing substantial change to the mortgaged property or allowing a third party to do so. Mortgages contain these covenants to protect and preserve the lender's collateral. By all accounts, most homeowner-borrowers did not obtain lender consent prior to entering into their gas lease. While the secondary mortgage market relies upon loan servicers to collect payments of principal and interest and tax and insurance escrow payments if the loan so requires, property site visits are not part of the loan servicer's job description. Further, no one at the FHFA, Fannie

---

[21] *Residential and Commercial Property Assessed Clean Energy (PACE) Financing in California,* supra at page 54.

Mae, Freddie Mac, or otherwise performs home visits to check on the mortgaged collateral. Yet, nationwide, an unknown portion of mortgaged homes with unconventional drilling operations also serve as collateral for residential mortgage backed securities in the secondary mortgage market.[22]

Fracking operations introduce risks to the secondary mortgage market and threaten the goals expressed in the FHFA's strategic plan by creating liens, which interfere with homeowner mortgage financing and supersede the mortgage lien. To finance unconventional drilling operations, multi-million dollar loans are secured by blanket mortgages, which appear in the land records as property liens against the multitude of homeowners who leased their land to the oil and gas company obtaining the commercial loan. In one example, among others, Chesapeake Energy pledged its mineral interests granted through multiple gas leases to obtain a $500 million line of credit. The commercial lender recorded the multimillion dollar lien in the public records against the homeowners who granted their mineral rights to Chesapeake Energy. This eye-popping lien caused three different banks to deny an affected Ohio homeowner the right to refinance his mortgage loan, even though Chesapeake's commercial loan was secured by a different property interest on the same land.[23] In another example, between October and December 2012, separate mechanics liens representing more than $3,384,000.00 in unpaid bills by Chesapeake Energy were filed by its contractors against the residential land of private property owners in Bradford County, Pennsylvania, who previously granted the mineral rights under their land to Chesapeake Energy for drilling operations.[24] These contractors' mechanics liens, which were filed against the residential land, gained a priority position superseding each affected homeowner's mortgage lien.

---

[22] *See* Elisabeth N. Radow, *At the Intersection of Wall Street and Main: Impacts of Hydraulic Fracturing on Residential Property Interests, Risk Allocation, and Implications for the Secondary Mortgage Market*, ALBANY LAW REVIEW 77 ALBANY LAW REVIEW 673 (2014), at pages 688-698, http://www.albanylawreview.org/issues/pages/articleinformation.aspx?volume=77&issue=2&page=673.

[23] Rachel Morgan, *Liens could be obstacles for gas leaseholders*, SHALEREPORTER, Jan. 6, 2014, http://www.shalereporter.com/industry/article_c1359fc4-768a-11e3-b790-001a4bcf6878.html.

[24] Brian Grow, Anna Driver and Joshua Schneyer, *Special Report: Chesapeake, McClendon endure rocky year; more uncertainty ahead*, REUTERS, Dec.27, 2012, http://www.reuters.com/article/2012/12/27/us-chesapeake-mcclendon-idUSBRE8BQ0IO20121227.

Fracking operations are linked to water contamination and structural damage, both of which adversely impact property value[25] and can affect a homeowner's ability to repay the loan, conditions that the FHFA seeks to protect against. For example, between January and August 11, 2014, the state of Oklahoma experienced 292 earthquakes measuring 3.0 or more on the Richter scale (nearly triple the amount from 2013), which were linked to deep well injection of hydraulic fracturing waste and caused residential structural damage.[26] Other states experiencing manmade earthquakes, resulting in property damage from hydraulic fracturing operations, include Arkansas, Colorado, Ohio, and Texas.[27] Homeowner's insurance does not cover risks from manmade earthquakes or water pollution from hydraulic fracturing operations; special endorsements to insurance will not necessarily cover these risks either.[28] Insurance companies such as Nationwide have publicly stated that they reserve the right not to renew homeowners' insurance policies with unconventional drilling operations. A home without homeowner's insurance represents an incurable mortgage default. This can lead to foreclosure.

The secondary mortgage market holds collateral located in every state with unconventional drilling operations. Recent studies have found that these heavy industrial operations can cause more than a 22% loss of property value for homes relying on well water located in proximity to drilling operations, *before* contamination occurs.[29] Once a home's well water becomes contaminated or the residence becomes structurally damaged as a result of one or another phase of unconventional drilling operations, the home

[25] For a compendium of articles addressing impacts to property value, see, generally, Drilling vs. the American Dream: Fracking impacts on property rights and home values, March 14, 2014, available at http://www.resource-media.org/drilling-vs-the-american-dream-fracking-impacts-on-property-rights-and-home-values/#.U_OXx2d0zIU.

[26] Carey Gillam, *Houses are bouncing; quakes trigger controls on Oklahoma industry*, Aug 11, 2014 www.reuters.com/.../us-usa-earthquakes-oklahoma-idUSKBN0G9/.

[27] Induced Earthquakes, USGS, http://earthquake.usgs.gov/research/induced/ (Last visited August 20, 2014).

[28] Patti Conley, *Insurers: Fracking-Related Damages Not Covered*, TIMESONLINE, Feb. 28, 2013, http://www.timesonline.com/special_sections/progress_edition/insurers-fracking-related-damages-not-covered-by-standard-policies/article_ca7228ac-ff24-5764-9019-2c3ddd94f291.html?_dc=35695646656.677124&cbst=79.

[29] Lucija Muehlenbachs, Elisheba Spiller & Christopher Timmins, *The Housing Market Impacts of Shale Gas Development 37–39 (Nat'l Bureau Econ. Research, Working Paper No. 19796,* (2014), http://goo.gl/TuoZ6R. *See also* Michelle Conlin, *Analysis: U.S. drilling boom leaves some homeowners in a big hole* REUTERS, Dec. 12, 2013, http://www.reuters.com/article/2013/12/12/us-fracking-homeowners-analysis-idUSBRE9BB0GS20131212.

can lose substantially all of its market value. A majority of potential homeowners surveyed would prefer to avoid altogether locating their families where unconventional drilling occurs.[30]

In the event a homeowner whose water supply or residence has been adversely impacted by unconventional drilling operations subsequently defaults on their mortgage loan, a foreclosure sale of the impacted property may fail to reap enough funds to repay the outstanding debt. These adverse property impacts can affect a borrower who signed a gas lease and a borrower who elected not to sign a gas lease but who lives next to someone who did. In contrast, a home that uses residential PACE financing to fund water or energy efficiency upgrades or renewable energy retrofits is likely to benefit from lower operating expenses, an increase in market value of the residence, and reduced GHG emissions and produces no negative impacts to the community.

MAKING RESIDENTIAL PACE FINANCING WORK

Millions more of Americans can conserve on water and energy consumption and power their daily activities from renewable energy sources once the federal policies associated with residential mortgage financing align with these conservation goals and permit residential PACE financing within reasonable parameters. Such an alignment would simultaneously enable borrowers to improve their property performance, property value, and cash flow and likewise make private capital investment in the secondary mortgage market more attractive, all of which would serve to fulfill the mission of the FHFA, as expressed in its 2014 strategic plan.

Coming full circle, federal residential mortgage financing policy decisions can produce powerful ripple effects, both positive and negative, in America's energy future. Aligning the strategic goals of the secondary mortgage market with the imperatives presented by climate change can result in direct economic benefits to homeowners and foster indirect but essential economic benefits to all Americans who otherwise face the escalating shared cost of lost crops from droughts and property damage and loss of life from storms, floods, and sea-level rise, among other potential catastrophic effects.

---

[30] *Drilling vs. the American Dream: Fracking impacts on property rights and home values,* RESOURCE MEDIA, March 14, 2014, http://www.resource-media.org/drilling-vs-the-american-dream-fracking-impacts-on-property-rights-and-home-values/#. UoAwr3DkuSr.

Avoidable water and energy consumption requires prompt correction. By enabling homeowners to access funds without the closing costs associated with secured financing and gradually pay for energy efficiency upgrades and renewable energy retrofits at a low interest rate, the residential PACE financing model can help American homeowners to phase out household-generated GHG emissions more effectively than other current financing options. Enabling homeowners to make shifts in water and energy consumption and minimize GHG emissions can likewise influence awareness and inspire similar corrective action in other sectors of life such as the workplace, retail, and recreational settings. What begins at home and benefits the homeowner can benefit the secondary mortgage market and the nation as a whole.

Smart Explorer's Wheel: Accelerating Innovation Integration in the Green Industrial Revolution

Julian Gresser[1,2]

[1]Alliances for Discovery, is an international attorney, inventor, social entrepreneur, and professional negotiator. He was twice Mitsubishi visiting professor at the Harvard Law School and has been a visiting professor at MIT, Beijing University, and Doshisha University in Kyoto, Japan. He has served as a senior advisor to the US State Department, the Prime Minister's Office of Japan, the European Commission (where he coached its Japanese negotiating teams), and the World Bank. During the past 25 years, he has trained hundreds of senior executives around the world in his integrity-based system of artful negotiation, described in his recent book *Piloting Through Chaos—The Explorer's Mind (Bridge21 Publications, August 2013)*. Julian Gresser is the author of *Environmental Law in Japan* (MIT Press, 1981) and *Partners in Prosperity—Strategic Industries for the United States and Japan* (McGraw Hill 1985), among other works. He is deeply involved in developing a global response to the unfolding tragedy from the nuclear accident at Fukushima, Japan (www.explorerswheel.com). One of his most exciting current ventures is to establish a unique action-oriented twenty-first-century think tank: the United States-China center on trade and innovation.

[2]© Copyright, Julian Gresser, August 2014, Santa Barbara, California; All rights reserved.

OVERVIEW

In the *green industrial revolution*, Woodrow Clark and Grant Cooke described the environmental, social, economic, and political forces and some of the critical technologies that are driving a third global industrial transformation. These technologies will shape the direction of planet Earth in the twenty-first century. In an earlier book, *Partners in Prosperity—Strategic Industries for the United States and Japan* (McGraw Hill, 1985), I documented how at specific historical times and places certain economically "strategic" technologies and industries served as catalysts to accelerate innovation, raise productivity, stimulate economic growth, and produce new and more creative jobs. The green industrial revolution is constellating, with compounding impact and breakthroughs in science and technology that have the potential to support a new era of global prosperity.

One intersection between *the green industrial revolution* and my present work is the question, how to raise global literacy and consciousness about the green industrial revolution? Another is how to build a corps of millions of green innovators and entrepreneurs who can imaginatively solve global problems collaboratively? A Smart Explorer's Wheel introduced in my 2013 book *Piloting Through Chaos—The Explorer's Mind* (www.explorerswheel.com) offers one useful approach. In fact, the convergence of advanced information communications technologies through this new learning-problem-solving platform will introduce an additional strategic capability that can accelerate breakthroughs in the green industrial revolution. Throughout history, many other strategic industries in their time, for example, displayed a similar converge with communications technologies. The canals and railroads in eighteenth- and nineteenth-century America and Japan's post-WWII "trigger industries" (semiconductor-computer-telecommunications-robotics) are good examples.

THE FUTURE EMBEDDED IN THE PRESENT

The Smart Explorer's Wheel platform is best viewed from several perspectives. As depicted in Illustration 1, the wheel gathers and integrates the creative energy of an Explorer's Community around a central theme or inventive challenge. In the present example, the focal theme could be measurably and rapidly to enable millions of people around the world almost overnight to achieve literacy about regarding all the critical challenges, issues, and available resources for the green industrial revolution. The center of the wheel—the "Explorer's Mind"—is surrounded by eight realms or prisms. They are the entry points for the Explorer's Journey. The realms are the following: the Past, Wisdom, Beauty, Life Force, Discovery-Invention-Innovation, Humanity, Networked Brain, and Future. The explorer enters within any realm that most strikes his or her fancy. As the wheel begins to spin, the explorer starts to discover interesting "intertidal" connections between the various realms. The "intertidal" is derived from the zone of greatest biological fertility where the ocean's tides flow in and out.

EXPLORER'S WORLD ™

FUTURE

BEAUTY

WISDOM

LIFE FORCE

DISCOVERY
INVENTION
INNOVATION

NETWORKED
BRAIN

HUMANITY

PAST

August 3, 2013

A More Abundant Life

The Explorer's Wheel--Creating Your Own Luck--A Complete Navigational System
For Immediate Distribution—Bridge 21 Publications Announces the Release of *Piloting Through Chaos—The Explorer's Mind* by Julian Gresser

The pioneering British geneticist and evolutionary biologist, J.B. S. Haldane famously observed, ""My suspicion is that the universe is not only queerer than we suppose, but queerer than we can suppose."

Julian Gresser's new book, *Piloting Through Chaos—The Explorer's Mind*, released today by Bridge 21 Publications, introduces explorers to the last unchartered frontier, at least by modern people: the vast terrain

s m a r t BOOK

BRING THIS BOOK TO LIFE
Scan here with your SmartPhone

In my work globally (see short bio below) over the last four decades, I have discovered this same liminal zone at the border of consciousness and the unconscious processes, where the barriers between conventional disciplines and fields collapse. This is also a fertile terrain of the human mind, where new perspectives, discoveries, and inventions can suddenly appear. The Explorer's Wheel helps to train the Explorer's Mind through the iterative process of consciously and continuously discovering often surprising connections.

Recent scientific discoveries suggest that this explorers' process of finding connections may itself actually generate new neuropathways. The human brain is an extraordinary neuroplastic organ, and its potential for systematic applications to the frontier of a green industrial revolution is virgin territory. The logical implication is that each member of a GIR Explorer's Community, and the Explorer's Community as a whole, will become smarter, more creative, resourceful, and resilient. At least, this seems a plausible hypothesis. It invites rigorous scientific confirmation.

TODAY'S APPLICATIONS OF THE EXPLORER'S WHEEL

I believe the most important discoveries and inventions in the green industrial revolution will be generated by systematically and collaboratively harvesting the intertidal connections between narrowly siloed conventional fields and academic disciplines. A particularly important application of this principle will be to accelerate breakthroughs at "choke points" between important areas of concern—for example, food security, clean water, and sustainable energy—in order to ensure that the solutions in one domain do not create secondary negative consequences in another.

For example, if not approached systemically, a breakthrough in food security could result in wasting huge amounts of available freshwater; a breakthrough in desalination could result in pollution of coastal waters and the consumption of huge amounts of energy. The Explorer's Wheel encourages systemic thinking and systemic solutions. As the explorer ventures more deeply, he or she may reach an interesting frontier of discovery: the realms are sufficiently broad while bounded that they provide an effective framework and process to connect every issue of importance in the green industrial revolution with every other.

The other important dimension of the Explorer's Wheel platform is that it is "smart." In other words, it learns 24/7 along with and about its community of explorers. Storytelling is the engine of an Explorer's Wheel. The Explorer's Wheel platform harvests these stories along with the embedded metaphors, images, and idiosyncratic ways of thinking of the millions of members of GIR-Smart Collaborative Innovation Network (GIR-SCOIN). The more actively engaged the members, the smarter the GIR-SCOIN becomes. The result is a new kind of "psychographic database," which can be licensed as a valuable asset to charter sponsors who share the vision and values of the GIR. The Explorer's Wheel in this way offers a powerful means to contribute to the financial sustainability of GIR-SCOIN.

The GIR-SCOIN platform is designed to incorporate additional elements. Virtually all of these elements are already developed, tested, and proved and await integration. They include the following:

- New processes for discovery and invention engineering and collaborative innovation embodied in a discovery-invention-innovation engine, which can accelerate breakthroughs for humanity virtually on demand (http://www.explorerswheel.com/blog/inventing-humanitya-collaborative-strategy-global-survival)
- A collaborative IP model to facilitate collaboration including advanced processes and frameworks for the negotiation, structuring, management, measurement, and conflict resolution of strategic alliances and networks
- A new model of artful negotiation based on the *principle of integrity*
- A translation engine, which enables domain experts and creative amateurs ("citizen scientists") to communicate and work effectively together
- A trust engine, which enables recipients to assess the degree of reliability of the data they are receiving
- A collaborative competition platform that will support XPRIZE-like competitions for breakthroughs for humanity
- Advanced Smart Search components so that the platform learns about each participant and supports his or her discovery process
- A wisdom expert system, which enables users to encode the "wisdom genome" of the great figures of history living and dead across all fields and turn them into mentors for our daily lives
- A way of integrating Big Data, Big Mechanism, and Big Heart (described below) with particular interest in humanitarian applications
- A new approach to online learning

Within three years or less, the Explorer's Wheel platform can become the online brain of the green industrial revolution.

A POST-GREEN INDUSTRIAL REVOLUTION FOR THE TWENTY-FIRST CENTURY: THE ALIGNMENT AND INTEGRATION OF THE HEART, MIND, AND HAND

The green industrial revolution will catalyze breakthroughs in science and technology that hold the promise, at least in theory, to usher in an unprecedented era of global abundance and prosperity. However, my observations over the past fifty years raise one profound concern: I do not believe that technology alone divorced from an expanded consciousness will achieve this result. According to visionaries like Ray Kurzweil (*The Singularity is Near*) and Michio

Kaku (*The Future of the Mind*) within the next 30-50 years, the human brain and its artifact, the mind, will merge with computers, conferring countless benefits on suffering humanity, including a form of bionic and digital immortality. I do not accept the reductionist proposition that all we dream and are can be reduced to bits and bytes governed by the implacable force of Moore's law.

In fact, the deep problems of human hatred, greed, arrogance, cruelty, and selfishness will remain; and even worse, the new green technologies with all their promise may well, if not tempered by the angels of our better nature—generosity, kindness, and unconditional love—easily be turned to devilish purpose. Today, the world must deal with the "devil's equation": $\text{nuclear arms} + \text{hatred} + \text{time acceleration} > \text{attention deficit} = \text{apocalypse}$. Is this not the present challenge presented by despotic regimes like ISIS and North Korea?

The critical question then is, how most effectively to release the Promethean power of mind but temper and guide it by the wisdom, foresight, balance, resilience, and generosity of the heart? Will not all the scientific and technological wonders humanity has thus far produced pale in comparison when, after so many millennia, we can discover a way to reconcile and harness the power of the mind and heart together? In my view, the green industrial revolution, which is still in its infancy, is ideally suited to build a foundation for what may prove the fourth great advance in human evolution: the alignment and integration of the heart, mind, and hand.

SMART EXPLORER'S WHEELS: NEXT STEPS

Currently, our team is well along in designing and planning the launch of several pilot Smart Explorer's Wheels during the next twenty-four months. Some applications most relevant to the green industrial revolution include the following:

- Smart creative sustainable cities
- Accelerating breakthroughs at the choke points of clean water, food security, and sustainable energy
- A California-China Solar Partnership (http://www.explorerswheel. com/blog/california-china-solar-partnership)

Consider one application in particular that is rapidly being implemented (2014-2015): a smart platform for a new United States-China trade and innovation center. This will become the preeminent action-oriented think tank of its kind in the world focusing on food security, clean tech, water, sustainable energy, investment liberalization, trade facilitation, and e-commerce.

This platform will include a breakthrough in instantaneous voice recognition in English/Chinese available via Skype and smartphones that will enable millions of green innovators and entrepreneurs inside and outside China to collaborate effectively together. With adequate financing and the effective collaboration of the major companies that have already developed the core technical components, the app can be delivered to market within the next 18 months (mid-2015) (http://www.explorerswheel.com/blog/how-integration-can-drive-trade-facilitation-several-languages; http://www.explorerswheel.com/blog/breaking-language-barrier-between-english-and-chinese-and-its-implications-world).

SUMMARY: NOT THE END AND MORE THAN A BEGINNING

A Smart Explorer's Wheel can contribute significantly to the realization of the goals of the global industrial revolution in the following ways:

- It will provide a platform to raise global literacy and consciousness by integrating the next-generation tools, methods, and resources of social media, smart search, crowd intelligence, Big Data, and Big Mechanism in a new model of online learning and education.
- It will help to accelerate breakthrough discoveries, inventions, and innovations by providing a collaborative platform for technology developers, users, financiers, and many others who will play a critical role in delivering effective innovations to the marketplace.
- In time, it will combine instantaneous voice recognition technology, which will dismantle the language and cultural barriers between nations and ultimately change forever the master–slave relationship between humans and machines.
- It will provide a crucible for the heart to play a crucially needed role in people-to-people (heart-to-heart) exchanges both online and in the outside world.
- It offers a new revenue model deriving from license fees based on a continuously enhanced psychographic data, generated by the storytelling of members of the GIR-SCOIN.
- As Woodrow Clark and Grant Cooke ably document, the green industrial revolution is already upon us. Millions of global citizens from every nation and culture are exuberant to engage in this new age of green exploration and innovation. When the heart becomes the source code of this revolution, the great next transformation will occur, for it must as the hourglass of our suffering planet to change course is rapidly running out.

INDEX

Note: Page numbers followed by *f* indicate figures and *t* indicate tables.

A

Agile system, 259–261
Algae fuel, 185–186
America
  California Global Warming Solution Act, 214
  electricity generation, 213–214
  Energy Efficiency Strategic Plan, 212
  energy policy, 215
  Energy Star program, 214
  Environmental Protection Agency, 211
  hydrogen fuel cells, 212
  issues, 213
  LEED, 215
  renewable auction mechanism, 213
  state legislation, 214
  USGBC, 215
American Association for the Advancement of Science (AAAS), 84
Asia, 205–206
Australia, 219–220

B

Biofuels
  algae fuel, 185–186
  gasoline and diesel, 185
  plants, 186–187
  products and sources, 184
  renewable energy, 184
  sugarcane, 184
Bitumen lake
  advanced distillation technique, 40
  Ancient Mesopotamia, 39
  fossils, 39
  Greek fire, 40
  medical applications, 40–41
  mummification process, 39
  muslim forces, 40
  natural gas, 40
  prehistoric humans, 39
  stone slabs, 39

C

California Energy Commission (CEC), 159–160, 159*f*
California Global Warming Solution Act, 214
Canada, 215–216
Cap-and-trade systems
  arguments, 243
  carbon taxes, 242, 243
  definition, 241–242
  economic model, 242
  ETS, 242
Capture and store the carbon (CCS), 73, 75
CEC. *See* California Energy Commission (CEC)
Central America, 216–218
China's central planning model
  economic growth, 237
  energy economics, 236
  five-year plans, 236–237
  international policies and programs, 236
  long-term planning and financing, 235–236
  post-Mao era, 236–237
  social capitalism, 235–236
China's leading smart grid
  attributes, 170
  benefits, 169–170
  Chinese-style power grid, 167–168
  five-year plan, 167, 169
  high-profile demonstration project, 168
  large-scale, 168
  on-site power, 170
  power industry, 167
  renewable energy, 167
  robust and low-cost smart grid, 170
  Zhangbei project, 168
Climate change
  additive manufacturing, 113–114
  animal protein, 106
  anthropogenic impacts, 106–107
  Arctic climate change, 99, 100

Climate change (*Continued*)
   Atlantic sea surface temperature, 97–98
   automobiles, 103
   cap-and-trade system (*see* Cap-and-trade systems)
   carbon emission, 100
   carbonic acid, 100
   clothes dryers, 113
   community-based and on-site energy generation, 111–112
   droughts, 98
   economic growth, 116–117, 240
   efficiency, effectiveness and fairness, 241
   Energy Star, 112
   food chain, 100
   fossil fuels, 103, 104
   fuel cell storage, 115–116
   GIR economics, 243
   global meat production, 106
   global warming, 98
   green jobs, 119
   greenhouse gases, 101, 104–105, 105*f*
   human/anthropogenic activity, 96, 241
   Hurricane Sandy, 96–97, 96*f*, 101–102
   hydrogen power, 115–116
   internal combustion engine, 103
   LED, 114–115
   livestock, 105–106
   N_2O, 106
   operating costs, 118
   polar vortex, 99–100, 99*f*
   population growth, 110–111
   scientific and technological change, 112
   seawater heat pumps, 112
   smog-covered community, 100–101
   snowstorm, 98
   society, 103
   solar installation, 119
   sustainability, 118
   tobacco smoking taxes, 241
   tourist-driven market, 118
   Typhoon Haiyan, 97, 101–102
   UN IPCC, 108–110
   United Parcel Service, 118–119
   urbanization, 104
   warmest period, 102–103
   Watt's steam engine, 103
   zero emission, 117–118
Coal
   Appalachian Mountain region, 74–75
   carbon emission, 72
   carbon-capture technology, 73
   CCS, 73, 75
   China's fossil fuel, 72–73
   energy statistics, 71–72
   environmental damage, 75–76
   fossil fuel-burning process, 74
   integrated gasification combined cycle, 73
   Schwarze Pumpe plant, 73
   sediment, 72
   underground coal gasification, 73
   ZEBS, 73
Combined heat and power (CHP) system. *See* Frederikshavn
Communist Party of China (CPC), 193–194
Concentrating solar power (CSP), 152–153
Cool roofs, 177–178

D
DARA report, 225–226
Deepwater Horizon oil spill, Gulf of Mexico, 229–230
Denmark
   base load power, 259
   buildings, 259
   certification process, 259
   components, 258–259
   energy efficiency, 257–258
   green jobs, 258
   wind turbines, 258
Diablo Canyon earthquake, 88, 88*f*
Distributed energy generation
   biofuels, 6
   carbonless technology, 2
   children education, 1–2
   climate change, 9
   CO_2-consuming algae, 2
   cutting-edge technology, 6
   ecosystems, 9
   energy storage, 7
   finance services-based economy, 3
   flexible energy sharing, 7
   government policy, 3
   government support and financing, 8–9
   humanitarian organization, 9
   hybrid technology, 4

hydrogen highway, 2
hydrogen-propelled vehicles, 4
integrated transportation, 7
magnetic force, 4
oil industry, 5
regenerative braking system, 4
renewable energy, 3, 6
solar energy, 1
steam-driven first industrial revolution, 3
Tesla Model S, 4, 5f
Drilling technique
    bamboo pipelines, 46, 47
    Daqing oil field, 48
    domestic oil production, 50
    drying process, 47–48
    energy demand, 50
    export/import equation, 49
    foreign technology, 48–49
    greenhouse gases and carbon
        emission, 51
    Kang Pen drum, 48
    open-door policy, 48
    refining and plant equipment, 48
    renewable energy, 51
    Sichuan salt farmers, 46
    Sinopec controls, 49
    State Planning Commission, 49
    state-owned companies, 50
    stockpiling plan, 50
Duke Energy ash coal spill, 228–229

E

Economics
    China (see China's central planning
        model)
    Chipotle, 227
    climate change effects, 226
    Climate Vulnerable Forum, 225
    cost of oil (see True cost of oil)
    DARA report, 225–226
    Duke Energy, 228–229
    energy generation, 227
    FiT model (see Feed-in-tariff (FiT)
        model)
    Forbes magazine, 227–228
    free-market (see Free-market
        economics)
    gasoline prices, 228

global temperatures, 226
    global weather patterns, 227
    Google (see Google investment)
    green jobs, 243–246
    IEA, 226–227
    megadrought, California, 227
    oil companies, 228
    oil cost (see True cost of oil)
    private investment, 246–247
    public land loopholes, 229
    subsidized railroads, 229
    tax breaks, 229
    US DOD, 225
Electricity
    coal-generated electricity, 125
    data centers, 125
    definition, 123
    electromechanical generators, 124
    kite experiment, 123–124
    mechanical transformers, 124
    natural gas generation, 124–125
Emerging GIR technology
    biofuels (see Biofuels)
    CHP system, 181–182
    commercial technology, 188–189
    cool roofs, 177–178
    Energy Star, 173–174
    FastOx, 187–188
    government-led initiatives, 173–174
    GSHP, 182–183
    high-speed rail train, 183–184
    lighting (see Lighting industry)
    long transmission lines, 173
    Maglev trains, 184
    nanotechnology, 178–179
    on-site energy generation, 173
    peak-load management, 176–177
    regeneration braking (see Regeneration
        braking systems)
    seawater heat pump, 183
Emission Trading System (ETS), 242
Energy deflation, 280–282
Energy Renewable Sources Act, 238
Energy return on investment (EROI), 68
Europe
    Act of Granting Priority, 209
    aging grid structure, 210
    Arab oil embargoes, 208

Europe (*Continued*)
  Clean Development Mechanism, 209–210
  energy efficiency and conservation, 211
  Green Technology Park, 210
  green/environment-first political
    movement, 208
  North Sea oil deposits, 208
  ROCs, 209
  social, political, and environmental issues,
    209
  trading system, 209
European Commission and International
    Energy Agency, 239
European FiT program, 33
Exxon Valdez, Alaska, 229–230
ExxonMobil Corporation, 227–228

F

Feed-in-tariff (FiT) model
  California, 213, 239–240
  China, 133, 200
  European, 33, 239
  Germany, 24, 129, 237, 238, 239
Flywheel energy storage (FES)
  battery systems, 152
  Eco-Gen, 152
  factors, 151–152
  high-strength carbon filaments, 151
  LLNL, 151
  potter's wheel, 151
  transportation systems, 152
  vehicles and power plants, 151
Fracturing Responsibility and Awareness of
    Chemicals (FRAC) Act, 80–81
Frederikshavn
  biogas plant, 268
  biomass boiler burning straw, 266
  emerging GIR, 181–182
  geothermal, 268
  heat pumps, 268
  heating grid, 267–268
  methanol production, 268
  transportation, 268
  wind power, 268
Free-market economics
  2IR result, 234
  draft animal powered economy, 233
  dramatic change, 234

Earth and inhabitants, 233
economist, 232
fossil fuel-powered economy, 233
global economic implosion, 232
historical precedence, 234
information and digital technology, 233
infrastructure components, 235
Middle East, 233
rampant economic growth, 234
renewable energy generation, 235

G

Geothermal heat pumps (GSHP), 182–183
German FiT model, 238, 239
GHG (*see* Greenhouse gases GHG)
Gigafactory, 153
Global Wind Energy Council projects, 200
Google investment
  2IR economy, 248
  data center and infrastructure spending, 247
  Ernst and Young report, 248–249
  fossil fuel and centralized utility industries,
    249
  renewable energy, 247, 248
  VB/Research, 248
Green industrial revolution (GIR)
  Arab Spring, 23
  big oil, 22
  carbon-based 2IR, 22
  carbon-intensive lifestyle, 22–23
  digital age, 23
  digital communication, 21
  economics (*see* Social capitalism)
  energy-independent community, 24–25
  energy-secure, 22
  environmental collapse, 23
  feed-in tariff program, 24
  government support, 29–30
  large-scale effort, 24
  natural resources, 21–22
  nuclear power, 24
  oil and natural gas industry, 24
  population, 23
  renewable energy generation, 22
  sustainable community (*see* Renewable
    energy)
  The European Dream, 21
Green jobs, 243–246

Greenhouse gases (GHG), 101, 104–105, 105*f*
GSHP. *See* Geothermal heat pumps (GSHP)

H

High-profile demonstration project, 168
Hurricane Katrina, 226–227
Hurricane Sandy, 96–97, 96*f*, 101–102
Husk Power System, 146
Hydraulic fracturing
    AAAS, 84
    Cabot Oil, 82
    cattle grazing, 83
    environmental and health damage, 80
    evaporation pits, 83
    farmers and nut growers, 82
    FRAC Act, 80–81
    groundwater system and air pollution, 83
    Halliburton Company, 79
    lateral drilling, 76–77
    Marcellus Shale natural gas, 81
    mineral wealth, 84
    public pressure, 83–84
    residential areas, 85
    TEDX, 83
    toxic chemicals, 79, 81–82
Hydrogen fuel cell
    biomass, 161
    Bloom Energy, 157
    California Energy Commission, 159–160,
        159*f*
    declining production costs, 161
    electric current, 156, 156*f*
    electrolyte, 157
    fossil fuel-burning vehicles, 158
    H2 Mobility plan, 160–161, 160*f*
    Honda FCX Clarity, 158–159
    in United States, 158
    reactions, 156
    Sweden, 161
    transportation, industry and homes, 157
    zero-emission engine, 158

I

IEA. *See* International Energy Agency (IEA)
India, 218
Industrial revolution
    Age of Enlightenment, 14
    analog communication, 18

Boulton-Watt double-acting engine, 14,
    15*f*
    electricity, 16
    energy consumption, 19
    energy-intensive lifestyle, 19
    GIR (*see* Green industrial revolution
        (GIR))
    goods and products, 17
    Gutenberg's printing press, 13–14, 13*f*
    internal combustion engine, 17
    lightning, 16
    literacy, 14
    mass communication, 14
    mechanical movable-type printing, 14
    miniexplosion, 17
    oil demand, 19–21
    personal transportation vehicle, 18, 18*f*
    petroleum, 16–17
    power demand, 19
    salt mines, 16–17
    steam engine, 16
    urbanization, 19
Inner Mongolia Autonomous Region
    (IMAR), 194–195, 199
Integral Fast Reactor, 88
Intermittent energy generation, 126
International Energy Agency (IEA), 77, 78,
    226
International Monetary Fund reports, 231

L

Lawrence Berkeley National Laboratory
    (LBNL), 177–178
Lawrence Livermore National Laboratory
    (LLNL), 151
Light emitting diode (LED), 114–115, 177
Lighting industry
    building design and task-ambient lighting,
        174–175
    digital communications and Internet
        connectivity, 174–175
    economic growth, 175
    energy and utility industries, 175
    fluorescent tubes, 174
    high-humidity locations, 175–176
    HVAC systems, 175–176
    incandescent bulbs, 174
    LED, 177

Lighting industry (*Continued*)
  mud-walled shack, 177
  smart sensors, 175

M

Maglev technology, 184
Mandatory Renewable Energy Target
  (MRET), 220
Mayak reactor, Russia, 91–92
Megadrought, California, 227
Middle East and North Africa (MENA),
  220–222

N

Nanotechnology, 178–179
National Energy Bureau, 199–200
Natural gas
  carbon deposits, 76
  carbon pollution, 77
  condensation/absorption, 76
  energy efficiency, 78
  environmental destruction, 77
  fracking (*see* Hydraulic fracturing)
  greenhouse gases, 77
  IEA, 77, 78
New Jersey Turnpike, 163
New zealand, 219–220
Nitrous oxide (N_2O), 106
Norway, oil-producing nation, 256
Nuclear power
  condenser, 87
  control rods, 87
  cylindrical cooling towers, 86, 86*f*
  Diablo Canyon earthquake, 88, 88*f*
  fuel rods, 87
  Integral Fast Reactor, 88
  International Atomic Energy Agency
    report, 86
  Mayak reactor, 91–92
  moderator, 87
  nuclear disasters, 89, 91*f*
  nuclear fission, 87, 88
  nuclear reaction, 86, 87
  nuclear waste, 89
  radioactivity, 87
  start-up costs, 88
  steam turbine, 87

O

Obninsk Nuclear Power Station, 85–86
Oil industry
  big oil wages, 62
  coal, 42
  Deepwater Horizon oil rig, 63
  downstream process, 41–42
  drilling technique, 46–51
  Foreign Policy, 43
  fossil fuel subsidies, 65
  global business analysis, 41
  greenhouse gases, 64–65
  hydrocarbons, 38
  impacts, 65
  kerosene/paraffin, 37
  lobby industry, 65–66
  Marshall Islands, 63
  Middle East nations, 43–44
  natural gas/unusable material, 38
  New York Times, 63
  oil embargo (*see* Organization of
    Petroleum Exporting Countries
    (OPEC))
  oil reserves (*see* Oil reserves)
  oil seeps, 44
  oil spills, 59–61
  petroleum industry, 42
  political pressure, 63
  potential elimination, 64
  prehistoric algae, 38
  rock oil (*see* Standard oil)
  Russian power struggle, 42
  smallish sludge/tar lakes (*see* Bitumen
    lake)
  state-owned companies, 41
  subsidies and politics, 59
  tax breaks, 64
  ThinkProgress, 66
  upstream process, 41
  US taxpayer, 63
Oil reserves
  EROI, 68
  financial and energy cost, 68
  global conventional oil production, 69
  oil data, 66–67
  oil demand, 68
  Oxford report, 67
  peak oil, 67

TED Conference, 67
unconventional oil resources, 68–69
Oil spills, 59–61
Organization of Petroleum Exporting
    Countries (OPEC)
  big oil, 54–55
  cartel members, 58
  cheap oil, 53
  energy costs, 57
  energy crisis, 54
  exploration techniques, 54
  fair price, 57
  fossil fuel fortunes, 56–57
  geopolitics, 58
  global market, 57
  globalization, 54
  goals, 51–53
  oil demand, 53
  oil embargo, 53
  oil spikes, 57
  personnel transportation, 58
  production level, 58
  recovery technology, 54
  resource extraction, 56
  tools, 55
  Yom Kippur war, 53

P

Peak-load management, 176–177
Pumped hydroelectric storage (PHS), 150–151

R

RAM. *See* Renewable Auction Mechanism
    (RAM)
Raworth system, 180
Regeneration braking system
  AMC Amitron, United States, 181
  distributed energy generation, 4
  electric railways, 180–181
  kinetic energy, 179–180
  Raworth system, 180
  stored energy, 179–180
  Toyota, Japan, 181
  tramcar motors, 180
Renewable Auction Mechanism (RAM),
    239–240
Renewable energy
  benefits, 147

biogenerator, 28–29
biomass, 26–27, 143–144
cell phones, 146
climate change, 25
Denmark, 125–126
digestive processes, 26
electricity, 123–125
farms/central plants, 126
fossil fuel, 25
fuel cells, 28
geothermal energy, 141–143
geothermal power, 27, 27f
global economy, 145
grid distribution, 145, 146
Husk Power System, 146
intermittent energy generation, 126
non-fossil fuel transportation, 25
nonhazardous microbial bacteria, 28
ocean and tidal waves, 27, 28
recycled and reusable generation,
    143–144
run-of-the-river system, 28
solar (*see* Solar)
solar generation system, 26
solar panels, 146
storage devices, 144
water, 139–141
wind generation, 26
windmills (*see* Wind power)
Renewable Obligation Certificates (ROCs),
    209
Rising Tigers
  advantages, 202
  benefits, 200–201
  fragile ecosystem, 203
  government policies and investments, 201
  market dominance, 203
  production costs, 201
  quality products, 201
  Silicon Valley, 201–203
  tax incentives, 201
  Vestas, 202
Roscoe Wind Farm, 129
Russia's Mayak nuclear reactor, 91–92

S

Schwarze Pumpe plant, 73
Seawater heat pump, 183

Sherman Antitrust Act, 45
Sino-Singapore Tianjin Eco-city (SSTEC)
    C40 program, 272
    eco-solution, 271
    Eco-Valley, 272
    Hurricane Sandy, 273
    hybrid cars and buses, 271
    population, 271
    sound water management system, 271
    wastewater pond, 270
Smart grid system
    AC electricity, 163–164
    Big Oil, 277
    cap-and-trade program, 279
    carbon emitters, 278–279
    central power plant, 163
    China's leading (see China's leading smart
        grid)
    Climate Vulnerable Forum, 279
    coal industry, 278
    DC electricity, 163–164
    Deutsche Bank, 276
    distribution bus, 164
    dot-com era, 164
    electricity generation, 278
    energy deflation, 280–282
    Europe's parallel lines, 166–167
    fossil fuels, 165
    grid parity, 279–280
    human affairs, 277
    industrial revolution, 275–276
    Internet communication, 165
    noncarbon energy sources, 278
    power transmission, 164
    real-time data, 165
    Russian natural gas, 277
    smart meter, 164–165
    solar power, 276
    solar PV and wind power, 165
    tax payer support, 277–278
    Tesla Model S, 278
    zero marginal cost, 282–285
Smart meter, 164–165
Social capitalism
    accelerated cooperation, 33
    animal-powered economy, 31
    climate change, 32, 33
    economic implosion, 30–31

energy independence, 31
environmental factors, 31
European FiT program, 33
fossil fuel-powered economy, 31
free-market capitalism, 31
global warming, 33
irreversible environmental damage, 30
national energy policy, 32
political issues, 31
Sociopolitical toxicity, 230
Solar
    peak power, 135
    photovoltaic systems, 134
    Sharp Corporation, 135
    silicon crystal, 135
    Solar Valley City, China, 136
    thin-film technologies, 135
    US industry, 137–139
Solar Valley City, China, 136, 197–198
Solar-Powered Rooftops Plan, 197–198
South America, 216–218
South Korea, 206–207
Southeast Asia, 218–219
Standard oil
    Allied forces, 46
    control pricing, 45
    military transportation, 46
    Model T automobile, 45, 46
    oil and kerosene business, 45
    oil patch, 44
    refining capacity, 45
    Sherman Antitrust Act, 45
Storage technology
    electric cars, 154–156
    energy storage applications, 149, 150t,
        153
    flywheels, 151–152
    Gigafactory, 153, 154
    grid storage devices, 150
    high-temperature batteries, 153
    liquid electrolyte batteries, 153
    PHS, 150–151
    TES, 152–153
    V2G power storage, 154–156
    ZAFC technology, 154
Sustainable community
    agile system, 259–261
    Arab oil embargo, 255

automakers, 261
biomass generation plants, 256
Brundtland Commission, 257
car-centric sustainable community,
    254–255
CHP system (*see* Frederikshavn)
codes and standards, 265
congestion and smog, 264
conservation, 256
Denmark (*see* Denmark)
diesel, 261
economic sustainability, 255
environmental sustainability, 255
gasoline-powered engines, 262
Harmonies (*see* Sino-Singapore Tianjin
    Eco-city (SSTEC))
hydroelectric dams, 264
hydrogen/electric cars, 264
mobility, 261
Norway, oil-producing nation, 256
public policies, 265
renewable energy, 264–265
smart green community, 256–257
tax breaks, 262
Tesla Model S, 263
true game changer, 263
West Village, 269–270
wind farms, 256

T
Technology, Entertainment and Design
    (TED) Conference, 67
Tesla Model S, 4, 5*f*, 263
Thanet Wind Farm, 129
Thermal energy storage (TES)
    CSP, 152–153
    hot and cold storage, 153
True cost of oil
    Alaska spill, 230
    climate change, 230
    Copulos' estimation, 231, 232
    Deepwater Horizon oil spill, 229–230
    global emissions, 232
    International Monetary Fund reports, 231
    oil production, 231
    sociopolitical toxicity, 230
Turbines, 26

Twenty-First-Century Green Powerhouse,
    China
    12th Five-Year Plan, 191, 194
    13th Five-Year Plan, 191–192
    climate change, 193
    distinction, 192–193
    green technology tigers (*see* Rising
        Tigers)
    gross domestic product, 192
    joint venture business model, 192
    long-term planning, 194
    massive energy, 196–197
    Olympics, 192
    renewable energy, 193
    social capitalism, 191–192, 193
    Solar Valley City, China, 197–198
    sustainable development, 194–196
    Vestas, 193
    wind power (*see* Wind power)
Typhoon Haiyan, Philippines, 97, 101–102,
    226–227

U
United Nations Intergovernmental Panel on
    Climate Change (UN IPCC), 108–110
University of California, Los Angeles
    (UCLA), 166–167
US Department of Defense (US DOD), 225
US Environmental Protection Agency, 158
US Federal Energy Regulatory Commission
    (FERC), 154
US Green Building Council (USGBC), 215
US Green Building Council's Leadership in
    Energy and Environmental Design
    (LEED), 173–174

V
Venture Business Research (VB/Research),
    248
Vestas
    benefits, 132
    Chinese National Energy Bureau,
        132–133
    commercialization, 134
    emission reduction, 131
    FiT system, 133
    government policies, 131
    offshore wind energy, 132

Vestas (*Continued*)
  Pew Charitable Trusts, 130–131
  pricing, 133
  smart gird technology, 131
  state-owned power supply, 132
  turbine construction, 133–134, 193
  wind base, 132
  wind farms, 133

W

West Village, 269–270
Wind power
  Arab oil embargo, 128
  California, 128
  challenges, 200
  Chile, 129–130
  China's wind power (*see* Vestas)
  Cleveland, 127–128
  commercial wind turbine market, 128
  emission reduction, 198
  energy production, 198
  equipment manufacturing, 199
  fan-type windmill, 127
  FiT program, 129, 200
  joint ventures and collaborations, 198
  National Energy Bureau, 199–200
  Offshore project, 199
  post-mills, European, 127
  power motors, 127
  pricing, 200
  Roscoe Wind Farm, 129
  sail-rotor windmills, 126–127
  state-owned power supply, 199
  Thanet Wind Farm, 129
  tower mill, Dutch, 127
  wind turbines, 127

Z

Zero Emission Boiler System
  (ZEBS), 73
Zero marginal cost
  Airbnb, 283
  carbonless energy generation, 284
  climate change, 284
  declining prices, 284
  elements, 283
  environmental benefits, 284–285
  file sharing technology, 282
  Internet, 283
  Massive Open Online Course, 283
  sustainability, 284
Zhangbei project, 168
Zinc–air fuel cell (ZAFC) technology, 154

Printed in the United States
By Bookmasters